GENETIC RECOMBINATION RESEARCH PROGRESS

GENETIC RECOMBINATION RESEARCH PROGRESS

JACOB H. SCHULZ
EDITOR

Nova Biomedical Books
New York

For permission to use material from this book please contact us:
Telephone 631-231-7269; Fax 631-231-8175
Web Site: http://www.novapublishers.com

NOTICE TO THE READER

The Publisher has taken reasonable care in the preparation of this book, but makes no expressed or implied warranty of any kind and assumes no responsibility for any errors or omissions. No liability is assumed for incidental or consequential damages in connection with or arising out of information contained in this book. The Publisher shall not be liable for any special, consequential, or exemplary damages resulting, in whole or in part, from the readers' use of, or reliance upon, this material. Any parts of this book based on government reports are so indicated and copyright is claimed for those parts to the extent applicable to compilations of such works.

Independent verification should be sought for any data, advice or recommendations contained in this book. In addition, no responsibility is assumed by the publisher for any injury and/or damage to persons or property arising from any methods, products, instructions, ideas or otherwise contained in this publication.

This publication is designed to provide accurate and authoritative information with regard to the subject matter covered herein. It is sold with the clear understanding that the Publisher is not engaged in rendering legal or any other professional services. If legal or any other expert assistance is required, the services of a competent person should be sought. FROM A DECLARATION OF PARTICIPANTS JOINTLY ADOPTED BY A COMMITTEE OF THE AMERICAN BAR ASSOCIATION AND A COMMITTEE OF PUBLISHERS.

Library of Congress Cataloging-in-Publication Data

Genetic recombination research progress / Jacob H. Schulz (editor).
      p. ; cm.
  Includes bibliographical references and index.
  ISBN 978-1-60456-482-2 (hardcover)
  1. Genetic recombination. I. Schulz, Jacob H.
  [DNLM: 1. Recombination, Genetic. 2. Research. QU 475 G328 2008]
  QH443.G456   2008
  572.8'77--dc22   2008007080

Published by Nova Science Publishers, Inc. ✦ *New York*

Contents

Preface

Genetic recombination is the process by which a strand of genetic material (usually DNA; but can also be RNA) is broken and then joined to a different DNA molecule. In eukaryotes recombination commonly occurs during meiosis as chromosomal crossover between paired chromosomes. This process leads to offsprings having different combinations of genes from their parents and can produce new chimeric alleles. In evolutionary biology this shuffling of genes is thought to have many advantages, including that of allowing sexually reproducing organisms to avoid Muller's ratchet. However, a recombination pathway in DNA is any way by which a broken DNA molecule is reconnected to form a whole DNA strand.

In molecular biology "recombination" can also refer to artificial and deliberate recombination of disparate pieces of DNA, often from different organisms, creating what is called recombinant DNA.

Enzymes called recombinases catalyze natural recombination reactions. RecA, the recombinase found in E. coli, is responsible for the repair of DNA double strand breaks (DSBs). In yeast and other eukaryotic organisms there are two recombinases required for repairing DSBs. The RAD51 protein is required for mitotic and meiotic recombination and the DMC1 protein is specific to meiotic recombination.

This new book presents the latest research in the field.

In response to DNA damage, the cell institutes an orchestrated and measured response. Central to this is the canonical DNA damage response pathway. The components of this pathway are many and include ATM, MRE11, RAD50, NBS1, CHEK2, p53, GADD45a, p21, BRCA1, BRCA2, FA proteins, BLM and WRN. Depending on cell cycle and damage type, different combinations of these components are utilized to affect an appropriate cellular response, including DNA repair. Absence of any one of these components is largely associated with clinical diseases, severe developmental defects, an increased propensity to develop cancer at an early age, and/or aging defects. Homologous recombination repair is a DNA repair mechanism that is capable of repairing a wide range of damages. It has been determined that homologous recombination repair is altered in the absence of almost every component of the damage response pathway in either a direct or indirect fashion or by functional implication. In Chapter 1 the authors will present the evidence for the intimate relationship of the damage response pathway, cell cycle and homologous recombination repair based on *in vivo* models and discuss the implications of their intertwining in mammalian systems.

Inducible gene expression systems offer great potential toward achieving a wide variety of basic and applied biomedical research goals. Current inducible gene technologies are based on binary transgenic systems in which the expression of a target gene is dependent upon the activity of a unique transcriptional activator in the presence or absence of its drug ligand.

Despite the conceptual simplicity, most gene inducible strategies suffer from a number of limitations, such as leaky basal transcriptional activity when uninduced and poor dynamic control of the system. Because of these shortcomings, regulated gene expression technologies are still being evaluated using stringent genetic studies and none are currently used in human gene therapy. To overcome some of the disadvantages of the current systems, the authors describe the development of a next-generation inducible gene expression system, which relies upon *multiple* heterologous transactivators and their respective ligands. Their system would provide negative/positive selection for transgene expression in its induced state that would create virtually no transgene expression when not induced (i.e., no leakage). In Chapter 2 the authors propose a strategy to engineer expression systems that include heterodimeric hormone receptors combined with other heterologous transactivators.

Genetic recombination, originating from the occurrences of crossovers during meiosis, may be seen as the foundation of genetic variation, i.e. underlying both population diversities and general evolution in itself. Moreover, it is the driving force behind the original formation of genetic material forming the genetic components corresponding to, human or other, diseases. The recombination process generally imply a low level of recombinations, given a fixed number of meiotic generations, with respect to closely located loci; being the basic mechanism facilitating gene mapping studies through using collected data in the form of genetic information on multigenerational families (pedigrees). A major role behind such mapping is played by a relevant set of statistical techniques, methods and algorithms; developed using tools and theory from, for instance, mathematics, computer science and genetics.

In linkage analysis or, in a wider sense, gene mapping one searches for disease loci along a genome. This is done by observing so called marker genotypes and phenotypes of a pedigree set, i.e. a set of pedigrees, in order to locate the loci corresponding to the underlying disease genes or, at least, to narrow down the interesting genome regions. In this context the key concept is the genetic inheritance of alleles and its correlation to phenotype outcomes; which is informative according to genetic recombination. A significant deviation from what is expected under random inheritance is taken as statistical evidence of existing genetic components suggested to be located at, or close to, the loci giving significant results.

Chapter 3 mainly serves as a (subjective) review on gene mapping in general and nonparametric linkage analysis in particular, but also includes some material on recent or new research. The authors will follow a path vaguely outlined as: Firstly, they begin by outlining the needed genetical foundation of statistical genetics as well as some basic concepts, for instance, noting on the process of allelic inheritance, the genetic model, the pedigree set, the inheritance vector and various types of genetic information. Secondly, they give an introduction to one-locus nonparametric linkage (NPL) analysis focusing on significance calculations of NPL scores. Corresponding approaches are based either on analytical approximations or Monte Carlo simulations. Thirdly, the authors make some comments on the generalizations to two-locus procedures; both on the unconditional (simultaneous) and conditional (sequential) search for two disease loci. Along the way they also at some length discuss, for instance, parametric linkage analysis, fuzzy significance, multiple testing and evidential interpretation of significant findings. Fourthly, they very briefly discuss some competing and complementary subfields within the context of gene mapping-based statistical genetics.

Discovering the impact of biological processes, such as lateral gene transfer, has significantly transformed our understanding of microbial evolution and our views of the natural genetic relationships between diverse life forms. As the diversity of evolutionary processes are revealed, real natural connections appear to be much more complex than initially believed at the time traditional molecular phylogenetics assigned itself the task to reconstruct the universal Tree of life, aka the unique genealogy of species, based on the use of carefully selected genes. Both the notions of species and of a unique inclusive hierarchy of living beings needs in fact to be reevaluated. Importantly, such a reevaluation calls for a substantial renewal of our phylogenetic practices and focuses, opening new perspectives to the whole field of evolutionary biology. Here, these changes are justified by recalling an important recent lesson in the philosophy of biology: how the consideration of evolutionary processes, necessary for species definition, indicates that many incompatible but legitimate definitions of species taxa and taxonomies are more realistic when it comes to represent life's natural relationships. In Chapter 4 the authors discuss why, in addition, many other evolutionary units, smaller or larger than the usual species taxa, have also to be considered as real, because they too emerge from the evolutionary process. Thus, even though the trajectories of these multiple evolutionary units may conflict, they deserve equally to be fully investigated by phylogenetics. Instead of the use of a unique tree, the authors explain how the consideration of multiple databases could help properly systematize a portion of the real biological diversity. In addition, consideration of networks could partly help investigate the dynamics sustaining the natural genetic connections between all the evolutionary units. Such analyses go beyond the scope of traditional phylogenetics, yet they matter, because, while some evolutionary units are "closed", others are "open", vastly changing, thus presenting fuzzy rather than precisely definable boundaries. They argue why allowing such a distinction at all evolutionary levels has important bearing on our ability and ways to describe life's evolution, which can not be accounted for by a unique model. Finally, after encouraging the development of additional evolutionary metaphors to describe the true course of evolution, the authors briefly present implications for biotechnologies and conservation biology that makes this pluralistic phylogenetics particularly valuable and promising.

Chapter 5 describes the molecular mechanisms which underly joining of human and animal retroviral DNA to host cell DNA. The authors discuss contributions of the viral protein integrase, as well as cellular co-factors, to the efficiency of integration and selection of integration sites in the host cell genome. They also address the questions related to the treatment of HIV (human immunodeficiency virus) infection, since the HIV-1 integrase is an attractive target for the development of anti-HIV-1 therapeutics. Finally, the authors present opportunities as well as the broblems associated with the integration of retroviral vectors, which are used in gene therapy applications.

φC31 integrase is a site specific recombinase derived from the *Streptomyces* phage. In the phage lifecycle, the enzyme mediates lysogeny by mediating recombination between specific sequences termed *attB* (present in the bacterial DNA) and *attP* (present in the phage genome). Screening the enzyme activity in mammalian cells provided positive results and also showed that the enzyme retained its property of site specific recombination into mammalian genomes. Mammalian genomes have been shown to contain sequences that are similar to the wild type *attP* sequence of the *Streptomyces* phage genome and experiments with the integrase in mammalian cells showed that it could mediate recombination and subsequent integration of any DNA bearing an *attB* site into these pseudo*attP* sites. These properties have made φC31 integrase emerge as a promising tool for nonviral gene therapy to achieve long-term gene expression in different tissues *in vitro* and *in vivo*. Chapter 6 introduces the different recombination mediating enzymes but focuses primarily on the activity of φC31 integrase in different tissues. Results of this enzyme system in the lung tissue and hematopoietic cells are also discussed. In murine and human lung cell lines, it could be shown that the φC31 integrase mediates specific recombination between *attB* and *attP* and subsequently long-term gene expression could be achieved without any selection pressure. The results were further validated *in vivo* in mice. However, when integration into a specific site "mpsL1" was investigated, it was revealed that this site was targeted only in 50% of the mice. This was in contrast to the reports from other tissues where all the mice showed integration at mpsL1. These results indicated that the activity of this enzyme may actually be tissue dependent.

Tissue specific efficiency of the φC31 integrase was confirmed in the human hematopoietic system. The activity of the φC31 integrase was extremely reduced in CD34$^+$ hematopoietic stem cells, primary T lymphocytes and T cell derived cell lines in comparison with mesenchymal stem cells and cell lines derived from lung -, liver - and cervix tissue. No enhanced long-term expression mediated by the φC31 integrase could be observed in T cell lines. A direct comparison of hematopoietic Jurkat T and A549 alveolar type-II lung cell lines indicated up to a 100-fold higher activity of the φC31 integrase in lung tissue compared to hematopoietic cells. Looking for possible mechanisms responsible for this discrepancy of φC31 integrase activity in different tissues, revealed that Jurkat cells contain significantly higher amounts of DAXX protein compared to A549 cells. As DAXX has been reported to interact with the φC31 integrase and inhibit recombination, higher levels in hematopoietic cells may be one reason for low activity of φC31 integrase in these cells.

These studies in lungs and the hematopoietic cells provide evidence for the tissue specificity of the φC31 integrase system and also raise the need for development of novel strategies for achieving recombination and long-term expression.

Recombination is a key evolutionary mechanism that should not be ignored. It is directly related with the amount of linkage disequilibrium, which is important in order to characterize populations from an evolutionary point of view and to localize genes in humans and other organisms. Recombination effects are not independent of other evolutionary forces, such as natural selection and genetic drift. Recently, new methods became available to infer positive selection in the presence of recombination and *viceversa*. These new methods are computationally very intensive and as data sets become larger, their analysis becomes unavoidable. Therefore, faster and efficient algorithms are needed to estimate recombination at the fine-scale and genomic level. Such approaches need also to disentangle recombination and natural selection signals on DNA. Additionally, it is usually difficult to evaluate the adequacy of the methods since there are few simulation tools capable to produce data under complex evolutionary models. Consequently, developing programs that allow simulating both, natural selection and recombination, is also necessary.

In what follows in Chapter 7 the authors will review recently developed algorithms to estimate recombination at the fine scale. In addition, they will also consider approaches that allow the simultaneous estimation of recombination and selection. The authors will explore possible future directions for recombination and selection estimation research. They will stress the importance of incorporating recombination jointly with other genetic factors both in evolutionary and epidemiological contexts, e.g. to model drug resistance emergence. In this framework, they will present a new result concerning the faster evolution of resistance favoured by the minimum co-infection rate combined with the higher recombination value. Finally, the authors will consider simulation tools. They will briefly point out how to simulate DNA sequences under a specific nonsynonymous/synonymous (dN/dS) ratio and recombination rate (inter and intracodon level). In doing this they will introduce a new method of forward simulation and an efficient way of simulating different substitution models forward in time.

As explained in Chapter 8, direct observations of chiasmata showed that there were chromosome intervals where only single chiasmata appeared and other intervals where rather chiasmata appeared which were accompanied by other chiasmata elsewhere. Particularly, double chiasmata were involved in both chromosome ends, but single chiasmata only in one end. There is no model of a chiasma process developed so far which is appropriate to such events. Therefore, a new model was developed here to infer the phenomenon. It takes into account suppression interference and allows the determination of the chiasma number distribution and the distribution of the chiasma locations for each number of appearing chiasmata. It was shown with an example that the novel model explains the chiasma formation process much better than other models.

Chapter 9 presents the disease Fanconi anemia (FA) is a cancer predisposition disorder involving progressive anemia, caused by deficiency in any of 13 known (FANC) genes. At the cellular level FA is a classical chromosomal instability disease associated with sensitivity to DNA damage, particularly interstrand crosslinks (ICLs). Unlike classical DNA repair pathways, it is not understood how most of the FANC proteins function to maintain chromosome stability, but many of them are necessary for the monoubiquitylation of two of the FANC proteins, FANCD2 and FANCI, both posttranslational modifications that are considered necessary for a functional FA "pathway". Through genetic and biochemical studies, this pathway is variously implicated in a number of key S-phase events such prevention of replication fork breaks by promotion of translesion synthesis (TLS), and the repair of broken chromatids through both of the double-strand break (DSB) repair pathways: homologous recombination repair (HRR) and non-homologous end joining (NHEJ). However, many published results concerning the functional links between the FA pathway and the two mutually exclusive DSB repair processes are, not surprisingly, contradictory. Recent use of the mutagenesis model system of hamster CHO cells, with knock-out clones defective in one of the FANC proteins (FANCG), or an HRR protein (Rad51D), have been highly informative in distinguishing functional differences between the FA and HRR pathways. Importantly, the experimental precision of CHO cells has clarified seemingly conflicting mutagenesis measurements in human FA cells, and implies contribution of the FA pathway in promoting all three replication-associated damage-recovery processes: TLS, HRR, and NHEJ. FA cells, therefore, are not completely deficient in any one of these processes, but are less efficient in all of them, suggesting a role for the FA proteins in co-ordination or optimization of these genome stabilizing processes, rather than direct participation in any one of them.

Chapter 10 discusses recombination, which might be an evolutionary development as ancient as the origin of life. In spite of numerous data on the mechanisms of genetic recombination in living organisms since the discovery of Rec mutations and of the existence of genetic exchanges in bacteria, some genetic mechanisms in these organisms remain not fully understood, notably in relation to the genetic diversity of bacteria and to their position with respect to the very first organisms having appeared on earth, ancestors of both eukaryotes and prokaryotes. Spontaneous zygogenesis (or Z-mating), recently discovered in *Escherichia coli*, is intriguing, as it resembles fusion of gametes in eukaryotes in that it involves complete genetic mixing. It is also question on how genetic exchanges can enable bacteria having undergone severe deletions or temporary chromosomal inactivation to survive and overcome what could be a selective disadvantage.

In the past, studying of the gene was very difficult. Genetic laboratory seems to be a complicated and mysterious field. However, the blooming of molecular biology leads to many simplified techniques for genetic studying. Finding the amino acid sequence of a gene can be easily performed by basic sequencing technique. At the end of the 20's century, the completion of the genome project lead to a new era of post genomics. The in silico laboratory is an advent in the post genomics era. Based on in silico techniques, manipulation on a genetic sequence is easy. This can help us to better understand the genetic recombination phenomenon. To study a genetic recombination, in silico mutating and docking can help create a genetic recombinant. In addition, in silico gene expression analysis such as gene ontology techniques can help identify the changes in phenotypic expression of a designed recombinant. In Chapter 11, principles and interesting examples of in silico genetic recombinant studies will be briefly discussed.

Pseudotyping lentiviral vector with other viral surface proteins could be applied for treating genetic anomalies in human skin. In Chapter 12, the modification of HIV vector tropism by pseudotyping with the envelope glycoprotein from vesicular stomatitis virus (VSV), the Zaire Ebola (EboZ) virus, murine leukemia virus (MuLV), lymphocytic choriomeningitis virus (LCMV), Rabies, or the rabies-related Mokola virus encoding *LacZ* as a reporter gene was evaluated qualitatively and quantitatively in human skin xenografts. High transgene expression was detected in dermal fibroblasts transduced with VSV-G-, EboZ-, or MuLV-pseudotyped HIV vector with tissue irregularities in the dermal compartments following repeated injections of EboZ- or LCMV-pseudotyped vectors. Four weeks after transduction, double-labeling immunofluorescence of β-galactosidase and involucrin or integrin β1 demonstrated that VSV-G-, EboZ-, or MuLV-pseudotyped HIV vector effectively targeted quiescent epidermal stem cells and their progenies, which were expressed dorsally and underwent terminal differentiation. Among the six different pseudotyped HIV-based vectors evaluated, VSV-G-pseudotyped vector was found to be the most efficient viral glycoprotein for cutaneous transduction as demonstrated by the highest level attained in β-galactosidase activity and genome copy number evaluated by *Taq*Man PCR.

The central nervous system comprises neuron networks, in which enormously diversified neurons connect and interact with each other. For decades, the diversity of such neurons has been hypothetically ascribed to a gene diversification mechanism similar to that of the antigen receptor genes in the immune system. Synaptogenesis, memory, and odoreceptor diversification have been raised as candidate representatives of the neuronal functions involved in DNA rearrangements mediated by the hypothetical somatic DNA recombination mechanism in the brain. Some reports have described the somatic DNA recombination activity and the possible rearranging gene loci in association with these neuronal functions in the brain. In spite of every effort to search for the associated gene loci for traces of gene rearrangement, no physiologically functional gene rearrangement has yet been identified. Two rearranged genomic regions were, however, identified in the brain by a rearranged genomic region-oriented approach. A repetitive genomic region (LINE) and a non-repetitive genomic region (BC-1) are the only known examples to undergo genomic rearrangement in the brain, thus far. Both of the regions have been observed to undergo DNA rearrangement not only in the brain and but also in the lens, thus implying that these DNA rearrangements are associated with ectodermal development. The relationship between the neighboring gene function and the rearrangement of the non-repetitive genomic region is discussed in Chapter 13.

As explained Innocuous loss of intron or *in situ* loss of intron is common during evolution. Together with intron gain, loss of intron has re-shaped many genomes dramatically, and has changed many gene structures together with intron gain. Five modes of intron loss can be defined in a multiple-intron gene: complete loss of all introns, 3'-biased loss, concerted loss of several internal adjacent introns, intron exclusion and multiple intron exclusion. The cDMHR/DSBR model presented in Chapter 14, which the authors recently established, can accommodate all the five patterns of intron losses. In this model, cDNA undergoes homologous recombination (HR) with its parent intron-containing genomic copy. This cDNA recombination is facilitated by the HR repair machinery of DSB in the cell. This process can be triggered by a DSB in a specific intron, which results in the loss of the very intron suffering the DSB. The reverse transcriptase activity could be from retrotransposon such as the yeast Ty1 element and possibly the mammalian LINE. DSB in intron, retrotransposon-encoded reverse transcriptase and homologous repair machinery might be the long searched driving forces for *in situ* intron elimination. This model is strongly supported by the independent experimental data from yeast in which cDNA recombination with the corresponding genomic copy is directly demonstrated, and this process is dependent on the HR protein RAD52. The widely used gap-repair technique also supports this model.

Genetic recombination is an important phenomenon in gene medicine. This phenomenon can lead to a new genotype and phenotype. Genetic recombination can be either natural or artificial. Natural genetic recombination is an important contributing factor to genetic shift and drift. In medicine, the genetic recombination in pathogenesis of an existing disease is important. Considering three epidemiological determinants, genetic recombination of agent is easier than that of host and environment. For pathogens, natural genetic recombination can result in a change in virulence and susceptibility. Natural genetic recombination in the virus is well characterized in clinical microbiology. In Chapter 15, the author performed a literature review on the natural genetic recombination of pathogens. Examples of important genetic recombination of pathogens and their clinical correlations are presented.

Chapter 16 discusses a randomized Spruce Budworm model with Holling III Functional Response . The authors show that the positive solution of the associated stochastic differential equation does not explode to infinity in a finite time. The authors proof the existence and uniqueness of the positive solutions.In addition, Uniformly Continuous of solution is studied.

In: Genetic Recombination Research Progress
Editor: Jacob H. Schulz, pp. 1-68

Chapter 1

The Intertwining of DNA Damage Response Pathway Components and Homologous Recombination Repair

Adam D. Brown[1,2,*], *Bijal Karia*[1,2*], *Amy M. Wiles*[2]
and Alexander J.R. Bishop[1,2,†]

[1]Department of Cellular and Structural Biology, University of Texas Health Science
Center at San Antonio, 7703 Floyd Curl Drive, San Antonio, TX 78229, USA
[2]Greehey Children's Cancer Research Institute, University of Texas Health Science
Center at San Antonio, 8403 Floyd Curl Drive, San Antonio, TX 78229, USA

Abstract

In response to DNA damage, the cell institutes an orchestrated and measured response. Central to this is the canonical DNA damage response pathway. The components of this pathway are many and include ATM, MRE11, RAD50, NBS1, CHEK2, p53, GADD45a, p21, BRCA1, BRCA2, FA proteins, BLM and WRN. Depending on cell cycle and damage type, different combinations of these components are utilized to affect an appropriate cellular response, including DNA repair. Absence of any one of these components is largely associated with clinical diseases, severe developmental defects, an increased propensity to develop cancer at an early age, and/or aging defects. Homologous recombination repair is a DNA repair mechanism that is capable of repairing a wide range of damages. It has been determined that homologous recombination repair is altered in the absence of almost every component of the damage response pathway in either a direct or indirect fashion or by functional implication. Here we will present the evidence for the intimate relationship of the damage response pathway, cell cycle and homologous recombination repair based on *in vivo* models and discuss the implications of their intertwining in mammalian systems.

[*] Authors contributed equally
[†] E-mail address: bishopa@uthscsa.edu (To whom correspondence should be addressed)

Introduction

The genome is under constant insult from both endogenous and exogenous sources, and yet genomic stability and integrity are essential for cellular viability. Mammalian cells have been estimated to have a spontaneous loss of purines at a rate of 10,000 per cell generation [1], and a human cell is estimated to acquire 10,000-150,000 oxidative damaged DNA sites per day [2]. Ionizing radiation (IR) studies using mammalian cell lines have shown that approximately 1,000 single strand breaks (SSBs) are generated per cell per 1 Gy [3] and 10-70 double strand breaks (DSBs) [4, 5]. If left unrepaired, the fate of these (and other) lesions would be either loss of enough genetic material to result in cell death, or misreplication and generation of mutations that would be fixed in somatic tissues or the germ line. To counteract these genomic damages, the cell has numerous DNA repair mechanisms. Amongst the various DNA repair systems, Homologous Recombination Repair (HRR) has the ability to correct a wide variety of DNA damages [6]. As such, HRR is an excellent indicator of the capacity of a cell to respond to DNA damage and reduce any resultant genomic instability.

In response to the constant barrage that the cell encounters, an elegant yet elaborate signaling and response system has evolved. The response to damage is pleiotropic, where, depending on the type and extent of damage, different components of the DNA damage response pathway (DDR) are utilized to elicit responses that include cell cycle arrest, transcriptional and translation changes, apoptosis, senescence, and DNA repair. We, and many others, have used HRR as an indictor of genomic instability and have taken advantage of this phenotype to dissect the relationship between components of DDR. Here we outline several of the central components of DDR, describing their relationship to one another and their likelihood for influencing HRR in response to particular types of damage.

Overview of Homologous Recombination

A DSB of a DNA molecule is considered to be the most toxic form of DNA damage due to the potential for excessive genomic instability leading to possible gross chromosomal rearrangements, and/or apoptosis. The contemporary view is that a DSB is the initiating event in HRR, though exposure to agents that result in different types of DNA lesions also induce HRR events [6]. At least five different pathways of homology-directed repair have been shown to comprise HRR: double Holliday junctions (HJ) for DSBs, single strand annealing (SSA), break induced repair (BIR), synthesis dependent strand annealing (SDSA), stalled replication fork repair. These pathways are briefly described below and are outlined in figures 1 and 2 [7].

Despite the different HRR pathway models, there is thought to be a conserved core set of events. Excluding SSA, HRR begins with the resection of double-stranded (dsDNA) from the DSB site in a 5'-3' direction, resulting in a 3' single-strand (ssDNA) tail. MRE11 of the MRE11, RAD50, NBS1 (MRN) complex has been proposed to possibly carry out this exonuclease activity [8]. To prevent reannealing of the ssDNA to other regions, replication protein A (RPA) binds to the 3' tail, thereby serving as a mediator and regulator of HRR [9]. The eukaryotic RecA homolog, RAD51, then coats the ssDNA while stretching out the DNA molecule to form a nucleofilament. RAD51 promotes strand invasion, allowing for DNA sequence homology searching. In addition to RAD51, there are other components that are thought to facilitate this process, including the RAD51 paralogs, RAD51B, RAD51C, RAD51D, x-ray repair complementing 2(XRCC2) and XRCC3, and several associated proteins, such as breast cancer 2, early onset (BRCA2), RAD52, RAD54 and RAD54B. RAD51C has been shown to complex with XRCC3 and has roles in filament formation [10, 11], strand pairing, and HJ resolution [12]. RAD51C and RAD51B can complex to mediate ssDNA-RAD51 filament formation [13]. RAD51B has also been shown to preferentially bind HJ and to possess a kinase activity [14, 15]. Lastly, RAD51D complexes with XRCC2 and, when bound to ssDNA, has an ATPase activity [16]. From this step the HRR pathways deviate yet still have the common goal of synthesizing DNA past the point of damage. The specific differences between the HRR pathways are briefly described below and are depicted in figures 1 and 2.

A. Double Holliday junction (Figure 1a). For the canonical double HJ model, ligation of the newly synthesized DNA to its original strand produces a second HJ. Following branching and migration, HRR is completed by HJ resolution by an as yet unidentified resolvase in mammalian cells. This model accounts for both gene conversion and crossing over of flanking markers, but studies suggest that crossover products are rarely observed in mammalian cells following the formation of a DSB [17].

B. Single strand annealing (SSA) (Figure 1b). SSA repair is an error prone pathway that results from the hybridization of homologous regions between repeated regions. The intervening regions that are deleted produce "flapped" DNA that must be excised before proper ligation. The removal of the genetic material between the homologous sequences is potentially deleterious.

C. Break-induced repair (BIR) (Figure 1c). BIR has been suggested as one form of repair used for telomere maintenance [18]. Similarly to SDSA, a second HJ does not form, allowing the displaced strand to serve as a template for lagging strand synthesis and the start of a new replication fork. This semiconservative replication could then be carried on to the end of the chromosome.

D. Synthesis dependent strand annealing (SDSA) (Figure 1d). In SDSA, branch migration through the break site releases the 3' invading strain, and the displaced strand reanneals to its complement. This type of repair leads to gene conversion with no crossing over of flanking markers.

E. Stalled replication fork repair (Figure 2). Mechanistically, repair at a stalled replication fork is similar to the double HJ model with the exception that only one HJ is formed. The resolution of this HJ structure permits the restart of a replication fork while allowing the exchange of genetic material between sister chromatids [19].

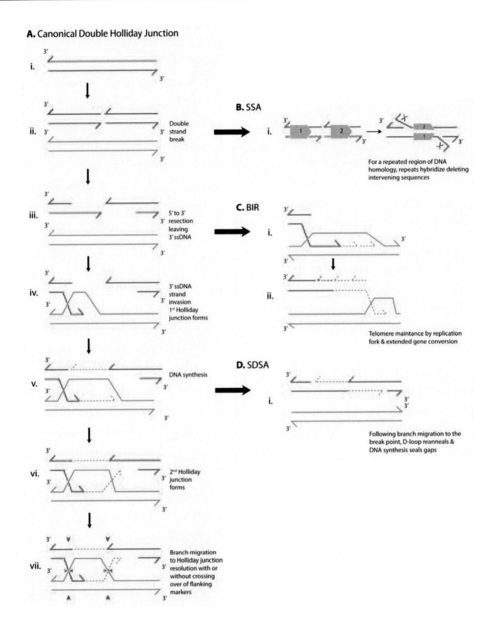

Figure 1. General Double Strand Break Repair Models. A. (i) The top solid blue lines depict a double stranded molecule of one chromosome with arrows showing direction of DNA synthesis. (ii) A region of homology, as shown in a solid red line, can be either intra- or inter-chromatid or on the same DNA molecule. (iii) 5' resection of the broken strand then leaves 3' overhangs, (iv) 3' single stranded DNA invasion and D-loop formation creates the first Holliday junction (HJ). (v) Nascent DNA synthesis occurs (red dashed lines) and (vi) forms the second HJ. (vii) Resolution (green arrowheads) of the HJs following branch migration leads to gene conversion with and without crossover products. B. (i) In lieu of 5' resection of the broken strand, repetitive DNA elements (orange pentagons) can anneal to correct the damage. Excision (black scissors) of intervening DNA sequence between the repeat elements potentially leads to mutations. C. (i) Following 3' single stranded DNA invasion, (ii) a replication fork can restart, leading to extended gene conversion. D. (i) As an alternative to forming a second HJ, branch migration to the site of the break will release the invading strain allowing the D-loop to reanneal to its sister chromatid.

Stalled Replication Fork Repair

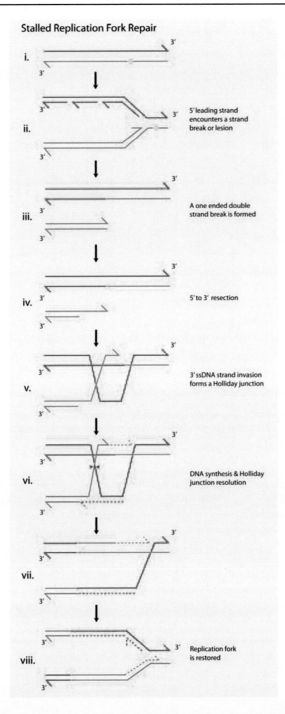

Figure 2. Double Strand Break Repair and Repair at Stalled Replication Forks. (i) Solid blue and red lines depict a DNA duplex, with arrows representing synthesis direction. Here, parental DNA is shown with a site of damage (yellow circle). (ii) As DNA synthesis occurs, the leading strand (solid red line) encounters the site of damage while the lagging strand (solid blue lines) continues synthesis; (iii) this results in a stalled replication fork and a one-ended DSB. (iv) 5' resection of the broken strand occurs. (v) Displacement of the invaded strand creates a D-loop forming a Holliday junction. (vi) Resolution (green arrowheads) following nascent DNA synthesis (red and blue dashed lines) (vii) creates a new replication fork, and (viii) DNA replication is restarted.

Common *in vivo* Systems Used to Measure HRR Events

There are a number of methods by which HRR events are often measured. One of the most widely used methods is the examination of damage-induced relocalization of proteins known to be involved in HRR. A number of DNA repair proteins are known to relocalize into distinct nuclear locations, or foci, and can be detected by immunofluorescence, including HRR components RAD51, RAD52, and RAD54 [20]. The frequency and timing of these foci are thought to relate to the amount of HRR that occurs. In addition, co-localization is used as evidence for protein interaction. Another method commonly used to determine if a protein has a role in HRR is by genetic analysis. Here investigators measure whether phenotypes associated with two mutants are epistatic to determine whether two genes lie on the same pathway. Work with the chicken DT40 cell line in particular takes advantage of this method [21]. The third method is to directly measure HRR events.

Numerous systems to directly measure HRR frequency have been developed over the years. It has largely been accepted that systems that are integrated within the chromosome, as opposed to episomal plasmids, best recapitulate normal genomic events. With regard to analyses of DDR, there are three systems of note. Two are tissue culture based systems that are derived from plasmid integration and are initiated by a restriction enzyme. These two systems take advantage of the unique 18 nt recognition sequence of the *Saccharomyces cerevisiae* (*S. cerevisiae*) intron encoded endonuclease I-*Sce*I. The third system is an *in vivo* mouse model that takes advantage of a spontaneous genomic mutation, pink-eyed unstable (p^{un}), and this system can measure both spontaneous and damage induced HRR events. Each system is described briefly.

A. Inverted repeat I-*Sce*I. The first system involves a stable integration of a construct composed of two inactive copies of the *E. coli* xanthine-guanine phosphoribosyl transferase gene (*gpt*) in an inverted tandem repeat configuration [22]. The upstream copy of *gpt* is inactivated by the insertion of the 18 nt I-*Sce*I recognition sequence and the other by deletion of the 3′ end. HRR is measured by growing cells in XHATM (xanthine, hypoxanthine, aminopterin, thymidine, and mycophenolic acid) selection media and quantifying mycophenolic acid-resistant colonies, that is, those colonies where functional *gpt* has been reconstituted by HRR. This system measures HRR gene conversion events resultant from intrachromatid pairing.

B. Direct repeat I-*Sce*I. The second system differs from the first in that it consists of a direct repeat of two inactive neomycin (*neo*) resistance genes separated by a hygromycin resistance gene [23]. The first *neo* is inactivated by a 5′ truncation (*3′neo*), the second inactivated by the insertion of the I-*Sce*I recognition sequences (*S2neo*). Colonies that are G418 resistant have undergone an HRR event, a gene conversion from either the same chromatid or a sister chromatid. Depending on the number of copies integrated, reconstitution of *neo* could also arise from unequal sister chromatid exchange (SCE).

C. Pink-eyed unstable (p^{un}) mouse model. The third system of note for the measurement of HRR events is the p^{un} mouse model [24, 25]. This system relies on the identification of cells or clones of cells that have arisen in the animal after they have undergone an HRR event within the p^{un} gene, reconstituting it to a functional gene, pink-eyed dilution (p). In order to achieve this gene reconstitution, the HRR event must correctly delete one copy of a tandem 65 kb repeat that encompasses exons 6 – 18 (i.e. the p^{un} allele consists of exons 1-18;6-23 compared to the p allele with exons 1-23). The functional allele resultant from an HRR event is phenotypically observable by the presence of pigmentation in melanocytes found in the fur [24, 26] and in cells of the retinal pigment epithelium (RPE) [27-29]. The HRR event that reconstitutes the p gene can be either SSA or unequal crossing over either between sister chromatids or between homologues. The resulting fur- or eye-spots are increased in frequency after exposure to different DNA damaging agents [26, 30-32] as well as in different genetic backgrounds [33-35]. In addition, we have demonstrated that due to the well-defined developmental pattern of the RPE [36], we are able to relate the time of exposure to a specific region within the RPE where HRR events are induced, providing a highly sensitive *in vivo* HRR assay [37].

The DNA Damage Response System and Cell Cycle Checkpoints

The DDR pathway is a highly complex, cellular signaling network that incorporates DNA damage sensing, the regulation of context appropriate response, the relaying of signals to transducer proteins, and the coordinated activation of a multitude of effector proteins that execute cellular responses [38]. The most notable and best-understood cellular responses to a DNA damage signal are cell cycle arrest, apoptosis, senescence, and DNA repair. Though much of this pathway has been elucidated, many of the details are still actively being researched. A complete discussion of this pathway is beyond the scope of this review, therefore we will only present some of the major components of the pathway and some detail of cell cycle arrest, mainly highlighting those features that have a potential relationship to HRR.

A proliferating cell undergoes a four-phase cycle of events: Gap 1 (G_1), DNA synthesis (S), Gap 2 (G_2) and mitosis (M) [39]. The fidelity of cell division is important for normal development, genomic stability and suppression of tumorigenesis. One mechanism to facilitate the fidelity of this process is the use of cell cycle checkpoints (Figure 3). The progression of the cell cycle is catalyzed by cyclin-dependent kinases (CDKs) and their activity requires the association with cyclins. The cell cycle is regulated by oscillating levels of cyclins and their interactions with CDKs, in that cyclin:CDK complexes phosporylate proteins necessary to progress through the various phases of the cell cycle.

A cell cycle checkpoint is the temporary arrest of cell cycle progression to correct any issues that are encountered, including DNA damage. There are a number of defined points within the cell cycle when a cell may execute a checkpoint: at the G_1/S border, intra S-phase, at the G_2/M border, and the spindle checkpoint. With regard to DNA damage, an arrest would allow the cell time to repair the damage, ensuring accurate replication of the genome [40]. If the damage is too severe, the cell will undergo programmed cell death, thereby avoiding the transmission of the damage and possible DNA mutations to its daughter cells. The G_1/S checkpoint, or restriction point, is the point during cell cycle when the cell becomes committed to progress into cellular division, delay division, or enter a resting or "quiescent" state. Once past this border, initiation of replication begins and the cell enters S-phase. Many intra S-phase checkpoints exist to ensure accurate replication of the chromosomes. In general, a cell cycle arrest during S-phase is the result of a stalled replication fork. Because a variety of DNA damaging agents, including IR, ultraviolet radiation (UV), hydroxyurea (HU), mitomycin C (MMC), cisplatin, and methylmethane sulfonate (MMS), can result in a stalled replication fork, it is not surprising that all these agents will induce arrest at this checkpoint [41]. The G_2/M checkpoint ensures accurate and complete replication of chromosomes before entering mitosis. The cell proceeds to mitosis when one last checkpoint, the spindle checkpoint, ensures proper segregation of chromosomes into daughter cells. The details of each checkpoint along with respective sensor, transducer, and effector proteins are discussed below.

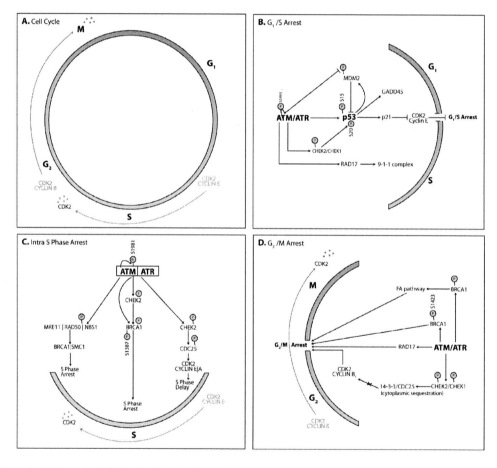

Figure 3. DDR and Cell Cycle Checkpoints. This schematic represents the main protein interactions (arrows), inhibitions (blunt ended lines) and phosphorylation events (P) involved in the response to DNA damage at various stages of the cell cycle. A. A simple representation of the phases of the cell cycle G_1, S, G_2 and M along with the relevant cyclins and cyclin-dependent kinases (CDK). The degradation of cyclins is represented by colored dots, purple for cyclin E and blue for cyclin B. Detail is added for each checkpoint: B. G_1/S arrest, C. intra S-phase arrest and D. G_2/M arrest.

Ataxia telangiectasia mutated (ATM) and ataxia telangiectasia and RAD3-related (ATR) are the two main transducer protein kinases that orchestrate a variety of effector proteins in response to damage. Both proteins act very early in the damage response pathway, apparently with many overlapping functions. The largest distinction between the roles of these proteins is the type of damage to which they respond. ATM responds mostly to DSBs induced by oxidative stress or IR and is thought to rapidly phosphorylate effector proteins, while, following IR, ATR is thought to play a delayed role, maintaining the phosphorylation of effector substrates. In contrast, ATR is also activated in response to a wide variety of damages, stalled replication forks in particular, as well as UV light damage and hypoxia [42-46].

Once activated by damage, ATM and ATR are responsible for triggering several cell cycle checkpoints. The major cellular checkpoint ATM and ATR elicit is at the G_1/S border and is by phosphorylation and activation of p53 [46-49], resulting in the expression of p21. p21 directly inhibits cyclinE:CDK2 activity [50] thereby inhibiting the G_1/S transition. In addition to the G_1/S checkpoint, ATM and ATR can mediate S-phase and G_2/M checkpoints via phosphorylation of BRCA1 [51-53]. RAD17 is also phosphorylated by ATM and possibly by ATR, and this may be important for both the G_1/S and G_2/M checkpoints. In addition, ATM and ATR activate checkpoint kinase 2 (CHEK2) and CHEK1, respectively [54-56], and deactivate the phosphotase cell division cycle 25 (CDC25C) [57]. CDC25C normally regulates the progression of G_2 by dephosphorylating and activating CDK1. Phosphorylation of CDC25C allows the 14-3-3 protein to bind and sequester these now complexed proteins in the cytoplasm, thereby activating the G_2/M checkpoint [58-60].

In summary, the DDR system, in combination with DNA damage checkpoints, is a highly complex, interconnected, self-reinforcing system. As such, this system is thought to play a crucial role in the coordination of DNA repair, allowing replication of the genome with the highest possible fidelity.

Central Components of the Damage Response Pathway

The DDR system is composed of many proteins and interactions. Here we outline some of the central components of DDR, describing each protein, the clinical manifestations of any associated diseases, and what is understood about the molecular roles of each protein with regard to DDR, cell cycle control, and HRR. Of note, we have not included a detailed description of ATR and CHEK1 since these proteins are a little less well characterized, are thought to be mainly involved in response to damages other than DSBs, and at least with regard to their molecular function, are likely to be largely similar to ATM and CHEK2, respectively.

ATM

Ataxia telangiectasia mutated (ATM) is a tumor suppressor gene that is responsible for the rare autosomal disease ataxia telangiectasia (A-T). ATM is a member of a conserved family of phosphatidylinositol 3-kinase (PI3K) related proteins that has an intricate role in

DDR, especially to DSBs [61]. A-T patients display profound cerebellar ataxia, marked dilation of small blood vessels (telangiectasia), increased and early onset of typically lymphoreticular cancer, and severe sensitivity to IR [62]. In the absence of ATM, cellular phenotypes include exquisite sensitivity to IR and radiomimetic agents; radioresistant DNA synthesis (RDS); defective cell cycle checkpoints following irradiation, such as prolonged S-phase, defective G_1/S and G_2/M, and DNA repair deficiencies; and chromosomal instability [38].

ATM contains a number of structurally important domains [63]. These include a nuclear localization signal (NLS), leucine zipper motif, proline rich motif, FRAP/ATM/TRRAP (FAT) domain, kinase (PI3K) domain, and FAT c-terminal (FATC) domain (Figure 4). ATM is an early component of the DDR signaling pathway with many downstream substrates including itself. ATM is located primarily in the nucleus with a small amount being cytoplasmic. In its inactive form, ATM is found as a dimeric structure. Bakkenist and Kastan very elegantly demonstrated that after DNA damage ATM undergoes an inter-chain autophosphorylation on Ser[1981] (Figure 4) [64], which disrupts the dimeric structure and releases the active monomers. Since then, several other autophosphorylation sites have been discovered on ATM including Ser[367], Ser[440], and Ser[1893] (Figure 4) [65]. In response to UV induced stalled replication forks, ATM undergoes an ATR-dependent ATM autophosphorylation on Ser[1981]. This autophosphorylation does not require the MRN complex but does result in normal phosphorylation of downstream targets such as CHEK2 [66]. The exact sequence of events required for ATM activation is still being elucidated, though it is known that the autophosphorylation is not dependent on the direct binding of DNA but accumulation of ATM at sites of damage does occur, an event apparently enhanced by the MRN complex [67-70]. It does appear though, that the interaction between ATM and chromatin state is important for the autophosphorylation event [71, 72].

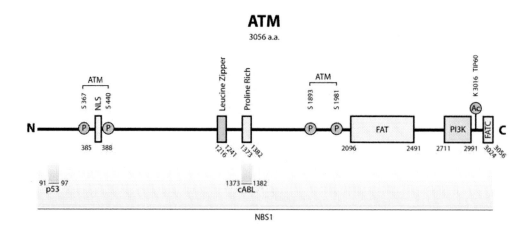

Figure 4. ATM Protein. ATM is a large protein composed of 3056 amino acid residues. Domains are depicted in colored boxes, including a nuclear localization signal (NLS), FRAP/ATM/TRRAP (FAT) domain, phosphatidylinositol 3-kinase (PI3K) and FAT C-terminal (FATC) domain. Phosphorylation events (P), by ATM, and acetylation events (Ac), by TIP60, are indicated, and relevant protein interactions are highlighted below the protein.

ATM has a number of targets of which only the central components of the DDR pathway are presented here. Two of the most rapidly phosphorylated targets of activated ATM include the alternative histone γ-H2AX and Nijmegen breakage syndrome 1 (NBS1) of the MRN complex. ATM phosphorylates γ-H2AX on Ser^{139} [73] and NBS1 on Ser^{278} and Ser^{343} [74-78]. In addition to phosphorylating NBS1, ATM also directly interacts with the C-terminus of NBS1 [78, 79]. Although the two phosphorylation events of NBS1 are independent of each other, both are necessary for proper *in vivo* DDR.

Activated ATM has many other downstream targets thought to regulate cell cycle, DNA repair and apoptosis. An interesting aspect of ATM target proteins is that a number of them can phosphorylate each other, reinforcing DDR signals. As described earlier, ATM directly phosphorylates CHEK2 on Thr^{68} [80, 81] and p53 on Ser^{15} [49]. The activated form of CHEK2 phosphorylates p53 on Ser^{20}, interrupting the ability of double minute 2 (MDM2) to bind with p53. ATM also directly phosphorylates MDM2 on Ser^{396} after exposure to IR [82, 83], decreasing its ability to shuttle p53 out of the nucleus into the cytoplasm for degradation. Thus, in a multiply reinforced manner, ATM both activates and stabilizes p53. Additional substrates of ATM that are thought to be involved in DDR and DNA repair include the Fanconia anemia complementation group D (FA-D2) on Ser^{222} [84]; the breast cancer susceptibility gene BRCA1 on Ser^{1457} [85], Ser^{1387} during S-phase [51], and Ser^{1423} during G_2/M-phase [52]; Abelson murine leukemia (cABL) on Ser^{465} [86]; BLM (gene mutated in Bloom syndrome) on Thr^{99} and Thr^{122} [87]; and the p34 subunit of RPA [88].

Three mouse models of A-T have been produced to recapitulate the human disease [89-91]. All three models display early onset of cancer, usually lymphoma and exquisite sensitivity to IR when homozygous mutant. No overt phenotype has been described for the heterozygous mutant animals.

The role of ATM in HRR has been indicated from its interactions, genetics, and by direct measurement. Some of the substrates of ATM are thought to play a direct role in HRR, for example BRCA1, BLM and FA-D2. Meyn was the first to demonstrate that cells from A-T patients have high spontaneous recombination rates and that these rates are probably due to an aberrant cell cycle regulation, allowing progression with damaged DNA [92]. ATM phosphorylates, as well as physically interacts with cABL [93]. Biochemical data have shown that ATM phosphorylation is necessary for cABL to phosphorylate and associate with members of the HRR mechanism (e.g. RAD51 and RAD52) following exposure to IR [94]. Using the *p53* null DT40 chicken cell line, Morrison *et al.* found that ATM was epistatic with RAD54, displaying no greater sensitivity to IR when these two mutations were combined [95]. The authors demonstrated that following IR exposure there was an increase in RAD51 and RAD54 nuclear foci in $ATM^{-/-}$ cells (see below). Bolderson *et al.* demonstrated that ATM is activated following thymidine-induced stalled replication forks; that this thymidine exposure also induced HRR events in an ATM dependent manner by using the direct repeat *I-SceI* assay system [96]. *In vivo*, using the p^{un} mouse system, Bishop *et al.* observed that $Atm^{-/-}$ mice had increased levels of spontaneous intrachromosomal HRR [97] and this increase was latter shown to be suppressed decreasing oxidative stress [98]. In addition to this finding, Bishop *et al.* went on to show that the increased HRR in the absence of Atm only occurs at later times during murine embryonic development [35]. Overall, it appears that ATM is involved in both suppressing spontaneous levels of HRR and controlling HRR following exposure to certain forms of damage.

MRE11/RAD50/NBS1 (MRN) Complex

The MRN complex is composed of three proteins, MRE11, RAD50 and NBS1. We present each of these three proteins in one section since their major role is thought to be while they are acting as the MRN complex, each component providing specific functionality to the complex. This combined functionality does not exclude that they may have additional activities outside of the complex.

MRE11

The core of the MRN complex is MRE11 and, when defective, results in the rare, autosomal recessive disease that is similar to A-T called A-T-like disorder (A-TLD). A-TLD patients share the neurological degeneration phenotype of A-T patients, but not the ocular telangiectasis [99]. Also reminiscent of A-T, cells from A-TLD patients display chromosomal instability, often involving chromosomes 7 and 14, and increased sensitivity to IR, resulting in chromosomal breakage and RDS [99]. These cellular phenotypes suggest a role for MRE11 in the DDR and in DNA repair.

MRE11 has been shown to possess multiple activities, which together suggest an intimate role in sensing damage and its initial processing. MRE11 contains two functional regions, including four conserved phosphoesterase motifs located at the N-terminus and two DNA binding domains, one of which is located at the N- and the other at the C-terminus [100] (Figure 5a). The MRN complex has been shown to bind cooperatively to DNA via the DNA binding domains of MRE11 [101], binding both dsDNA and ssDNA, with a preference for ssDNA [102]. Following a DSB, the MRN complex has been postulated to bring together two DNA ends, acting as "molecular Velcro," a bridging structure thought to be involved in DNA repair [103]. This bridging may facilitate the ability of MRE11 to promote the joining of non-complementary ends by annealing short regions of homologies in a manner similar to non-homologous end joining (NHEJ), an activity that MRE11 has *in vitro* [8, 104]. Another activity of MRE11 is 3'-5' exonuclease. This exonuclease activity is increased when MRE11 is complexed with RAD50; additionally, MRE11 has a 3'-5' endonuclease activity [8, 104] that is ATP-dependent and requires both RAD50 and NBS1 [101]. Interestingly, the MRE11 nuclease activities appear to pause when the protein encounters regions of sequence microhomology [105], again suggesting a potential role in NHEJ. Furthermore, MRE11 has been shown to possess a strand-annealing activity towards complementary ssDNA molecules, and this activity, unlike RAD52 in HRR, is abrogated by RPA binding to ssDNA [102].

MRE11 has several protein:protein interactions that are integral to DDR, the most notable of which are its associations within the MRN complex. MRE11 has been shown to bind to itself [8], with RAD50 [8, 106] and with NBS1 [107, 108]. NBS1 is thought to be required for both damage induced focus formation and nuclear localization of the complex [107, 108]. Based on co-localization of damage induced immunofluorescent foci, MRE11 is believed to also associate with BRCA1, BLM, proliferating cell nuclear antigen (PCNA) and other DNA repair proteins that make up the BRCA1-associated genome surveillance complex (BASC) [109]. An independent study showed that BRCA1 interacts both *in vitro* and *in vivo* with RAD50, leading to an indirect association between BRCA1 and MRE11 [110]. Furthermore, the binding of BRCA1 to DNA inhibits the nucleolytic activities of MRN [111]. Additional foci data indicated that p53 also localizes with MRE11 following an induced replication fork arrest [112].

Within the nucleus, MRN normally resides in promyelocytic leukemia (PML) nuclear bodies in the absence of damage, but shortly after IR treatment this complex leaves the PML bodies [113, 114]. Within 30 minutes of DNA damage, MRE11 and RAD50 are recruited to sites of damage independent of RAD51, suggesting a distinct role for these repair proteins [115]. Following IR, DNA repair deficient cells had an increase in MRE11:RAD50 foci when compared to repair proficient cells [116], and additional work showed a marked reduction of MRE11:RAD50 foci in A-T cells. Subsequent studies demonstrated that the posttranslational modification of MRN via ATM phosphorylation of NBS1 was not required for MRN to associate with damaged DNA [113]. These studies suggest that MRE11 complexed with RAD50 and NBS1 acts in DDR in a partially ATM-dependent manner at best. Another important association of the MRN complex is with PCNA and E2F family members via NBS1 at replication forks during S-phase [117]. Altogether it appears that the MRN complex is involved with both DDR and intra S-phase checkpoint to suppress genomic instability.

Work by Xiao and Weaver was the first to show that mammalian MRE11 is essential, with its loss resulting in embryonic lethality [118]. To better recapitulate A-TLD, Theunissen *et al.* have since developed a hypomorphic viable mouse ($MRE11^{ATLD}$) with an amino acid change (R633 to STOP), resulting in a 75-residue truncation [119]. Mouse embryonic fibroblasts derived from the $MRE11^{ATLD/ATLD}$ mouse recapitulate the A-TLD cellular phenotype in that the MRN complex forms, albeit at lower levels, and these cells display IR sensitivity and lack MRN foci following IR treatment. $MRE11^{ATLD/ATLD}$ cells display RDS (a defect in intra S-phase checkpoint) and a defect in G_2/M checkpoint compared to wild-type cells [119]. Although MRE11 is found at stalled replication forks, is associated with PCNA, and possess exonuclease activity, there have been no studies that have been published that directly link MRE11 with HRR in mammalian cells. Interestingly though, studies using yeast have shown an association between MRE11 and meiotic recombination [120, 121] and damage induced mitotic recombination [122, 123].

RAD50

RAD50 is a component of the MRN complex with an ATPase activity that facilitates DNA binding [124]. Patients who have been identified with germline mutations in RAD50 have cellular radiosensitivity and microcephaly, although no apparent immunodeficiency is present [125]. These phenotypes are reminiscent of cells with defects in other components of DDR and DNA repair.

RAD50 has 1328 amino acid residues (Figure 5b) and is member to the SMC (structural maintenance of chromosomes) family of proteins [124]. It consists of two ATP-binding cassettes (ABC) at either terminal of the protein, separated by two, long, heptad repeat regions [126]. The N-terminal ABC harbors a Walker A motif; the C-terminal ABC harbors a Walker B motif. When correctly folded, the RAD50 protein folds back onto itself, allowing the two ATP domains to join at the apex of the structure. The central region of the protein is also a conserved motif (Cys-X-X-Cys) that forms the zinc-hook binding coiled-coil dimerization motif. The heptad repeats that flank this central region fold into an anti-parallel coiled-coil domain. The MRN complex is formed when two MRE11 proteins interact with RAD50 by binding each of the ABCs, followed by NBS1 interaction with MRE11 [126].

The MRN complex is critical for mediating diverse functions in DSB repair. Mutation of any one of the MRN complex components results in embryonic lethality [118, 127-129]. Bender *et al.* recently described a hypomorphic *Rad50* mutant mouse ($Rad50^{S/S}$) that is analogous to the *S. cerevisiae rad50S* allele [130]. In *S. cerevisiae rad50S* mutants, the meiotic HRR initiating DSB is produced, but is not fully processed [131]. The $Rad50^{S/S}$ is partially embryonic lethal, with a loss of male germ cells and cells in the hematopoietic lineages. Most of the mice that do survive development die by three months of age from hematopoietic failure, although a few are longer-lived with a high predisposition to lymphomas. p53 deficiency rescued the $Rad50^{S/S}$ hematopoietic and testicular attrition, whereas the tumor latency in $p53^{-/-}$ and $p53^{+/-}$ mice was reduced by $Rad50^{S/S}$. Compared to $p53^{-/-}$ tumors, lymphomas in the double mutant mice exhibited increased chromosomal rearrangements attributable to genomic instability [130]. Primary cell cultures from $Rad50^{S/S}$ mutant mice exhibit chronic genotoxic stress and have chromosome rearrangements though no checkpoint deficiency or impairment in DSB recognition by the MRN complex is observed [130].

$Rad50^{S/S}$ mice exhibit constitutive signaling to ATM, leading to increased levels of ATM autophosphorylation [132]. This chronic stimulation of ATM leads to apoptosis in hematopoietic and spermatogenic lineages and early death of these mice due to severe anemia. $Rad50^{S/S}$ along with deficiencies in *Atm*, *Chek2* and *p53* attenuated this apoptotic response indicating that the Rad50S allele up regulated an Atm-Chek2-p53 dependent apoptosis pathway. Smc1 is also a substrate of Atm, important in intra S-phase cell cycle checkpoint activation, and therefore was constitutively phosphorylated in $Rad50^{S/S}$ mice. Smc1 deficiency in $Smc1^{2SA/2SA}$ mice, however, had no effect on the $Rad50^{S/S}$ phenotype [132]. From these results, Morales *et al.* proposed a second and parallel endpoint to the MRN complex-ATM DNA damage response pathway by demonstrating a separation of the influence of MRN on apoptosis from checkpoint activation [132]. Thus, studies with the *Rad50S* allele have provided strong support for the MRN complex as a DNA damage sensor, in its requirement for the activation of ATM-dependent apoptosis, and the elucidation of two parallel outcomes in response to DNA damage (apoptosis and cell cycle checkpoint activation) governed by the MRN complex and ATM [133].

RAD50 is a critical component of the cellular response to DSBs including HRR. Use of immunofluorescence, Maser *et al.* observed that MRE11 and RAD50 co-localize in response to IR induced DNA damage. Interestingly, foci formation was cell line specific; foci formation was drastically reduced in cells with *ATM* mutations. This links the MRN complex in DSB repair regulated by the ATM-dependent DDR pathway [116].

NBS1

The *NBS1* gene product is a member of the MRN complex. Mutation of *NBS1* is responsible for Nijmegen breakage syndrome (NBS), a disease that shares many clinical symptoms with A-T. Clinical phenotypes of NBS patients include characteristic faces, microcephaly, mental retardation, some immunodeficiency, stunted growth and an increased propensity for cancer onset at an early age, mainly hematolymphoid with some cases of rhabdomyosarcoma and neuroblastoma [134-136]. Unlike A-T, NBS does not result in increased α-feto protein levels, telangiectasia, or any abnormal neurological characteristics [137]. On the cellular level, NBS and A-T do share some characteristics, chromosomal instability in particular. Cells from NBS patients have increased spontaneous chromosomal instability and fragility that often involves chromosomes 7 and 14 [138-140]. In addition, similar to A-T patient cells, NBS patient cells are hypersensitive to IR or radio-mimetic drug induced chromosomal breaks [140-142]. NBS patient cells also display accelerated telomere shortening, cell cycle checkpoint defects and RDS [138, 143, 144].

NBS1 mRNA is ubiquitously expressed to produce a 754-residue protein. NBS1 is comprised of three regions, the N-terminus region, central region, and C-terminus region; each region has a unique role in NBS1 action (outlined below) [144]. Within the N-terminal region, or damage recognition portion of the protein, lies the fork-head associated (FHA) domain, a BRCA1 carboxyl-terminal (BRCT) domain and a second putative BRCT domain (Figure 5c). The central region, or the signal transduction region, contains two serine/glutamine (SQ) motifs that are sites for phosphorylation. The C-terminal region of NBS1, also known as the interaction domain, contains a MRE11 binding domain and a PI3K-interacting motif (Figure 5c) [63].

In order to better understand the *in vivo* role of NBS1, a number of different mouse models have been developed. Zhu *et al.* first showed NBS1 as essential when disruption of exon and intron 1 caused embryonic lethality [129]. Two additional models later demonstrated that N-terminal truncation of NBS1 recapitulated the symptoms found in NBS patients; one model being a hypomorph recapitulating the 657Δ5 allele found in approximately 95% of NBS patients [145, 146]. A fourth model, which is heterozygous for NBS1, also has clinical phenotypes similar to the other models, a predisposition to develop lymphomas and a sensitivity to IR exposure, suggests that haploinsufficiency is the mechanism responsible for tumor development [147].

The role of NBS1 in DDR and HRR is inferred by its interaction with various proteins involved in both processes. MRE11 and RAD50 interact, and the cloning of p95 (*NBS1*) revealed that NBS1 is also a member of this complex [108, 148]. Since its discovery, NBS1 has been shown to interact and associate with many other proteins involved in DDR and DNA repair.

The complexity of how ATM, γ-H2AX and the MRN complex act on damaged DNA is still largely unclear. Nonetheless, these three components are necessary for the activation of DNA repair, cell cycle checkpoints and apoptosis mechanisms. To summarize the current models, DNA undergoes damage and causes a change in chromatin modeling. ATM phosphorylates γ-H2AX, and NBS1 independently recruits MRE11 and RAD50 to the nucleus via interactions at the C-terminus of NBS1, forming the MRN complex [149]. At the site of damage, the MRN complex, via the NBS1 FHA domain, interacts with phosphorylated γ-H2AX [150]. ATM can both phosphorylate NBS1 at Ser^{278} and Ser^{343}, thus activating the protein, as well as interact directly with the PI3K interaction motif within the C-terminus of NBS1 [74-79]. The activity and interactions of other ATM targets either requires or is enhanced by phosphorylated NBS1 [72, 144, 151]. Altogether, it appears that NBS1 is required for MRN recruitment and the interaction of ATM at sites of DNA damage.

NBS1 appears to facilitate cellular functions in addition to damage response through interactions with a number of other proteins. For example, the MRN complex via NBS1 has been observed at both origins of replication and at replication forks. This localization is mediated by an interaction between NBS1 and E2F1 and was demonstrated by both co-localization at immunofluorescent foci and co-immunoprecipitation [117]. NBS1 has also been shown to interact with MDM2 in a p53-independent manner, and though the exact role of this interaction is not understood, it does appear to inhibit DSB repair [152]. Additionally, NBS1 has been shown to localize to telomeres during S-phase. Considering the accelerated telomere shortening observed in NBS cells, it is possible that NBS1 has a role in telomere maintenance. In support of this, NBS1 has been demonstrated to interact with telomeric repeat binding factor (TERF1), a telomere-capping protein [153, 154]. Furthermore, WRN (gene mutated in Werner sydrome), interacts with MRN via NBS1 following DNA damage, IR and cross-linking damage in particular. NBS1 apparently regulates WRN helicase activity, though the role of WRN in repair or DDR is poorly understood. It is known that WRN deficient cells have shortened telomeres and increased genomic instability (see below regarding WRN) [155].

NBS1 is believed to have both a direct and an indirect role in controlling HRR. Indirectly NBS1 affects HRR in an ATM-dependent mechanism as outlined above. The MRN complex is also a member BASC and has been shown to co-localize with BRCA1 by immunofluorescent foci formation following damage [109]. Studies using DNA damaging agents other than IR, namely UV and HU, have shown that ATR also phosphorylates NBS1 at Ser^{343} (Figure 5c), and that this phosphorylation event is responsible for RPA2 to mediate the S-phase checkpoint [156]. Similar studies using MMC to cause DNA damage demonstrated that NBS1 and FA-D2 co-localize in nuclear foci [157].

Direct evidence for NBS1 involved in HRR comes from a number of independent studies. NBS1 mutant chicken DT40 cells have an increased level of SCE after treatment with MMC. Tauchi et al. demonstrated that this increase was due to HRR and not NHEJ by using the direct repeat I-SceI assay system [158]. Using a similar I-SceI assay, Sakamoto et al. determined that human NBS fibroblasts are involved in regulating I-SceI induced HRR frequencies depending on the interaction that is disrupted [159]. The I-SceI induced HRR frequencies were unaltered when the ATM interaction was disrupted but were reduced upon disruption of the MRE11 interaction. This decrease in HRR frequency was also recapitulated using cells from a Cre/lox conditional mouse model, where NBS1 exon 6 is deleted. This study also utilized the direct repeat I-SceI system to quantify HRR events as well as RAD51 and BRCA1 foci formation assays [160].

Clearly, the MRN complex, through the individual protein functions of MRE11, RAD50, and NBS1, is directly involved in DDR. Though NBS1 has been the only member of the complex shown to have a direct role in HRR, the involvement of the MRN complex as a whole in mammalian systems is highly implicated. Furthermore, the orchestrated signaling pathway that is thought to involve the MRN complex proteins and ATM following DNA damage is unresolved.

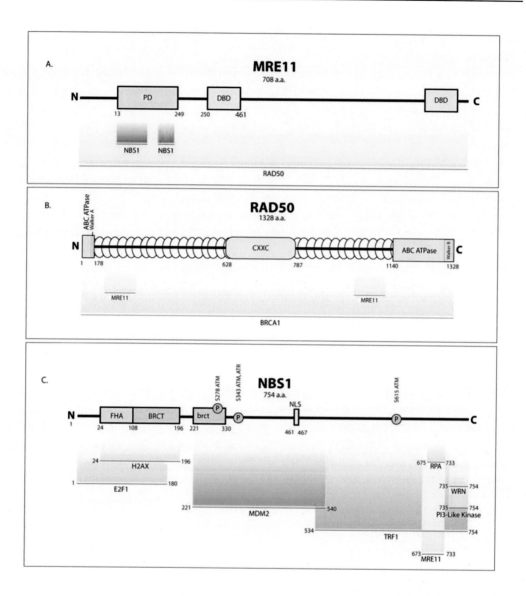

Figure 5. MRN Complex. The MRN complex is composed of MRE11, RAD50 and NBS1. Domains are depicted by colored boxes, and relevant protein interactions are highlighted below each protein. A. MRE11 domains include a phosphoesterase domain (PD) and two DNA binding domains (DBD) B. RAD50 include two ATP binding cassette (ABC) ATPase domains and a central Cys-X-X-Cys (CXXC) zinc finger hook domain. C. NBS1 domains include forkhead-associated (FHA), followed by two BRCA1 C-terminal (BRCT) repeats, the second being putative repeat (brct) and a nuclear localization signal (NLS). Phosphorylation events (P), by ATM and ATR, are noted.

CHEK2

Checkpoint kinase 2 (CHEK2) is a tumor suppressor involved in the DDR signaling cascade, controlling cell cycle and possibly apoptosis. Mutation in CHEK2 is associated with numerous clinical cases of various cancer types. The inheritance of a single nucleotide deletion within the kinase domain (1100ΔC) was found in a patient diagnosed with Li-Fraumeni syndrome, normally the result of inheriting a p53 mutation [161] and suggests a functional link between these two proteins. This same mutation is also found in breast cancer tumors [162-164], as well as in ovarian [165], prostate [166, 167] and colorectal cancer [168].

The gene product of *CHEK2* is a 543-residue protein that is a member of a conserved group of serine/threonine kinases. Through evolution, this group of kinases has retained three conserved elements [169]: an N-terminal serine or threonine followed by a glutamine (SQ/TQ) rich motif, a FHA domain and a kinase domain (Figure 6). CHEK2 activity is regulated by post-translational phosphorylation, but a single phosphorylation event is insufficient for activation. Instead a series of phosphorylation events are required for complete activation, the details of which are an area of active investigation. The first step of CHEK2 activation is the phosphorylation of Thr^{68} within the SQ/TQ domain. Phosphorylation at Thr^{68} by ATM and ATR [80, 81] was the first insight into the role of CHEK2 in DDR. More recent data have shown that DNA-dependent protein kinase catalytic subunit (DNA-PKcs) [170] and the MRN complex [171] could possibly also phosphorylate CHEK2 on Thr^{68}. Lou *et al.* demonstrated an interaction with mediator of DNA damage checkpoint (MDC1) and suggested the possibility that MDC1 facilitates the interaction between CHEK2 and upstream activating proteins (e.g. ATM and MRN) [172]. In contrast, wild-type p53-induced phosphatase 1 (WIP1), a known phosphatase, has been shown to directly interact with the SQ/TQ motif (Figure 6) [173], thereby functioning as a negative regulator of CHEK2 by dephosphorylating Thr^{68} in response to DNA damage [174]. Once Thr^{68} is phosphorylated, two CHEK2 molecules dimerize via their FHA domains. This in turn, facilitates their complete activation through *trans*-autophosphoylation on residues Thr$^{383/387}$, followed by CHEK2 dimer dissociation to release active monomers [175]. An interaction between PML bodies and CHEK2 has been shown to mediate this autophosphorylation step [176].

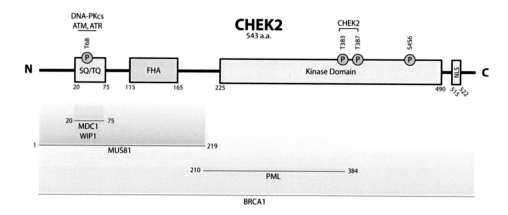

Figure 6. CHEK2 Protein. CHEK2 is a relatively small protein consisting of 543 amino acid residues. Domains are depicted in colored boxes: serine/glutamine or threonine/glutamine (SQ/TQ), forkhead-associated domain (FHA) and a nuclear localization signal (NLS). Phosphorylation events (P), by DNA-PKcs, ATM, ATR and CHEK2, are indicated, and relevant protein interactions are depicted below the protein.

CHEK2 has many substrates, the best characterized of which appear to be involved in regulating cell cycle. CHEK2 directly phosphorylates CDC25A on Ser^{123}, blocking the G_1/S-phase transition and transient S-phase block [177]; CDC25C on Ser^{216}, blocking the G_2/M-phase transition [178]; and p53 on Ser^{20}, sustaining G_1 and G_2 arrest [58, 59, 179]. In addition to this role in cell cycle control, CHEK2 regulates apoptosis through its interaction with p53, as shown *in vivo* using two knockout mouse models. Hirao *et al.* concluded that in response to ionizing radiation, Chek2 is required for apoptosis of thymocytes in a p53-mediated manner, independent of the Atm-Chek2 cell cycle arrest pathway [180]. A second mouse model by Motoyama and colleagues also demonstrated a role for Chek2 in p53-mediated apoptosis after IR treatment [180]. The direct phosphorylation of Brca1 on Ser^{988} by Chek2 suggests an involvement in DNA repair regulation [181]. Chek2 directly interacts with Brca1, but this interaction is disrupted by the Ser^{988} phosphorylation event [181]. BRCA1 has been implicated as having a role in DNA repair, particularly in HRR (see below). A third mouse model, a knock-in of the *CHEK2*(*)1100ΔC mutation, has recently been developed to understand the role of CHEK2 in breast cancer maturation and HRR. Mouse embryonic fibroblasts (MEFs) from homozygous knock-in mice have abnormal cell cycle profiles, multinucleated cells and increased genomic instability, indicated by an elevation of constitutive levels of DNA damage as observed by γ-H2AX foci [182].

CHEK2 is an important mediator of the DDR signaling cascade, playing a role in controlling cell cycle progression. Recent studies though have shown a more direct role for CHEK2 in regulating HRR. Zhang et al., used the inverted repeat I-*Sce*I system to show that BRCA1 regulation on DSB repair is dependent on CHEK2 phosphorylation at Ser^{988} [22]. CHEK2 has also been shown to interact at HJ with MUS81, which is a protein with known resolvase activity believed to be involved in HRR [183].

p53

TP53 (p53) is a tumor suppressor, mutated in at least half of all human cancers of all varieties [184]. p53 appears to be important for tumor development and in protecting the integrity of the genome [185] with a strong selective pressure for its loss. Further, individuals with Li-Fraumeni syndrome, patients who commonly inherit a mutant *TP53* gene, are highly predisposed to cancer development. These patients experience a spectrum of cancers, usually at an early age, that include breast carcinomas, soft tissue carcinomas, brain tumors, osteosarcomas, leukemia and adrenocortical carcinomas [186].

p53 is central to the DDR pathway. p53 deficient primary cells and tumor cells exhibit enhanced proliferation and increased levels of chromosomal instability. They also have centrosome duplication abnormalities, which could contribute to the aneuploidy phenotypes seen in these cells [187]. A key phenotype of these cells is that they exhibit a p53 dose dependent resistance to DNA damaging agents, with p53 null cells more resistant to DNA damage than p53 wild-type cells, and p53 heterozygous cells having an intermediate resistance to DNA damaging agents [188-190]. Using human colon carcinoma cells, it has been demonstrated that loss of p53 results in a reduced ability to repair UV induced DNA damage, suggesting a role for p53 in controlling some DNA repair reactions [191]. The importance of p53 not only in DDR but also in control of DNA repair is clear, but much of the details of the mechanisms are still being elucidated.

p53 is a 393 amino acid protein with several well-defined domains (Figure 7). These include two transactivation domains (TAD), followed by a DNA binding core region, which contains an NLS. In addition, there is the tetramerization domain (TET), which contains the nuclear export sequence (NES). Finally, near the C-terminus is the regulatory domain (RD) that binds chromatin and sequence independent single and double stranded nucleic acids [192].

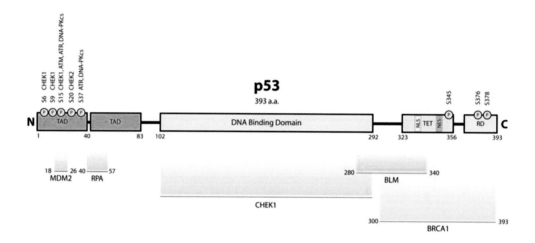

Figure 7. p53 Protein. p53 consists of 393 amino acids and has conserved domains represented by colored boxes: two transactivation domains (TAD), a DNA binding domain, a tetramerization domain (TET) that houses a nuclear localization signal (NLS) and a nuclear export signal (NES), and a regulatory domain (RD). Phosphorylation sites (P), by CHEK1, CHEK2, ATM, ATR and DNA-PKcs are indicated, as well as p53 interactions highlighted below the protein.

p53 is activated by a variety of stress signals including DNA damage, hypoxia and aberrant proliferation. The p53 response includes stabilization, accumulation and then execution of activity, such as induction of cell cycle arrest, senescence, or if the damage is severe enough, apoptosis. p53 mediates its action by either protein:protein interactions, interacting with other DDR proteins (e.g. ATM [49], CHEK2 [59, 193] and Jun-N (amino)-terminal kinase (JNK) [194]), cell cycle proteins (e.g. Polo-like kinase 1 (PLK1) [195]), anti-apoptotic proteins (e.g. B-cell leukemia/lymphoma x (BCL-XL) [196]), transcription coactivators (e.g. p300 and CREB binding protein (CBP) [197]), structural proteins (e.g. PML [198]) and DNA repair proteins (e.g. poly (ADP-ribose) polymerase 1 (PARP1) [199], WRN [200] and APEX nuclease (APE1) [201]) or by transcriptional activation. Targets for transcriptional activation have been much investigated [202-205] and certainly include cell cycle control genes (e.g. p21 [206] and 14-3-3σ [207]), pro-apoptotic genes (e.g. BCL2, Bcl2-associated X (BAX) [208] and TNF receptor superfamily (FAS/APO1) [209]), angiogenic genes (e.g. tumor suppressor region 1 (TSP1) [210]), as well as DNA repair genes (eg. Xerderma Pigmentosum, complementation group E (XP-E) [211] and DNA-Polβ [212]). It is beyond the scope of this review to discuss all these interactions (for a more complete review see [213]), thus we will summarize those most pertinent to DDR in general and HRR in specific.

p53 is normally present within the resting cell at low levels. The control of p53 level is mediated by MDM2, a p53 transcriptional target, thus forming a negative autoregulatory loop [214]. The MDM2 protein binds with the N-terminus of p53, thereby preventing its interaction with the transcriptional machinery of the cell [215]. In unstressed cells the amount of p53 is kept low by its rate of degradation and lack of p53 mRNA translation. Upon damage MDM2 is phosphorylated by ATM on Ser^{395}, disrupting its interaction with p53 and allowing p53 to stabilize in response to stress [82].

The activity of p53 is regulated by a variety of posttranslational modifications, including phosphorylation, acetylation, glycosylation, ribosylation, ubiquitination, or sumoylation [216]. These posttranslational modifications are induced in response to a variety of stimuli, including IR and other DNA damaging agents. In response to IR exposure or DSBs, ATM phosphorylates p53 on Ser^{15} [49], where ATR performs the same modification in response to UV. This modification is important to the efficiency of the overall damage response [217] because it renders MDM2 less capable of shuttling p53 to the cytoplasm for degradation [83]. ATM also phosphorylates p53 at additional sites in response to IR, including Ser^6, Ser^9, Thr^{18} and Ser^{20} (Figure 6) [218]. In comparison, ATR and DNA-PKcs, phosphorylate p53 on both Ser^{15} and Ser^{37} [219]. Overall, ATR and DNA-PKcs, in addition to other protein kinases, are important in stabilizing p53 in response to other types of DNA damage such as UV damage [220]. Kinetically, the Ser^{15} phosphorylation is rapid following damage induction, with Thr^{18} and Ser^{20} being phosphorylated subsequently. The major kinases for these subsequent events appear to be CHEK2 [221] and CHEK1 [59], with the Ser^{20} modification being important for p53 stability [222]. Once Thr^{18} and Ser^{20} are phosphorylated, p300, CBP and p300/CBP-associated factor (P/CAF) are recruited and acetylate the C-terminus of p53, preventing its ubiquitination and degradation [223]. Another important p53 modification in response to IR damage is the dephosphorylation of the constitutively phosphorylated Ser^{376} and Ser^{378} [224]. The Ser^{376} dephosphorylation in particular allows 14-3-3σ binding, thereby increasing the affinity of p53 for sequence-specific DNA binding sites [224].

There are a number of additional p53 protein interactions of note, including those with CHEK1, BRCA1, and BRCA2. Not only is CHEK1 capable of phosphorylating p53 at Ser^{20} [59], but once CHEK1 is phosphorylated at Ser^{345} following damage, it can bind to the central DNA-binding domain of p53, thereby potentially regulating downstream responses of p53 [225]. The C-terminal region of p53 also forms a stable complex with BRCA1. This p53:BRCA1 interaction has been postulated to activate p53 as well as bridge p53 to the basal transcriptional machinery [226]. Finally, it should be noted that BRCA2 has also been shown to interact with p53 and inhibit its transcriptional activity; this inhibition is enhanced by RAD51 [227]. The effect of the interaction with p53 on BRCA2 and RAD51 functionality, and possibly on HRR, has not been examined to date.

The conclusive demonstration that p53 inactivation promotes tumorigenesis comes from p53 mouse models. *p53* knockout mice are viable, except for a small subset of females that fail to complete neural tube closure during embryogenesis [228, 229]. The p53 mutant mice that do survive to adulthood are highly prone to developing cancer, primarily lymphomas and sarcomas, within two to nine months after birth. In contrast, p53 heterozygous mice have a longer tumor latency with about half of the tested mice developing tumors by 18 months of age [230-233]. One of the first targeted mutations (L22Q and W26S; the $p53^{QS}$ allele) tested was in the TAD domain at the N-terminus of p53, where p53 interacts with MDM2. $p53^{QS}$ mutant protein, though very stable owing to decreased MDM2 binding, fails to recruit histone acetyl transferases and therefore does not induce the transcription of most p53 target genes after DNA damage [234-237]. Conditional knock-out mice have also been created to study tumorigenesis in non-lymphoid organs in order to bypass the rapid onset of lymphomas and resulting mortality in p53 deficient mice. Following LoxP mutation of the p53 allele in mammary tissue, tumors were seen with a slow latency; this was dramatically increased when both Brca2 and p53 were conditionally removed [238]. The role for p53 in regulating cell cycle was elucidated by knock-in mice ($p53^{R172P}$ and $p53^{R175P}$) in which only the p53 arrest function and not the apoptosis function is preserved [239]. Many additional knock-out, knock-in, conditional, and transgenic p53 mice have provided important insights into p53 activities in development, tumorigenesis, DNA repair and aging [184].

The involvement of p53 in HRR has been extensively studied because p53 is known to be involved in maintaining the stability of the genome [240-248]. At early passages, fibroblasts from $Trp53^{-/-}$ mice develop several chromosomal abnormalities [249]. Tumors from $Trp53^{-/-}$ mice are often aneuploid, and there has been some evidence of chromosomal instability [232, 250]. Many early papers found conflicting HRR effects, though most showed that cells lacking p53 have a higher than normal frequency of HRR [244-246, 251-253]. The current support for the involvement of p53 in HRR comes from its direct protein:protein interactions, its transcriptional regulations, as well as by direct measurement. We will only highlight some of the most pertinent studies.

A suggestive interaction, implicating a possible mechanism of p53 regulating HRR, is with BLM and WRN [254]. Both of these RecQ homologues are DNA-structure specific helicases that unwind HJ (see sections below). Recombinant p53 was shown to bind BLM and WRN and attenuate their ability to unwind HJ. This activity was blocked by phosphorylation of p53 residues Ser^{376} and Ser^{378} and by mutant forms of p53 ($p53^{248W}$ and $p53^{273H}$). In addition, p53 has been shown to co-localize in vivo with BLM and RAD51 at sites of stalled replication forks and HJ in S-phase arrested cells [254].

RPA also interacts with p53 near the amino terminus and prevents RPA from binding ssDNA, important for HR processes. Romanov et al., employed a plasmid-based assay in which p53, p53 mutant and N-terminal p53 mutant plasmids carried a nonfunctional gpt gene copy with a 5' and 3' deletion. Homologous recombination allowed reconstitution of gpt gene expression and conferred cellular resistance to XHATM in p53-null H1299 lung carcinoma cells. They demonstrated that the transactivation-deficient N-terminal p53 mutant that was capable of binding RPA suppressed HR, similar to wild-type. However, transactivation-proficient and p53 mutants failed to interact with RPA and also failed to inhibit HR [255]. These results suggested the direct interaction of p53 and RPA is necessary for suppression of HRR.

There have been a number of co-localization studies that further implicate an involvement of p53 with HRR. Restle et al. used CV1 green monkey cells to analyze the spatial-temporal association of p53pSer[15] with HRR proteins after replication fork stalling. Aphidicolin treatment induced phosphorylation of p53 on Ser[15] was detectable and located within discrete nuclear foci, while total p53 protein showed both focal and diffuse nuclear staining. p53pSer[15] was localized to 38% and 36% of RAD51 foci in S-phase cells after aphidicolin or IR, respectively. After induction of replication fork stalling, p53pSer[15] also co-localized with MRE11 and RAD54 foci, 56% and 19%, respectively [112]. p53pSer[15] did not co-localize with RAD52 foci in aphidicolin treated cells, suggesting an involvement of p53 in RAD51-dependent gene conversion rather than RAD52 mediated SSA [256, 257].

Interestingly, p53pSer15 co-localization with RAD51, MRE11, and RAD54 was reduced to background levels within six hours post aphidicolin treatment, whereas co-localization with 72% of BLM foci remained stably associated long after six hours [112]. This suggests p53pSer[15] coincides with HRR proteins early on after replication inhibition. p53pSer[15] may play a role with the MRN complex in executing the earliest steps of recognizing DNA lesions and possibly in processing and tethering the ends of exchange substrates in HRR [258]. RAD51 performs the initial and continued strand exchange while RAD54 accelerates the search for homologous regions [259, 260]. p53pSer[15] was found to localize to a lesser extent at sites of RAD54 that is loaded to RAD51 foci and thus may monitor the correct placement of the 3′ invading strand during replication-associated HRR processes [261]. The evidence of sustained association of p53pSer[15] and BLM foci suggests cooperation at early stages of repair surveillance and later in correction of errors in HRR repair synthesis, since BLM directly interacts with mismatch repair proteins and the mutS homolog 2 (MSH2) and MSH6 domains on p53 enhance binding to HJ [262]. It is more likely that BLM recruits p53 to repair foci, leading to a net increase in p53pSer[15], as confirmed by HRR measurements that BLM and p53 act on separate regulatory pathways [263, 264].

Recent work has suggested a role for p53 in regulating HRR through its transcriptional functionality. p53 has been shown to down regulate the mRNA expression of RAD51 [265] in contrast to the direct role of p53 promoting HRR as discussed above. The interpretation of this difference is that if the damage is severe, p53 inhibits inconvenient DNA repair and initiates apoptosis. Arias-Lopez et al. have shown that p53 binds to the RAD51 promoter, leading to a reduction in RAD51 mRNA and protein and therefore inhibition of RAD51 foci formation in response to DSBs [266]. Ivanov et al. suggest that the induction of apoptosis in cells with an overload of DNA damage requires the repression of HRR, thus revealing another potential role for p53 in controlling HRR [267].

The involvement of p53 in HRR has been examined by directly examining HRR frequency. p53 has been shown in a number of studies to play a role in the homology directed repair of double strand breaks, suppressing HRR frequency [35, 245, 251, 252, 261, 268-270]. More elegant studies with over expression of specific p53 mutations in either wild-type [256, 271] or p53 null cells [245] using the inverted I-SceI assay system have demonstrated that p53 suppression of HRR is not dependent upon its transcriptional activation activity. The ability of p53 to suppress HRR was subsequently recapitulated in vivo using the p^{un} mouse model [35]. Similar to the effect of Atm deficiency, the role of p53 in suppressing HRR was largely temporally restricted. In contrast to Atm though, the temporal restriction was during early development. Using the p^{un} system, it was also shown that benzo[a]pyrene (B[a]P) induced HRR events irrespective of p53 genotype whereas X-ray induction required p53 [272]. These results suggest an interesting contradiction, where p53 suppresses spontaneous levels of HRR yet is required for HRR induction in response to particular forms of damage. Clearly, the relationship between DDR and HRR is complex.

GADD45a

The growth arrest and DNA damage-inducible (GADD) gene *GADD45a* is member to a group of genes that are induced by DNA damaging agents and growth arrest signals. The gene was first isolated in a large screen of Chinese Hamster Ovarian cell (CHO) cDNA clones after treatment with UV [273-276]. *GADD45a* was the first well defined *p53* target gene and is known for its roles in the G$_2$/M cell cycle checkpoint [277-280], DNA repair [281-283] and induction of cell death [284-286].

Hollander *et al.* reported severe genomic instability in *Gadd45a* null MEFs, including aneuploidy, chromosomal aberrations, gene amplifications and abnormal mitosis and cytokinesis. *Gadd45a* null keratinocytes contained amplified centrosomes with multiple spindle poles, which may have been a consequence of unbalanced chromosome segregation. IR-induction (DNA strand break) failed to induce G$_2$/M arrest in *Gadd45a* null lymphocytes, but arrest occurred with the use of DNA base damaging agents such as UV [287].

The *GADD45a* gene encodes a 165-residue protein (Figure 8) and can be found in the heart, skeletal muscle and kidneys [284]. Two oligomerization domains exist to facilitate GADD45a binding to DNA damage and to interact with other proteins in response to specific stresses [288]. Upon IR exposure, GADD45a interacts with PCNA, a normal component of cyclin-dependent kinase complexes, DNA replication and DNA repair [281]. The various CDK complexes and their interactions with different cyclins are key for cell progression [289]. GADD45a binding to PCNA has been postulated to displace PCNA from CDK complexes, thus inhibiting entry of cells into S-phase. This interaction with PCNA is thought to also regulate nucleotide excision repair (NER) [281]. The importance of another interaction, that of GADD45a with the p53 target gene p21, is unclear, though it underscores a possible relationship between GADD45a and cell cycle progression [290] since p21 can institute G$_1$/S [50, 291-293] and G$_2$/M [294] cell cycle arrest. GADD45a directly binds CDK1DC2 and inhibits the kinase activity of the CDK1:cyclin B1 complex, important for G$_2$/M transition [278, 295-297]. This interaction has also been shown to be important in GADD45a-mediated activation of the G$_2$/M checkpoint in response to UV, IR and MMS in human fibroblasts [277]. A final GADD45a interaction of note is with mitogen-activated protein kinase kinase kinase 4 (MEKK4/MTK1), an upstream kinase of the p38/JNK MAP kinase pathway implicated in stress-induced responses [298].

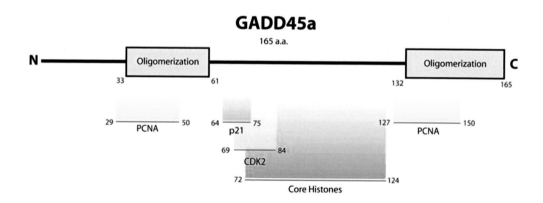

Figure 8. GADD45a Protein: GADD45a is comprised of 165 amino acid residues. Two oligomerization domains are boxed and protein interactions are indicated below.

GADD45a becomes transcriptionally active following exposure to a variety of DNA damaging agents [273]. Overexpression of GADD45a results in a G$_2$/M checkpoint that is dependent upon the presence of functional p53 [277]. Gadd45a deficient murine lymphocytes display a defective G_2/M checkpoint in response to UV but not to IR [277], similar to p53 deficient cells. Thus a p53:GADD45a-mediated response may be necessary for UV or MMS induction of G_2/M but is not required for G_2/M arrest after an IR induction [277].

Hollander *et al.* used gene targeting to create viable *Gadd45a* null mice that appeared grossly normal with a low frequency of exencephaly comparable to what is seen in *p53* null mice. Irradiated *Gadd45a* null mice showed a three-fold increase in tumor susceptibility compared to wild-type, with a majority of the tumors being lymphomas of thymic origin. These mice also had a great susceptibility to DNA damage, with induction of tumors by agents such as UV and dimethylbenzanthracene (DMBA) in addition to IR [287]. In contrast to IR induced lymphomas, DMBA treated *Gadd45a* null mice had an increased incidence of vascular tumors, the females also having an increase in ovarian tumors and males an increased frequency of hepatocellular tumors [283].

GADD45a plays a role in the DNA damage response pathway, and its absence leads to an increased level of genomic instability. GADD45a has been implicated in NER after UV damage and base excision repair (BER) after MMS treatment [299]. Bishop *et al.*, using the *in vivo* p^{un} system, determined that *Gadd45a* deficient mice have an increased frequency of HRR [35]. Further, the observed increase was determined to be temporally restricted to early development, much like the observation with p53 deficient mice. Though the mechanism by which either GADD45a or p53 might suppress HRR is currently unknown, these data are suggestive that they work together, similar to their coordination of G_2/M checkpoint response.

p21

p21/WAF1/CIP1/SDI1/MDA6 (p21) is a potent inhibitor of many cyclin-dependent kinases [50, 292] and is required for a p53-dependent G_1/S cell cycle checkpoint after IR [300]. p21 is a well-established component of the DDR pathway, and its transcription is upregulated by p53 following exposure to a variety of damaging agents, such as X-rays, MMS, UV and IR [206, 293, 301-304]. As might be expected from the cell cycle role of p21, cells deficient in this protein exhibit RDS and defective G_1/S checkpoint control [293, 301].

p21 binds to proteins involved in both DNA replication as well as cell cycle progression, thereby eliciting a potent G_1/S cell cycle checkpoint. A zinc finger domain exists in the amino terminus followed by a NLS near the carboxy terminus. Through its C-terminal region (Figure 9), p21 binds PCNA and inhibits PCNA-induced DNA replication [305, 306]. A large component of the N-terminus of p21 is used to bind cyclins and CDKs. Specifically, residues 16-24 bind cyclin E, cyclin A, and the cyclin D1:CDK4, whereas residues 45-65 bind CDK [307] with a second cyclin binding site near residues 152-158 [307]. Finally, the C-terminus of p21 interacts with GADD45a [290].

p21 deficient mice are developmentally normal, suggesting p21 is not essential in development and differentiation [293, 301]. Early studies with p21 deficient mice suggested that p21 is not a tumor suppressor due to the tumor incidence in these mice being no different than wild-type [293]. Interestingly, loss of p21 in Atm deficient mice resulted in a 50-day delay in the onset of lymphomas, with high levels of apoptotic cells not observed when null for *atm* alone [308].

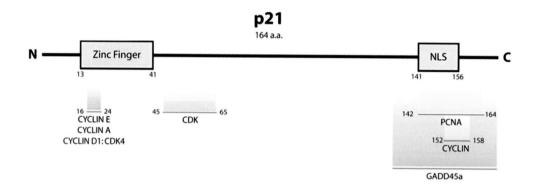

Figure 9. p21 Protein. p21 consists of 164 amino acid residues and two main domains: a zinc finger domain and a nuclear localization signal (NLS). Protein interactions are highlighted below the protein.

Bishop *et al.* used the p^{un} system in the RPE to investigate whether p21 plays a role in genomic stability during development by controlling HRR [309]. No increase in spontaneous HRR frequency was observed in these mutant mice. This does not, however, rule out the possibility that p21 is involved in damage induced HRR events in a DDR-dependent manner.

BRCA1

Breast cancer 1, early onset (*BRCA1*) was the first identified and cloned breast cancer gene [310]. Hereditary breast cancers comprise 5% to 10% of all breast cancers. Of these, mutations of the *BRCA1* gene account for 40% to 45% of breast cancers and 80% of breast and ovarian cancers from predisposed families [311-313]. *BRCA1* is therefore considered to be a tumor suppressor gene that is involved in cell cycle checkpoints, transcription, ubiquitin-ligation, apoptosis and DNA repair. Surprisingly, *BRCA1* gene mutations are very rare in sporadic breast and ovarian cancers, but the BRCA1 protein has often been noted to be reduced or absent in these sporadic cases [314, 315]. The *BRCA1* gene product consists of 1863 amino acid residues with a RING (Really Interesting New Gene) domain, followed by two NLS domains and a BRCT repeat domain at the C-terminal region (Figure 10) [316].

BRCA1 is thought to play a central role in DDR and multiple cell cycle checkpoints. In response to IR exposure, ATM phosphorylates BRCA1 at Ser^{1387} to induce an S-phase checkpoint [51] and Ser^{1423} to induce a G_2/M checkpoint [52]. BRCA1 interacts with large number of proteins, as first alluded to by the isolation of the BASC [109]. The list of interacting proteins has grown significantly, thus only those implicated in the major components of DDR or HRR are highlighted here [317]. A key protein interaction for mediating the effects of BRCA1 is BRCA1 associated RING domain 1 (BARD1), a ubiquitin-ligase that interacts via the RING finger consensus motif located near the N-terminus of BRCA1 [316]. Cellular localization studies have shown that BARD1 and BRCA1 localize to discrete nuclear foci during S-phase and to sites of damaged replicating DNA structures [318]. Another key interaction placing BRCA1 in a central position of DDR is its interaction with RAD50, and presumably with the MRN complex, in IR induced nuclear foci [110]. BRCA1 has also been implicated by co-immunoprecipitation of both proteins *in vitro* as a co-activator of p53 transcriptional activity through domains at both the N- and C-termini. The transcriptional transactivation may then be through the interaction and recruitment of p300 or CBP and their associated acetyl transferase activities [319, 320].

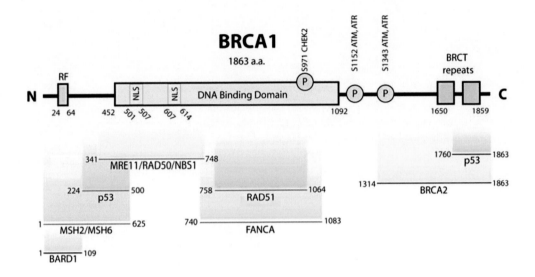

Figure 10. BRCA1 Protein. BRCA1 consists of 1863 amino acid residues. The main domains are an N-terminal ring finger (RF) domain, a large DNA binding domain including two nuclear localization signals (NLS) and two BRCA1 C-terminal (BRCT) repeats. Phosphorylation sites (P), by CHEK2, ATM and ATR are indicated. The main protein interactions of BRCA1 are highlighted below.

BRCA1 is involved in DNA repair. In particular, BRCA1 is strongly implicated in DSB repair through its interaction with BRCA2 [321] as well as its interaction with RAD51 during S-phase and at meiotic synaptonemal complexes. Together, this suggests that BRCA1 is involved in the control of HRR and genome integrity [321, 322]. BRCA1 also interacts with FA-A. Interestingly, this BRCA1:FA-A interaction does not depend on damage, that is, these proteins constitutively interact, which implies that BRCA1 is intimately involved in the FA pathway of DNA repair [323]. Another DNA repair pathway that BRCA1 appears to be involved with is mismatch repair through its interaction with MSH2 and MSH6 [324]. BRCA1 therefore plays a major role in both DDR and DNA repair, though these roles can be resolved by single amino acid mutations. S988 of BRCA1 is a phosphorylation site for CHEK2 and is important for the release of CHEK2 from BRCA1 following IR exposure; an S988A mutation results in normal S-phase checkpoint but severe defects in HRR [181]. In contrast, a S1423A mutation on BRCA1, the phosphorylation site used by ATM to induce G_2/M checkpoint [52], has normal DNA repair but is defective at the G_2/M checkpoint [22].

There have only been a limited number of investigations into the role of BRCA1 in HRR. The involvement of BRCA1 in DSB repair was first implicated when cells deficient for BRCA1 were found to be sensitive to IR and other DNA damaging agents that cause DSBs [325]. Using the observation of RAD51 foci, this sensitivity was likely due to a deficiency in HRR [318]. More directly, Moynahan *et al.* demonstrated that *Brca1* deficient embryonic stem (ES) cells had a deficiency in HRR using the direct repeat I-*Sce*I system [326]. In addition, the authors reported both a profound sensitivity of these cells to DNA interstrand cross-links and a high level of gross chromosomal aberrations in these cells, although no aneuploidy was detected as is common in other homology-directed mutant cells [326]. Overall *BRCA1* deficiency leads to an accumulation of DNA damage and the added insult of defective repair, HRR in particular, may be the mechanism for the observed genomic instability.

BRCA2

Breast cancer 2, early onset (*BRCA2*) is a prototypical tumor suppressor gene. Patients who carry germline mutations in *BRCA2* are predisposed to breast cancer and, less commonly, to ovarian cancer [327]. *BRCA2* mutations are also seen in prostatic, pancreatic, gastric and colon cancers, as well as melanoma [327]. The *BRCA2* mutation may also confer a higher risk for male breast cancer [328]. Finally, a subset of BRCA2 patients manifests an FA phenotype that includes early onset of tumors, high rates of leukemia and specific solid tumors. Actually, the *FA-D1* gene is *BRCA2* [329]; the role of BRCA2 in the FA pathway is discussed in the next section.

The *BRCA2* gene encodes a 3418 amino acid protein of approximately 385 kDa (Figure 11). The N-terminus contains a transcription activation domain [330], and there is an NLS towards the C-terminus [331]. The *BRCA2* gene is structurally interesting in that it contains a very large exon (exon 11) that encodes eight BRC repeat motifs, some of which are required to interact with RAD51 to accomplish error-free HRR in response to DBSs, as discussed below. With regard to its regulation following damage, it has been shown that ATR directly phosphorylates BRCA2 at Ser^{2156} [219]. More recently, phosphorylation of Ser^{3291}, a modification that blocks the BRCA2:RAD51 interaction, has been shown to be dependent upon CDKs and correlates with cell cycle progress. The level of BRCA2 Ser^{3291} is lowest in S-phase and increases with progression to mitosis [332].

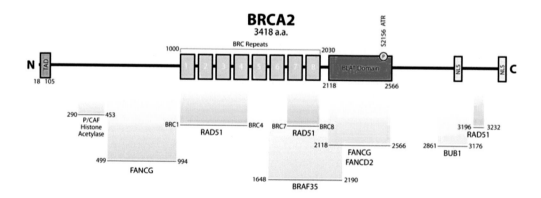

Figure 11. BRCA2 Protein. BRCA2 is a large protein, consisting of 3418 amino acid residues. At the N-terminus there is a transactivation domain (TAD) followed by eight BRCA2 (BRC) repeats. The C-terminus is comprised of a BRCA2-motif in *Leishmania*, *Arabidopsis*, and *Trypanosoma* (BLAT) domain followed by two nuclear localization signals (NLS). Phosphorylation sites (P), by ATR, are indicated and protein interactions with BRCA2 are highlighted below.

Several mouse models for BRCA2 have been created [333, 334]. In mice, *Brca2* is expressed in the adult thymus and in the testis and ovaries of midgestation embryos. Both *Brca1* and *Brca2* are expressed in highly proliferating cells, such as mammary epithelial cells, with peak expression at the G_1/S boundary of the cell cycle [335]. The importance of BRCA2 in cell cycle progression is particularly obvious with the mouse models since *Brca2* null embryos die at approximately embryonic day 6.5 from a severe growth defect [333]. *Brca2* heterozygotes, though, are indistinguishable from wild-type littermates [333]. As might be expected from the role of BRCA2 in DDR, *Brca2* mutant embryos have an increased sensitivity to IR, resulting in a strong apoptotic response [336]. Jonkers et al. created a conditional *Brca2* mouse model in mammary gland epithelium to model human hereditary breast cancer. No tumors were observed in mice carrying only the conditional *Brca2* alleles but females with combined p53 inactivation depicted mammary and skin tumors. Thus, the disruption of p53 is crucial in BRCA2-associated tumorigenesis [238].

Human BRCA2-defective pancreatic cancer cells (CAPAN-1) are commonly used tissue culture models for BRCA2 mutations. In addition to a homozygous truncation of BRCA2, it should be noted that these cells also carry mutations in p53, p16 (an inhibitor of CDK4 and thus G_1 progression), and Rb (a protein that binds E2F to allow G_1 progression into S-phase, before it is deactivated by the actions of Cyclin D, CDK4, and CDK6) [337, 338]. CAPAN-1 cells have a marked defect in DSB repair and display radiation hypersensitivity both *in vivo* and *in vitro* [339]. With the evidence of fluorescence *in situ* hybridization (FISH) data, CAPAN-1 cells exhibit extremely abnormal karyotypes, with 70% of chromosomes being rearranged and 20% of those being multiple translocations [340, 341].

RCA2 is involved in the DRR pathway as indicated by its direct regulation by ATR, its interaction with other components of DDR, its integration in cell cycle checkpoint signals and its role in transcriptional and chromatin regulation. Here we briefly outline some of these roles. BRCA2 is a member of the BRCA1-BRCA2-containing complex (BRCC) along with BRCA1 and BARD1 [342]. BRCC directly ubiquitinates the central DDR component p53 and interacts with the HRR component RAD51, suggesting this complex is important in both DDR and HRR. BRCA2 also interacts directly with p53, limiting its transcriptional activity, an inhibition that is enhanced by RAD51 [227],. Perhaps, given this evidence, successful HRR response might have role in modulating DDR. BRCA2 also has an involvement in cell cycle progression derived from its interaction with PLK1. BRCA2 undergoes enhanced phosphorylation by PLK1 during normal G_2/M cell cycle transition, while this phosphorylation is inhibited upon DNA damage [343]. Another indication of involvement of BRCA2 in cell cycle is through its interaction with BRCA2-associated factor 35 (BRAF35), which binds BRCA2 at the BRC repeat region [344] (Figure 11). The BRCA2:BRAF35 complex plays a regulatory role in mitotic progression, modulating the components of chromosome condensation and segregation machinery as well as being recruited to sites of unusual DNA architecture such as cruciforms, a structure that resembles an HJ [344]. Further, BRCA2 also interacts with budding uninhibited by benzimidazoles 1 (BUB1), a spindle-damage checkpoint protein in M-phase. BUB1, through its kinase activity, is thought to phosphorylate BRCA2 upon sustained microtubule damage as seen when expressed in COS7 cells, the African Green Monkey SV40-transformed kidney fibroblast cell line, after treatment with the microtubule disrupting drug vincristine [345]. Finally, BRCA2 may act as a transcription regulator possibly through the recruitment of the histone modifying activity of the P/CAF co-activator [346] (Figure 11). Overall, BRCA2 appears to be intimately involved in both central DDR and cell cycle progression, possibly acting as one bridge to the HRR machinery.

BRCA2 plays a functional role in the repair of DSBs by HRR [347]. RAD51 is directly bound by BRCA2 BRC repeats BRC1-4, BRC-7, and BRC-8, the most highly conserved of the BRC repeats within BRCA2 [348] (Figure 11). It has been postulated that this interaction is required for the transport of RAD51 into the nucleus and to sites of DNA damage because no NLS has been found in RAD51 and because in the absence of the BRCA2 NLS sequences, both proteins are found in the cytoplasm [331]. Once inside the nucleus, the RAD51:BRCA2 complex is thought to disengage, allowing RAD51 to form a nucleoprotein filament at the initiation of HRR.

Direct analysis of HRR events has been conducted in cells deficient for BRCA2, as reported in two independent studies. The first study used the $Brca2^{lex1/lex2}$ mutant embryonic stem cell line, where the terminal exon 27 is deleted in *lex1* and both exon 27 and a portion of exon 26 are deleted in *lex2*. These *Brca2* mutation alleles both retain the Rad51 interacting BRC repeats and the NLS. Using the direct repeat I-*Sce*I assay system, a five-fold reduction in HRR was observed [347]. The second study used a mouse embryonic stem cell line that expressed two truncated forms of Brca2. One allele retains the ability to bind Rad51 while the other allele has lost the C terminal Rad51 binding domain of *Brca2* ($Brca2^{Tr/\Delta Ex27}$) [349]. Again, the gene products from these alleles were shown to interact with Rad51, but the cells had impaired Rad51 nuclear foci formation following IR exposure [349]. Using the direct repeat I-*Sce*I system as well, a 42% reduction in HRR was observed. Overall, it appears that BRCA2 facilitates at least induced DSB HRR repair events, presumably through its direct interaction with RAD51.

Fanconi Anemia Proteins

Fanconi anemia (FA) is an autosomal recessive, congenital disease caused by a mutation in any of a number of FA genes, each named after the complementation group that was originally identified. The clinical symptoms are diverse and include skeletal abnormalities (radial ray defects in the hip, thumb, arm, and rib, and scoliosis), hypopigmentation and café-au-lait spots, and short stature [350]. FA patients have bone marrow failure and early onset of pancytopenia, the resultant anemic state leading to an increased cancer predisposition with the greatest risk being leukemias, especially acute myeloid leukemia [351]. Cells from FA patients have increased spontaneous chromosomal aberrations and are hypersensitive to DNA inter- and intra-strand crosslinking (ICL) agents like MMC, diepoxybutane (DEB) and cisplatin [352-355]. These cells also show increased sensitivity to chromosomal breaks and apoptosis following exposure to ICL agents, as well as UV, IR and reactive oxygen species (ROS) [352, 356, 357]. FA patient cells also display altered karyotypes, with sub-4N DNA and prolonged G_2/M or delayed S-phase after exposure to ICL agents [358].

To date, 13 different FA complementation groups have been identified: FA-A, -B, -C, -D1, -D2, -E, -F, -G, -I, -J, -L, -M and –N [359]. Below is a short description of what is known about the FA genes. The different FA genes can be categorized into three groups based on the proposed cellular pathway (Figure 12): the FA core complex, the ID group, and the third FA group. The FA core complex includes FA-A, -B, -C, -E, -F, -G, -L, and -M [358, 360-365]. The FA ID group, identified by both direct interaction and posttranslational modifications, is composed of FA-I and FA-D2 [366, 367]. The third FA group contains FA-D1 (BRCA2), -J, and -N [359]. This third group links the FA pathway with DDR and HRR.

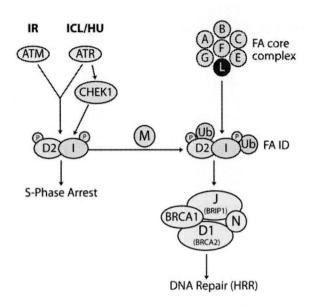

Figure 12. Fanconi Anemia (FA) Pathway. The FA pathway is involved in S- phase arrest and HR. In response to various DNA damaging agents FA-D2 (D2) and FA-I (I) (blue ovals) are phosphorylated (P) by upstream kinases (green ovals). The activated FA core complex (brown circles), through FA-M, binds DNA and FA-L (dark brown) monoubiquinates (Ub) FA-D2 and FA-I. Monoubiquinated FA-D2 and FA-I, in conjunction with FA-J, FA-N, FA-D1 and BRCA1 (yellow ovals), act in DNA repair.

Many of the FA proteins do not have well defined functional domains, though those that have been identified, which are mainly in the core complex, are described here. The primary function of the core complex is the monoubiquitination of FA-D2 and the physical interaction with DNA. FA-M contains a DEAH (Asp-Glu-Ala-His)-helicase, XPF-like endonuclease domain, and ATP-dependent DNA translocase activity [368]. Although neither the helicase nor the endonuclease domains appear to be functionally active, the core complex is believed to physically associate with DNA via the translocase activity of FA-M [368]. The remaining members of this complex either directly or indirectly serve to monoubiquitinate FA-D2 [369, 370]. An E3 ubiquitin ligase is required to ubiquinate a protein; FA-L contains three WD40 repeats (40 amino acid repeat terminating in Trp-Asp) and a PHD (plant homeobox domain) or RING-finger-type ubiquitin-ligase motif and is responsible for the monoubiquitination of FA-D2 [371, 372]. FA-A, -E, -G, and -M contain phosphorylation sites that could be substrates for ATM, ATR and CHEK1 mediated activation [360, 373-376].

FA-D2 and FA-I are paralogs of one another. They both contain Armadillo repeats, as well as phosphorylation and ubiquitination sites, posttranslational modifications that dictate the FA pathway response to different types of damage. Following IR exposure, FA-D2 cells have a similar S-phase checkpoint response to ATM null cells [84], and indeed, after IR, ATM phosphorylates FA-D2 on Ser[222]. Following exposure to MMC or HU, FA-D2 is phosphorylated even in the absence of ATM [42, 84, 377]. In addition, FA-D2 is monoubiquitinated (FA-D2L) on Lys^{561} during S-phase and following exposure to ICL agents [378, 379]. Once monoubiquinated, FA-D2L relocalizes to the nucleus, is observed in nuclear foci, and has an increased affinity for DNA ends and HJs [379, 380]. In a similar fashion, FA-I is also monoubiquitinated, with the monoubquination of both proteins FA-D2 and FA-I being interdependent [367, 381]. Most importantly, these monoubiquitinations are independent of the ATM phosphorylation S-phase arrest, and monoubiquitinated FA-D2 and FA-I foci still occur in ATM^{-} cells [358].

A majority of interactions with the FA proteins have been linked to DDR and DNA repair components, suggesting their relationship. With regard to DDR, following MMC exposure, FA-D2L and NBS1 form nuclear foci in wild-type cells. In the absence of NBS1, ATM-dependent FA-D2 phosphorylation is defective [157]. The association of FA to DNA repair is supported by the identification of the BRAFT (BLM, RPA, FA, topoisomerase (TOPO) IIIα) complex. This multi-protein complex contains 17 polypeptides and joins together two distinct genetic diseases: Bloom's Syndrome (see below) and FA [382]. As additional evidence for the involvement of the FA pathway in DNA repair, it was shown that FA inactive DT40 cells showed an increased level of SCE, a feature similar to the diagnostic feature of BS [383]. BASC also provides a link between the FA proteins and BLM, as well as with other cancer susceptibility genes like BRCA1 [109]. Furthermore, BRCA1 directly interacts with FA-A [323] and FA-J, also known as BRCA1-interacting protein 1 (BRIP1) [384, 385]. Damage induced FA-D2L foci co-localize with BRCA1 and RAD51 [378, 379]. Subsequent studies have shown that monoubiquitination of FA-D2 is necessary for these damage induced foci [386]. Considering this evidence, FA is clearly involved in both DDR and HRR, at least with regard to the intra S-phase checkpoint.

The role of the FA pathway in HRR is suggested by both the interactions of its protein components and the cellular phenotypes of FA cells. Recent studies have directly examined the role of FA in HRR. Using a cell-free plasmid system to measure HRR from FA patient cells, Thyagarajan and Campbell showed that HRR was 20- to 100-fold higher than in wild-type cells [387]. Yamamoto *et al.*, in two independent studies using DT40 cells deficient for FA-D2, showed decreased levels of gene targeting efficiency compared to wild-type. Similarly, FA-G deficient cells had a modest decrease in gene targeting [388, 389]. Using the direct repeat I-*Sce*I assay system, these authors went on to show an approximately 40-fold and 9-fold reduction in HRR in FA-D2 and FA-G deficient cells, respectively [388]. Further, upon exposure to ICL agents, FA-C has been shown to be epistatic with HRR mutations, as well as a number of translesion synthesis (TLS) polymerases [390]. Lastly, Nakanishi *et al.*, also using the direct I-*Sce*I plasmid system in either mouse or human cells, demonstrated that FA-A and FA-C (FA core complex members) promote activation of SSA and HRR, yet FA-D1, which acts downstream of the monoubiquitination of FA-D2, promotes HRR while suppressing SSA [391]. Overall, the FA pathway has been shown to have a role in an ATM-dependent S-phase arrest and HRR at both stalled replication forks and induced DSB.

BLM

BLM, encoding a RecQ helicase, is the gene mutated in Bloom's Syndrome (BS). BS is a rare, autosomal recessive disorder that has an increased propensity of a broad spectrum of cancers. Clinical features include dwarfing, facial abnormalities, fertility complications, and an early onset of cancer. Clinical diagnosis of BS is based on its hallmark chromosomal instability phenotype, including symmetric, multiradial rearrangements, chromosomal breaks, and increased SCE [392, 393]. The elevation in DNA breaks and possible increase of unequal exchange of genetic material resultant of the increased levels of SCE can lead to mutation and loss of heterozygosity, mechanisms thought to predispose to cancer onset. Four mouse models have been developed; each recapitulated different clinical phenotypes of BS. Blm^{cin} was designed to recapitulate the Blm^{Ash} allele that is most prevalent in the Ashkenazi Jewish population. This model had an increase in lymphoma development earlier than wild-type controls, as well as an increase of intestinal tumors [394]. The $Blm^{tm1Ches}$ model disrupted exon 8, resulting in a null allele. Homozygous mutants of this model are embryonic lethal, dying at embryonic day 13.5 [395]. These mutant embryos displayed elevated levels of SCE, were developmentally delayed, had increased amounts of apoptosis and had anemia. The Blm^{tm1Brd} model disrupted BLM function by a duplication of exon 4, resulting in a hypomorphic allele; homozygous mutants are viable and predisposed to a number of cancers [396]. $Blm^{tm3Ches}$ is the fourth mouse model and also demonstrated a correlation between the amount of BLM present and levels of chromosomal instability and tumor presence [397]. Using tumor cell lines derived from a BLM conditional model, Chester *et al.* supported the idea that chromosomal instability gives rise to cancer development and that the loss of BLM is responsible [398].

The *BLM* gene encodes for a 1,417-residue protein with three evolutionarily conserved domains including the DEAH helicase domain, RecQ C-terminal (RecQCt) domain and the Helicase and RNaseD C-terminal (HRDC) domain (Figure 13) [399]. Biochemically, BLM has an ATP-dependent DNA helicase activity that unwinds DNA in the 3'-5' direction; a number of BS patients carry mutations that disable this helicase function suggesting it importance [400].

BLM is regulated at both the transcriptional and post-translational levels. In mitoticly dividing cells, the highest expression of *BLM* is during S-phase [401]. BLM contains a C-terminal NLS (Figure 13), and once recruited inside the nucleus, it localizes to PML nuclear bodies [402], a subnuclear localization possibly mediated by a BLM:p53 physical interaction [403]. Once within the PML bodies, BLM co-localizes with topo III-α [404]. Additional studies demonstrated that topo III-α interacts with BLM on two independent sites (Figure 13) [405]. Even though the subcellular localization of BLM is known, the activation of BLM is still poorly understood. In response to IR damage, ATM phosphorylates BLM at two independent sites, Thr^{99} and Thr^{122} (Figure 13) [87], and this ATM-dependent phosphorylation suggests that BLM reacts to such damage as a downstream target of ATM [406]. CDK1 has also been shown to phosphorylate BLM (Ser^{714} and Thr^{766}) at two sites that are independent of the sites phosphorylated by ATM [407]. Considering this CDK1 phosphorylation of BLM and the high expression of BLM during S-phase, it has been suggested that if DNA damage persists during S-phase, cyclinB and CDK2 will dissociate, thus preventing progression of the cell cycle into mitosis and freeing CDK1 to phosphorylate BLM (Figure 13) to act at sites of DNA damage [407].

BLM has been implicated as having a role in both DDR and DNA repair. Foci studies using laser light IR in DT40 cells has shown BLM to be an early responder to DNA DSBs [408]. Additional experiments using BLM mutant proteins further demonstrated that this response appears to be mediated through its HRDC domain and is independent of other

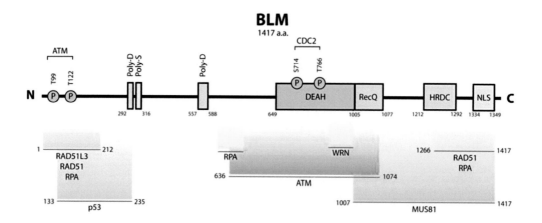

Figure 13. BLM Protein. The BLM protein consists of 1417 amino acid residues. Colored boxes depict domains: Asp-Glu-Ala-His (DEAH) domain, RecQ C-terminal helicase (RecQ), and RNaseD C-terminal (HRDC) domain as well as a nuclear localization signal (NLS). Phosphorylation events (P), by ATM and CDC2, are indicated, and protein interactions are highlighted below.

components of DDR, such as ATM, RAD17 and RAD52 [408]. BLM is also found at stalled replication forks and that its presence at sites of damage is necessary for the rapid localization of BRCA1 [263]. Using DNA combing to measure replication origins, Rao *et al.* found that BLM deficient cells have an increase in spontaneous replication fork arrests and that ATM, CHEK2, and phosphorylated γ-H2AX co-localized with these sites [409]. Further support for the involvement of BLM in DDR includes it being a member of BASC [109] and its physical interactions with the MRN complex [410] and members of the FA pathway [411] (see below).

The major function of BLM is thought to be DNA repair. Biochemical studies using synthetic oligonucleotide structures (e.g. blunt end double-stranded DNA with a bubble, HJ, G-quadruplexes or forked structures) demonstrated that BLM acts on specific DNA structures [412], promotes branch migration [413], suppresses crossing over during HRR [414], and promotes annealing of complementary ssDNA [415]. In addition to these activities, BLM interacts with a number of other DNA repair proteins. RPA directly interacts with BLM, stimulating BLM helicase activity and leading to the unwinding of DNA duplexes (Figure 13) [416]. It has also been shown that BLM interacts with RAD51 [417] and the RAD51 paralog RAD51L3 [418] (Figure 13). These interacting proteins, RPA, Rad51 and RAD51L3, are all involved in HRR. The final steps of HRR often leave DNA substrates that require a nuclease activity (see Figure 1 and 2); BLM may promote this nuclease activity by stimulating MUS81 endonuclease activity, an enzyme that has been shown to act on 3' flaps and nicked HJ *in vitro* [419], as well as the endonucleolytic and exonucleolytic cleavage activity of flap structure-specific endonuclease 1 (FEN-1) during either DNA repair or Okazaki fragment removal [420]. BLM has been shown to interact with another RecQ helicase, WRN, at sites of DNA damage. The role of WRN at sites of damage is poorly understood (see below), though the BLM interaction results in an inhibition of the exonuclease activity unique to WRN [421]. As for other types of DNA repair, BLM associates with DNA repair proteins involved in mismatch repair (MLH1 [422] and MSH2 and MSH6 [423]) and NHEJ (DNA ligase IV [424]).

In addition to the role BLM plays in inhibiting SCE frequency, an event thought to be mediated by HRR, and the biochemical evidence for its involvement in HRR, there has been a limited amount of work using *Drosophila* that directly examines the effect of BLM on HRR frequency *in vivo*. The first model utilized a transposable element in *Drosophila* to induce defined breaks. A majority of the DSBs that arise in *Drosophila* are corrected by HRR, specifically SDSA. In the absence of Mus309, the *Drosophila* BLM ortholog (DmBlm), DSBs induced by *P elements* resulted in an elevation of deleted genomic DNA suggesting that Mus309 cooperated in SDSA HRR [425]. Mus309 mutants were unable to repair site-specific DNA damage induced from the cleavage and mobilization of the transposable *P element* as well as exogenous agents, specifically MMS [426]. A second set of experiments using Mus309 mutants and the *P element* system supported the role for DmBlm in chromosomal stability as shown by its role in embryonic development and meiotic recombination [427]. Together these studies very elegantly demonstrated an *in vivo* function for DmBlm with HRR.

BLM certainly appears be involved in both DDR and HRR, particularly during S-phase and at stalled replication forks. Due to p53-dependent apoptosis and the BLM:p53 interaction at PML nuclear bodies, Wang *et al.* postulated that their interaction might attenuate apoptosis and explain the high predisposition to cancer in BS patients [403]. Alternatively, p53 may play a different role, since once p53 is phosphorylated on Ser[15], BLM and phosphorylated p53 co-localize with RAD51 in nuclear foci [254]. These co-localized proteins are thought to represent sites of stalled DNA replication forks [264], possibly during HRR [428]. Biochemically, BLM appears to be involved in the later stages of HRR, either in branch migration or HJ resolution. Whatever the exact task of BLM, it is an important segue between DDR and HRR, functioning in both pathways, determining the outcome of both.

WRN

Werner syndrome (WS) is a rare disorder caused by defects in the RecQ helicase gene *WRN* [429]. WS patients exhibit pleiotropic phenotypes, including premature aging phenotypes, increased incidence of malignancy, osteoporosis, atherosclerosis, Type II Diabetes, early graying, hair loss, skin atrophy, and cataracts [430, 431]. Cells obtained from these patients display reduced replicative lifespan and increased genomic instability [432]. WRN has two biochemically characterized activities, a 3'-5' DNA helicase [433] and a 3'-5' DNA exonuclease activity [434, 435]. The NLS is at the C-terminal end of the protein, thus because the majority of WS patients produce truncated protein, the syndrome is most likely caused by mislocalized WRN, if any stable product exists. Similar to BLM, WRN contains a DEAD (Asp-Glu-Ala-Asp) helicase domain, whereas BLM has a DEAH helicase. In connection with DDR, it is known that WRN is phosphorylated by the PI3Ks ATM and ATR, in response to replication fork blockage [436], and DNA-PKcs [437]. Phosphorylation of WRN does not appear to be necessary for its relocalization to replication fork blocks or its co-localization with ATR [436]. Additional proteins with which WRN either interacts with or co-localizes with following damage exposure are p53 [254, 438-440], H2AX, NBS1, MRE11, RAD51, RAD52, RAD54, and RAD54B [155, 441-443] (Figure 14).

In order to facilitate studies into the function of WRN and to model the disease, various mouse models have been generated, including a knock-out of the *Wrn* gene, a deletion of the helicase region of *Wrn*, and a transgenic insertion of a dominant negative human allele of *WRN*. The first model generated was a deletion of part of the helicase domain (*Wrn*[helΔ]); a stable mutant protein containing an internal deletion is expressed [444]. The second mutation was a deletion of the entire *Wrn* gene [445]. Finally, a dominant negative transgenic mouse model with a point mutation from a human WS allele (K577M) has also been reported [446]. Fibroblasts from *Wrn*^{Δhel} and *Wrn*⁻ mice displayed a reduced ability to proliferate as compared to wild-type mouse fibroblasts, as well as displaying increased genomic instability [444, 445], phenotypes comparable to cells from human WS patients. Additionally, *Wrn*^{Δhel} cells are more sensitive to DNA damaging agents such as the topoisomerase inhibitors etoposide and camptothecin [444], again comparable to human patients, unlike *Wrn*^{-/-} cells, which are not [445].

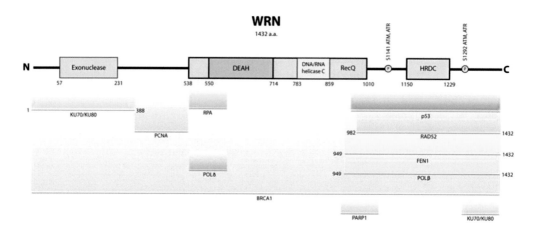

Figure 14. WRN Protein. The WRN protein consists of 1432 amino acid residues. Domains are indicated by colored boxes: Asp-Glu-Ala-His (DEAH) domain, RecQ helicase, and RNaseD C-terminal (HRDC) domain. ATM and ATR phosphorylation events (P) are indicated. Many relevant proteins that interact with WRN are highlighted below the protein.

The exact molecular function of WRN is not well understood, though it has been implicated in a number of DNA repair processes. It is not possible to review all of the work here, but in summary, WRN has been reported as having roles in long patch BER through DNA polβ and PARP-1 [447]; in NER, both directly and through FEN-1; in single-strand break repair (SSBR) through PARP-1 [448]; ICL-repair, through BRCA1 [449]; in NHEJ through KU86 and DNA-PKcs [450, 451]; in HRR, by direct activity and through RPA, Rad51, Rad52, RAD54 and RAD54B with a possible regulation by PARP-1 [442, 448, 452-455]; in repair of stalled replication forks by direct activity [456]; and in DNA replication, both directly and through PCNA, RPA and DNA polδ [457, 458]. Molecular analysis of the processing of a DSB on an episomal plasmid demonstrated that extensive deletion of the plasmid DNA occurs in the absence of WRN [459]. Similarly, the Tg.AC mouse model was found to display increased deletion when placed on the Werner helicase mutant background *Wrn*^{Δhel} [460]. In summary, WRN has been implicated in almost every DNA repair mechanism and appears to be intricately linked to components of DDR [461, 462]. Through each of these interactions, models can be hypothesized whereby WRN can directly or indirectly promote or compete with HRR. Irrespective of all the reported activities of WRN, it has been established that Wrn functions to suppress some HRR events as determined by the *p*^{un} assay [463].

Summary

The DDR system forms an intricate signaling web to control a number of responses, including cell cycle and possibly HRR. From the simplified descriptions of the core components that we have described here, the DDR involves cross-talk, signal reinforcement, feedback loops, cellular re-localization, posttranslational modification, protein:protein interaction, protein degradation and transcriptional control. The components of the DDR that are utilized and the response elicited depend on the type of damage and the stage of the cell cycle. Throughout this review, we have highlighted the roles of each component of the DDR in cell cycle progression and how each component is related to one another. We have also summarized the potential role for each of these proteins in either directly or indirectly controlling HRR.

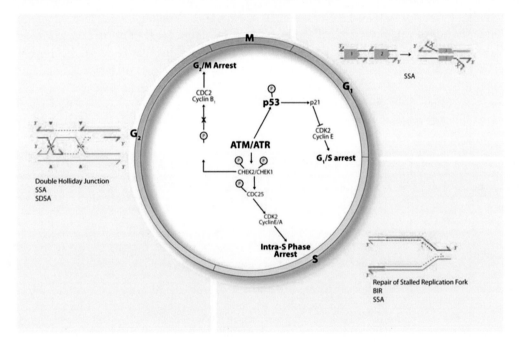

Figure 15. Overlay of Cell Cycle and HRR Models. The main components of the DDR cell cycle checkpoints are illustrated; central in the response are the transducer proteins ATM and ATR. Phosphorylation events (P) are indicated as well as protein interactions (arrows) and inhibitory interactions (blunt ended lines). During G_1, the cells assess damage by inducing G_1/S arrest in order to repair lesions by single strand annealing (SSA). Lesions in S-phase can have the added complexity of stalled replication forks; the nascent strand, however, provides a homologous template that can be used in HRR stalled replication fork repair, break induced repair (BIR) or simply single strand annealing (SSA). DNA damage during G_2 will induce G_2/M arrest, where repair mechanisms include HRR mediated resolution of Holliday junctions, synthesis dependent strand annealing (SDSA) or SSA. Blue and red lines represent homologous sections of DNA, and arrows represent the direction of DNA synthesis. Solid lines represent existing DNA, while dashed lines represent nascent DNA synthesis. Green arrowheads depict the possible sites of resolution of the Holliday structures, and blocked orange arrows represent sections of repetitive DNA, with scissors depicting cleavage of excess overhanging DNA.

HRR plays a crucial rule in repairing spontaneous or exogenously induced errors that may involve DSBs, stalled replication forks, lesion bypass or cohesion dependent repair [464]. We have described the major models for HRR (Figures 1 and 2). As described, each model has prerequisites, such as homology being present on the same chromosome (i.e. SSA), initiation by stalled replication fork, or the presence of either sister chromatids or the chromosomal homologue. These considerations relate cell cycle progression with the type of HRR that can occur (Figure 15). When combined with the cell cycle dependent response of DDR (Figure 3) and the potential interrelationship of DDR and HRR, it can be seen that this is a very complex system.

It is clear that there is a relationship between DDR and HRR. Each of the central components of DDR that we have highlighted, with the exception of p21, has been shown to be involved in the regulation of HRR. Though neither DDR nor HRR are fully understood, there are at least three HRR components that appear to be directly influenced by DDR. In particular, these are BLM, the MRN complex and RAD51. A summary of these interactions, without the added complexity of cell cycle, is given in Figure 16. RAD51 is clearly known to be a part of the HRR machinery. Biochemical evidence suggests that BLM is also important for HRR. With regard to MRN, its role in HRR is still unclear, at least in mammalian systems. Certainly MRN does influence HRR and it is suggested that it is an active participant in the process. Clearly there is much work left to do to unravel DDR and HRR intertwining and determine whether their relationship can be used to therapeutic advantage.

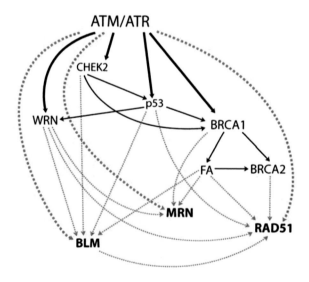

Figure 16. Summary of DDR Interactions with HRR Components. This schematic depicts the overall view of the relationship between DDR and HRR. Thick lines represent interactions with ATM/ATR. Thin lines represent interactions between proteins. Red lines are direct links to HRR (proteins in bold).

Acknowledgements

Supported by the NIEHS (K22-ES12264; A.J.R.B.) and NIA (T32AG021890-05; A.D.B.). We gratefully acknowledge the graphical support provided by David Rodriguez.

References

[1]   Lindahl T. DNA glycosylases, endonucleases for apurinic/apyrimidinic sites, and base excision-repair. *Prog Nucleic Acid Res Mol Biol* **22**(135-92, 1979.

[2]   Beckman KB, Ames BN. Oxidative decay of DNA. *J Biol Chem* **272**(32): 19633-6, 1997.

[3]   Ward JF. The radiation-induced lesions which trigger the bystander effect. *Mutat Res* **499**(2): 151-4, 2002.

[4]   Charlton DE, Nikjoo H, Humm JL. Calculation of initial yields of single- and double-strand breaks in cell nuclei from electrons, protons and alpha particles. *Int J Radiat Biol* **56**(1): 1-19, 1989.

[5]   Ward JF. The yield of DNA double-strand breaks produced intracellularly by ionizing radiation: a review. *Int J Radiat Biol* **57**(6): 1141-50, 1990.

[6]   Bishop AJ, Schiestl RH. Homologous recombination as a mechanism of carcinogenesis. *Biochim Biophys Acta* **1471**(3): M109-21, 2001.

[7]   Helleday T, Lo J, van Gent DC, Engelward BP. DNA double-strand break repair: from mechanistic understanding to cancer treatment. *DNA Repair (Amst)* 6(7): 923-35, 2007.

[8]   Paull TT, Gellert M. The 3' to 5' exonuclease activity of Mre 11 facilitates repair of DNA double-strand breaks. *Mol Cell* **1**(7): 969-79, 1998.

[9]   Sung P, Klein H. Mechanism of homologous recombination: mediators and helicases take on regulatory functions. *Nat Rev Mol Cell Biol* **7**(10): 739-50, 2006.

[10] Kurumizaka H, Ikawa S, Nakada M, Eda K, Kagawa W, Takata M, Takeda S, Yokoyama S, Shibata T. Homologous-pairing activity of the human DNA-repair proteins Xrcc3.Rad51C. *Proc Natl Acad Sci U S A* **98**(10): 5538-43, 2001.

[11] Masson JY, Stasiak AZ, Stasiak A, Benson FE, West SC. Complex formation by the human RAD51C and XRCC3 recombination repair proteins. *Proc Natl Acad Sci U S A* **98**(15): 8440-6, 2001.

[12] Liu Y, Masson JY, Shah R, O'Regan P, West SC. RAD51C is required for Holliday junction processing in mammalian cells. *Science* **303**(5655): 243-6, 2004.

[13] Sigurdsson S, Van Komen S, Bussen W, Schild D, Albala JS, Sung P. Mediator function of the human Rad51B-Rad51C complex in Rad51/RPA-catalyzed DNA strand exchange. *Genes Dev* **15**(24): 3308-18, 2001.

[14] Yokoyama H, Kurumizaka H, Ikawa S, Yokoyama S, Shibata T. Holliday junction binding activity of the human Rad51B protein. *J Biol Chem* **278**(4): 2767-72, 2003.

[15] Havre PA, Rice M, Ramos R, Kmiec EB. HsRec2/Rad51L1, a protein influencing cell cycle progression, has protein kinase activity. *Exp Cell Res* **254**(1): 33-44, 2000.

[16] Braybrooke JP, Spink KG, Thacker J, Hickson ID. The RAD51 family member, RAD51L3, is a DNA-stimulated ATPase that forms a complex with XRCC2. *J Biol Chem* **275**(37): 29100-6, 2000.

[17] Johnson RD, Jasin M. Sister chromatid gene conversion is a prominent double-strand break repair pathway in mammalian cells. *Embo J* **19**(13): 3398-407, 2000.

[18] Kraus E, Leung WY, Haber JE. Break-induced replication: a review and an example in budding yeast. *Proc Natl Acad Sci U S A* **98**(15): 8255-62, 2001.

[19] Arnaudeau C, Lundin C, Helleday T. DNA double-strand breaks associated with replication forks are predominantly repaired by homologous recombination involving an exchange mechanism in mammalian cells. *J Mol Biol* **307**(5): 1235-45, 2001.

[20] Haaf T, Golub EI, Reddy G, Radding CM, Ward DC. Nuclear foci of mammalian Rad51 recombination protein in somatic cells after DNA damage and its localization in synaptonemal complexes. *Proc Natl Acad Sci U S A* **92**(6): 2298-302, 1995.

[21] Hochegger H, Takeda S. Phenotypic analysis of cellular responses to DNA damage. *Subcell Biochem* **40**(313-25, 2006.

[22] Zhang J, Willers H, Feng Z, Ghosh JC, Kim S, Weaver DT, Chung JH, Powell SN, Xia F. Chk2 phosphorylation of BRCA1 regulates DNA double-strand break repair. *Mol Cell Biol* **24**(2): 708-18, 2004.

[23] Johnson RD, Liu N, Jasin M. Mammalian XRCC2 promotes the repair of DNA double-strand breaks by homologous recombination. *Nature* **401**(6751): 397-9, 1999.

[24] Brilliant MH, Gondo Y, Eicher EM. Direct molecular identification of the mouse pink-eyed unstable mutation by genome scanning. *Science* **252**(5005): 566-569, 1991.

[25] Gondo Y, Gardner JM, Nakatsu Y, Durham-Pierre D, Deveau SA, Kuper C, Brilliant MH. High-frequency genetic reversion mediated by a DNA duplication: the mouse pink-eyed unstable mutation. *Proc Natl Acad Sci U S A* **90**(1): 297-301, 1993.

[26] Schiestl RH, Aubrecht J, Khogali F, Carls N. Carcinogens induce reversion of the mouse pink-eyed unstable mutation. *Proc Natl Acad Sci U S A* **94**(9): 4576-4581, 1997.

[27] Searle AG. The use of pigment loci for detecting reverse mutations in somatic cells of mice. *Arch Toxicol* **38**(1-2): 105-108, 1977.

[28] Deol MS, Truslove GM. The effects of the pink-eyed unstable gene on the retinal pigment epithelium of the mouse. *J Embryol Exp Morphol* **78**(291-298, 1983.

[29] Bodenstein L, Sidman RL. Growth and development of the mouse retinal pigment epithelium. II. *Cell* patterning in experimental chimaeras and mosaics. *Developmental Biology* **121**(1): 205-219, 1987.

[30] Schiestl RH, Khogali F, Carls N. Reversion of the mouse pink-eyed unstable mutation induced by low doses of x-rays. *Science* **266**(5190): 1573-6, 1994.

[31] Jalili T, Murthy GG, Schiestl RH. Cigarette smoke induces DNA deletions in the mouse embryo. *Cancer Res.* **58**(12): 2633-2638, 1998.

[32] Bishop AJR, Kosaras B, Sidman RL, Schiestl RH. Benzo(a)pyrene and X-rays induce reversions of the pink-eyed unstable mutation in the retinal pigment epithelium of mice. *Mutation Research* **457**(1-2): 31-40, 2000.

[33] Bishop AJ, Barlow C, Wynshaw-Boris AJ, Schiestl RH. Atm deficiency causes an increased frequency of intrachromosomal homologous recombination in mice. *Cancer Res* **60**(2): 395-9., 2000.

[34] Lebel M. Increased frequency of DNA deletions in pink-eyed unstable mice carrying a mutation in the Werner syndrome gene homologue. *Carcinogenesis* **23**(1): 213-6., 2002.

[35] Bishop AJ, Hollander MC, Kosaras B, Sidman RL, Fornace AJ, Jr., Schiestl RH. Atm-, p53-, and Gadd45a-deficient mice show an increased frequency of homologous recombination at different stages during development. *Cancer Res* **63**(17): 5335-43, 2003.

[36]   Bodenstein L, Sidman RL. Growth and development of the mouse retinal pigment epithelium. I. *Cell* and tissue morphometrics and topography of mitotic activity. *Developmental Biology* **121**(1): 192-204, 1987.

[37]   Bishop AJ, Kosaras B, Carls N, Sidman RL, Schiestl RH. Susceptibility of proliferating cells to benzo[a]pyrene-induced homologous recombination in mice. *Carcinogenesis* **22**(4): 641-9., 2001.

[38]   Shiloh Y. ATM and related protein kinases: safeguarding genome integrity. *Nat Rev Cancer* **3**(3): 155-68, 2003.

[39]   Giacinti C, Giordano A. RB and cell cycle progression. *Oncogene* **25**(38): 5220-7, 2006.

[40]   Kastan MB, Bartek J. *Cell*-cycle checkpoints and cancer. *Nature* **432**(7015): 316-23, 2004.

[41]   Tourriere H, Pasero P. Maintenance of fork integrity at damaged DNA and natural pause sites. *DNA Repair (Amst)* **6**(7): 900-13, 2007.

[42]   Shiloh Y. ATM and ATR: networking cellular responses to DNA damage. *Curr Opin Genet Dev* **11**(1): 71-7, 2001.

[43]   Abraham RT. *Cell* cycle checkpoint signaling through the ATM and ATR kinases. *Genes Dev* **15**(17): 2177-96, 2001.

[44]   Hammond EM, Denko NC, Dorie MJ, Abraham RT, Giaccia AJ. Hypoxia links ATR and p53 through replication arrest. *Mol Cell Biol* **22**(6): 1834-43, 2002.

[45]   Heffernan TP, Simpson DA, Frank AR, Heinloth AN, Paules RS, Cordeiro-Stone M, Kaufmann WK. An ATR- and Chk1-dependent S checkpoint inhibits replicon initiation following UVC-induced DNA damage. *Mol Cell Biol* **22**(24): 8552-61, 2002.

[46]   Tibbetts RS, Brumbaugh KM, Williams JM, Sarkaria JN, Cliby WA, Shieh SY, Taya Y, Prives C, Abraham RT. A role for ATR in the DNA damage-induced phosphorylation of p53. *Genes Dev* **13**(2): 152-7, 1999.

[47]   Banin S, Moyal L, Shieh S, Taya Y, Anderson CW, Chessa L, Smorodinsky NI, Prives C, Reiss Y, Shiloh Y, Ziv Y. Enhanced phosphorylation of p53 by ATM in response to DNA damage. *Science* **281**(5383): 1674-7, 1998.

[48]   Canman CE, Lim DS, Cimprich KA, Taya Y, Tamai K, Sakaguchi K, Appella E, Kastan MB, Siliciano JD. Activation of the ATM kinase by ionizing radiation and phosphorylation of p53. *Science* **281**(5383): 1677-9, 1998.

[49]   Khanna KK, Keating KE, Kozlov S, Scott S, Gatei M, Hobson K, Taya Y, Gabrielli B, Chan D, Lees-Miller SP, Lavin MF. ATM associates with and phosphorylates p53: mapping the region of interaction. *Nat Genet* **20**(4): 398-400, 1998.

[50]   Harper JW, Adami GR, Wei N, Keyomarsi K, Elledge SJ. The p21 Cdk-interacting protein Cip1 is a potent inhibitor of G1 cyclin-dependent kinases. *Cell* **75**(4): 805-16, 1993.

[51]   Xu B, O'Donnell AH, Kim ST, Kastan MB. Phosphorylation of serine 1387 in Brca1 is specifically required for the Atm-mediated S-phase checkpoint after ionizing irradiation. *Cancer Res* **62**(16): 4588-91, 2002.

[52]   Xu B, Kim S, Kastan MB. Involvement of Brca1 in S-phase and G(2)-phase checkpoints after ionizing irradiation. *Mol Cell Biol* **21**(10): 3445-50, 2001.

[53]   Tibbetts RS, Cortez D, Brumbaugh KM, Scully R, Livingston D, Elledge SJ, Abraham RT. Functional interactions between BRCA1 and the checkpoint kinase ATR during genotoxic stress. *Genes Dev* **14**(23): 2989-3002, 2000.

[54]   Zhao H, Piwnica-Worms H. ATR-mediated checkpoint pathways regulate phosphorylation and activation of human Chk1. *Mol Cell Biol* **21**(13): 4129-39, 2001.

[55]   Guo Z, Kumagai A, Wang SX, Dunphy WG. Requirement for Atr in phosphorylation of Chk1 and cell cycle regulation in response to DNA replication blocks and UV-damaged DNA in Xenopus egg extracts. *Genes Dev* **14**(21): 2745-56, 2000.

[56]   Lopez-Girona A, Tanaka K, Chen XB, Baber BA, McGowan CH, Russell P. Serine-345 is required for Rad3-dependent phosphorylation and function of checkpoint kinase Chk1 in fission yeast. *Proc Natl Acad Sci U S A* **98**(20): 11289-94, 2001.

[57]   Lam MH, Rosen JM. Chk1 versus Cdc25: chking one's levels of cellular proliferation. *Cell Cycle* **3**(11): 1355-7, 2004.

[58]   Chehab NH, Malikzay A, Appel M, Halazonetis TD. Chk2/hCds1 functions as a DNA damage checkpoint in G(1) by stabilizing p53. *Genes Dev* **14**(3): 278-88, 2000.

[59]   Shieh SY, Ahn J, Tamai K, Taya Y, Prives C. The human homologs of checkpoint kinases Chk1 and Cds1 (Chk2) phosphorylate p53 at multiple DNA damage-inducible sites. *Genes Dev* **14**(3): 289-300, 2000.

[60]   O'Connell MJ, Walworth NC, Carr AM. The G2-phase DNA-damage checkpoint. *Trends Cell Biol* **10**(7): 296-303, 2000.

[61]   Harper JW, Elledge SJ. The DNA damage response: ten years after. *Mol Cell* **28**(5): 739-45, 2007.

[62]   Shiloh Y. Ataxia-telangiectasia, ATM and genomic stability: maintaining a delicate balance. Two international workshops on ataxia-telangiectasia, related disorders and the ATM protein. *Biochim Biophys Acta* **1378**(2): R11-8, 1998.

[63]   Lee JH, Paull TT. Activation and regulation of ATM kinase activity in response to DNA double-strand breaks. *Oncogene* **26**(56): 7741-8, 2007.

[64]   Bakkenist CJ, Kastan MB. DNA damage activates ATM through intermolecular autophosphorylation and dimer dissociation. *Nature* **421**(6922): 499-506, 2003.

[65]   Kozlov SV, Graham ME, Peng C, Chen P, Robinson PJ, Lavin MF. Involvement of novel autophosphorylation sites in ATM activation. *Embo J* **25**(15): 3504-14, 2006.

[66]   Stiff T, Walker SA, Cerosaletti K, Goodarzi AA, Petermann E, Concannon P, O'Driscoll M, Jeggo PA. ATR-dependent phosphorylation and activation of ATM in response to UV treatment or replication fork stalling. *Embo J* **25**(24): 5775-82, 2006.

[67]   Kitagawa R, Bakkenist CJ, McKinnon PJ, Kastan MB. Phosphorylation of SMC1 is a critical downstream event in the ATM-NBS1-BRCA1 pathway. *Genes Dev* **18**(12): 1423-38, 2004.

[68]   Uziel T, Lerenthal Y, Moyal L, Andegeko Y, Mittelman L, Shiloh Y. Requirement of the MRN complex for ATM activation by DNA damage. *Embo J* **22**(20): 5612-21, 2003.

[69]   Carson CT, Schwartz RA, Stracker TH, Lilley CE, Lee DV, Weitzman MD. The Mre11 complex is required for ATM activation and the G2/M checkpoint. *Embo J* **22**(24): 6610-20, 2003.

[70]   Horejsi Z, Falck J, Bakkenist CJ, Kastan MB, Lukas J, Bartek J. Distinct functional domains of Nbs1 modulate the timing and magnitude of ATM activation after low doses of ionizing radiation. *Oncogene* **23**(17): 3122-7, 2004.

[71]   Sun Y, Xu Y, Roy K, Price BD. DNA damage-induced acetylation of lysine 3016 of ATM activates ATM kinase activity. *Mol Cell Biol* **27**(24): 8502-9, 2007.

[72] Berkovich E, Monnat RJ, Jr., Kastan MB. Roles of ATM and NBS1 in chromatin structure modulation and DNA double-strand break repair. *Nat Cell Biol* **9**(6): 683-90, 2007.

[73] Burma S, Chen BP, Murphy M, Kurimasa A, Chen DJ. ATM phosphorylates histone H2AX in response to DNA double-strand breaks. *J Biol Chem* **276**(45): 42462-7, 2001.

[74] Lim DS, Kim ST, Xu B, Maser RS, Lin J, Petrini JH, Kastan MB. ATM phosphorylates p95/nbs1 in an S-phase checkpoint pathway. *Nature* **404**(6778): 613-7, 2000.

[75] Zhao S, Weng YC, Yuan SS, Lin YT, Hsu HC, Lin SC, Gerbino E, Song MH, Zdzienicka MZ, Gatti RA, Shay JW, Ziv Y, Shiloh Y, Lee EY. Functional link between ataxia-telangiectasia and Nijmegen breakage syndrome gene products. *Nature* **405**(6785): 473-7, 2000.

[76] Wu X, Ranganathan V, Weisman DS, Heine WF, Ciccone DN, O'Neill TB, Crick KE, Pierce KA, Lane WS, Rathbun G, Livingston DM, Weaver DT. ATM phosphorylation of Nijmegen breakage syndrome protein is required in a DNA damage response. *Nature* **405**(6785): 477-82, 2000.

[77] Lee JH, Xu B, Lee CH, Ahn JY, Song MS, Lee H, Canman CE, Lee JS, Kastan MB, Lim DS. Distinct functions of Nijmegen breakage syndrome in ataxia telangiectasia mutated-dependent responses to DNA damage. *Mol Cancer Res* **1**(9): 674-81, 2003.

[78] Falck J, Coates J, Jackson SP. Conserved modes of recruitment of ATM, ATR and DNA-PKcs to sites of DNA damage. *Nature* **434**(7033): 605-11, 2005.

[79] Gatei M, Young D, Cerosaletti KM, Desai-Mehta A, Spring K, Kozlov S, Lavin MF, Gatti RA, Concannon P, Khanna K. ATM-dependent phosphorylation of nibrin in response to radiation exposure. *Nat Genet* **25**(1): 115-9, 2000.

[80] Matsuoka S, Rotman G, Ogawa A, Shiloh Y, Tamai K, Elledge SJ. Ataxia telangiectasia-mutated phosphorylates Chk2 in vivo and in vitro. *Proc Natl Acad Sci U S A* **97**(19): 10389-94, 2000.

[81] Ahn JY, Li X, Davis HL, Canman CE. Phosphorylation of threonine 68 promotes oligomerization and autophosphorylation of the Chk2 protein kinase via the forkhead-associated domain. *J Biol Chem* **277**(22): 19389-95, 2002.

[82] Khosravi R, Maya R, Gottlieb T, Oren M, Shiloh Y, Shkedy D. Rapid ATM-dependent phosphorylation of MDM2 precedes p53 accumulation in response to DNA damage. *Proc Natl Acad Sci U S A* **96**(26): 14973-7, 1999.

[83] Maya R, Balass M, Kim ST, Shkedy D, Leal JF, Shifman O, Moas M, Buschmann T, Ronai Z, Shiloh Y, Kastan MB, Katzir E, Oren M. ATM-dependent phosphorylation of Mdm2 on serine 395: role in p53 activation by DNA damage. *Genes Dev* **15**(9): 1067-77, 2001.

[84] Taniguchi T, Garcia-Higuera I, Xu B, Andreassen PR, Gregory RC, Kim ST, Lane WS, Kastan MB, D'Andrea AD. Convergence of the fanconi anemia and ataxia telangiectasia signaling pathways. *Cell* **109**(4): 459-72, 2002.

[85] Gatei M, Scott SP, Filippovitch I, Soronika N, Lavin MF, Weber B, Khanna KK. Role for ATM in DNA damage-induced phosphorylation of BRCA1. *Cancer Res* **60**(12): 3299-304, 2000.

[86] Baskaran R, Wood LD, Whitaker LL, Canman CE, Morgan SE, Xu Y, Barlow C, Baltimore D, Wynshaw-Boris A, Kastan MB, Wang JY. Ataxia telangiectasia mutant

protein activates c-Abl tyrosine kinase in response to ionizing radiation. *Nature* **387**(6632): 516-9, 1997.

[87] Beamish H, Kedar P, Kaneko H, Chen P, Fukao T, Peng C, Beresten S, Gueven N, Purdie D, Lees-Miller S, Ellis N, Kondo N, Lavin MF. Functional link between BLM defective in Bloom's syndrome and the ataxia-telangiectasia-mutated protein, *ATM. J Biol Chem* **277**(34): 30515-23, 2002.

[88] Gately DP, Hittle JC, Chan GK, Yen TJ. Characterization of ATM expression, localization, and associated DNA-dependent protein kinase activity. *Mol Biol Cell* **9**(9): 2361-74, 1998.

[89] Barlow C, Hirotsune S, Paylor R, Liyanage M, Eckhaus M, Collins F, Shiloh Y, Crawley JN, Ried T, Tagle D, Wynshaw-Boris A. Atm-deficient mice: a paradigm of ataxia telangiectasia. *Cell* **86**(1): 159-71, 1996.

[90] Xu Y, Ashley T, Brainerd EE, Bronson RT, Meyn MS, Baltimore D. Targeted disruption of ATM leads to growth retardation, chromosomal fragmentation during meiosis, immune defects, and thymic lymphoma. *Genes Dev* **10**(19): 2411-22, 1996.

[91] Elson A, Wang Y, Daugherty CJ, Morton CC, Zhou F, Campos-Torres J, Leder P. Pleiotropic defects in ataxia-telangiectasia protein-deficient mice. *Proc Natl Acad Sci U S A* **93**(23): 13084-9, 1996.

[92] Meyn MS. High spontaneous intrachromosomal recombination rates in ataxia-telangiectasia. *Science* **260**(5112): 1327-30, 1993.

[93] Shafman T, Khanna KK, Kedar P, Spring K, Kozlov S, Yen T, Hobson K, Gatei M, Zhang N, Watters D, Egerton M, Shiloh Y, Kharbanda S, Kufe D, Lavin MF. Interaction between ATM protein and c-Abl in response to DNA damage. *Nature* **387**(6632): 520-3, 1997.

[94] Chen G, Yuan SS, Liu W, Xu Y, Trujillo K, Song B, Cong F, Goff SP, Wu Y, Arlinghaus R, Baltimore D, Gasser PJ, Park MS, Sung P, Lee EY. Radiation-induced assembly of Rad51 and Rad52 recombination complex requires ATM and c-Abl. *J Biol Chem* **274**(18): 12748-52, 1999.

[95] Morrison C, Sonoda E, Takao N, Shinohara A, Yamamoto K, Takeda S. The controlling role of ATM in homologous recombinational repair of DNA damage. *Embo J* 19(3): 463-71, 2000.

[96] Bolderson E, Scorah J, Helleday T, Smythe C, Meuth M. ATM is required for the cellular response to thymidine induced replication fork stress. *Hum Mol Genet* **13**(23): 2937-45, 2004.

[97] Bishop AJ, Barlow C, Wynshaw-Boris AJ, Schiestl RH. Atm deficiency causes an increased frequency of intrachromosomal homologous recombination in mice. *Cancer Res* **60**(2): 395-9, 2000.

[98] Reliene R, Fischer E, Schiestl RH. Effect of N-acetyl cysteine on oxidative DNA damage and the frequency of DNA deletions in atm-deficient mice. *Cancer Res* **64**(15): 5148-53, 2004.

[99] Stewart GS, Maser RS, Stankovic T, Bressan DA, Kaplan MI, Jaspers NG, Raams A, Byrd PJ, Petrini JH, Taylor AM. The DNA double-strand break repair gene hMRE11 is mutated in individuals with an ataxia-telangiectasia-like disorder. *Cell* **99**(6): 577-87, 1999.

[100] D'Amours D, Jackson SP. The Mre11 complex: at the crossroads of dna repair and checkpoint signalling. *Nat Rev Mol Cell Biol* **3**(5): 317-27, 2002.

[101] Paull TT, Gellert M. Nbs1 potentiates ATP-driven DNA unwinding and endonuclease cleavage by the Mre11/Rad50 complex. *Genes Dev* **13**(10): 1276-88, 1999.

[102] de Jager M, Dronkert ML, Modesti M, Beerens CE, Kanaar R, van Gent DC. DNA-binding and strand-annealing activities of human Mre11: implications for its roles in DNA double-strand break repair pathways. *Nucleic Acids Res* **29**(6): 1317-25, 2001.

[103] de Jager M, van Noort J, van Gent DC, Dekker C, Kanaar R, Wyman C. Human Rad50/Mre11 is a flexible complex that can tether DNA ends. *Mol Cell* **8**(5): 1129-35, 2001.

[104] Trujillo KM, Yuan SS, Lee EY, Sung P. Nuclease activities in a complex of human recombination and DNA repair factors Rad50, Mre11, and p95. *J Biol Chem* **273**(34): 21447-50, 1998.

[105] Paull TT, Gellert M. A mechanistic basis for Mre11-directed DNA joining at microhomologies. *Proc Natl Acad Sci U S A* **97**(12): 6409-14, 2000.

[106] Dolganov GM, Maser RS, Novikov A, Tosto L, Chong S, Bressan DA, Petrini JH. Human Rad50 is physically associated with human Mre11: identification of a conserved multiprotein complex implicated in recombinational DNA repair. *Mol Cell Biol* **16**(9): 4832-41, 1996.

[107] Desai-Mehta A, Cerosaletti KM, Concannon P. Distinct functional domains of nibrin mediate Mre11 binding, focus formation, and nuclear localization. *Mol Cell Biol* **21**(6): 2184-91, 2001.

[108] Carney JP, Maser RS, Olivares H, Davis EM, Le Beau M, Yates JR, 3rd, Hays L, Morgan WF, Petrini JH. The hMre11/hRad50 protein complex and Nijmegen breakage syndrome: linkage of double-strand break repair to the cellular DNA damage response. *Cell* **93**(3): 477-86, 1998.

[109] Wang Y, Cortez D, Yazdi P, Neff N, Elledge SJ, Qin J. BASC, a super complex of BRCA1-associated proteins involved in the recognition and repair of aberrant DNA structures. *Genes Dev* **14**(8): 927-39, 2000.

[110] Zhong Q, Chen CF, Li S, Chen Y, Wang CC, Xiao J, Chen PL, Sharp ZD, Lee WH. Association of BRCA1 with the hRad50-hMre11-p95 complex and the DNA damage response. *Science* **285**(5428): 747-50, 1999.

[111] Paull TT, Cortez D, Bowers B, Elledge SJ, Gellert M. Direct DNA binding by Brca1. *Proc Natl Acad Sci U S A* **98**(11): 6086-91, 2001.

[112] Restle A, Janz C, Wiesmuller L. Differences in the association of p53 phosphorylated on serine 15 and key enzymes of homologous recombination. *Oncogene* **24**(27): 4380-7, 2005.

[113] Mirzoeva OK, Petrini JH. DNA damage-dependent nuclear dynamics of the Mre11 complex. *Mol Cell Biol* **21**(1): 281-8, 2001.

[114] Lombard DB, Guarente L. Nijmegen breakage syndrome disease protein and MRE11 at PML nuclear bodies and meiotic telomeres. *Cancer Res* **60**(9): 2331-4, 2000.

[115] Nelms BE, Maser RS, MacKay JF, Lagally MG, Petrini JH. In situ visualization of DNA double-strand break repair in human fibroblasts. *Science* **280**(5363): 590-2, 1998.

[116] Maser RS, Monsen KJ, Nelms BE, Petrini JH. hMre11 and hRad50 nuclear foci are induced during the normal cellular response to DNA double-strand breaks. *Mol Cell Biol* **17**(10): 6087-96, 1997.

[117] Maser RS, Mirzoeva OK, Wells J, Olivares H, Williams BR, Zinkel RA, Farnham PJ, Petrini JH. Mre11 complex and DNA replication: linkage to E2F and sites of DNA synthesis. *Mol Cell Biol* **21**(17): 6006-16, 2001.

[118] Xiao Y, Weaver DT. Conditional gene targeted deletion by Cre recombinase demonstrates the requirement for the double-strand break repair Mre11 protein in murine embryonic stem cells. *Nucleic Acids Res* **25**(15): 2985-91, 1997.

[119] Theunissen JW, Kaplan MI, Hunt PA, Williams BR, Ferguson DO, Alt FW, Petrini JH. Checkpoint failure and chromosomal instability without lymphomagenesis in Mre11(ATLD1/ATLD1) mice. *Mol Cell* **12**(6): 1511-23, 2003.

[120] Ajimura M, Leem SH, Ogawa H. Identification of new genes required for meiotic recombination in Saccharomyces cerevisiae. *Genetics* **133**(1): 51-66, 1993.

[121] Ivanov EL, Korolev VG, Fabre F. XRS2, a DNA repair gene of Saccharomyces cerevisiae, is needed for meiotic recombination. *Genetics* **132**(3): 651-64, 1992.

[122] Saeki T, Machida I, Nakai S. Genetic control of diploid recovery after gamma-irradiation in the yeast Saccharomyces cerevisiae. *Mutat Res* **73**(2): 251-65, 1980.

[123] Zakharov IA, Kasinova GV, Koval'tsova SV. [Intragenic mitotic recombination induced by ultraviolet and gamma rays in radiosensitive mutants of Saccharomyces cerevisiae yeasts]. *Genetika* **19**(1): 49-57, 1983.

[124] Aravind L, Walker DR, Koonin EV. Conserved domains in DNA repair proteins and evolution of repair systems. *Nucleic Acids Res* **27**(5): 1223-42, 1999.

[125] Gennery AR. Primary immunodeficiency syndromes associated with defective DNA double-strand break repair. *Br Med Bull* **77-78**(71-85, 2006.

[126] Hopfner KP, Tainer JA. Rad50/SMC proteins and ABC transporters: unifying concepts from high-resolution structures. *Curr Opin Struct Biol* **13**(2): 249-55, 2003.

[127] Luo G, Yao MS, Bender CF, Mills M, Bladl AR, Bradley A, Petrini JH. Disruption of mRad50 causes embryonic stem cell lethality, abnormal embryonic development, and sensitivity to ionizing radiation. *Proc Natl Acad Sci U S A* **96**(13): 7376-81, 1999.

[128] Yamaguchi-Iwai Y, Sonoda E, Sasaki MS, Morrison C, Haraguchi T, Hiraoka Y, Yamashita YM, Yagi T, Takata M, Price C, Kakazu N, Takeda S. Mre11 is essential for the maintenance of chromosomal DNA in vertebrate cells. *Embo J* **18**(23): 6619-29, 1999.

[129] Zhu J, Petersen S, Tessarollo L, Nussenzweig A. Targeted disruption of the Nijmegen breakage syndrome gene NBS1 leads to early embryonic lethality in mice. *Curr Biol* **11**(2): 105-9, 2001.

[130] Bender CF, Sikes ML, Sullivan R, Huye LE, Le Beau MM, Roth DB, Mirzoeva OK, Oltz EM, Petrini JH. Cancer predisposition and hematopoietic failure in Rad50(S/S) mice. *Genes Dev* **16**(17): 2237-51, 2002.

[131] Alani E, Padmore R, Kleckner N. Analysis of wild-type and rad50 mutants of yeast suggests an intimate relationship between meiotic chromosome synapsis and recombination. *Cell* **61**(3): 419-36, 1990.

[132] Morales M, Theunissen JW, Kim CF, Kitagawa R, Kastan MB, Petrini JH. The Rad50S allele promotes ATM-dependent DNA damage responses and suppresses ATM deficiency: implications for the Mre11 complex as a DNA damage sensor. *Genes Dev* **19**(24): 3043-54, 2005.

[133] Usui T, Petrini JH, Morales M. Rad50S alleles of the Mre11 complex: questions answered and questions raised. *Exp Cell Res* **312**(14): 2694-9, 2006.

[134] Weemaes CM, Hustinx TW, Scheres JM, van Munster PJ, Bakkeren JA, Taalman RD. A new chromosomal instability disorder: the Nijmegen breakage syndrome. *Acta Paediatr Scand* **70**(4): 557-64, 1981.

[135] van der Burgt I, Chrzanowska KH, Smeets D, Weemaes C. Nijmegen breakage syndrome. *J Med Genet* **33**(2): 153-6, 1996.

[136] Seemanova E. An increased risk for malignant neoplasms in heterozygotes for a syndrome of microcephaly, normal intelligence, growth retardation, remarkable facies, immunodeficiency and chromosomal instability. *Mutat Res* **238**(3): 321-4, 1990.

[137] Shiloh Y. Ataxia-telangiectasia and the Nijmegen breakage syndrome: related disorders but genes apart. *Annu Rev Genet* **31**(635-62, 1997.

[138] Chrzanowska KH, Kleijer WJ, Krajewska-Walasek M, Bialecka M, Gutkowska A, Goryluk-Kozakiewicz B, Michalkiewicz J, Stachowski J, Gregorek H, Lyson-Wojciechowska G, et al. Eleven Polish patients with microcephaly, immunodeficiency, and chromosomal instability: the Nijmegen breakage syndrome. *Am J Med Genet* **57**(3): 462-71, 1995.

[139] Conley ME, Spinner NB, Emanuel BS, Nowell PC, Nichols WW. A chromosomal breakage syndrome with profound immunodeficiency. *Blood* **67**(5): 1251-6, 1986.

[140] Taalman RD, Hustinx TW, Weemaes CM, Seemanova E, Schmidt A, Passarge E, Scheres JM. Further delineation of the Nijmegen breakage syndrome. *Am J Med Genet* **32**(3): 425-31, 1989.

[141] Tupler R, Marseglia GL, Stefanini M, Prosperi E, Chessa L, Nardo T, Marchi A, Maraschio P. A variant of the Nijmegen breakage syndrome with unusual cytogenetic features and intermediate cellular radiosensitivity. *J Med Genet* **34**(3): 196-202, 1997.

[142] Jaspers NG, Gatti RA, Baan C, Linssen PC, Bootsma D. Genetic complementation analysis of ataxia telangiectasia and Nijmegen breakage syndrome: a survey of 50 patients. *Cytogenet Cell Genet* **49**(4): 259-63, 1988.

[143] Zhang Y, Zhou J, Lim CU. The role of NBS1 in DNA double strand break repair, telomere stability, and cell cycle checkpoint control. *Cell Res* **16**(1): 45-54, 2006.

[144] Tauchi H, Matsuura S, Kobayashi J, Sakamoto S, Komatsu K. Nijmegen breakage syndrome gene, NBS1, and molecular links to factors for genome stability. *Oncogene* **21**(58): 8967-80, 2002.

[145] Kang J, Bronson RT, Xu Y. Targeted disruption of NBS1 reveals its roles in mouse development and DNA repair. *Embo J* **21**(6): 1447-55, 2002.

[146] Williams BR, Mirzoeva OK, Morgan WF, Lin J, Dunnick W, Petrini JH. A murine model of Nijmegen breakage syndrome. *Curr Biol* **12**(8): 648-53, 2002.

[147] Dumon-Jones V, Frappart PO, Tong WM, Sajithlal G, Hulla W, Schmid G, Herceg Z, Digweed M, Wang ZQ. Nbn heterozygosity renders mice susceptible to tumor formation and ionizing radiation-induced tumorigenesis. *Cancer Res* **63**(21): 7263-9, 2003.

[148] Varon R, Vissinga C, Platzer M, Cerosaletti KM, Chrzanowska KH, Saar K, Beckmann G, Seemanova E, Cooper PR, Nowak NJ, Stumm M, Weemaes CM, Gatti RA, Wilson RK, Digweed M, Rosenthal A, Sperling K, Concannon P, Reis A. Nibrin, a novel DNA double-strand break repair protein, is mutated in Nijmegen breakage syndrome. *Cell* **93**(3): 467-76, 1998.

[149] Tauchi H, Kobayashi J, Morishima K, Matsuura S, Nakamura A, Shiraishi T, Ito E, Masnada D, Delia D, Komatsu K. The forkhead-associated domain of NBS1 is essential

for nuclear foci formation after irradiation but not essential for hRAD50[middle dot]hMRE11[middle dot]NBS1 complex DNA repair activity. *J Biol Chem* **276**(1): 12-5, 2001.

[150] Kobayashi J, Tauchi H, Sakamoto S, Nakamura A, Morishima K, Matsuura S, Kobayashi T, Tamai K, Tanimoto K, Komatsu K. NBS1 localizes to gamma-H2AX foci through interaction with the FHA/BRCT domain. *Curr Biol* **12**(21): 1846-51, 2002.

[151] Kobayashi J. Molecular mechanism of the recruitment of NBS1/hMRE11/hRAD50 complex to DNA double-strand breaks: NBS1 binds to gamma-H2AX through FHA/BRCT domain. *J Radiat Res (Tokyo)* **45**(4): 473-8, 2004.

[152] Alt JR, Bouska A, Fernandez MR, Cerny RL, Xiao H, Eischen CM. Mdm2 binds to Nbs1 at sites of DNA damage and regulates double strand break repair. *J Biol Chem* **280**(19): 18771-81, 2005.

[153] Wu G, Lee WH, Chen PL. NBS1 and TRF1 colocalize at promyelocytic leukemia bodies during late S/G2 phases in immortalized telomerase-negative cells. Implication of NBS1 in alternative lengthening of telomeres. *J Biol Chem* **275**(39): 30618-22, 2000.

[154] Zhu XD, Kuster B, Mann M, Petrini JH, de Lange T. *Cell*-cycle-regulated association of RAD50/MRE11/NBS1 with TRF2 and human telomeres. *Nat Genet* **25**(3): 347-52, 2000.

[155] Cheng WH, von Kobbe C, Opresko PL, Arthur LM, Komatsu K, Seidman MM, Carney JP, Bohr VA. Linkage between Werner syndrome protein and the Mre11 complex via Nbs1. *J Biol Chem* **279**(20): 21169-76, 2004.

[156] Olson E, Nievera CJ, Lee AY, Chen L, Wu X. The Mre11-Rad50-Nbs1 complex acts both upstream and downstream of ataxia telangiectasia mutated and Rad3-related protein (ATR) to regulate the S-phase checkpoint following UV treatment. *J Biol Chem* **282**(31): 22939-52, 2007.

[157] Nakanishi K, Taniguchi T, Ranganathan V, New HV, Moreau LA, Stotsky M, Mathew CG, Kastan MB, Weaver DT, D'Andrea AD. Interaction of FANCD2 and NBS1 in the DNA damage response. *Nat Cell Biol* **4**(12): 913-20, 2002.

[158] Tauchi H, Kobayashi J, Morishima K, van Gent DC, Shiraishi T, Verkaik NS, vanHeems D, Ito E, Nakamura A, Sonoda E, Takata M, Takeda S, Matsuura S, Komatsu K. Nbs1 is essential for DNA repair by homologous recombination in higher vertebrate cells. *Nature* **420**(6911): 93-8, 2002.

[159] Sakamoto S, Iijima K, Mochizuki D, Nakamura K, Teshigawara K, Kobayashi J, Matsuura S, Tauchi H, Komatsu K. Homologous recombination repair is regulated by domains at the N- and C-terminus of NBS1 and is dissociated with ATM functions. *Oncogene* **26**(41): 6002-9, 2007.

[160] Yang YG, Saidi A, Frappart PO, Min W, Barrucand C, Dumon-Jones V, Michelon J, Herceg Z, Wang ZQ. Conditional deletion of Nbs1 in murine cells reveals its role in branching repair pathways of DNA double-strand breaks. *Embo J* 25(23): 5527-38, 2006.

[161] Bell DW, Varley JM, Szydlo TE, Kang DH, Wahrer DC, Shannon KE, Lubratovich M, Verselis SJ, Isselbacher KJ, Fraumeni JF, Birch JM, Li FP, Garber JE, Haber DA. Heterozygous germ line hCHK2 mutations in Li-Fraumeni syndrome. *Science* **286**(5449): 2528-31, 1999.

[162] Vahteristo P, Bartkova J, Eerola H, Syrjakoski K, Ojala S, Kilpivaara O, Tamminen A, Kononen J, Aittomaki K, Heikkila P, Holli K, Blomqvist C, Bartek J, Kallioniemi OP,

Nevanlinna H. A CHEK2 genetic variant contributing to a substantial fraction of familial breast cancer. *Am J Hum Genet* **71**(2): 432-8, 2002.

[163] Meijers-Heijboer H, van den Ouweland A, Klijn J, Wasielewski M, de Snoo A, Oldenburg R, Hollestelle A, Houben M, Crepin E, van Veghel-Plandsoen M, Elstrodt F, van Duijn C, Bartels C, Meijers C, Schutte M, McGuffog L, Thompson D, Easton D, Sodha N, Seal S, Barfoot R, Mangion J, Chang-Claude J, Eccles D, Eeles R, Evans DG, Houlston R, Murday V, Narod S, Peretz T, Peto J, Phelan C, Zhang HX, Szabo C, Devilee P, Goldgar D, Futreal PA, Nathanson KL, Weber B, Rahman N, Stratton MR. Low-penetrance susceptibility to breast cancer due to CHEK2(*)1100delC in noncarriers of BRCA1 or BRCA2 mutations. *Nat Genet* **31**(1): 55-9, 2002.

[164] Meijers-Heijboer H, Wijnen J, Vasen H, Wasielewski M, Wagner A, Hollestelle A, Elstrodt F, van den Bos R, de Snoo A, Fat GT, Brekelmans C, Jagmohan S, Franken P, Verkuijlen P, van den Ouweland A, Chapman P, Tops C, Moslein G, Burn J, Lynch H, Klijn J, Fodde R, Schutte M. The CHEK2 1100delC mutation identifies families with a hereditary breast and colorectal cancer phenotype. Am J Hum Genet **72**(5): 1308-14, 2003.

[165] Baysal BE, DeLoia JA, Willett-Brozick JE, Goodman MT, Brady MF, Modugno F, Lynch HT, Conley YP, Watson P, Gallion HH. Analysis of CHEK2 gene for ovarian cancer susceptibility. *Gynecol Oncol* **95**(1): 62-9, 2004.

[166] Dong X, Wang L, Taniguchi K, Wang X, Cunningham JM, McDonnell SK, Qian C, Marks AF, Slager SL, Peterson BJ, Smith DI, Cheville JC, Blute ML, Jacobsen SJ, Schaid DJ, Tindall DJ, Thibodeau SN, Liu W. Mutations in CHEK2 associated with prostate cancer risk. Am J Hum Genet **72**(2): 270-80, 2003.

[167] Seppala EH, Ikonen T, Mononen N, Autio V, Rokman A, Matikainen MP, Tammela TL, Schleutker J. CHEK2 variants associate with hereditary prostate cancer. *Br J Cancer* **89**(10): 1966-70, 2003.

[168] Lipton L, Fleischmann C, Sieber OM, Thomas HJ, Hodgson SV, Tomlinson IP, Houlston RS. Contribution of the CHEK2 1100delC variant to risk of multiple colorectal adenoma and carcinoma. *Cancer Lett* **200**(2): 149-52, 2003.

[169] Bartek J, Falck J, Lukas J. CHK2 kinase--a busy messenger. *Nat Rev Mol Cell Biol* **2**(12): 877-86, 2001.

[170] Li J, Stern DF. Regulation of CHK2 by DNA-dependent protein kinase. *J Biol Chem* **280**(12): 12041-50, 2005.

[171] Takemura H, Rao VA, Sordet O, Furuta T, Miao ZH, Meng L, Zhang H, Pommier Y. Defective Mre11-dependent activation of Chk2 by ataxia telangiectasia mutated in colorectal carcinoma cells in response to replication-dependent DNA double strand breaks. *J Biol Chem* **281**(41): 30814-23, 2006.

[172] Lou Z, Minter-Dykhouse K, Wu X, Chen J. MDC1 is coupled to activated CHK2 in mammalian DNA damage response pathways. *Nature* **421**(6926): 957-61, 2003.

[173] Yoda A, Xu XZ, Onishi N, Toyoshima K, Fujimoto H, Kato N, Oishi I, Kondo T, Minami Y. Intrinsic kinase activity and SQ/TQ domain of Chk2 kinase as well as N-terminal domain of Wip1 phosphatase are required for regulation of Chk2 by Wip1. *J Biol Chem* **281**(34): 24847-62, 2006.

[174] Fujimoto H, Onishi N, Kato N, Takekawa M, Xu XZ, Kosugi A, Kondo T, Imamura M, Oishi I, Yoda A, Minami Y. Regulation of the antioncogenic Chk2 kinase by the oncogenic Wip1 phosphatase. *Cell* Death Differ 13(7): 1170-80, 2006.

[175] Lee CH, Chung JH. The hCds1 (Chk2)-FHA domain is essential for a chain of phosphorylation events on hCds1 that is induced by ionizing radiation. *J Biol Chem* **276**(32): 30537-41, 2001.

[176] Yang S, Jeong JH, Brown AL, Lee CH, Pandolfi PP, Chung JH, Kim MK. Promyelocytic leukemia activates Chk2 by mediating Chk2 autophosphorylation. *J Biol Chem* **281**(36): 26645-54, 2006.

[177] Falck J, Mailand N, Syljuasen RG, Bartek J, Lukas J. The ATM-Chk2-Cdc25A checkpoint pathway guards against radioresistant DNA synthesis. *Nature* **410**(6830): 842-7, 2001.

[178] Peng CY, Graves PR, Thoma RS, Wu Z, Shaw AS, Piwnica-Worms H. Mitotic and G2 checkpoint control: regulation of 14-3-3 protein binding by phosphorylation of Cdc25C on serine-216. *Science* **277**(5331): 1501-5, 1997.

[179] Chehab NH, Malikzay A, Stavridi ES, Halazonetis TD. Phosphorylation of Ser-20 mediates stabilization of human p53 in response to DNA damage. *Proc Natl Acad Sci U S A* **96**(24): 13777-82, 1999.

[180] Hirao A, Cheung A, Duncan G, Girard PM, Elia AJ, Wakeham A, Okada H, Sarkissian T, Wong JA, Sakai T, De Stanchina E, Bristow RG, Suda T, Lowe SW, Jeggo PA, Elledge SJ, Mak TW. Chk2 is a tumor suppressor that regulates apoptosis in both an ataxia telangiectasia mutated (ATM)-dependent and an ATM-independent manner. *Mol Cell Biol* **22**(18): 6521-32, 2002.

[181] Lee JS, Collins KM, Brown AL, Lee CH, Chung JH. hCds1-mediated phosphorylation of BRCA1 regulates the DNA damage response. *Nature* **404**(6774): 201-4, 2000.

[182] Bahassi el M, Penner CG, Robbins SB, Tichy E, Feliciano E, Yin M, Liang L, Deng L, Tischfield JA, Stambrook PJ. The breast cancer susceptibility allele CHEK2*1100delC promotes genomic instability in a knock-in mouse model. *Mutat Res* **616**(1-2): 201-9, 2007.

[183] McGowan CH. Checking in on Cds1 (Chk2): A checkpoint kinase and tumor suppressor. *Bioessays* **24**(6): 502-11, 2002.

[184] Attardi LD, Donehower LA. Probing p53 biological functions through the use of genetically engineered mouse models. *Mutat Res* **576**(1-2): 4-21, 2005.

[185] Oren M. The p53 saga: the good, the bad, and the dead. *Harvey Lect* **97**(57-82, 2001.

[186] Malkin D, Li FP, Strong LC, Fraumeni JF, Jr., Nelson CE, Kim DH, Kassel J, Gryka MA, Bischoff FZ, Tainsky MA, et al. Germ line p53 mutations in a familial syndrome of breast cancer, sarcomas, and other neoplasms. *Science* **250**(4985): 1233-8, 1990.

[187] Fukasawa K, Choi T, Kuriyama R, Rulong S, Vande Woude GF. Abnormal centrosome amplification in the absence of p53. *Science* **271**(5256): 1744-7, 1996.

[188] Clarke AR, Purdie CA, Harrison DJ, Morris RG, Bird CC, Hooper ML, Wyllie AH. Thymocyte apoptosis induced by p53-dependent and independent pathways. *Nature* **362**(6423): 849-52, 1993.

[189] Lowe SW, Schmitt EM, Smith SW, Osborne BA, Jacks T. p53 is required for radiation-induced apoptosis in mouse thymocytes. *Nature* **362**(6423): 847-9, 1993.

[190] Ziegler A, Jonason AS, Leffell DJ, Simon JA, Sharma HW, Kimmelman J, Remington L, Jacks T, Brash DE. Sunburn and p53 in the onset of skin cancer. *Nature* **372**(6508): 773-6, 1994.

[191] Smith ML, Chen IT, Zhan Q, O'Connor PM, Fornace AJ, Jr. Involvement of the p53 tumor suppressor in repair of u.v.-type DNA damage. *Oncogene* **10**(6): 1053-9, 1995.

[192] Toledo F, Wahl GM. Regulating the p53 pathway: in vitro hypotheses, in vivo veritas. *Nat Rev Cancer* **6**(12): 909-23, 2006.

[193] Berger M, Stahl N, Del Sal G, Haupt Y. Mutations in proline 82 of p53 impair its activation by Pin1 and Chk2 in response to DNA damage. *Mol Cell Biol* **25**(13): 5380-8, 2005.

[194] Fuchs SY, Adler V, Buschmann T, Yin Z, Wu X, Jones SN, Ronai Z. JNK targets p53 ubiquitination and degradation in nonstressed cells. *Genes Dev* **12**(17): 2658-63, 1998.

[195] Ando K, Ozaki T, Yamamoto H, Furuya K, Hosoda M, Hayashi S, Fukuzawa M, Nakagawara A. Polo-like kinase 1 (Plk1) inhibits p53 function by physical interaction and phosphorylation. *J Biol Chem* **279**(24): 25549-61, 2004.

[196] Mihara M, Erster S, Zaika A, Petrenko O, Chittenden T, Pancoska P, Moll UM. p53 has a direct apoptogenic role at the mitochondria. *Mol Cell* **11**(3): 577-90, 2003.

[197] Ito A, Lai CH, Zhao X, Saito S, Hamilton MH, Appella E, Yao TP. p300/CBP-mediated p53 acetylation is commonly induced by p53-activating agents and inhibited by MDM2. *Embo J* **20**(6): 1331-40, 2001.

[198] Guo A, Salomoni P, Luo J, Shih A, Zhong S, Gu W, Pandolfi PP. The function of PML in p53-dependent apoptosis. *Nat Cell Biol* **2**(10): 730-6, 2000.

[199] Wesierska-Gadek J, Wojciechowski J, Schmid G. Phosphorylation regulates the interaction and complex formation between wt p53 protein and PARP-1. *J Cell Biochem* **89**(6): 1260-84, 2003.

[200] Sommers JA, Sharma S, Doherty KM, Karmakar P, Yang Q, Kenny MK, Harris CC, Brosh RM, Jr. p53 modulates RPA-dependent and RPA-independent WRN helicase activity. *Cancer Res* **65**(4): 1223-33, 2005.

[201] Hanson S, Kim E, Deppert W. Redox factor 1 (Ref-1) enhances specific DNA binding of p53 by promoting p53 tetramerization. *Oncogene* **24**(9): 1641-7, 2005.

[202] Yu J, Zhang L, Hwang PM, Rago C, Kinzler KW, Vogelstein B. Identification and classification of p53-regulated genes. *Proc Natl Acad Sci U S A* **96**(25): 14517-22, 1999.

[203] Zhao R, Gish K, Murphy M, Yin Y, Notterman D, Hoffman WH, Tom E, Mack DH, Levine AJ. Analysis of p53-regulated gene expression patterns using oligonucleotide arrays. *Genes Dev* **14**(8): 981-93, 2000.

[204] Kannan K, Amariglio N, Rechavi G, Jakob-Hirsch J, Kela I, Kaminski N, Getz G, Domany E, Givol D. DNA microarrays identification of primary and secondary target genes regulated by p53. *Oncogene* **20**(18): 2225-34, 2001.

[205] Kho PS, Wang Z, Zhuang L, Li Y, Chew JL, Ng HH, Liu ET, Yu Q. p53-regulated transcriptional program associated with genotoxic stress-induced apoptosis. *J Biol Chem* **279**(20): 21183-92, 2004.

[206] el-Deiry WS, Tokino T, Velculescu VE, Levy DB, Parsons R, Trent JM, Lin D, Mercer WE, Kinzler KW, Vogelstein B. WAF1, a potential mediator of p53 tumor suppression. *Cell* **75**(4): 817-25, 1993.

[207] Hermeking H, Lengauer C, Polyak K, He TC, Zhang L, Thiagalingam S, Kinzler KW, Vogelstein B. 14-3-3 sigma is a p53-regulated inhibitor of G2/M progression. *Mol Cell* **1**(1): 3-11, 1997.

[208] Miyashita T, Krajewski S, Krajewska M, Wang HG, Lin HK, Liebermann DA, Hoffman B, Reed JC. Tumor suppressor p53 is a regulator of bcl-2 and bax gene expression in vitro and in vivo. *Oncogene* **9**(6): 1799-805, 1994.

[209] Owen-Schaub LB, Zhang W, Cusack JC, Angelo LS, Santee SM, Fujiwara T, Roth JA, Deisseroth AB, Zhang WW, Kruzel E, et al. Wild-type human p53 and a temperature-sensitive mutant induce Fas/APO-1 expression. *Mol Cell Biol* **15**(6): 3032-40, 1995.

[210] Dameron KM, Volpert OV, Tainsky MA, Bouck N. Control of angiogenesis in fibroblasts by p53 regulation of thrombospondin-1. *Science* **265**(5178): 1582-4, 1994.

[211] Hwang BJ, Ford JM, Hanawalt PC, Chu G. Expression of the p48 xeroderma pigmentosum gene is p53-dependent and is involved in global genomic repair. *Proc Natl Acad Sci U S A* **96**(2): 424-8, 1999.

[212] Seo YR, Fishel ML, Amundson S, Kelley MR, Smith ML. Implication of p53 in base excision DNA repair: in vivo evidence. *Oncogene* **21**(5): 731-7, 2002.

[213] Levine AJ, Hu W, Feng Z. The P53 pathway: what questions remain to be explored? Cell Death Differ **13**(6): 1027-36, 2006.

[214] Wu X, Bayle JH, Olson D, Levine AJ. The p53-mdm-2 autoregulatory feedback loop. *Genes Dev* **7**(7A): 1126-32, 1993.

[215] Chen J, Lin J, Levine AJ. Regulation of transcription functions of the p53 tumor suppressor by the mdm-2 *oncogene*. *Mol Med* **1**(2): 142-52, 1995.

[216] Bode AM, Dong Z. Post-translational modification of p53 in tumorigenesis. *Nat Rev Cancer* **4**(10): 793-805, 2004.

[217] Fuchs B, O'Connor D, Fallis L, Scheidtmann KH, Lu X. p53 phosphorylation mutants retain transcription activity. *Oncogene* **10**(4): 789-93, 1995.

[218] Saito S, Goodarzi AA, Higashimoto Y, Noda Y, Lees-Miller SP, Appella E, Anderson CW. ATM mediates phosphorylation at multiple p53 sites, including Ser(46), in response to ionizing radiation. *J Biol Chem* **277**(15): 12491-4, 2002.

[219] Kim ST, Lim DS, Canman CE, Kastan MB. Substrate specificities and identification of putative substrates of ATM kinase family members. *J Biol Chem* **274**(53): 37538-43, 1999.

[220] Lavin MF, Gueven N. The complexity of p53 stabilization and activation. *Cell Death Differ* **13**(6): 941-50, 2006.

[221] Saito S, Yamaguchi H, Higashimoto Y, Chao C, Xu Y, Fornace AJ, Jr., Appella E, Anderson CW. Phosphorylation site interdependence of human p53 post-translational modifications in response to stress. *J Biol Chem* **278**(39): 37536-44, 2003.

[222] Unger T, Juven-Gershon T, Moallem E, Berger M, Vogt Sionov R, Lozano G, Oren M, Haupt Y. Critical role for Ser20 of human p53 in the negative regulation of p53 by Mdm2. *Embo J* **18**(7): 1805-14, 1999.

[223] Li M, Luo J, Brooks CL, Gu W. Acetylation of p53 inhibits its ubiquitination by Mdm2. *J Biol Chem* **277**(52): 50607-11, 2002.

[224] Waterman MJ, Stavridi ES, Waterman JL, Halazonetis TD. ATM-dependent activation of p53 involves dephosphorylation and association with 14-3-3 proteins. *Nat Genet* **19**(2): 175-8, 1998.

[225] Tian H, Faje AT, Lee SL, Jorgensen TJ. Radiation-induced phosphorylation of Chk1 at S345 is associated with p53-dependent cell cycle arrest pathways. *Neoplasia* **4**(2): 171-80, 2002.

[226] Zhang H, Somasundaram K, Peng Y, Tian H, Zhang H, Bi D, Weber BL, El-Deiry WS. BRCA1 physically associates with p53 and stimulates its transcriptional activity. *Oncogene* **16**(13): 1713-21, 1998.

[227] Marmorstein LY, Ouchi T, Aaronson SA. The BRCA2 gene product functionally interacts with p53 and RAD51. *Proc Natl Acad Sci U S A* **95**(23): 13869-74, 1998.

[228] Armstrong JF, Kaufman MH, Harrison DJ, Clarke AR. High-frequency developmental abnormalities in p53-deficient mice. *Curr Biol* **5**(8): 931-6, 1995.

[229] Sah VP, Attardi LD, Mulligan GJ, Williams BO, Bronson RT, Jacks T. A subset of p53-deficient embryos exhibit exencephaly. *Nat Genet* **10**(2): 175-80, 1995.

[230] Donehower LA, Harvey M, Slagle BL, McArthur MJ, Montgomery CA, Jr., Butel JS, Bradley A. Mice deficient for p53 are developmentally normal but susceptible to spontaneous tumours. *Nature* **356**(6366): 215-21, 1992.

[231] Jacks T, Remington L, Williams BO, Schmitt EM, Halachmi S, Bronson RT, Weinberg RA. Tumor spectrum analysis in p53-mutant mice. *Curr Biol* **4**(1): 1-7, 1994.

[232] Purdie CA, Harrison DJ, Peter A, Dobbie L, White S, Howie SE, Salter DM, Bird CC, Wyllie AH, Hooper ML, et al. Tumour incidence, spectrum and ploidy in mice with a large deletion in the p53 gene. *Oncogene* **9**(2): 603-9, 1994.

[233] Harvey M, McArthur MJ, Montgomery CA, Jr., Butel JS, Bradley A, Donehower LA. Spontaneous and carcinogen-induced tumorigenesis in p53-deficient mice. *Nat Genet* **5**(3): 225-9, 1993.

[234] Jimenez GS, Nister M, Stommel JM, Beeche M, Barcarse EA, Zhang XQ, O'Gorman S, Wahl GM. A transactivation-deficient mouse model provides insights into Trp53 regulation and function. *Nat Genet* **26**(1): 37-43, 2000.

[235] Chao C, Saito S, Kang J, Anderson CW, Appella E, Xu Y. p53 transcriptional activity is essential for p53-dependent apoptosis following DNA damage. *Embo J* **19**(18): 4967-75, 2000.

[236] Lin J, Chen J, Elenbaas B, Levine AJ. Several hydrophobic amino acids in the p53 amino-terminal domain are required for transcriptional activation, binding to mdm-2 and the adenovirus 5 E1B 55-kD protein. *Genes Dev* **8**(10): 1235-46, 1994.

[237] Johnson TM, Hammond EM, Giaccia A, Attardi LD. The p53QS transactivation-deficient mutant shows stress-specific apoptotic activity and induces embryonic lethality. *Nat Genet* **37**(2): 145-52, 2005.

[238] Jonkers J, Meuwissen R, van der Gulden H, Peterse H, van der Valk M, Berns A. Synergistic tumor suppressor activity of BRCA2 and p53 in a conditional mouse model for breast cancer. *Nat Genet* **29**(4): 418-25, 2001.

[239] Liu G, Parant JM, Lang G, Chau P, Chavez-Reyes A, El-Naggar AK, Multani A, Chang S, Lozano G. Chromosome stability, in the absence of apoptosis, is critical for suppression of tumorigenesis in Trp53 mutant mice. *Nat Genet* **36**(1): 63-8, 2004.

[240] Levine AJ. p53, the cellular gatekeeper for growth and division. *Cell* **88**(3): 323-31, 1997.

[241] Agarwal ML, Taylor WR, Chernov MV, Chernova OB, Stark GR. The p53 network. *J Biol Chem* **273**(1): 1-4, 1998.

[242] Wang XW, Yeh H, Schaeffer L, Roy R, Moncollin V, Egly JM, Wang Z, Freidberg EC, Evans MK, Taffe BG, et al. p53 modulation of TFIIH-associated nucleotide excision repair activity. *Nat Genet* **10**(2): 188-95, 1995.

[243] Offer H, Wolkowicz R, Matas D, Blumenstein S, Livneh Z, Rotter V. Direct involvement of p53 in the base excision repair pathway of the DNA repair machinery. *FEBS Lett* **450**(3): 197-204, 1999.

[244] Sturzbecher HW, Donzelmann B, Henning W, Knippschild U, Buchhop S. p53 is linked directly to homologous recombination processes via RAD51/RecA protein interaction. *Embo J* **15**(8): 1992-2002, 1996.

[245] Willers H, McCarthy EE, Wu B, Wunsch H, Tang W, Taghian DG, Xia F, Powell SN. Dissociation of p53-mediated suppression of homologous recombination from G1/S cell cycle checkpoint control. *Oncogene* **19**(5): 632-9, 2000.

[246] Gebow D, Miselis N, Liber HL. Homologous and nonhomologous recombination resulting in deletion: effects of p53 status, microhomology, and repetitive DNA length and orientation. *Mol Cell Biol* **20**(11): 4028-35, 2000.

[247] Livingstone LR, White A, Sprouse J, Livanos E, Jacks T, Tlsty TD. Altered cell cycle arrest and gene amplification potential accompany loss of wild-type p53. *Cell* **70**(6): 923-935, 1992.

[248] Yin Y, Tainsky MA, Bischoff FZ, Strong LC, Wahl GM. Wild-type p53 restores cell cycle control and inhibits gene amplification in cells with mutant p53 alleles. *Cell* **70**(6): 937-48, 1992.

[249] Buchhop S, Gibson MK, Wang XW, Wagner P, Sturzbecher HW, Harris CC. Interaction of p53 with the human Rad51 protein. *Nucleic Acids Res* **25**(19): 3868-74, 1997.

[250] Donehower LA, Godley LA, Aldaz CM, Pyle R, Shi YP, Pinkel D, Gray J, Bradley A, Medina D, Varmus HE. Deficiency of p53 accelerates mammary tumorigenesis in Wnt-1 transgenic mice and promotes chromosomal instability. *Genes Dev* **9**(7): 882-95, 1995.

[251] Bertrand P, Rouillard D, Boulet A, Levalois C, Soussi T, Lopez BS. Increase of spontaneous intrachromosomal homologous recombination in mammalian cells expressing a mutant p53 protein. *Oncogene* **14**(9): 1117-22, 1997.

[252] Mekeel KL, Tang W, Kachnic LA, Luo CM, DeFrank JS, Powell SN. Inactivation of p53 results in high rates of homologous recombination. *Oncogene* **14**(15): 1847-57, 1997.

[253] Aubrecht J, Secretan MB, Bishop AJ, Schiestl RH. Involvement of p53 in X-ray induced intrachromosomal recombination in mice. *Carcinogenesis* **20**(12): 2229-2236, 1999.

[254] Yang Q, Zhang R, Wang XW, Spillare EA, Linke SP, Subramanian D, Griffith JD, Li JL, Hickson ID, Shen JC, Loeb LA, Mazur SJ, Appella E, Brosh RM, Jr., Karmakar P, Bohr VA, Harris CC. The processing of Holliday junctions by BLM and WRN helicases is regulated by p53. *J Biol Chem* **277**(35): 31980-7, 2002.

[255] Romanova LY, Willers H, Blagosklonny MV, Powell SN. The interaction of p53 with replication protein A mediates suppression of homologous recombination. *Oncogene* **23**(56): 9025-33, 2004.

[256] Saintigny Y, Rouillard D, Chaput B, Soussi T, Lopez BS. Mutant p53 proteins stimulate spontaneous and radiation-induced intrachromosomal homologous recombination independently of the alteration of the transactivation activity and of the G1 checkpoint. *Oncogene* **18**(24): 3553-63, 1999.

[257] Akyuz N, Boehden GS, Susse S, Rimek A, Preuss U, Scheidtmann KH, Wiesmuller L. DNA substrate dependence of p53-mediated regulation of double-strand break repair. *Mol Cell Biol* **22**(17): 6306-17, 2002.

[258] Hopfner KP, Craig L, Moncalian G, Zinkel RA, Usui T, Owen BA, Karcher A, Henderson B, Bodmer JL, McMurray CT, Carney JP, Petrini JH, Tainer JA. The Rad50 zinc-hook is a structure joining Mre11 complexes in DNA recombination and repair. *Nature* **418**(6897): 562-6, 2002.

[259] Krejci L, Chen L, Van Komen S, Sung P, Tomkinson A. Mending the break: two DNA double-strand break repair machines in eukaryotes. *Prog Nucleic Acid Res Mol Biol* **74**(159-201, 2003.

[260] West SC. Molecular views of recombination proteins and their control. *Nat Rev Mol Cell Biol* **4**(6): 435-45, 2003.

[261] Dudenhoffer C, Rohaly G, Will K, Deppert W, Wiesmuller L. Specific mismatch recognition in heteroduplex intermediates by p53 suggests a role in fidelity control of homologous recombination. *Mol Cell Biol* **18**(9): 5332-42, 1998.

[262] Subramanian D, Griffith JD. Interactions between p53, hMSH2-hMSH6 and HMG I(Y) on Holliday junctions and bulged bases. *Nucleic Acids Res* **30**(11): 2427-34, 2002.

[263] Davalos AR, Campisi J. Bloom syndrome cells undergo p53-dependent apoptosis and delayed assembly of BRCA1 and NBS1 repair complexes at stalled replication forks. *J Cell Biol* **162**(7): 1197-209, 2003.

[264] Sengupta S, Linke SP, Pedeux R, Yang Q, Farnsworth J, Garfield SH, Valerie K, Shay JW, Ellis NA, Wasylyk B, Harris CC. BLM helicase-dependent transport of p53 to sites of stalled DNA replication forks modulates homologous recombination. *Embo J* **22**(5): 1210-22, 2003.

[265] Zhivotovsky B, Kroemer G. Apoptosis and genomic instability. *Nat Rev Mol Cell Biol* **5**(9): 752-62, 2004.

[266] Arias-Lopez C, Lazaro-Trueba I, Kerr P, Lord CJ, Dexter T, Iravani M, Ashworth A, Silva A. p53 modulates homologous recombination by transcriptional regulation of the RAD51 gene. *EMBO Rep* **7**(2): 219-24, 2006.

[267] Ivanov A, Cragg MS, Erenpreisa J, Emzinsh D, Lukman H, Illidge TM. Endopolyploid cells produced after severe genotoxic damage have the potential to repair DNA double strand breaks. *J Cell Sci* **116**(Pt 20): 4095-106, 2003.

[268] Meyn MS, Strasfeld L, Allen C. Testing the role of p53 in the expression of genetic instability and apoptosis in ataxia-telangiectasia. *Int J Radiat Biol* **66**(6 Suppl): S141-9, 1994.

[269] Wiesmuller L, Cammenga J, Deppert WW. In vivo assay of p53 function in homologous recombination between simian virus 40 chromosomes. *J Virol* **70**(2): 737-44, 1996.

[270] Xia RY, Luo YX, Wang TP. Operational techniques and combination treatment for the recurrent sacro-coccygeal tumor. *J Tongji Med Univ* **14**(4): 245-8, 1994.

[271] Dudenhoffer C, Kurth M, Janus F, Deppert W, Wiesmuller L. Dissociation of the recombination control and the sequence-specific transactivation function of P53. *Oncogene* **18**(42): 5773-84, 1999.

[272] Aubrecht J, Secretan MB, Bishop AJ, Schiestl RH. Involvement of p53 in X-ray induced intrachromosomal recombination in mice. *Carcinogenesis* **20**(12): 2229-36, 1999.

[273] Fornace AJ, Jr., Alamo I, Jr., Hollander MC. DNA damage-inducible transcripts in mammalian cells. *Proc Natl Acad Sci U S A* **85**(23): 8800-4, 1988.

[274] Papathanasiou MA, Fornace AJ, Jr. DNA-damage inducible genes. *Cancer Treat Res* **57**(13-36, 1991.

[275] Papathanasiou MA, Kerr NC, Robbins JH, McBride OW, Alamo I, Jr., Barrett SF, Hickson ID, Fornace AJ, Jr. Induction by ionizing radiation of the gadd45 gene in cultured human cells: lack of mediation by protein kinase C. *Mol Cell Biol* **11**(2): 1009-16, 1991.

[276] Fornace AJ, Jr., Jackman J, Hollander MC, Hoffman-Liebermann B, Liebermann DA. Genotoxic-stress-response genes and growth-arrest genes. gadd, MyD, and other genes induced by treatments eliciting growth arrest. *Ann N Y Acad Sci* **663**(139-53, 1992.

[277] Wang XW, Zhan Q, Coursen JD, Khan MA, Kontny HU, Yu L, Hollander MC, O'Connor PM, Fornace AJ, Jr., Harris CC. GADD45 induction of a G2/M cell cycle checkpoint. *Proc Natl Acad Sci U S A* **96**(7): 3706-11, 1999.

[278] Zhan Q, Antinore MJ, Wang XW, Carrier F, Smith ML, Harris CC, Fornace AJ, Jr. Association with Cdc2 and inhibition of Cdc2/Cyclin B1 kinase activity by the p53-regulated protein Gadd45. *Oncogene* **18**(18): 2892-900, 1999.

[279] Jin S, Antinore MJ, Lung FD, Dong X, Zhao H, Fan F, Colchagie AB, Blanck P, Roller PP, Fornace AJ, Jr., Zhan Q. The GADD45 inhibition of Cdc2 kinase correlates with GADD45-mediated growth suppression. *J Biol Chem* **275**(22): 16602-8, 2000.

[280] Jin S, Tong T, Fan W, Fan F, Antinore MJ, Zhu X, Mazzacurati L, Li X, Petrik KL, Rajasekaran B, Wu M, Zhan Q. GADD45-induced cell cycle G2-M arrest associates with altered subcellular distribution of cyclin B1 and is independent of p38 kinase activity. *Oncogene* **21**(57): 8696-704, 2002.

[281] Smith ML, Chen IT, Zhan Q, Bae I, Chen CY, Gilmer TM, Kastan MB, O'Connor PM, Fornace AJ, Jr. Interaction of the p53-regulated protein Gadd45 with proliferating cell nuclear antigen. *Science* **266**(5189): 1376-80, 1994.

[282] Smith ML, Kontny HU, Zhan Q, Sreenath A, O'Connor PM, Fornace AJ, Jr. Antisense GADD45 expression results in decreased DNA repair and sensitizes cells to u.v.-irradiation or cisplatin. *Oncogene* **13**(10): 2255-63, 1996.

[283] Hollander MC, Kovalsky O, Salvador JM, Kim KE, Patterson AD, Haines DC, Fornace AJ, Jr. Dimethylbenzanthracene *carcinogenesis* in Gadd45a-null mice is associated with decreased DNA repair and increased mutation frequency. *Cancer Res* **61**(6): 2487-91, 2001.

[284] Takekawa M, Saito H. A family of stress-inducible GADD45-like proteins mediate activation of the stress-responsive MTK1/MEKK4 MAPKKK. *Cell* **95**(4): 521-30, 1998.

[285] Harkin DP, Bean JM, Miklos D, Song YH, Truong VB, Englert C, Christians FC, Ellisen LW, Maheswaran S, Oliner JD, Haber DA. Induction of GADD45 and JNK/SAPK-dependent apoptosis following inducible expression of BRCA1. *Cell* **97**(5): 575-86, 1999.

[286] Zhan Q, Jin S, Ng B, Plisket J, Shangary S, Rathi A, Brown KD, Baskaran R. Caspase-3 mediated cleavage of BRCA1 during UV-induced apoptosis. *Oncogene* **21**(34): 5335-45, 2002.

[287] Hollander MC, Sheikh MS, Bulavin DV, Lundgren K, Augeri-Henmueller L, Shehee R, Molinaro TA, Kim KE, Tolosa E, Ashwell JD, Rosenberg MP, Zhan Q, Fernandez-Salguero PM, Morgan WF, Deng CX, Fornace AJ, Jr. Genomic instability in Gadd45a-deficient mice. *Nat Genet* **23**(2): 176-84, 1999.

[288] Kovalsky O, Lung FD, Roller PP, Fornace AJ, Jr. Oligomerization of human Gadd45a protein. *J Biol Chem* **276**(42): 39330-9, 2001.

[289] John PC, Mews M, Moore R. Cyclin/Cdk complexes: their involvement in cell cycle progression and mitotic division. *Protoplasma* **216**(3-4): 119-42, 2001.

[290] Kearsey JM, Coates PJ, Prescott AR, Warbrick E, Hall PA. Gadd45 is a nuclear cell cycle regulated protein which interacts with p21Cip1. *Oncogene* **11**(9): 1675-83, 1995.

[291] Zhang H, Xiong Y, Beach D. Proliferating cell nuclear antigen and p21 are components of multiple cell cycle kinase complexes. *Mol Biol Cell* **4**(9): 897-906, 1993.

[292] Xiong Y, Hannon GJ, Zhang H, Casso D, Kobayashi R, Beach D. p21 is a universal inhibitor of cyclin kinases. *Nature* **366**(6456): 701-4, 1993.

[293] Deng C, Zhang P, Harper JW, Elledge SJ, Leder P. Mice lacking p21CIP1/WAF1 undergo normal development, but are defective in G1 checkpoint control. *Cell* **82**(4): 675-84, 1995.

[294] Chan TA, Hwang PM, Hermeking H, Kinzler KW, Vogelstein B. Cooperative effects of genes controlling the G(2)/M checkpoint. *Genes Dev* **14**(13): 1584-8, 2000.

[295] Vairapandi M, Balliet AG, Hoffman B, Liebermann DA. GADD45b and GADD45g are cdc2/cyclinB1 kinase inhibitors with a role in S and G2/M cell cycle checkpoints induced by genotoxic stress. *J Cell Physiol* **192**(3): 327-38, 2002.

[296] Zhao H, Jin S, Antinore MJ, Lung FD, Fan F, Blanck P, Roller P, Fornace AJ, Jr., Zhan Q. The central region of Gadd45 is required for its interaction with p21/WAF1. *Exp Cell Res* **258**(1): 92-100, 2000.

[297] Yang Q, Manicone A, Coursen JD, Linke SP, Nagashima M, Forgues M, Wang XW. Identification of a functional domain in a GADD45-mediated G2/M checkpoint. *J Biol Chem* **275**(47): 36892-8, 2000.

[298] Zhang W, Hoffman B, Liebermann DA. Ectopic expression of MyD118/Gadd45/CR6 (Gadd45beta/alpha/gamma) sensitizes neoplastic cells to genotoxic stress-induced apoptosis. *Int J Oncol* **18**(4): 749-57, 2001.

[299] Jung HJ, Kim EH, Mun JY, Park S, Smith ML, Han SS, Seo YR. Base excision DNA repair defect in Gadd45a-deficient cells. *Oncogene* **26**(54): 7517-25, 2007.

[300] Waldman T, Kinzler KW, Vogelstein B. p21 is necessary for the p53-mediated G1 arrest in human cancer cells. *Cancer Res* **55**(22): 5187-90, 1995.

[301] Brugarolas J, Chandrasekaran C, Gordon JI, Beach D, Jacks T, Hannon GJ. Radiation-induced cell cycle arrest compromised by p21 deficiency. *Nature* **377**(6549): 552-7, 1995.

[302] Di Leonardo A, Linke SP, Clarkin K, Wahl GM. DNA damage triggers a prolonged p53-dependent G1 arrest and long-term induction of Cip1 in normal human fibroblasts. *Genes Dev* **8**(21): 2540-51, 1994.

[303] Kuo ML, Chou YW, Chau YP, Huang TS. Resistance to apoptosis induced by alkylating agents in v-Ha-ras-transformed cells due to defect in p53 function. *Mol Carcinog* **18**(4): 221-31, 1997.

[304] Yuan ZM, Huang Y, Fan MM, Sawyers C, Kharbanda S, Kufe D. Genotoxic drugs induce interaction of the c-Abl tyrosine kinase and the tumor suppressor protein p53. *J Biol Chem* **271**(43): 26457-60, 1996.

[305] Li R, Waga S, Hannon GJ, Beach D, Stillman B. Differential effects by the p21 CDK inhibitor on PCNA-dependent DNA replication and repair. *Nature* **371**(6497): 534-7, 1994.

[306] Waga S, Hannon GJ, Beach D, Stillman B. The p21 inhibitor of cyclin-dependent kinases controls DNA replication by interaction with PCNA. *Nature* **369**(6481): 574-8, 1994.

[307] Poon RY, Hunter T. Expression of a novel form of p21Cip1/Waf1 in UV-irradiated and transformed cells. *Oncogene* **16**(10): 1333-43, 1998.

[308] Wang YA, Elson A, Leder P. Loss of p21 increases sensitivity to ionizing radiation and delays the onset of lymphoma in atm-deficient mice. *Proc Natl Acad Sci U S A* **94**(26): 14590-5, 1997.

[309] Bishop AJ, Kosaras B, Hollander MC, Fornace A, Jr., Sidman RL, Schiestl RH. p21 controls patterning but not homologous recombination in RPE development. *DNA Repair (Amst)* **5**(1): 111-20, 2006.

[310] Miki Y, Swensen J, Shattuck-Eidens D, Futreal PA, Harshman K, Tavtigian S, Liu Q, Cochran C, Bennett LM, Ding W, et al. A strong candidate for the breast and ovarian cancer susceptibility gene BRCA1. *Science* **266**(5182): 66-71, 1994.

[311] Easton DF, Ford D, Bishop DT. Breast and ovarian cancer incidence in BRCA1-mutation carriers. Breast Cancer Linkage Consortium. *Am J Hum Genet* **56**(1): 265-71, 1995.

[312] Ford D, Easton DF, Bishop DT, Narod SA, Goldgar DE. Risks of cancer in BRCA1-mutation carriers. Breast Cancer Linkage Consortium. *Lancet* **343**(8899): 692-5, 1994.

[313] Hall JM, Lee MK, Newman B, Morrow JE, Anderson LA, Huey B, King MC. Linkage of early-onset familial breast cancer to chromosome 17q21. *Science* **250**(4988): 1684-9, 1990.

[314] Futreal PA, Liu Q, Shattuck-Eidens D, Cochran C, Harshman K, Tavtigian S, Bennett LM, Haugen-Strano A, Swensen J, Miki Y, et al. BRCA1 mutations in primary breast and ovarian carcinomas. *Science* **266**(5182): 120-2, 1994.

[315] McCoy ML, Mueller CR, Roskelley CD. The role of the breast cancer susceptibility gene 1 (BRCA1) in sporadic epithelial ovarian cancer. *Reprod Biol Endocrinol* **1**(72, 2003.

[316] Meza JE, Brzovic PS, King MC, Klevit RE. Mapping the functional domains of BRCA1. Interaction of the ring finger domains of BRCA1 and BARD1. *J Biol Chem* **274**(9): 5659-65, 1999.

[317] Boulton SJ. Cellular functions of the BRCA tumour-suppressor proteins. *Biochem Soc Trans* **34**(Pt 5): 633-45, 2006.

[318] Scully R, Chen J, Ochs RL, Keegan K, Hoekstra M, Feunteun J, Livingston DM. Dynamic changes of BRCA1 subnuclear location and phosphorylation state are initiated by DNA damage. *Cell* **90**(3): 425-35, 1997.

[319] Ouchi T, Monteiro AN, August A, Aaronson SA, Hanafusa H. BRCA1 regulates p53-dependent gene expression. *Proc Natl Acad Sci U S A* **95**(5): 2302-6, 1998.

[320] Chai YL, Cui J, Shao N, Shyam E, Reddy P, Rao VN. The second BRCT domain of BRCA1 proteins interacts with p53 and stimulates transcription from the p21WAF1/CIP1 promoter. *Oncogene* **18**(1): 263-8, 1999.

[321] Chen J, Silver DP, Walpita D, Cantor SB, Gazdar AF, Tomlinson G, Couch FJ, Weber BL, Ashley T, Livingston DM, Scully R. Stable interaction between the products of the BRCA1 and BRCA2 tumor suppressor genes in mitotic and meiotic cells. *Mol Cell* **2**(3): 317-28, 1998.

[322] Scully R, Chen J, Plug A, Xiao Y, Weaver D, Feunteun J, Ashley T, Livingston DM. Association of BRCA1 with Rad51 in mitotic and meiotic cells. *Cell* **88**(2): 265-75, 1997.

[323] Folias A, Matkovic M, Bruun D, Reid S, Hejna J, Grompe M, D'Andrea A, Moses R. BRCA1 interacts directly with the Fanconi anemia protein FANCA. *Hum Mol Genet* **11**(21): 2591-7, 2002.

[324] Wang Q, Zhang H, Guerrette S, Chen J, Mazurek A, Wilson T, Slupianek A, Skorski T, Fishel R, Greene MI. Adenosine nucleotide modulates the physical interaction between hMSH2 and BRCA1. *Oncogene* **20**(34): 4640-9, 2001.

[325] Foray N, Randrianarison V, Marot D, Perricaudet M, Lenoir G, Feunteun J. Gamma-rays-induced death of human cells carrying mutations of BRCA1 or BRCA2. *Oncogene* **18**(51): 7334-42, 1999.

[326] Moynahan ME, Cui TY, Jasin M. Homology-directed dna repair, mitomycin-c resistance, and chromosome stability is restored with correction of a Brca1 mutation. *Cancer Res* **61**(12): 4842-50, 2001.

[327] Rahman N, Stratton MR. The genetics of breast cancer susceptibility. *Annu Rev Genet* **32**(95-121, 1998.

[328] Ford D, Easton DF, Stratton M, Narod S, Goldgar D, Devilee P, Bishop DT, Weber B, Lenoir G, Chang-Claude J, Sobol H, Teare MD, Struewing J, Arason A, Scherneck S, Peto J, Rebbeck TR, Tonin P, Neuhausen S, Barkardottir R, Eyfjord J, Lynch H, Ponder BA, Gayther SA, Zelada-Hedman M, et al. Genetic heterogeneity and penetrance analysis of the BRCA1 and BRCA2 genes in breast cancer families. The Breast Cancer Linkage Consortium. *Am J Hum Genet* **62**(3): 676-89, 1998.

[329] Howlett NG, Taniguchi T, Olson S, Cox B, Waisfisz Q, De Die-Smulders C, Persky N, Grompe M, Joenje H, Pals G, Ikeda H, Fox EA, D'Andrea AD. Biallelic inactivation of BRCA2 in Fanconi anemia. *Science* **297**(5581): 606-9, 2002.

[330] Milner J, Ponder B, Hughes-Davies L, Seltmann M, Kouzarides T. Transcriptional activation functions in BRCA2. *Nature* **386**(6627): 772-3, 1997.

[331] Spain BH, Larson CJ, Shihabuddin LS, Gage FH, Verma IM. Truncated BRCA2 is cytoplasmic: implications for cancer-linked mutations. *Proc Natl Acad Sci U S A* **96**(24): 13920-5, 1999.

[332] Esashi F, Christ N, Gannon J, Liu Y, Hunt T, Jasin M, West SC. CDK-dependent phosphorylation of BRCA2 as a regulatory mechanism for recombinational repair. *Nature* **434**(7033): 598-604, 2005.

[333] Ludwig T, Chapman DL, Papaioannou VE, Efstratiadis A. Targeted mutations of breast cancer susceptibility gene homologs in mice: lethal phenotypes of Brca1, Brca2, Brca1/Brca2, Brca1/p53, and Brca2/p53 nullizygous embryos. *Genes Dev* **11**(10): 1226-41, 1997.

[334] Suzuki A, de la Pompa JL, Hakem R, Elia A, Yoshida R, Mo R, Nishina H, Chuang T, Wakeham A, Itie A, Koo W, Billia P, Ho A, Fukumoto M, Hui CC, Mak TW. Brca2 is required for embryonic cellular proliferation in the mouse. *Genes Dev* **11**(10): 1242-52, 1997.

[335] Rajan JV, Wang M, Marquis ST, Chodosh LA. Brca2 is coordinately regulated with Brca1 during proliferation and differentiation in mammary epithelial cells. *Proc Natl Acad Sci U S A* **93**(23): 13078-83, 1996.

[336] Sharan SK, Morimatsu M, Albrecht U, Lim DS, Regel E, Dinh C, Sands A, Eichele G, Hasty P, Bradley A. Embryonic lethality and radiation hypersensitivity mediated by Rad51 in mice lacking Brca2. *Nature* **386**(6627): 804-10, 1997.

[337] Schutte M, Hruban RH, Geradts J, Maynard R, Hilgers W, Rabindran SK, Moskaluk CA, Hahn SA, Schwarte-Waldhoff I, Schmiegel W, Baylin SB, Kern SE, Herman JG. Abrogation of the Rb/p16 tumor-suppressive pathway in virtually all pancreatic carcinomas. *Cancer Res* **57**(15): 3126-30, 1997.

[338] Redston MS, Caldas C, Seymour AB, Hruban RH, da Costa L, Yeo CJ, Kern SE. p53 mutations in pancreatic carcinoma and evidence of common involvement of homocopolymer tracts in DNA microdeletions. *Cancer Res* **54**(11): 3025-33, 1994.

[339] Abbott DW, Freeman ML, Holt JT. Double-strand break repair deficiency and radiation sensitivity in BRCA2 mutant cancer cells. *J Natl Cancer Inst* **90**(13): 978-85, 1998.

[340] Ghadimi BM, Schrock E, Walker RL, Wangsa D, Jauho A, Meltzer PS, Ried T. Specific chromosomal aberrations and amplification of the AIB1 nuclear receptor coactivator gene in pancreatic carcinomas. *Am J Pathol* **154**(2): 525-36, 1999.

[341] Grigorova M, Staines JM, Ozdag H, Caldas C, Edwards PA. Possible causes of chromosome instability: comparison of chromosomal abnormalities in cancer cell lines with mutations in BRCA1, BRCA2, CHK2 and BUB1. *Cytogenet Genome Res* **104**(1-4): 333-40, 2004.

[342] Dong Y, Hakimi MA, Chen X, Kumaraswamy E, Cooch NS, Godwin AK, Shiekhattar R. Regulation of BRCC, a holoenzyme complex containing BRCA1 and BRCA2, by a signalosome-like subunit and its role in DNA repair. *Mol Cell* **12**(5): 1087-99, 2003.

[343] Lee M, Daniels MJ, Venkitaraman AR. Phosphorylation of BRCA2 by the Polo-like kinase Plk1 is regulated by DNA damage and mitotic progression. *Oncogene* **23**(4): 865-72, 2004.

[344] Marmorstein LY, Kinev AV, Chan GK, Bochar DA, Beniya H, Epstein JA, Yen TJ, Shiekhattar R. A human BRCA2 complex containing a structural DNA binding component influences cell cycle progression. *Cell* **104**(2): 247-57, 2001.

[345] Futamura M, Arakawa H, Matsuda K, Katagiri T, Saji S, Miki Y, Nakamura Y. Potential role of BRCA2 in a mitotic checkpoint after phosphorylation by hBUBR1. *Cancer Res* **60**(6): 1531-5, 2000.

[346] Fuks F, Milner J, Kouzarides T. BRCA2 associates with acetyltransferase activity when bound to P/CAF. *Oncogene* **17**(19): 2531-4, 1998.

[347] Moynahan ME, Pierce AJ, Jasin M. BRCA2 is required for homology-directed repair of chromosomal breaks. *Mol Cell* **7**(2): 263-72, 2001.

[348] Bignell G, Micklem G, Stratton MR, Ashworth A, Wooster R. The BRC repeats are conserved in mammalian BRCA2 proteins. *Hum Mol Genet* **6**(1): 53-8, 1997.

[349] Tutt A, Bertwistle D, Valentine J, Gabriel A, Swift S, Ross G, Griffin C, Thacker J, Ashworth A. Mutation in Brca2 stimulates error-prone homology-directed repair of DNA double-strand breaks occurring between repeated sequences. *Embo J* **20**(17): 4704-16, 2001.

[350] Tischkowitz M, Dokal I. Fanconi anaemia and leukaemia - clinical and molecular aspects. *Br J Haematol* **126**(2): 176-91, 2004.

[351] Alter BP. Cancer in Fanconi anemia, 1927-2001. *Cancer* **97**(2): 425-40, 2003.

[352] Sasaki MS, Tonomura A. A high susceptibility of Fanconi's anemia to chromosome breakage by DNA cross-linking agents. *Cancer Res* **33**(8): 1829-36, 1973.

[353] Kato H, Stich HF. Sister chromatid exchanges in ageing and repair-deficient human fibroblasts. *Nature* **260**(5550): 447-8, 1976.

[354] Auerbach AD, Wolman SR. Susceptibility of Fanconi's anaemia fibroblasts to chromosome damage by carcinogens. *Nature* **261**(5560): 494-6, 1976.

[355] Auerbach AD, Wolman SR. Carcinogen-induced chromosome breakage in Fanconi's anaemia heterozygous cells. *Nature* **271**(5640): 69-71, 1978.

[356] Alter BP. Radiosensitivity in Fanconi's anemia patients. *Radiother Oncol* **62**(3): 345-7, 2002.

[357] Gatti RA. The inherited basis of human radiosensitivity. *Acta Oncol* **40**(6): 702-11, 2001.

[358] D'Andrea AD, Grompe M. The Fanconi anaemia/BRCA pathway. *Nat Rev Cancer* **3**(1): 23-34, 2003.

[359] Wang W. Emergence of a DNA-damage response network consisting of Fanconi anaemia and BRCA proteins. *Nat Rev Genet* **8**(10): 735-48, 2007.

[360] Yamashita T, Kupfer GM, Naf D, Suliman A, Joenje H, Asano S, D'Andrea AD. The fanconi anemia pathway requires FAA phosphorylation and FAA/FAC nuclear accumulation. *Proc Natl Acad Sci U S A* **95**(22): 13085-90, 1998.

[361] Medhurst AL, Huber PA, Waisfisz Q, de Winter JP, Mathew CG. Direct interactions of the five known Fanconi anaemia proteins suggest a common functional pathway. *Hum Mol Genet* **10**(4): 423-9, 2001.

[362] Kupfer GM, Naf D, Suliman A, Pulsipher M, D'Andrea AD. The Fanconi anaemia proteins, FAA and FAC, interact to form a nuclear complex. *Nat Genet* **17**(4): 487-90, 1997.

[363] Garcia-Higuera I, Kuang Y, Denham J, D'Andrea AD. The fanconi anemia proteins FANCA and FANCG stabilize each other and promote the nuclear accumulation of the Fanconi anemia complex. *Blood* **96**(9): 3224-30, 2000.

[364] Garcia-Higuera I, Kuang Y, Naf D, Wasik J, D'Andrea AD. Fanconi anemia proteins FANCA, FANCC, and FANCG/XRCC9 interact in a functional nuclear complex. *Mol Cell Biol* **19**(7): 4866-73, 1999.

[365] Demuth I, Wlodarski M, Tipping AJ, Morgan NV, de Winter JP, Thiel M, Grasl S, Schindler D, D'Andrea AD, Altay C, Kayserili H, Zatterale A, Kunze J, Ebell W, Mathew CG, Joenje H, Sperling K, Digweed M. Spectrum of mutations in the Fanconi anaemia group G gene, FANCG/XRCC9. *Eur J Hum Genet* **8**(11): 861-8, 2000.

[366] Nijman SM, Huang TT, Dirac AM, Brummelkamp TR, Kerkhoven RM, D'Andrea AD, Bernards R. The deubiquitinating enzyme USP1 regulates the Fanconi anemia pathway. *Mol Cell* **17**(3): 331-9, 2005.

[367] Smogorzewska A, Matsuoka S, Vinciguerra P, McDonald ER, 3rd, Hurov KE, Luo J, Ballif BA, Gygi SP, Hofmann K, D'Andrea AD, Elledge SJ. Identification of the FANCI protein, a monoubiquitinated FANCD2 paralog required for DNA repair. *Cell* **129**(2): 289-301, 2007.

[368] Meetei AR, Medhurst AL, Ling C, Xue Y, Singh TR, Bier P, Steltenpool J, Stone S, Dokal I, Mathew CG, Hoatlin M, Joenje H, de Winter JP, Wang W. A human ortholog of archaeal DNA repair protein Hef is defective in Fanconi anemia complementation group M. *Nat Genet* **37**(9): 958-63, 2005.

[369] Meetei AR, Levitus M, Xue Y, Medhurst AL, Zwaan M, Ling C, Rooimans MA, Bier P, Hoatlin M, Pals G, de Winter JP, Wang W, Joenje H. X-linked inheritance of Fanconi anemia complementation group B. *Nat Genet* **36**(11): 1219-24, 2004.

[370] Ling C, Ishiai M, Ali AM, Medhurst AL, Neveling K, Kalb R, Yan Z, Xue Y, Oostra AB, Auerbach AD, Hoatlin ME, Schindler D, Joenje H, de Winter JP, Takata M, Meetei AR, Wang W. FAAP100 is essential for activation of the Fanconi anemia-associated DNA damage response pathway. *Embo J* **26**(8): 2104-14, 2007.

[371] Gurtan AM, Stuckert P, D'Andrea AD. The WD40 repeats of FANCL are required for Fanconi anemia core complex assembly. *J Biol Chem* **281**(16): 10896-905, 2006.

[372] Meetei AR, de Winter JP, Medhurst AL, Wallisch M, Waisfisz Q, van de Vrugt HJ, Oostra AB, Yan Z, Ling C, Bishop CE, Hoatlin ME, Joenje H, Wang W. A novel ubiquitin ligase is deficient in Fanconi anemia. *Nat Genet* **35**(2): 165-70, 2003.

[373] Matsuoka S, Ballif BA, Smogorzewska A, McDonald ER, 3rd, Hurov KE, Luo J, Bakalarski CE, Zhao Z, Solimini N, Lerenthal Y, Shiloh Y, Gygi SP, Elledge SJ. ATM and ATR substrate analysis reveals extensive protein networks responsive to DNA damage. *Science* **316**(5828): 1160-6, 2007.

[374] Wang X, Kennedy RD, Ray K, Stuckert P, Ellenberger T, D'Andrea AD. Chk1-mediated phosphorylation of FANCE is required for the Fanconi anemia/BRCA pathway. *Mol Cell Biol* **27**(8): 3098-108, 2007.

[375] Qiao F, Mi J, Wilson JB, Zhi G, Bucheimer NR, Jones NJ, Kupfer GM. Phosphorylation of fanconi anemia (FA) complementation group G protein, FANCG, at serine 7 is important for function of the FA pathway. *J Biol Chem* **279**(44): 46035-45, 2004.

[376] Mi J, Qiao F, Wilson JB, High AA, Schroeder MJ, Stukenberg PT, Moss A, Shabanowitz J, Hunt DF, Jones NJ, Kupfer GM. FANCG is phosphorylated at serines 383 and 387 during mitosis. *Mol Cell Biol* **24**(19): 8576-85, 2004.

[377] Andreassen PR, D'Andrea AD, Taniguchi T. ATR couples FANCD2 monoubiquitination to the DNA-damage response. *Genes Dev* **18**(16): 1958-63, 2004.

[378] Taniguchi T, Garcia-Higuera I, Andreassen PR, Gregory RC, Grompe M, D'Andrea AD. S-phase-specific interaction of the Fanconi anemia protein, FANCD2, with BRCA1 and RAD51. *Blood* **100**(7): 2414-20, 2002.

[379] Garcia-Higuera I, Taniguchi T, Ganesan S, Meyn MS, Timmers C, Hejna J, Grompe M, D'Andrea AD. Interaction of the Fanconi anemia proteins and BRCA1 in a common pathway. *Mol Cell* **7**(2): 249-62, 2001.

[380] Park WH, Margossian S, Horwitz AA, Simons AM, D'Andrea AD, Parvin JD. Direct DNA binding activity of the Fanconi anemia D2 protein. *J Biol Chem* **280**(25): 23593-8, 2005.

[381] Sims AE, Spiteri E, Sims RJ, 3rd, Arita AG, Lach FP, Landers T, Wurm M, Freund M, Neveling K, Hanenberg H, Auerbach AD, Huang TT. FANCI is a second monoubiquitinated member of the Fanconi anemia pathway. *Nat Struct Mol Biol* **14**(6): 564-7, 2007.

[382] Meetei AR, Sechi S, Wallisch M, Yang D, Young MK, Joenje H, Hoatlin ME, Wang W. A multiprotein nuclear complex connects Fanconi anemia and Bloom syndrome. *Mol Cell Biol* **23**(10): 3417-26, 2003.

[383] Hirano S, Yamamoto K, Ishiai M, Yamazoe M, Seki M, Matsushita N, Ohzeki M, Yamashita YM, Arakawa H, Buerstedde JM, Enomoto T, Takeda S, Thompson LH,

Takata M. Functional relationships of FANCC to homologous recombination, translesion synthesis, and BLM. *Embo J* **24**(2): 418-27, 2005.

[384] Litman R, Peng M, Jin Z, Zhang F, Zhang J, Powell S, Andreassen PR, Cantor SB. BACH1 is critical for homologous recombination and appears to be the Fanconi anemia gene product FANCJ. *Cancer Cell* **8**(3): 255-65, 2005.

[385] Levitus M, Waisfisz Q, Godthelp BC, de Vries Y, Hussain S, Wiegant WW, Elghalbzouri-Maghrani E, Steltenpool J, Rooimans MA, Pals G, Arwert F, Mathew CG, Zdzienicka MZ, Hiom K, De Winter JP, Joenje H. The DNA helicase BRIP1 is defective in Fanconi anemia complementation group J. *Nat Genet* **37**(9): 934-5, 2005.

[386] Montes de Oca R, Andreassen PR, Margossian SP, Gregory RC, Taniguchi T, Wang X, Houghtaling S, Grompe M, D'Andrea AD. Regulated interaction of the Fanconi anemia protein, FANCD2, with chromatin. *Blood* **105**(3): 1003-9, 2005.

[387] Thyagarajan B, Campbell C. Elevated homologous recombination activity in fanconi anemia fibroblasts. *J Biol Chem* **272**(37): 23328-33, 1997.

[388] Yamamoto K, Hirano S, Ishiai M, Morishima K, Kitao H, Namikoshi K, Kimura M, Matsushita N, Arakawa H, Buerstedde JM, Komatsu K, Thompson LH, Takata M. Fanconi anemia protein FANCD2 promotes immunoglobulin gene conversion and DNA repair through a mechanism related to homologous recombination. *Mol Cell Biol* **25**(1): 34-43, 2005.

[389] Yamamoto K, Ishiai M, Matsushita N, Arakawa H, Lamerdin JE, Buerstedde JM, Tanimoto M, Harada M, Thompson LH, Takata M. Fanconi anemia FANCG protein in mitigating radiation- and enzyme-induced DNA double-strand breaks by homologous recombination in vertebrate cells. *Mol Cell Biol* **23**(15): 5421-30, 2003.

[390] Niedzwiedz W, Mosedale G, Johnson M, Ong CY, Pace P, Patel KJ. The Fanconi anaemia gene FANCC promotes homologous recombination and error-prone DNA repair. *Mol Cell* **15**(4): 607-20, 2004.

[391] Nakanishi K, Yang YG, Pierce AJ, Taniguchi T, Digweed M, D'Andrea AD, Wang ZQ, Jasin M. Human Fanconi anemia monoubiquitination pathway promotes homologous DNA repair. *Proc Natl Acad Sci U S A* **102**(4): 1110-5, 2005.

[392] German J, Crippa LP, Bloom D. Bloom's syndrome. III. Analysis of the chromosome aberration characteristic of this disorder. *Chromosoma* **48**(4): 361-6, 1974.

[393] German J, Schonberg S, Louie E, Chaganti RS. Bloom's syndrome. IV. Sister-chromatid exchanges in lymphocytes. *Am J Hum Genet* **29**(3): 248-55, 1977.

[394] Goss KH, Risinger MA, Kordich JJ, Sanz MM, Straughen JE, Slovek LE, Capobianco AJ, German J, Boivin GP, Groden J. Enhanced tumor formation in mice heterozygous for Blm mutation. *Science* **297**(5589): 2051-3, 2002.

[395] Chester N, Kuo F, Kozak C, O'Hara CD, Leder P. Stage-specific apoptosis, developmental delay, and embryonic lethality in mice homozygous for a targeted disruption in the murine Bloom's syndrome gene. *Genes Dev* **12**(21): 3382-93, 1998.

[396] Luo G, Santoro IM, McDaniel LD, Nishijima I, Mills M, Youssoufian H, Vogel H, Schultz RA, Bradley A. Cancer predisposition caused by elevated mitotic recombination in Bloom mice. *Nat Genet* **26**(4): 424-9, 2000.

[397] McDaniel LD, Chester N, Watson M, Borowsky AD, Leder P, Schultz RA. Chromosome instability and tumor predisposition inversely correlate with BLM protein levels. *DNA Repair (Amst)* **2**(12): 1387-404, 2003.

[398] Chester N, Babbe H, Pinkas J, Manning C, Leder P. Mutation of the murine Bloom's syndrome gene produces global genome destabilization. *Mol Cell Biol* **26**(17): 6713-26, 2006.

[399] Hickson ID. RecQ helicases: caretakers of the genome. *Nat Rev Cancer* **3**(3): 169-78, 2003.

[400] Karow JK, Chakraverty RK, Hickson ID. The Bloom's syndrome gene product is a 3'-5' DNA helicase. *J Biol Chem* **272**(49): 30611-4, 1997.

[401] Kitao S, Ohsugi I, Ichikawa K, Goto M, Furuichi Y, Shimamoto A. Cloning of two new human helicase genes of the RecQ family: biological significance of multiple species in higher eukaryotes. *Genomics* **54**(3): 443-52, 1998.

[402] Bischof O, Kim SH, Irving J, Beresten S, Ellis NA, Campisi J. Regulation and localization of the Bloom syndrome protein in response to DNA damage. *J Cell Biol* **153**(2): 367-80, 2001.

[403] Wang XW, Tseng A, Ellis NA, Spillare EA, Linke SP, Robles AI, Seker H, Yang Q, Hu P, Beresten S, Bemmels NA, Garfield S, Harris CC. Functional interaction of p53 and BLM DNA helicase in apoptosis. *J Biol Chem* **276**(35): 32948-55, 2001.

[404] Johnson FB, Lombard DB, Neff NF, Mastrangelo MA, Dewolf W, Ellis NA, Marciniak RA, Yin Y, Jaenisch R, Guarente L. Association of the Bloom syndrome protein with topoisomerase IIIalpha in somatic and meiotic cells. *Cancer Res* **60**(5): 1162-7, 2000.

[405] Wu L, Davies SL, North PS, Goulaouic H, Riou JF, Turley H, Gatter KC, Hickson ID. The Bloom's syndrome gene product interacts with topoisomerase III. *J Biol Chem* **275**(13): 9636-44, 2000.

[406] Ababou M, Dutertre S, Lecluse Y, Onclercq R, Chatton B, Amor-Gueret M. ATM-dependent phosphorylation and accumulation of endogenous BLM protein in response to ionizing radiation. *Oncogene* **19**(52): 5955-63, 2000.

[407] Bayart E, Dutertre S, Jaulin C, Guo RB, Xi XG, Amor-Gueret M. The Bloom syndrome helicase is a substrate of the mitotic Cdc2 kinase. *Cell Cycle* **5**(15): 1681-6, 2006.

[408] Karmakar P, Seki M, Kanamori M, Hashiguchi K, Ohtsuki M, Murata E, Inoue E, Tada S, Lan L, Yasui A, Enomoto T. BLM is an early responder to DNA double-strand breaks. *Biochem Biophys Res Commun* **348**(1): 62-9, 2006.

[409] Rao VA, Conti C, Guirouilh-Barbat J, Nakamura A, Miao ZH, Davies SL, Sacca B, Hickson ID, Bensimon A, Pommier Y. Endogenous gamma-H2AX-ATM-Chk2 checkpoint activation in Bloom's syndrome helicase deficient cells is related to DNA replication arrested forks. *Mol Cancer Res* **5**(7): 713-24, 2007.

[410] Franchitto A, Pichierri P. Bloom's syndrome protein is required for correct relocalization of RAD50/MRE11/NBS1 complex after replication fork arrest. *J Cell Biol* **157**(1): 19-30, 2002.

[411] Pichierri P, Franchitto A, Rosselli F. BLM and the FANC proteins collaborate in a common pathway in response to stalled replication forks. *Embo J* **23**(15): 3154-63, 2004.

[412] Mohaghegh P, Karow JK, Brosh Jr RM, Jr., Bohr VA, Hickson ID. The Bloom's and Werner's syndrome proteins are DNA structure-specific helicases. *Nucleic Acids Res* **29**(13): 2843-9, 2001.

[413] Karow JK, Constantinou A, Li JL, West SC, Hickson ID. The Bloom's syndrome gene product promotes branch migration of holliday junctions. *Proc Natl Acad Sci U S A* **97**(12): 6504-8, 2000.

[414] Wu L, Hickson ID. The Bloom's syndrome helicase suppresses crossing over during homologous recombination. *Nature* **426**(6968): 870-4, 2003.

[415] Cheok CF, Wu L, Garcia PL, Janscak P, Hickson ID. The Bloom's syndrome helicase promotes the annealing of complementary single-stranded DNA. *Nucleic Acids Res* **33**(12): 3932-41, 2005.

[416] Brosh RM, Jr., Li JL, Kenny MK, Karow JK, Cooper MP, Kureekattil RP, Hickson ID, Bohr VA. Replication protein A physically interacts with the Bloom's syndrome protein and stimulates its helicase activity. *J Biol Chem* **275**(31): 23500-8, 2000.

[417] Wu L, Davies SL, Levitt NC, Hickson ID. Potential role for the BLM helicase in recombinational repair via a conserved interaction with RAD51. *J Biol Chem* **276**(22): 19375-81, 2001.

[418] Braybrooke JP, Li JL, Wu L, Caple F, Benson FE, Hickson ID. Functional interaction between the Bloom's syndrome helicase and the RAD51 paralog, RAD51L3 (RAD51D). *J Biol Chem* **278**(48): 48357-66, 2003.

[419] Zhang R, Sengupta S, Yang Q, Linke SP, Yanaihara N, Bradsher J, Blais V, McGowan CH, Harris CC. BLM helicase facilitates Mus81 endonuclease activity in human cells. *Cancer Res* **65**(7): 2526-31, 2005.

[420] Sharma S, Sommers JA, Wu L, Bohr VA, Hickson ID, Brosh RM, Jr. Stimulation of flap endonuclease-1 by the Bloom's syndrome protein. *J Biol Chem* **279**(11): 9847-56, 2004.

[421] von Kobbe C, Karmakar P, Dawut L, Opresko P, Zeng X, Brosh RM, Jr., Hickson ID, Bohr VA. Colocalization, physical, and functional interaction between Werner and Bloom syndrome proteins. *J Biol Chem* **277**(24): 22035-44, 2002.

[422] Pedrazzi G, Perrera C, Blaser H, Kuster P, Marra G, Davies SL, Ryu GH, Freire R, Hickson ID, Jiricny J, Stagljar I. Direct association of Bloom's syndrome gene product with the human mismatch repair protein MLH1. *Nucleic Acids Res* **29**(21): 4378-86, 2001.

[423] Yang Q, Zhang R, Wang XW, Linke SP, Sengupta S, Hickson ID, Pedrazzi G, Perrera C, Stagljar I, Littman SJ, Modrich P, Harris CC. The mismatch DNA repair heterodimer, hMSH2/6, regulates BLM helicase. *Oncogene* **23**(21): 3749-56, 2004.

[424] So S, Adachi N, Lieber MR, Koyama H. Genetic interactions between BLM and DNA ligase IV in human cells. *J Biol Chem* **279**(53): 55433-42, 2004.

[425] Adams MD, McVey M, Sekelsky JJ. Drosophila BLM in double-strand break repair by synthesis-dependent strand annealing. *Science* **299**(5604): 265-7, 2003.

[426] Beall EL, Rio DC. Drosophila IRBP/Ku p70 corresponds to the mutagen-sensitive mus309 gene and is involved in P-element excision in vivo. *Genes Dev* **10**(8): 921-33, 1996.

[427] McVey M, Andersen SL, Broze Y, Sekelsky J. Multiple functions of Drosophila BLM helicase in maintenance of genome stability. *Genetics* **176**(4): 1979-92, 2007.

[428] Slupianek A, Gurdek E, Koptyra M, Nowicki MO, Siddiqui KM, Groden J, Skorski T. BLM helicase is activated in BCR/ABL leukemia cells to modulate responses to cisplatin. *Oncogene* **24**(24): 3914-22, 2005.

[429] Yu CE, Oshima J, Fu YH, Wijsman EM, Hisama F, Alisch R, Matthews S, Nakura J, Miki T, Ouais S, Martin GM, Mulligan J, Schellenberg GD. Positional cloning of the Werner's syndrome gene. *Science* **272**(5259): 258-62, 1996.

[430] Epstein CJ, Martin GM, Schultz AL, Motulsky AG. Werner's syndrome a review of its symptomatology, natural history, pathologic features, genetics and relationship to the natural aging process. *Medicine (Baltimore)* **45**(3): 177-221, 1966.

[431] Goto M. Hierarchical deterioration of body systems in Werner's syndrome: implications for normal ageing. *Mech Ageing Dev* **98**(3): 239-54, 1997.

[432] Fukuchi K, Martin GM, Monnat RJ, Jr. Mutator phenotype of Werner syndrome is characterized by extensive deletions. *Proc. Natl. Acad. Sci. U. S. A.* **86**(15): 5893-5897, 1989.

[433] Gray MD, Shen J-C, Kamath-Loeb AS, Blank A, Sopher BL, Martin GM, Oshima J, Loeb LA. The Werner syndrome protein is a DNA helicase. *Nat Genet* **17**(1): 100, 1997.

[434] Huang S, Li B, Gray MD, Oshima J, Mian IS, Campisi J. The premature ageing syndrome protein, WRN, is a 3[prime][rarr]5[prime] exonuclease. *Nat Genet* **20**(2): 114, 1998.

[435] Shen J-C, Gray MD, Oshima J, Kamath-Loeb AS, Fry M, Loeb LA. Werner Syndrome Protein. I. DNA HELICASE AND DNA EXONUCLEASE RESIDE ON THE SAME POLYPEPTIDE. *J. Biol. Chem.* %R 10.1074/jbc.273.51.34139 **273**(51): 34139-34144, 1998.

[436] Pichierri P, Rosselli F, Franchitto A. Werner's syndrome protein is phosphorylated in an ATR/ATM-dependent manner following replication arrest and DNA damage induced during the S phase of the cell cycle. *Oncogene* **22**(10): 1491-500, 2003.

[437] Karmakar P, Piotrowski J, Brosh RM, Jr., Sommers JA, Miller SP, Cheng WH, Snowden CM, Ramsden DA, Bohr VA. Werner protein is a target of DNA-dependent protein kinase in vivo and in vitro, and its catalytic activities are regulated by phosphorylation. *J Biol Chem* **277**(21): 18291-302, 2002.

[438] Brosh RM, Jr., Karmakar P, Sommers JA, Yang Q, Wang XW, Spillare EA, Harris CC, Bohr VA. p53 Modulates the exonuclease activity of Werner syndrome protein. *J Biol Chem* **276**(37): 35093-102, 2001.

[439] Blander G, Kipnis J, Leal JF, Yu CE, Schellenberg GD, Oren M. Physical and functional interaction between p53 and the Werner's syndrome protein. *J Biol Chem* **274**(41): 29463-9, 1999.

[440] Blander G, Zalle N, Leal JF, Bar-Or RL, Yu CE, Oren M. The Werner syndrome protein contributes to induction of p53 by DNA damage. *Faseb J* **14**(14): 2138-40, 2000.

[441] Cheng WH, Sakamoto S, Fox JT, Komatsu K, Carney J, Bohr VA. Werner syndrome protein associates with gamma H2AX in a manner that depends upon Nbs1. *FEBS Lett* **579**(6): 1350-6, 2005.

[442] Baynton K, Otterlei M, Bjoras M, von Kobbe C, Bohr VA, Seeberg E. WRN interacts physically and functionally with the recombination mediator protein RAD52. *J Biol Chem* **278**(38): 36476-86, 2003.

[443] Otterlei M, Bruheim P, Ahn B, Bussen W, Karmakar P, Baynton K, Bohr VA. Werner syndrome protein participates in a complex with RAD51, RAD54, RAD54B and ATR in response to ICL-induced replication arrest. *J Cell Sci* **119**(Pt 24): 5137-46, 2006.

[444] Lebel M, Leder P. A deletion within the murine Werner syndrome helicase induces sensitivity to inhibitors of topoisomerase and loss of cellular proliferative capacity. *Proc Natl Acad Sci U S A* **95**(22): 13097-102, 1998.

[445] Lombard DB, Beard C, Johnson B, Marciniak RA, Dausman J, Bronson R, Buhlmann JE, Lipman R, Curry R, Sharpe A, Jaenisch R, Guarente L. Mutations in the WRN gene in mice accelerate mortality in a p53-null background. *Mol Cell Biol* **20**(9): 3286-91, 2000.

[446] Wang L, Ogburn CE, Ware CB, Ladiges WC, Youssoufian H, Martin GM, Oshima J. *Cell*ular Werner phenotypes in mice expressing a putative dominant-negative human WRN gene. *Genetics* **154**(1): 357-62, 2000.

[447] Harrigan JA, Wilson DM, 3rd, Prasad R, Opresko PL, Beck G, May A, Wilson SH, Bohr VA. The Werner syndrome protein operates in base excision repair and cooperates with DNA polymerase beta. *Nucleic Acids Res* **34**(2): 745-54, 2006.

[448] Lebel M, Lavoie J, Gaudreault I, Bronsard M, Drouin R. Genetic cooperation between the Werner syndrome protein and poly(ADP-ribose) polymerase-1 in preventing chromatid breaks, complex chromosomal rearrangements, and cancer in mice. *Am J Pathol* **162**(5): 1559-69, 2003.

[449] Cheng WH, Kusumoto R, Opresko PL, Sui X, Huang S, Nicolette ML, Paull TT, Campisi J, Seidman M, Bohr VA. Collaboration of Werner syndrome protein and BRCA1 in cellular responses to DNA interstrand cross-links. *Nucleic Acids Res* **34**(9): 2751-60, 2006.

[450] Li B, Comai L. Functional interaction between Ku and the werner syndrome protein in DNA end processing. *J Biol Chem* **275**(50): 39800, 2000.

[451] Orren DK, Machwe A, Karmakar P, Piotrowski J, Cooper MP, Bohr VA. A functional interaction of Ku with Werner exonuclease facilitates digestion of damaged DNA. *Nucleic Acids Res* **29**(9): 1926-34, 2001.

[452] Shen JC, Gray MD, Oshima J, Loeb LA. Characterization of Werner syndrome protein DNA helicase activity: directionality, substrate dependence and stimulation by replication protein A. *Nucleic Acids Res* **26**(12): 2879-85, 1998.

[453] von Kobbe C, Harrigan JA, Schreiber V, Stiegler P, Piotrowski J, Dawut L, Bohr VA. Poly(ADP-ribose) polymerase 1 regulates both the exonuclease and helicase activities of the Werner syndrome protein. *Nucleic Acids Res* **32**(13): 4003-14, 2004.

[454] Saintigny Y, Makienko K, Swanson C, Emond MJ, Monnat RJ, Jr. Homologous recombination resolution defect in werner syndrome. *Mol Cell Biol* **22**(20): 6971-8, 2002.

[455] Sakamoto S, Nishikawa K, Heo SJ, Goto M, Furuichi Y, Shimamoto A. Werner helicase relocates into nuclear foci in response to DNA damaging agents and co-localizes with RPA and Rad51. *Genes Cells* **6**(5): 421-30, 2001.

[456] Rodriguez-Lopez AM, Jackson DA, Iborra F, Cox LS. Asymmetry of DNA replication fork progression in Werner's syndrome. *Aging Cell* **1**(1): 30-9, 2002.

[457] Rodriguez-Lopez AM, Jackson DA, Nehlin JO, Iborra F, Warren AV, Cox LS. Characterisation of the interaction between WRN, the helicase/exonuclease defective in progeroid Werner's syndrome, and an essential replication factor, PCNA. *Mech Ageing Dev* **124**(2): 167-74, 2003.

[458] Agrelo R. A new molecular model of cellular aging based on Werner syndrome. *Med Hypotheses* 2006.

[459] Oshima J, Huang S, Pae C, Campisi J, Schiestl RH. Lack of WRN results in extensive deletion at nonhomologous joining ends. *Cancer Res* **62**(2): 547-551, 2002.

[460] Leder A, Lebel M, Zhou F, Fontaine K, Bishop A, Leder P. Genetic interaction between the unstable v-Ha-RAS transgene (Tg.AC) and the murine Werner syndrome gene: transgene instability and tumorigenesis. *Oncogene* **21**(43): 6657-68, 2002.

[461] Crabbe L, Verdun RE, Haggblom CI, Karlseder J. Defective telomere lagging strand synthesis in cells lacking WRN helicase activity. *Science* **306**(5703): 1951-3, 2004.

[462] Eller MS, Liao X, Liu S, Hanna K, Backvall H, Opresko PL, Bohr VA, Gilchrest BA. A role for WRN in telomere-based DNA damage responses. *Proc Natl Acad Sci U S A* **103**(41): 15073-8, 2006.

[463] Lebel M. Increased frequency of DNA deletions in pink-eyed unstable mice carrying a mutation in the Werner syndrome gene homologue. *Carcinogenesis* **23**(1): 213-6, 2002.

[464] Andreassen PR, Ho GP, D'Andrea AD. DNA damage responses and their many interactions with the replication fork. *Carcinogenesis* **27**(5): 883-92, 2006.

In: Genetic Recombination Research Progress
Editor: Jacob H. Schulz, pp. 69-83

ISBN: 978-1-60456-482-2
© 2008 Nova Science Publishers, Inc.

Chapter 2

Ecdysone-Based-Tet-on Expression System

Leonardo D'Aiuto[1,], Luigi Viggiano[2], Daniel Suter[1], and J. Richard Chaillet[1]*

[1]Department of Molecular Genetics and Biochemistry, University of Pittsburgh, Pittsburgh, Pennsylvania, 15213. USA
[2]Department Genetics and Microbiology, University of Bari, Italy, 70126

Abstract

Inducible gene expression systems offer great potential toward achieving a wide variety of basic and applied biomedical research goals. Current inducible gene technologies are based on binary transgenic systems in which the expression of a target gene is dependent upon the activity of a unique transcriptional activator in the presence or absence of its drug ligand.

Despite the conceptual simplicity, most gene inducible strategies suffer from a number of limitations, such as leaky basal transcriptional activity when uninduced and poor dynamic control of the system. Because of these shortcomings, regulated gene expression technologies are still being evaluated using stringent genetic studies and none are currently used in human gene therapy. To overcome some of the disadvantages of the current systems, we describe the development of a next-generation inducible gene expression system, which relies upon *multiple* heterologous transactivators and their respective ligands. Our system would provide negative/positive selection for transgene expression in its induced state that would create virtually no transgene expression when not induced (i.e., no leakage). We propose a strategy to engineer expression systems that include heterodimeric hormone receptors combined with other heterologous transactivators.

* E-mail addresses: daiutol@upmc.edu. (Corresponding author)

Introduction

Regulated gene expression technologies for proteins in mammalian cells are potent tools for studying the function of a gene and its corresponding protein as well as the production of difficult-to-express proteins. The importance of this technology is that the function of the protein can be analyzed in the absence of endogenous cellular controls. Establishment of a cell line that allows induction of the target genes' expression using these systems makes it possible to study the genes' function under highly reproducible conditions. Tight regulation of gene expression systems allows induction of high levels of proteins causing cell death [1] or cell growth-inhibitory molecules [2]. Furthermore, the use of these regulated expression systems allows the production of knockout mice for disease-causing genes that are lethal at embryonic and neonatal stages. The problem with current techniques, though, is that there are still relatively high levels of leakiness in non-induced conditions. If the product of a transgene is toxic or unwanted, then even a low level of leaky expression could be deleterious to embryos, preventing any further analysis of the potential phenotype during development, infancy, or as an adult.

Transcriptional regulation systems were initially based on the use of endocellular *cis*-genetic elements that respond to exogenous signals or stressors such as hormones [3], heat [4] hypoxia [5], cytokines [6] or metal ions [7]. The major limitation common to all these endogenous regulation systems was the pleiotropic effects that the inducer/inducing conditions may exercise. Such interference can lead to leakiness (detectable level of expression in absence of induction), low induction ratios, and induction of several cellular genes. For instance, repeated heat-shock of cells engineered with temperature-responsive expression systems results in growth arrest, increased glucose consumption, low product yield, and decreased cell viability [8].

A radical change in the development of expression inducible technologies evolved from attempts to adapt heterologous transcriptional control systems to mammalian cells. These heterologous gene regulation system include: i) Gal4-estrogen receptor-based technology [9]; ii) the mutant human progesterone receptor, RU486 inducible [10]; iii) the Tet-repressor-based system, tetracycline-inducible [11]; iv) chemically induced dimerization [12]; v) the Pip-repressor-based system, steptogramin-inducible [13]; and, vi) the ecdyson receptor (EcR)-based system, ecdyson inducible [14].

These technologies all involve the drug-dependent recruitment of a transcriptional activation domain to a minimal promoter driving the expression of the gene of interest, but differ in the mechanism of recruitment. A successful gene inducible system has two key features: i) basal expression is absent or very low while at the same time being inducible to high levels over a wide enough dose range to provide useful responsive control [15]; and ii) the regulatory protein(s) should not interfere with endogenous gene expression. A principal advantage of an inducible expression system is that it permits the adjustment of transgene expression to a desired level.

Here, we first give a brief introduction to the tetracycline- and ecdysone-dependent expression systems and focus on the major drawbacks of these technologies. Afterward, we will also propose a new strategy to achieve highly inducible gene expression with low basal activity.

Tetracycline-Inducible Expression Systems

Tetracycline/doxycycline expression systems are binary transgenic systems in which expression of a target transgene is driven by an inducible transcriptional activator. The tetracycline system (Tet) is based on the Tet repressor DNA binding protein (TetR) from the Tc resistance operon of *Escherichia coli* transposon Tn10. In the bacteria, TetR binds to the operator sequences (TetO) of the tet-resistance operon and represses gene expression [16].

In the Tet inducible system, TetR is fused to the C-terminal transactivation domain of the Herpes simplex virus protein VP16 [17], which interacts with transcription factors TFIIB, TBP and TAFII$_{40}$ to form an active transcription complex. Fusion of TetR to VP16 results in a tetracycline-controlled transactivator protein (*tTA*), which activates transcription through a tet-responsive promoter element (TRE) in the absence of tetracycline. The TRE consists of heptameric *tetO* repeats fused to a minimal promoter sequence derived from the human cytomegalovirus (hCMV) immediate-early promoter. Together, TetR and tTA form the Tet-OFF system: upon addition of tetracycline (Tc) or its analog Doxycycline (Dox), the antibiotics bind to TetR, inducing an allosteric change, and inducing the release of the tTA from the TRE operator, which as a result represses the transgene expression.

Binding of Tc to the TetR homodimer abolishes its molecular affinity to two palindromic operator sites on tetO. Each polypeptide chain of the functional TetR homodimer is folded into 10 α-helices (α1- α10 and α1'- α10'). Two N-terminal DNA-binding domains are formed by α1-α3 helices, where α2 and α3 constitute helix-turn-helix (HTH) motifs. Each HTH motif binds to a corresponding major groove of tetO. Helices α5 to α10 and α5' to α10' form the core of TetR where the binding of the Tc takes place (known as the regulatory domain). The N-terminus of helix α4 links the DNA binding domain to the regulatory domain. Binding of the Tc to TetR determines a shift of helix α6 and a consequent rotation of helix α4 and α4'. As a result of this motion the N terminal DNA-binding domains are shifted apart, increasing the separation of the helices α3 and α3' along the major groove of tetO. It is this shift that abolishes the affinity of the TetR for TetO [18].

The Tet-On system is based on a reverse tetracycline-controlled transactivator, *rtTA*, which differs from tTA by mutations of only four amino acids in TetR but as a result displays Doxycycline (Dox)-dependent tetO-binding. Thus, in the Tet-On system, transcription of the TRE-regulated target gene is stimulated by *rtTA* only in the presence of Dox (Figure1). These mutations have several major effects. First of all, they lock the DNA binding domains in the position necessary to bind the operator. Secondly, they restrict the TetR to a non-inducible conformation. After the initial discovery, mutation analysis allowed the identification of two novel rtTA alleles, called rtTA2^S-S2 and rtTA2^S-M2 which display a considerably lower basal activity in the absence of the inducer, and a higher sensitivity to Dox [19]. These rtTA variants contain a minimal transactivator domain named VP16-F3, consisting of three tandem repeats of a 12 amino- acid peptide derived from VP16.

Although the Tet systems are by far the most widely used technology for regulated gene expression in mammalian cells, they have two major drawbacks: leakiness and poor overall gene regulation. These drawbacks are actually related to one another, and both are common to most inducible gene expression systems. Leakiness is defined as gene expression in the absence of the inducer, and therefore usually in the absence of the operator-bound activator too. Leakiness can often contribute to poor overall regulation of gene expression by adversely affecting the fold increase in expression gained from induction. The Tet-Off system is preferred over the Tet-On system because the latter often exhibits leaky expression under noninduced conditions and requires a higher concentration of Dox for regulation of the gene expression [20]. However, many applications, such as gene therapy approaches, biotechnological production of recombinant proteins, or analysis of the role of specific genes in development require the expression of the gene of interest limited to a defined time window. The Tet-Off system is not suitable for gene therapy projects because prolonged exposure of the patients to tetracycline to repress or down-regulate the expression of the therapeutic transgene may elicit negative side-effects and poor induction kinetics [17,18]. Furthermore, the Tet-off system does not allow a homogeneous regulation of the tetracycline-responsive transgene in most cell types of transgenic mice [19]. The Tet-On system is conceptually ideal, but its activity is not under as tight genetic control. In this system, leakiness from the transgenes' minimal promoters may depend on the presence or absence of enhancers located near the site of integration [20]. On the other hand, the absence of leakiness may be a result of chromatin repression due to a "position effect" [21].

To mitigate the basal transgene leak in tet-based systems under uninduced conditions, the KRAB (Kruppel-associated box) repressor domain of the human Kox 1 zinc finger protein was fused to the tetracycline repressor (TetR). For better localization in the nucleus, a triple nuclear localization signal was used to link TetR to KRAB (tTS) [22, 23]. In the absence of Doxycycline, tTS binds to several tetO sequences of TRE and silences the downstream P_{CMV} promoter. As a result, the tTS alleviates the leaky expression under uninduced conditions. As the Dox is added, tTS dissociates from the target DNA while rtTA becomes active and triggers transcription. Recent work has demonstrated that tTS can also enhance the rtTA activity under induced conditions by protecting the rtTA from the ubiquitin-dependent proteosomal degradation pathway [24].

This enhanced rtTA activity, though, increases the risk of pleiotropic effects from squelching mediated by VP16 transcriptional activator. Squelching, or transcriptional interference, seems dependent on the competition between activating factors for a productive interaction with the limiting transcription factors [25]. In some cases, squelching can even be independent of the presence of a DNA binding domain (DBD) [26]. In gene inducible systems using VP16 as a transcriptional activator, squelching is thought to occur as a consequence of VP16 acting as a sink for cellular transcriptional machinery that is attracted and binds to it [27]. Furthermore, tTS not only represses basal transcription from tetO, but also efficiently down-regulates gene expression from other promoters [28].

Ecdysone-Inducible Systems

Tet-On systems typically encounter certain limitations, such as pleiotropic effects of the inducer, basal leakiness, toxicity of inducing agents and low levels of induced gene expression. A theoretical improvement in regulation of heterologous gene expression is represented by the ecdysone-inducible system. This system is based on insects' heterodimer ecdysone receptor (EcR) and their ultraspiracle protein (USP) [29]. EcR and USP mediate the morphological changes triggered by the hormone ecdysone during molting, metamorphic reproduction and other developmental processes in insects. To this date, EcR has been detected only in plants, insects and other invertebrates. Given the absence of EcR and its ligand in vertebrates, these components including the agonist RLS1 have been utilized to develop an inducible gene expression system for use in vertebrates. These inducible expression systems give negligible levels of basal expression in the absence of inducer, and ligands show no pleiotropic effects in mammalian cells [30].

The sensitivity of the initial version of the ecdysone-responsive expression technology has been increased by modifications to its two main components. The N-terminal transactivation domain has been replaced with the corresponding domain of the glucocorticoid receptor, and USP has been changed for its mammalian homologue, the retinoid X receptor (RxR). Subsequently, the system has subsisted several more modifications and the current version consists of a double hybrid format switch, in which the Gal4 DNA binding domain is fused to the ligand binding domain of *Choristneura fumiferana* EcR, and the VP16 activation domain to the EF domains of RxR (Figure1) [31].

Ecdysone-responsive expression systems show lower levels of basal expression in the absence of inducers when compared to the Tet-On system. Figure 2 compares the levels of basal expression of Tet-On and ecdysone-based (RheoGene) inducible systems in BHK-21 cells. Unfortunately although basal activity of the RheoGene is negligible, the ecdysone-responsive gene expression technologies have not given a solution to the problem of poor overall regulation performance.

To overcome some of the drawbacks associated with the inducible gene expression technologies, we propose the idea of developing a double selection tetracycline-ecdysone responsive regulatory system based on negative/positive selection that will isolate mammalian cells which displays high inducibility of the transgene along with low baseline activity.

Tetracycline-Ecdysone Responsive Regulatory System

Inducible gene expression technologies are invaluable tools for cell biologists, but their application is very demanding. In particular, the conventional screening strategies to isolate ideal, highly regulatable, and inducible clones are costly and time consuming.

An ideal inducible cell line or transgenic mouse should express the transgene only upon induction. The generation of such genetic systems would require the integration of the inducible transgene into a chromosomal region competent for transcription and devoid of genetic elements that could influence the activity of the inducible promoter. Selection of these sorts of tightly regulated and highly inducible clones from transfected cells would require a labor-intensive screening of hundreds of transformants. This problem could be resolved by leading the integration of the inducible transgene into specific chromosomal regions, but methods for site-specific integration into eukaryotic genomes are normally laborious and once again time consuming.

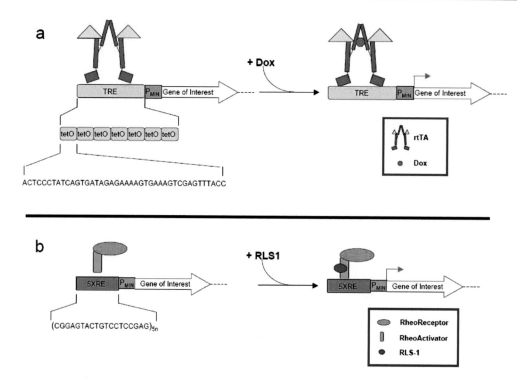

Figure 1. Schematic of gene regulation in Tet-On and ecdysone-based systems. a) In the Tet-On system, a reverse-controlled transactivator protein (rtTA) regulates the expression of a target gene that is under the control of a tetracycline-responsive promoter element (TRE), consisting of seven tandemly arranged Tet operator sequence fused to aminmal promoter (Pmin). In the absence of doxycycline (Dox) the transactivator in in a non-inducible conformation. Transcription of the tagret gene is stimulated by rtTA in the presence of Dox. b) Ecdysone-based RheoSwitch expression system requires a heterodimeric transactivator composed of two proteins, Rheoreceptor-1 and Rheoactivator. The gene of interest in under the control of a responsive target promoter, which is composed of five tandem repeats of Gal4 response element (5XRE) fused to a minimal promoter. In the absence of RLS1 ligand (a synthetic ecdysone antagonist), the transcription is repressed because the heterodimer binds to the 5XRE in a transcriptionally inactive onformation. In the presence of RLS1, the heterodimer binds to the 5XRE in an active conformation, and transcrition is stimulated.

The negative selection cassette contains the gene for Herpes simplex virus thymidine kinase (HSV-tk), which converts the nontoxic pro-drug gancyclovir (GCV) into a lethal compound that will eliminate cells expressing the gene [32]. The cytotoxic effect of the GCV results from its incorporation into DNA by a process starting with its phosphorylation by the viral enzyme HSV-tk to form the GCV-monophosphate [33]. Further phosphorylation by cellular kinases converts the GCV-monophosphate to GCV-triphosphate, which competitively inhibits incorporation of the endogenous GTP into DNA [34,36].

As a consequence, the GCV terminated strands can no longer be elongated, which leads to cell death. HSV-TK has one major disadvantage: the TK cassette prevents transmission of the male germ line to future generations [37]. The male sterility is caused by cryptic expression of a truncated form of tk in the testes [38]. The shorter transcript is initiated from a cryptic promoter that resides within the tk region. More recently, a carboxyl-terminal truncated version of HSV tk (Dtk) has been generated, which retains the tk function but does not cause male sterility [39].

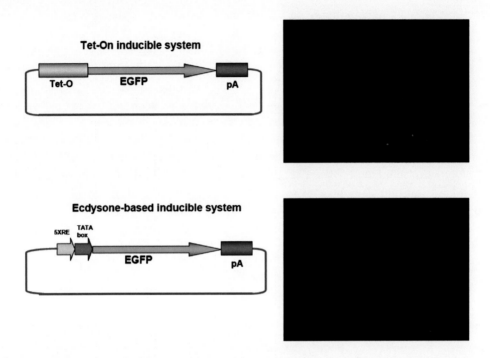

Figure 2. Leakage of Doxycicline and RSL1 responsive promoters in BHK-21 cells. BHK-21 cells were transfected with (a) pTetO-EGFP and (b) PG6-EGFP. EGFP expression was analyzed 16 hours after transfection using fluorescence microscopy.

Antibiotic Resistance Selectable Marker Allows Positive Selection Of Inducible Cells

The tetracycline-ecdysone responsive regulatory system is based on two autoregulated vectors: i) pEcdR-rtTA encoding for a synthetic ecdyson receptor and rtTA, and ii) pHSV-tk-biRheo-NFIRE-ARG which contains a bidirectional Rheogene operator flanked upstream by HSV-TK, and downstream by a multiple cloning site (MCS), followed by a NRF-IRES (NFIRE) element and an appropriate antibiotic resistance gene (Figure 3).

The pEcdR-rtTA vector encodes a synthetic receptor composed of two proteins, RheoReceptor-1 and RheoActivator, that dimerize to make a holoreceptor. In the pEcdR-rtTA vector, both proteins are expressed from a bidirectional promoter composed of a tetO$_7$ operator that is flanked by two divergently orientated minimal P$_{cmv}$ promoters. Synthesis coregulation of the RheoReceptor-1 and RheoActivator enables production of the two proteins in stoichiometric amounts, which is required. The gene to be expressed is cloned into the pHSV-tk-biRheo-NFIRE-ARG plasmid under control of five tandem repeats of the GAL4 response element (5XRE). pHSV-tk-biRheo-NFIRE-ARG contains a bidirectional 5XRE flanked upstream by HSV-TK, and downstream by a multiple cloning site (MCS) followed by an NRF-IRES element [40] that drives the expression of an antibiotic resistance gene (ARG).

Each element has an important role. HSV-TK and ARG are required respectively for negative and positive selection. The presence of the NFIRE element in the pHSV-tk-biRheo-NFIRE-ARG increases the translation efficiency of ARG and consequently increase the frequency of puromycin-resistant transformants. NFIRE actually corresponds to the 5' UTR of NRF (NF-kB repressing factor), a particularly long element (653 nucleotides) when compared to most vertebrate mRNAs (which have a 5'UTR of 20 to 100 nucleosides). NRF is thought to fold into a complex secondary structure with several stable hairpins, and is very ably translated in mammalian cells. When placed between two open reading frames in dicistronic reporter constructs, the 5' UTR of NRF acts as an internal ribosome entry site (IRES), which directs ribosomes to the downstream start codon by a cap-independent mechanism. The efficiency of translation initiation by NRF IRES is substantially higher than that of picornaviral IRESs. A comparison of the relative IRES activities in BHK-21 cells shows that the NRF IRES is 10-fold more active than the EMCV IRES (D'Aiuto et al., unpublished). Another advantage to using NRF IRES in bicistronic constructs is that the initiation codon of the downstream gene does not need to be placed precisely at the 3' end of the IRES.

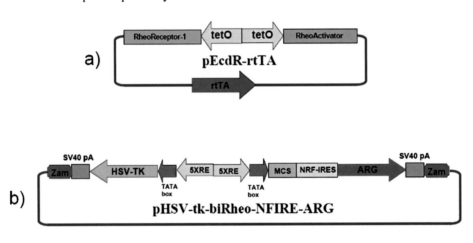

Figure 3. pEcdR and pHSV-tk-biRheo-NFIRE-puro plasmid maps. (a) pEcdR expresses a synthetic receptor composed of two proteins, RheoReceptor-1 and RheoActivator that dimerize to make a holoreceptor. Expression of both proteins is under the control of the tetracycline responsive promoter element (TRE) composed of a Tet operator sequence concatamers fused to a minimal CMV promoter. Transcription of RheoReceptor-1 and RheoActivator is stimulated by rtTA only in the presence of doxycicline (Dox). (b) pHSV-tk-biRheo-NFIRE-ARG has a bidirectional GAL4 response element (5X RE) flanked on one side by HSV-TK cDNA, and on the other side by a multiple cloning site followed by the NRF-IRES and puromycin resistance gene. To protect the expression of the transgene from position effect, two Zam insulator elements are placed downstream of the HSV-TK and puromycin resistance gene.

To select the highly inducible cells, the HSV-tk gene is introduced into pHSV-tk-biRheo-NFIRE-ARG to allow a suicide negative selection with Herpes simplex virus thymidine kinase and ganciclovir. After the negative selection, the choice of marker for positive selection largely depends on the cell line and the sequence of the gene of interest cloned upstream the IRES element. In fact, the protein expression from a bicistronic unit is normally quite unpredictable; the arrangement and composition of reading frames has a strong effect on the IRES-driven translation efficiency of the second cistron [41].

Negative/Positive Selection

After cotransfection with pEcdR-rtTA and pHSV-tk-biRheo-NFIRE-ARG carrying the gene of interest, the cells will be placed under negative selection using ganciclovir in uninduced conditions. The expression of HSV-TK requires the synthesis of the holoreceptor, which depends on the allosteric change of rtTA induced by the binding of Dox, RSL1, and the appropriate antibiotic. Cells expressing HSV-TK in the absence of Dox and RSL1 (leakage) will be selected against (negative selection) (Figure 4). After ganciclovir –mediated negative selection, positive selection is induced through replacement of the culture medium by a new medium containing Dox and RSL1 inducers. Under these conditions, rtTA is mobilized by Dox and triggers the transcription of the two holoreceptor subunits which are activated in the presence of RSL1 ligand. The activated holoreceptor and the VP16 activation domain bind 5XRE and starts the transcription of the bicistronic mRNA encoding both the gene of interest and the antibiotic resistance. Therefore, the presence of RSL1, Dox, and the appropriate antibiotics in the culture medium will allow isolation of inducible cell clones (positive selection) (Figure 5).

Use of the Zam Insulator to Prevent Leakiness or Silencing of the Transgene

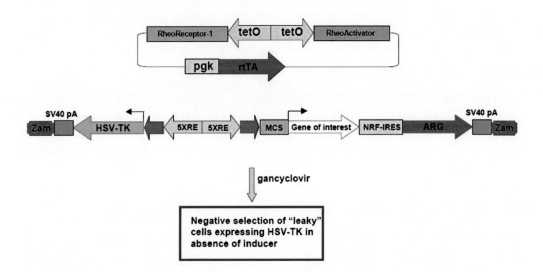

Figure 4. Positive selection. Transcription of the tetracycline responsive element (TRE)-regulated holoreceptor is stimulated by rtTA in the presence of Doxycicline (Dox). The holoreceptor is in an inactive conformation in the absence of the RSL1 ligand. In induced condition, the RSL1 binds and changes the conformation of the RheoReceptor protein, which stabilizes the holoreceptor on the 5XRE. The activated holoreceptor triggers the transcription of the bicistronic mRNAs while the ORFs (gene of interest and antibiotic resistance selectable marker) are separated by the NRF-IRES element. Therefore, after ganciclovir-mediated negative selection, addition of Dox, RSL1 and the appropriate antibiotic to the culture will allow the isolation of cells clones which display tightly regulated expression of the gene of interest.

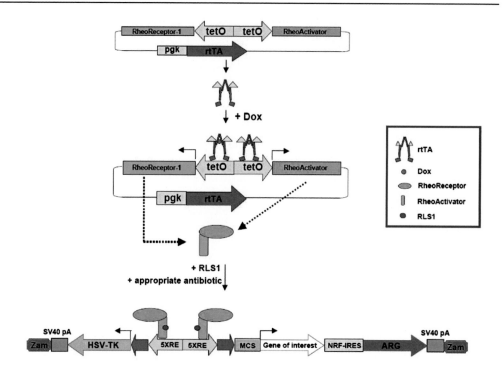

Figure 5. Ecdyson and Tet-on inducible expression systems. (a) The Ecdyson inducible system is composed of the VP16 activation domain fused to the EF domain of RxR and the Gal4 DBD fused to the binding domain of EcR. In the presence of Ecdyson, a conformational change is induced and transcription is activated. (b) The Tet-on system is based on the rtTA construct, composed of reverse TetR fused to the C-terminal domain of VP16, which in the presence of Doxycycline activates transcription through interactions with the TRE domain.

Crosstalk enhancer-promoters or chromatin-dependent repressive transgenic states can be overcome by flanking the inducible transgene with chromatin insulators [42]. These sequences are defined operationally by two characteristics: first, they interfere with enhancer-promoter interactions when located between them, and secondly they buffer transgenes against chromosomal position effects [43].

Several models have been proposed to explain how insulator sequences protect a transgene from chromosomal position effects and prevent extinction of its expression. All are based on the supposition that insulators play an active role in establishing chromatin domains. In these models, the insulator i) actively recruits histone acetylates and/or other proteins associated with active chromatin, maintaining an open chromatin configuration (histone acetylated/ DNA hypomethylated) within the domain; ii) captures or titrates proteins responsible for the closed chromatin configuration (heterochromatic proteins, DNA methyltransferases, histone deacetylases and methylases), avoiding gene silencing within the protected domain and thus preventing heterochromatin-mediated domain repression; and iii) locates the gene into an active subnuclear compartment [44]. We propose to use the ZAM insulator to reduce the probability of leakage or silencing of the transgene (Figure5). ZAM is a 8.4 kb retrotransposon determined to be closely related to gypsy both by comparative analysis of their pole genes and by the presence of a third ORF encoding for the ENVELOPE protein. These two characteristics make ZAM a member of the Ty3-gypsy RTE subset [45].

ZAM is mainly localized in the constitutive heterochromatin, even if there are some D. *melanogaster* strains with euchromatic copies (i.e. Charolles, Oregon-R) [46]. Its 5'-UTR carries a well-structured tandem repeat sequence and contains regulative sequences (enhancers) that are able to interact genetically with both transcription factors, such as ARAUCAN and COMBGAP, and chromatin protein such as HP1 [47]. The ZAM leader region contains a 699 bp region structured in 2.3 tandem repeats of 307 bp each that shows the "insulating" property that affects enhancer-promoter interaction. ZAM insulator activity was compared to the Su(Hw) binding sequence located in 5'UTR of the retrotransposon gypsy and was shown to have stronger activity.

Furthermore, the ZAM insulator does not show any silencer activity, indicating that it does not interfere with the correct regulation of the transgene (Viggiano et al. unpublished).

Isolation of Highly Inducible Embryonic Stem Cells to Generate Transgenic Mice

Despite several disadvantages, the Tet-On system has been successfully established in transgenic mice, with the responsive transgene controlled by addition of Dox to the drinking water or mouse chow. To utilize the Tet-On system in vivo, it is necessary to generate two sets of trangenic animals. One mouse line harbors rtTA under the control of a tissue-specific promoter. The other line carries the transgene of interest, whose expression is driven by the TRE element. Mating the two strains of mice allows spatiotemporal control of transgene expression or of the desired gene alteration [48]. However, leaky expression and inability to protect the transgene of interest from position effects greatly limits the use of Tet-based systems to generate inducible transgenic mice. These two problems are found at a frequency of around 1-5% in transgenic founders. This relatively high frequency necessitates the generation of several founders to identify those that express the transgene uniquely in an inducible manner. Consequently, a large number of oocyte microinjections are required, making this procedure very costly and time consuming. Creating inducible transgenic mice at a higher frequency could be achieved by using mouse embryonic stem (ES) cells. ES cells isolated directly from mouse embryos can be cultured for long periods of time in vitro and still resume normal embryonic development if returned to a carrier embryo. In the resulting chimeric mice, the ES cells contribute to the germ line in addition to the somatic tissues, which makes them very potent vectors in the generation of transgenic animals. They are amenable to a variety of genetic manipulations in tissue culture via electroporation, lipofection or infection with retroviral vectors. Genetically altered cells can then be transferred into the germ line of chimeric mice.

The use of the tetracycline-ecdysone responsive regulatory system has the potential to be a powerful tool to generate highly inducible transgenic mice, with a tight control of transgene expression. Mouse ES cells can first be transfected with pEcdR-rtTA and the gene of interest carried in pHSV-tk-biRheo-NFIRE-ARG vectors and then subjected to negative selection with ganciclovir. Once leaky expression is eliminated, positive selection with Dox, RLS1, and an antibiotic resistance selectable marker can be used to identify inducible transformants in which the transgene is expressed only in the induced state. Inducible ES cells showing no change of the induction/suppression phenotypes after several passages can be injected into

C57 BL/6 blastocysts, which will be implanted into foster mothers to derive somatic chimeras. Chimeric males can then be mated with females, and analysis of coat color of the offspring will indicate whether the germline had been transmitted.

Conclusions

Over the last few years, the potential of drug-regulated gene expression technologies in basic and applied biological research areas has been widely demonstrated. Although the conceptual simplicity of these systems suggests elegant solutions to gene therapy, functional genomics, and biopharmaceutical manufacturing, they are by no means easily achieved. The main bottleneck of these technologies is the lack of a high throughput screening strategy to isolate tightly regulatable and highly inducible cell clones. The traditional strategy to identify ideal inducible clones can cost months of tedious and laborious work. The tetracycline-ecdysone responsive regulatory system based on negative-positive selection would allow the isolation of nearly ideal candidate clones in just a few weeks.

Acknowledgments

We thank Marco Marzulli for critical reading of the manuscript.

References

[1]   Gil, J; Yamamoto, H; Zapata, JM; Reed, JC; Perucho, M. (1999). Related Impairment of the proapoptotic activity of Bax by missense mutations found in gastrointestinal cancers. *Cancer Res.*, **59**, 2034-7.

[2]   Wang Y, Blandino, G; Oren, M; Givol, D. (1998). Related Induced p53 expression in lung cancer cell line promotes cell senescence and differentially modifies the cytotoxicity of anti-cancer drugs. *Oncogene*, **17**, 1923-30.

[3]   Ko, MS; Takahashi, N; Sugiyama, N; Takano, T. (1989). An auto-inducible vector conferring high glucocorticoid inducibility upon stable transformant cells. *Gene*, **84**, 383-9.

[4]   Holmgren, R; Livak, K; Morimoto, R; Freund, R; Meselson, M. (1979). Studies of cloned sequences from four Drosophila heat shock loci. *Cell*, **18**, 1359-70.]

[5]   Rinsch, C; Regulier, E; Deglon, N; Dalle, B; Beuzard, YA; Aebischer, P. (1997). A gene therapy approach to regulated delivery of erythropoietin as a function of oxygen tension. *Hum Gene Ther.*, **8**, 1840-1.

[6]   Ryals, J; Dierks, P; Ragg, H; Weissmann, C. (1985). A 46-nucleotide promoter segment from an IFN-alpha gene renders an unrelated promoter inducible by virus. *Cell*, **41**, 497-507.

[7]   Mayo, KE; Warren, R; Palmiter, RD. (1982). The mouse metallothionein-I gene is transcriptionally regulated by cadmium following transfection into human or mouse cells. *Cell*, **29**, 701-10.

[8]   Hovey, A; Bebbington, CR; Jenkins, N. (1994) Control of growth and recombinant protein synthesis by heat-shock in a mutant mammalian cell line. *Biotechnol. Lett.*, **16**, 215-220.

[9]   Webster, NJ; Green, S; Jin, JR; Chambon, P. (1988). The hormone-binding domains of the estrogen and glucocorticoid receptors contain an inducible transcription activation function. *Cell*, **54**:199-207.

[10] Baulieu, EE. (1989). Contragestion and other clinical applications of RU 486, an antiprogesterone at the receptor. *Science*, **245**, 1351-7.

[11] Gossen, M; Bujard, H. (1992). Tight control of gene expression in mammalian cells by tetracycline-responsive promoters. *Proc. Natl. Acad .Sci. U S A*, **89**, 5547-51.

[12] Spencer, DM; Wandless, TJ; Schreiber, SL; Crabtree, GR. (1993). Controlling signal transduction with synthetic ligands. *Science*, **262**, 1019-24.

[13] Folcher, M; Morris, RP; Dale, G; Salah-Bey-Hocini, K; Viollier, PH; Thompson, CJ. (2001). A transcriptional regulator of a pristinamycin resistance gene in Streptomyces coelicolor. *J. Biol. Chem.*, **276**, 1479-85.

[14] Christopherson, KS; Mark, MR; Bajaj, V; Godowski, PJ. (1992). Ecdysteroid-dependent regulation of genes in mammalian cells by a Drosophila ecdysone receptor and chimeric transactivators. *Proc. Natl. Acad. Sci. U S A*, **89**, 6314-8.

[15] Rubanyi, GM. (2001). The future of human gene therapy. *Mol Aspects Med.*, 22, 113-42.

[16] Hillen, W; Berens, C. (1994). Mechanisms underlying expression of Tn10 encoded tetracycline resistance. *Annu. Rev. Microbiol.*, **48**, 345-69.

[17] Sadowski, I; Ma, J; Triezenberg, S; Ptashne, M. (1988). GAL4-VP16 is an unusually potent transcriptional activator. *Nature*, **335**, 563-4.

[18] Orth, P; Schnappinger, D; Hillen, W; Saenger, W; Hinrichs, W. (2000). Structural basis of gene regulation by the tetracycline inducible Tet repressor-operator system. *Nat. Struct. Biol.*, **7**-215-9.

[19] Urlinger, S; Baron, U; Thellmann, M; Hasan, MT; Bujard, H; Hillen W. (2000). Exploring the sequence space for tetracycline-dependent transcriptional activators: novel mutations yield expanded range and sensitivity. *Proc. Natl. Acad. Sci. U S A*, **97**, 7963-8.

[20] Mizuguchi H, Hayakawa T. (2002). The tet-off system is more effective than the tet-on system for regulating transgene expression in a single adenovirus vector. *J .Gene Med.* **4**:240-7.

[21] Cohlan, S. Q. (1997). Tetracycline staining of teeth. *Teratology*, **15**, 127-130.

[22] A. Mohammadi, S.; Alvarez-Vallina, L.; Ashworth, L. J.; Hawkins, R. E. (1997). Delay in resumption of the activity of tetracycline regulatable promoter following removal of tetracycline analogues. *GeneTher.*, **4**, 993-997.

[23] Bohl D., Heard J. M. (1997). Modulation of erythropoietin delivery from engineered muscles in mice. Hum. *GeneTher.*, **8**, 195-204.

[24] Razin, SV; Farrell, CM; Recillas-Targa, F. (2003). Genomic domains and regulatory elements operating at the domain level. *Int. Rev. Cytol.*, **226**, 63-125.

[25] Ellis, J. (2005). Silencing and variegation of gammaretrovirus and lentivirus vectors. *Hum. Gene Ther.*, **16**, 1241-6.

[26] Deuschle, U; Meyer, WK; Thiesen, HJ. (1995). Tetracycline-reversible silencing of eukaryotic promoters. *Mol. Cell Biol.*, **15**, 1907-14.

[27] Zhu, Z; Ma, B; Homer, RJ; Zheng, T; Elias, JA. (2001). Use of the tetracycline-controlled transcriptional silencer (tTS) to eliminate transgene leak in inducible overexpression transgenic mice. *J. Biol. Chem.*, **276**, 25222-9.

[28] Lai, JF; Cheng, HY; Cheng, TL; Lin, YY; Chen, LC; Lin, MT; Jou, TS. (2004). Doxycycline- and tetracycline-regulated transcriptional silencer enhance the expression level and transactivating performance of rtTA. *J. Gene Med.*, **6**:1403-13.

[29] Lewin, B. (1990). Commitment and activation at PolII promoters. *Cell*, 61, 1161-1164.

[30] Gill, G; Ptashne, M. (1988). Negative effect of the transcriptional activator GAL4. *Nature*, **334**, 721-724.

[31] Triezenberg, SJ; Kingsbury, RC; McKnight, SL. (1988). Functional dissection of VP16, the trans-activator of herpes simplex virus immediate early gene expression. *Genes Dev.*, **2**, 718-29.

[32] Ryu, JR; Olson, LK; Arnosti DN. (2001). Cell-type specificity of short-range transcriptional repressors. *Proc. Natl. Acad. Sci. U S A.*, **98**, 12960-5.

[33] Mangelsdorf, DJ; Evans, RM. (1995). The RXR heterodimers and orphan receptors. *Cell*, **83**, 841-50

[34] No, D; Yao, TP; Evans, RM. (1996). Ecdysone-inducible gene expression in mammalian cells and transgenic mice. *Proc. Natl. Acad. Sci. U S A..* **93**, 3346-51.

[35] Palli, SR; Kapitskaya, MZ; Kumar, MB; Cress, DE. (2003) Improved ecdysone receptor-based inducible gene regulation system. *Eur. J. Biochem.*, **270**, 1308-15.

[36] Shewach, DS; Zerbe, LK., Hughes, TL; Roessler, BJ; Breakefield, XO; Davidson, BL. (1994). Enhanced cytotoxicity of antiviral drugs mediated by adenovirus directed transfer of the herpes simplex virus thymidine kinase gene in rat glioma cells. *Cancer Gene Ther.*, **1**, 107–112.

[37] Keller, PM; Fyfe, JA; Beauchamp, L; Spector, T. (19810. Enzymatic phosphorylation of acyclic nucleoside analogs and correlations with antiherpetic activities. *Biochem. Pharmacol.*, **30**: 3071–3077.

[38] Frank, K B; Chiou, JF; Cheng, YC. (1984). Interaction of herpes simplex virus- induced DNA polymerase with 9-(1,3-dihydroxy-2-propoxymethyl)guanine triphosphate. *J. Biol. Chem.*, **259**, 1566 –1569.

[39] Mar, EC; Chiou, JF; Cheng, YC; Huang, ES. Inhibition of cellular DNA polymerase and human cytomegalovirus-induced DNA polymerase by the triphosphates of 9-(2-hydroxyethoxymethyl)guanineand9-(1,3-dihydroxy-2-propoxymeth-yl)guanine. (1985). *J. Virol.*, **53**(3): 776–780.

[40] Matthews, T; Boehme, R. (1997). Antiviral activity and mechanism of action of ganciclovir. *Rev. Infect. Dis.*, **S490 8,** 1825–1835.

[41] Al-Shawi, R; Burke, J; Jones, CT; Simons, JP; Bishop, JO. (1988). Mup promoter-thymidine kinase reporter gene shows relaxed tissue-specific expression and confers male sterility upon transgenic mice. *Mol. Cell. Biol.*, **8**, 4821-8.

[42] Al-Shawi, R; Burke, J; Wallace, H; Jones, C; Harrison, S; Buxton, D; Maley, S; Chandley, A; Bishop, JO. (1991). The herpes simplex virus type 1 thymidine kinase is expressed in the testes of transgenic mice under the control of a cryptic promoter. *Mol. Cell Biol..* **11**, 4207-16.

[43] Cohen, JL; Boyer, O; Salomon, B; Onclerco, R; Depetris, D; Lejeune, L; Dubus-Bonnet, V; Bruel, S; Charlotte, F; Mattei, MG; Klatzmann, D. (1998). Fertile homozygous

transgenic mice expressing a functional truncated herpes simplex thymidine kinase delta TK gene. *Transgenic Res.*, **7**, 321-30.

[44] Oumard, A; Hennecke, M; Hauser, H; Nourbakhsh. M. (2000). Translation of NRF mRNA is mediated by highly efficient internal ribosome entry. *Mol. Cell Biol.* **20**, 2755-9.

[45] Hennecke, M; Kwissa, M; Metzger, K; Oumard, A; Kroger, A; Schirmbeck, A; Reimann, J; Hauser, H. (2001) Composition and arrangement of genes define the strength of IRES-driven translation in bicistronic mRNAs. *Nucleic Acids Res.*, **29**, 3327-3334.

[46] Roy, S; Tan, YY; Hart, CM. (2007). A genetic screen supports a broad role for the Drosophila insulator proteins BEAF-32A and BEAF-32B in maintaining patterns of gene expression. *Mol. Genet. Genomics.* **277**, 273-86.

[47] Gdula, DA; Gerasimova, TI; Corces, VG. (1996). Genetic and molecular analysis of the gypsy chromatin insulator of Drosophila. *Proc. Natl. Acad. Sci. U S A.*, **93**, 9378-83.

[48] Recillas-Targa, F; Valadez-Graham, V; Farrell, CM. (2004) Prospects and implications of using chromatin insulators in gene therapy and transgenesis. *Bioessays*, **26**, 796-807.

[49] Leblanc, P; Desset, S; Dastugue, B; Vaury, C. (1997). Invertebrate retroviruses: ZAM a new candidate in D.melanogaster. *EMBO J.*, **16**, 7521-31.

[50] Baldrich, E; Dimitri, P; Desset, S; Leblanc, P; Codipietro, D; Vaury, C. (1997). Genomic distribution of the retrovirus-like element ZAM in Drosophila. *Genetica*, **100**, 131-40.

[51] Minervini, CF; Marsano, RM; Casieri, P; Fanti, L; Caizzi, R; Pimpinelli, S; Rocchi, M; Viggiano, L. (2007). Heterochromatin protein 1 interacts with 5'UTR of transposable element ZAM in a sequence-specific fashion. *Gene*, **393**:1-10.

[52] Jaisser, F. (2000). Inducible gene expression and gene modification in transgenic mice. *J. Am. Soc. Nephrol.*, **11**, S95-S100

In: Genetic Recombination Research Progress
Editor: Jacob H. Schulz, pp. 85–138

ISBN 978-1-60456-482-2
© 2008 Nova Science Publishers, Inc.

Chapter 3

A Somewhat Subjective Introduction to Gene Mapping Through Nonparametric Linkage Analysis

Lars Ängquist[*]

Department of Mathematical Statistics
Centre for Mathematical Sciences
Lund University, Sweden

Abstract

Genetic recombination, originating from the occurrences of crossovers during meiosis, may be seen as the foundation of genetic variation, i.e. underlying both population diversities and general evolution in itself. Moreover, it is the driving force behind the original formation of genetic material forming the genetic components corresponding to, human or other, diseases. The recombination process generally imply a low level of recombinations, given a fixed number of meiotic generations, with respect to closely located loci; being the basic mechanism facilitating gene mapping studies through using collected data in the form of genetic information on multigenerational families (pedigrees). A major role behind such mapping is played by a relevant set of statistical techniques, methods and algorithms; developed using tools and theory from, for instance, mathematics, computer science and genetics.

In linkage analysis or, in a wider sense, gene mapping one searches for disease loci along a genome. This is done by observing so called marker genotypes and phenotypes of a pedigree set, i.e. a set of pedigrees, in order to locate the loci corresponding to the underlying disease genes or, at least, to narrow down the interesting genome regions. In this context the key concept is the genetic inheritance of alleles and its correlation

---

[*]E-mail address: Lars.Angquist@matstat.lu.se or Lars.Angquist@telia.com

to phenotype outcomes; which is informative according to genetic recombination. A significant deviation from what is expected under random inheritance is taken as statistical evidence of existing genetic components suggested to be located at, or close to, the loci giving significant results.

This chapter mainly serves as a (subjective) review on gene mapping in general and nonparametric linkage analysis in particular, but also includes some material on recent or new research. We will follow a path vaguely outlined as: Firstly, we begin by outlining the needed genetical foundation of statistical genetics as well as some basic concepts, for instance, noting on the process of allelic inheritance, the genetic model, the pedigree set, the inheritance vector and various types of genetic information. Secondly, we give an introduction to one-locus nonparametric linkage (NPL) analysis focusing on significance calculations of NPL scores. Corresponding approaches are based either on analytical approximations or Monte Carlo simulations. Thirdly, we make some comments on the generalizations to two-locus procedures; both on the unconditional (simultaneous) and conditional (sequential) search for two disease loci. Along the way we also at some length discuss, for instance, parametric linkage analysis, fuzzy significance, multiple testing and evidential interpretation of significant findings. Fourthly, we very briefly discuss some competing and complementary subfields within the context of gene mapping-based statistical genetics.

1.   Human Genetics

In this introductory section we will present some background notation and information, hopefully laying the foundation for subsequent understanding, possibly increasing the utility, of the material to come. More thorough and detailed treatments of, all or some, included topics are e.g. given in the textbooks Sham (1998), Ott (1999), Almgren et al. (2003), Strachan and Read (2003), Thomas (2004), Haines and Pericak-Vance (2006), Ziegler and Koenig (2006) and Siegmund and Yakir (2007).

1.1.   Basic Notation and Key Concepts

The human *genome*, i.e. operating manual, consists of 23 pair of *chromosomes*. In total 22 pairs are so called *autosomes* which are structurally equal with respect to the sexes, whereas one single pair constitutes the *sex-chromosomes* which are content-wise sex-dependent. Throughout this work we will only consider autosomes. The main chemical structure of the chromosomes is the *deoxyribonucleic acid (DNA)*, which is based on units called *nucleotides* that each consists of a sugar (deoxyribonucleic), a phosphate and a nitrogeneous base. There are 4 possible bases available: adenine (A), cytosine (C), guanine (G) and thymine (T).

The physical structure of DNA is actually double-stranded, in the form of a double-helix, but since the two strands are strictly complementary[1] one might look at the chromosomal DNA as a single sequence of nucleotides represented by the underlying bases. An alternative, but essentially similar, view is to consider the whole (or parts of the) sequence as a word written using the $\{A, C, G, T\}$-alphabet.

---

[1]The genetic codes with respect to the two strands are *structurally equivalent* since the only admissible bonds between these strands are $A - T$ and $C - G$.

The genetic code is organized into nucleotide-triplets, *codons*, which code for specific *amino acids*. Obviously we may define $4^3 = 64$ different codons, but the code is degenerate in the sense that only 20 different amino acids may be produced. This follows since several codons may code for the same amino acid and some codons serve as code punctuation.[2]

In total the human genome consists of approximately 3×10^9 *base pairs (bp)*.[3] A *gene* is a sequence of DNA, i.e. a unique amino acid sequence, at a specific genome position which specifies the function and structure of a subunit in a *protein*. Some genome regions are in this sense informative coding regions (*exons*) and the space between them are noncoding regions (*introns*). Recent investigations specify the number of distinct genes to be about 20000-25000 (International Human Genome Sequencing Consortium, 2001, 2004). The actual nucleotide-length of genes shows great variation.

A well-defined position[4] on the genome is called a *locus* (pl. *loci*). Loci located on the same chromosome are *syntenic*. The opposite (complementary) term is *nonsyntenic*. At each locus a human-being hosts two *alleles*.[5] These alleles constitute the individual's *genotype*. Different nucleotide sequences at the same locus give rise to different *allelic variants* (polymorphisms). For a specific (homogeneous) population, the genotype probabilities associated with randomly picking an individual and finding these possible variants are referred to as the *gene frequencies* at the corresponding locus; noting that the same reasoning with respect to a single allele is giving the *allele frequencies*. Further, if a genotype consists of two copies of the same allelic variant it is called *homozygous*, otherwise it is *heterozygous*.

Formally, at a locus we may assume that the genetic variation is summarized by *a* different allelic variants $A_1, A_2, \ldots, A_a$. For an allele of a randomly chosen individual, these variants occur with gene frequencies (probabilities) $p_1, p_2, \ldots, p_a$. A common criterion of a *polymorphic locus* is that $a \geq 2$ and $p_i \leq 0.95$ ($\forall i$), but sometimes a slightly less restrictive constraint on the probabilities is used (e.g. $p_i \leq 0.99$). The number of distinct genotypes is $a(a+1)/2$.[6]

A general probabilistic assumption on the formation of distinct genotypes is the *Hardy-Weinberg equilibrium (HWE)*:

Definition 1. *HWE means that the gene frequencies are directly proportional to the two corresponding allele frequencies, i.e. $P(A_iA_i) = p_i^2$ and $P(A_iA_j) = 2p_ip_j$ ($i \neq j$).*

Considering a randomly chosen individual, this reflects a completely random formation of a genotype given the set of allele frequencies.[7]

---

[2]*Start* and *stop* codons, which tell the code interpreter to start or stop reading, i.e. to begin or end a coding region.

[3]One base-pair may be seen as one single position in the DNA sequence or, equivalently understood, as one letter in the genomewide genetic word, i.e. as one instance of either A, C, G or T.

[4]Should be understood as a small chromosomal segment.

[5]This makes the human organism a *diploid* species.

[6]Distinct here means that the unordered genotypes are different, i.e. the genotypes A_iA_j and A_jA_i are not distinct.

[7]One may view this as the genotype's second allele is not being dependent on the outcome of the first allele.

1.2. The Inheritance of Alleles

Of the 46 human chromosomes 23 are inherited from the father[8] (*paternal* chromosomes) and 23 are inherited from the mother (*maternal* chromosomes). This implies that for each individual, at each locus, the genotype consists of one paternal and one maternal allele. A simple inheritance example is given in Figure 1. Knowledge of the parental origins of an individual's alleles implies that one may form an *ordered genotype* where the so called *phase* is known. More generally, knowledge of an individual's ordered genotypes at several syntenic loci implies that one may infer the corresponding *haplotypes*; see Figure 2.

Example 1 (Inheritance at single locus in small pedigree). *Consider Figure 1. The unaffected female offspring to the left has inherited her paternal allele, B, from the pedigree (top-level) father and her maternal allele, C, from the pedigree mother.*

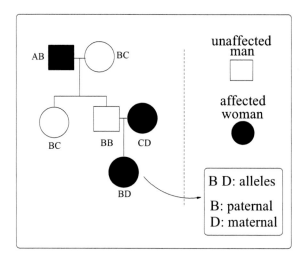

Figure 1. An example of the inheritance of alleles for a single small pedigree.

Example 2 (Inheritance at two loci in small pedigree). *Consider Figure 2. The unaffected female offspring to the left has inherited her paternal alleles, 2 and 3, from the pedigree father at the 1st and 2nd loci respectively; and her maternal alleles, 3 and 2, are similarly inherited from the pedigree mother.*

Each chromosome consists of mixed segments from the two corresponding grandparents. In other words there is a *blockwise chromosomal inheritance* of alleles interchangingly from both the grandparental chromosomes. The positions where one block ends and a new one starts are called *crossovers*. This behaviour is explained by the biological process of the formation of *gametes*, i.e. sperm and ovum cells, which is called *meiosis*.

During meiosis all the chromosomes are duplicated and then the homologous chromosomes pair up, i.e. we have initially an arrangement of four (2×2) DNA strands known as *chromatids*. Now, some physical contact between nonidentical chromatids may occur.

---

[8]One from each pair of *homologous* chromosomes.

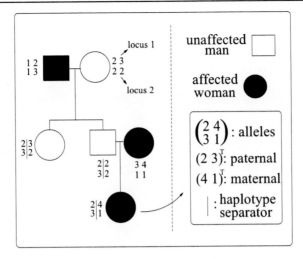

Figure 2. An example of the inheritance of haplotypes for a single small pedigree. We assume that the founder (top-level; see below) genotypes are phase-unknown. This is typically the case unless information from previous generations is available; and in such cases one could preferably extend the pedigree to include corresponding generations.

These positions are known as *chiasmata* and correspond to crossovers. Moreover, there is at least one chiasmata per chromosome pair. Finally, one of these genetically mixed chromatids is chosen, so to speak, for reproduction. This complex process is schematically shown in Figure 3.

Inheritance at different chromosomes is considered to be independent. For each chromosome, occurrence of a crossover at a locus usually lowers the probability of having a second crossover in close vicinity of this locus. This phenomenon is referred to as positive *chiasma interference*. Moreover, the human genome hosts both so called *hot spots* and *cold spots*, which are chromosomal regions with high and low intensity of crossovers respectively. Often it is a good approximation to ignore chiasma interference and assume that crossovers occur randomly according to a Poisson process.[9] If variation of spot temperature is ignored as well, the Poisson process has constant intensity.

Equipped with the concept of crossovers one may introduce a new measure of distance. The *genetic distance g* is measured in units of Morgans,[10] based on the concept of expected number of crossovers between loci. Formally, given two loci, l_1 and l_2, located 1 Morgan from each other, there is an expected number of 1 crossover for each meiosis, with respect to the actually inherited chromatid, between these loci. Since crossovers occurs with higher intensity for females than for males, this distance measure is really sex-dependent, but often one uses sex-averaged numbers. Adopting the latter approach gives us a total genetic length of 35.75 Morgans of the human autosomes (Collins et al., 1996). On average $1 cM = 0.01 M$ corresponds to a physical distance of $10^6 bp$ although the intensity of crossovers varies along

---

[9]Continuously oberverable phenomena that occurs one at a time, in space or time, with a fixed (or time- or space dependent) intensity may be modeled using *Poisson processes*. An introduction, presented at an intermediate level, is given in Gut (1995).

[10]The alternatively used *physical distance* is simply measured in base-pairs.

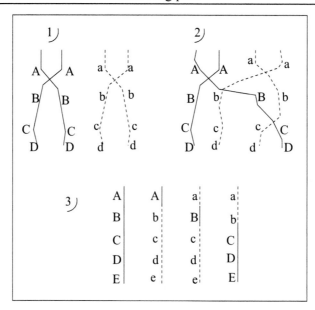

Figure 3. A simple overview of a single meiosis: 1) The replicated homologous chromosomes pair up. 2) Some physical contact at chiasmatas occur between the two pairs of chromatids. 3) Four mixed (recombinated) strands of DNA ready for reproduction.

a chromosome.

Alternatively one may define genomic distance using the concept of *recombination*. If the alleles transmitted by a parent at two loci, l_1 and l_2, are inherited from the same grandparent they are said to be *nonrecombinants*. Otherwise, they are referred to as *recombinants*. If a recombination has occurred between l_1 and l_2 we know, by definition, that there has been an odd number of crossovers between them. The probability of recombination between two arbitrary loci l_1 and l_2 is called the *recombination fraction* and is denoted by $\theta = \theta(l_1, l_2)$. Obviously, according to the blockwise inheritance, this parameter is an increasing function of the genetic distance between the loci.

Normally there is introduced a one-to-one function between the recombination fraction and the genetic distance. The link between these concepts is called the *map function*. Several suggested map functions exist in the literature, each choice corresponding to a specific way of modeling crossovers.

The most common one, which we exclusively use in this thesis, is the *Haldane map function* (Haldane, 1919), defined by,

$$g = -\frac{1}{2}\ln(1 - 2\theta), \qquad (1)$$

where g is the genetic distance. This function corresponds to lack of interference and, as pointed out above, for each chromosome yields a Poisson(g) distributed number of crossovers between l_1 and l_2. From this follows that the distance between crossovers with respect to a single meiosis is exponentially distributed.[11] Other well-known map functions

---

[11] With intensity and mean value 1 if measured in Morgans. If using centiMorgans these numbers will be

are defined by Morgan (1928) and Kosambi (1944). The *Morgan function*,

$$g = \theta,$$

is valid if excluding multiple crossovers (complete interference) and may therefore be used as an approximation over short distances. Somewhat more involved is the *Kosambi function*,

$$g = \frac{1}{4} \ln \left[\frac{1 + 2\theta}{1 - 2\theta} \right],$$

which models interference as being large at small distances, then decreasing with distance.

To prove (1), we notice that an odd number of crossovers c is needed between two loci for them to be recombinant. If the crossovers are purely random (no interference) they follow a Poisson process with expected value $g = E(c)$. This gives,

$$\theta = \sum_{i=1}^{\infty} P(c = 2i - 1 | g) = \sum_{i=1}^{\infty} \exp(-g) g^{2i-1} / (2i - 1)!$$

$$= \left[\frac{1 - \exp(-2g)}{2} \right] \quad \Rightarrow g = -\frac{1}{2} \ln(1 - 2\theta),$$

where we used that $\sum_{i=1}^{\infty} g^{2i-1} / (2i-1)! = \frac{1}{2} [\exp(g) - \exp(-g)]$, which is based on standard Taylor expansions of the exponential function $\exp(x)$.

Looking at (1), and most of its existing alternatives, one may note that $0 \leq \theta \leq 0.5$ is required.[12] This is no coincidence since generally two loci, l_1 and l_2, are considered to be *unlinked*, i.e. the inheritance at these loci are independent, if $\theta = 0.5$. This is interpreted as the distance between these loci being infinitely large, $g(l_1, l_2) = \infty$, or that there is an infinite expected number of crossover between them, meaning that they are located on different chromosomes. The other extreme case is $\theta = 0$ which corresponds to $l_1 = l_2$, i.e. inheritance at these loci is completely dependent or *linked*. Intuitively this seems surprising, but it makes sense when we compare a hypothetical disease locus with a so called marker locus in close vicinity.

1.3. Mendel, Markers and General Information

In one sense modern research in genetics started with the Austrian monk Johann Gregor Mendel's (1822-1884) publication on the inheritance in the pea plant (Mendel, 1866). He observed the behaviour of *random inheritance (RI)* and *independent assortment (IA)*, which today are known as the 1st and 2nd *Mendelian laws on inheritance*. Next, we will formally define these concepts and also introduce what is known as the assumption of *random mating (RM)*.

Definition 2. *RI means that the parental alleles are transmitted with equal probabilities to the offspring and this is done in an independent way when considering multiple offspring.*

0.01 and 100 respectively.

[12]In some cases with interference present, one may find $\theta > 0.5$. Generally this theoretical possibility is not considered to be of practical importance in the context of linkage analysis (Ott, 1999).

Definition 3. *IA means that normally the inheritance at distinct loci are performed in an independent way. Deviances from this rule is referred to as genetic (or physical) linkage.*

Definition 4. *RM means that the mating of parents is not dependent on genetic factors, i.e. the probability that a mating couple has genotypes G_P (paternal) and G_M (maternal) is $P(G_P)P(G_M)$.*

Consider $|l|$ different loci. If we denote the number of allelic variants at the ith locus with a_i, then we may form $h = a_1 a_2 \cdots a_{|l|}$ distinct haplotypes. This gives a total number of $h(h+1)/2$ different multi-loci genotypes and leads to the definition of so called *allelic association (AA)* or *linkage disequilibrium (LD)*.

Definition 5. *AA (or LD) means that the probability for at least one haplotype $A_{i_1} A_{i_2} \cdots A_{i_{|l|}}$ of a randomly chosen individual satisfies,*

$$P\left[A_{i_1}^1 A_{i_2}^2 \cdots A_{i_{|l|}}^{|l|} \right] \neq p_{i_1}^1 p_{i_2}^2 \cdots p_{i_{|l|}}^{|l|}.$$

Here $i_j \in \{1, 2, \ldots, a_j\}$ is an index number and $A_{i_j}^j$ the corresponding allelic variant at the jth locus which occurs with probability $p_{i_j}^j$.

Assume we want to perform a *genome scan* with respect to a genome region Ω. In order to do so we have to define *genetic markers* throughout Ω, facilitating the investigation of inheritance at these positions. A marker is a locus of known chromosomal position, where it is possible to measure allelic outcomes and where the population shows allelic variation, i.e. each marker locus is polymorphic.

In order to perform *linkage analysis* one needs to define polymorphic markers throughout Ω, estimate allele frequencies corresponding to all possible allelic variants at included markers and use this set of markers to produce a *marker map*.[13] This involves: (i) Ordering the markers with respect to chromosomal position. (ii) Specifying the distances between each consecutive pair of markers. If this is done using genetic distances one has produced a genetic marker map, whereas a physical marker map measures distances in base-pairs.[14]

The degree of polymorphism of a marker at locus $x \in \Omega$, with a allelic variants and corresponding allele frequencies $p_1, p_2, \ldots, p_a$, may be defined in different ways: (i) Through the *heterozygosity (H)* value,

$$H(x) = 1 - \sum_{i=1}^{a} p_i^2. \tag{2}$$

Note that this is the probability for an arbitrary individual of being heterozygote at locus x. (ii) Through the *polymorphism information content (PIC)* value,

$$PIC(x) = 1 - \sum_{i=1}^{a} p_i^2 - \sum_{i=1}^{a-1} \sum_{j=i+1}^{a} 2 p_i^2 p_j^2. \tag{3}$$

---

[13] Strictly speaking, a marker map is only needed when performing multipoint analyses; see below.

[14] Several different techniques for measuring or constructing genotyping markers exist. This includes, for example, *restriction fragment lengths polymorphisms*, *simple sequence repeats* (commonly called microsatellites) and *single nucleotide polymorphisms*. For instance, the latter is abbreviated as SNP and corresponds to allelic variants, at the marker locus, based on information (variation) regarding a single specific nucleotide (base-pair). See Strachan and Read (2003), Haines and Pericak-Vance (2006) and Siegmund and Yakir (2007).

Here the last sum is the probability that a child's genotype is heterozygous with unknown phase.[15] Both (2) and (3) quantify marker informativity on the population level.

1.4. The Genetic Model

Usually in linkage analysis one investigates inheritance of alleles with respect to a given disease. More generally one may divide the set of individuals in the study with respect to their *phenotypes*. This is a non-genetically observable quantity that may be qualitative or quantitative. Throughout this text we will only consider the binary qualitative phenotype of *affection status*. Typical examples of a quantitative phenotypes are body-mass-index (BMI), body weight and cholestrol level.

For an underlying disease to be genetically inheritable, i.e. to include a *genetic component*, some kind of correlation between the phenotype and the disease genotypes must exist. This is usually described by means of a *genetic model* λ. One may note that λ generally, at least to some extent, is unknown so, if needed, it is estimated prior to analysis using so called segregation analysis. Moreover, the disease may be governed by one or several different possibly interacting genes, *monogenic* and *polygenic* diseases respectively. The latter case is also referred to as a *complex disease*. If several distinct allelic variants at the same locus are susceptible with respect to the disease we speak of *allelic heterogeneity* and if more than one locus independently are susceptible to the disease we phrase this as *locus heterogeneity*.

The complete genetic model may be summarized as,

$$\lambda = (p, f, l), \tag{4}$$

where p is the set of *disease allele frequencies*, f is the set of *disease penetrance values*, describing the link between phenotypes and disease genotypes, and l defines the *disease loci positions*. We will now more formally describe these components for the one-locus (monogenic) case and then make some comments about the two-locus (instance of polygenic) case.

1.4.1. One-Locus Case

Generally one assumes a *biallelic* disease locus with disease allele D and normal (wild-type) allele d. The disease allele frequency is denoted by $p = P(D)$ and the normal allele frequency by $q = 1 - p = P(d)$.[16]

The probabilistic link between the disease phenotypes and genotypes is given by the *penetrance vector*,

$$f = (f_0, f_1, f_2), \tag{5}$$

---

[15]Each term corresponds to the probability that jointly: (i) An arbitrary pair of parents has the unordered-genotype mating-type $A_i A_j \times A_i A_j$ $(i < j)$. (ii) A corresponding offspring inherits the uninformative unordered-genotype $A_i A_j$.

[16]In general D and d may be thought of as collections of several different allelic variants with similar phenotypic effects.

where $f_i = P(\text{affected} \mid i \text{ disease alleles})$.[17] This formulates the disease structure, whereas the disease allele frequency, with respect to this structure, decides how common the disease will be. Adopting the common assumption of well-ordered penetrance vectors, i.e. that $f_0 \leq f_1 \leq f_2$, this is reflected through the *disease prevalence*

$$K = f_0 q^2 + f_1 2pq + f_2 p^2, \tag{6}$$

which is then an increasing function with p. Note that the prevalence K is the genetically-unconditional population-wise probability of disease, which oftenly is of great interest.

Example 3 (No genetic component). *If there is no genetic component of the disease what-soever,* $f = (f_0, f_1, f_2) = (K, K, K)$. *This follows since then* $f_0 = f_1 = f_2 = f_C$ *which, using (6), gives* $K = f_C$.[18]

Finally, in this case $l = l_1$ gives the actual location of the single disease locus.

1.4.2. Two-Locus Case

This is a straightforward generalization of the one-locus case. The disease alleles are denoted D_1 and D_2, where D_i is the disease allele at the ith disease locus, having disease allele frequency parameter $p = (p_1, p_2)$ with $p_1 = P(D_1)$ and $p_2 = P(D_2)$.

The penetrance vector in (5) is generalized to a 3×3 *penetrance matrix*,

$$f = \begin{bmatrix} f_{00} & f_{01} & f_{02} \\ f_{10} & f_{11} & f_{12} \\ f_{20} & f_{21} & f_{22} \end{bmatrix},$$

where $f_{ij} = P(\text{affected} \mid i \text{ of } D_1, j \text{ of } D_2)$.

Finally, $l = (l_1, l_2)$ denotes the chromosomal positions of the two disease loci. Often l_1 and l_2 are assumed to be nonsyntenic, i.e. $c(l_1) \neq c(l_2)$ where $c(x)$ is the chromosome where x is located.

1.5. The Pedigree Set and Allele-Sharing

We assume there is available data in the form of phenotype and genotype information from a given *pedigree set*, which is a set of (possibly) multigenerational families. See Figure 4 for an example.

Each pedigree may be divided into subsets of *founders* and *nonfounders*, where the parents of the founders are not included in the pedigree, whereas both of the parents of a nonfounder are.[19] The inheritance within a pedigree may be seen as the distribution of alleles from the founders to the corresponding nonfounders (descendants).

---

[17] If there is a penetrance difference whether a disease allele D is paternally or maternally inherited we face *parental imprinting*; leading to the penetrance vector $f = (f_0, f_{10}, f_{01}, f_2)$ with the two values in the middle reflecting these alternatives. This is not used further in this chapter; for further information see Strauch et al. (2000).

[18] More explicitly, $K = f_C q^2 + f_C 2pq + f_C p^2 = f_C(q^2 + 2pq + p^2) = f_C(p+q)^2 = f_C$.

[19] Having information only on one of the parents may easily be transformed to an equivalent pedigree following this standard through letting the uninformative parent be represented by a *pseudo-individual* having missing, phenotype and genotype, values only.

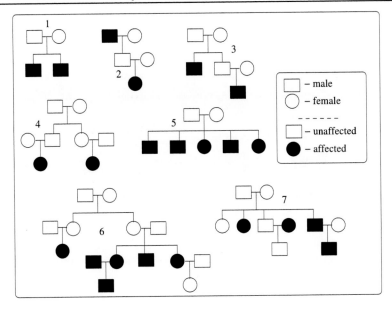

Figure 4. A pedigree set consisting of seven different pedigrees of varying pedigree structure and phenotypic configurations.

Family-based *gene mapping* is, explicitly or implicitly, based on allele-sharing between individuals in a pedigree. In this context, we say that: (i) Two individuals share an allele *identical-by-descent (IBD)* if they have both inherited exactly the same allele, i.e. an identical founder allele, from a common ancestor. (ii) Two individuals share an allele *identical-by-state (IBS)* if they have both inherited a common allelic variant.[20]

Obviously IBD is a stronger sharing property than IBS, since sharing an allele IBD implies sharing IBS as well. Throughout this text we will exclusively be interested in test statistics based on IBD-sharing since they are more efficient for testing genetic linkage.

Example 4 (A simple example of allele-sharing IBD and IBS in a small pedigree). *Consider Figure 5. Assuming no inbreeding the parents share no allele IBD, whereas one allele, 4, is shared IBS. The female offspring is homozygous with respect to the 4-allele implying that one of these must be IBD with the 4-allele of the male offspring. Note that a parent and a child always share one-allele IBD according to the rules of inheritance.*

Consider an *affected sib-pair (ASP)* pedigree; see Figure 5. In the one-locus case, the sib-pair may share either 0, 1 or 2 alleles IBD.[21] This may be summarized in the *IBD-*

---

[20]One may note that a genotype consisting of two IBD alleles are called *autozygous* whereas the complementary term (for the more common occurrence) is *allozygous*. The former case is important when including allowance for *inbreeding* in the statistical modeling; see Terwilliger and Ott (1994), Hössjer (2006) and Siegmund and Yakir (2007).

[21]This is of course logically possible to formulate, in the same way, in the general case of two individuals of arbitrary relatedness; the difference being that in most cases the probability of them sharing two alleles IBD (z_2 below) is 0. For unrelated individuals the probability of sharing one allele IBD (z_1 below) is 0 as well.

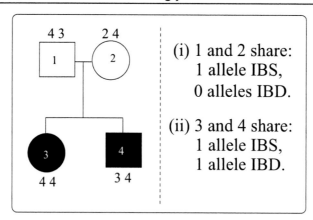

Figure 5. An allele-sharing example of an affected sib-pair (ASP) pedigree.

sharing vector,

$$z = (z_0, z_1, z_2); \quad \text{where} \quad \sum_{i=0}^{2} z_i = 1, \tag{7}$$

and z_i is the probability for the ASP to share i alleles IBD at the disease locus. In the two-locus case one may generalize (7) to the *IBD-sharing matrix,*

$$z = \begin{bmatrix} z_{00} & z_{01} & z_{02} \\ z_{10} & z_{11} & z_{12} \\ z_{20} & z_{21} & z_{22} \end{bmatrix}; \quad \text{where} \quad \sum_{i,j=0}^{2} z_{ij} = 1, \tag{8}$$

and z_{ij} is the probability for the ASP of sharing i and j alleles at the 1st and 2nd disease locus respectively. The IBD-sharing vector (matrix) z depends on the genetic model λ and the locus (loci) at which IBD-sharing is (are) evaluated. Discussions on further constraints on z are given, for instance, by Holmans (1993), Dudoit and Speed (1998) and Bengtsson (2001). Such constraints are induced by sets of valid, according to certain assumptions, disease models. A classic reference with respect to one-locus IBD-sharing is Suarez (1978).

Example 5 (A small example of IBD versus IBS sharing-probabilities). *Consider Figure 6. Introduce* $z' = (z'_0, z'_1, z'_2)$*, where* z'_i *is the probability for an ASP of sharing i alleles IBS under the null hypothesis of no linkage (random inheritance), conditioned on the founder alleles. From left to right, the IBS-sharing probabilities are* $z' = (0.25, 0.5, 0.25)$*,* $z' = (0.125, 0.5, 0.375)$ *and* $z' = (0, 0, 1)$ *respectively. Contrastingly, the corresponding IBD-sharing vector (7) equals* $z = (0.25, 0.5, 0.25)$ *for all three pedigrees.*

We end this subsection with a few words on a common abuse of notation. Generally one uses the notion allele for two distinct purposes: (i) For the *type* of allele, i.e. the actual allelic variant. Examples: *A*, *B*, 1 or 2. (ii) For the *origin* of the allele, i.e. the actual ancestral founder allele. Examples: 'The paternal allele of the third founder' or 'The fifth founder allele'. In this respect IBD and IBS analysis correspond to allele-sharing with respect to allelic origin and type respectively.

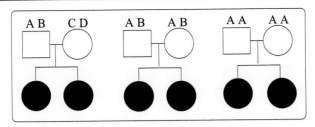

Figure 6. Three ASP pedigrees with different founder-allele configurations.

1.6. The Inheritance Vector and Entropy-Based Information Contents

Consider a pedigree consisting of n individuals, including f founders and $n - f$ non-founders. There are $m = 2(n - f)$ meioses associated with the pedigree, since each non-founder inherits one paternal and one maternal allele.

One of the core concepts in this thesis is the *inheritance vector* $v(x)$ (Donnelly, 1983). This binary 0-1 vector efficiently summarizes all the inheritance information at locus x for a single pedigree as,

$$v(x) = (p_1, m_1, p_2, m_2, \ldots, p_{n-f}, m_{n-f}), \tag{9}$$

where p_i and m_i correspond to the ith nonfounder's paternal and maternal allele respectively, i.e. each value is connected to a specific meiosis. Moreover, note that given the pedigree structure the inheritance vector completely determines the corresponding IBD-sharing through defining the exact distribution of founder alleles among the nonfounders. Somewhat more formally, given the inheritance vector and pedigree structure, the IBD sharing in the pedigree is unambiguous, i.e. known with probability one.

At a crossover point $v(x)$ will change since the corresponding meiosis, p_i or m_i for some i, switches between 0 and 1. We let 0 and 1 correspond to inheriting the grandpaternal and grandmaternal allele respectively. A schematic overview is given in Figure 7.[22]

Example 6 (Consistent inheritance vectors for a small pedigree). *Consider Figure 8. Assume that the founder alleles are phase-known. The origin of the paternal allele for the grandchild (individual 3) is ambiguous since the son (individual 2) is 1-allele homozygous. This leads to the two different (equally probable) consistent inheritance vectors displayed.*

Given data, measures on the locus-specific *information content* or *marker data information* may be based on the certainty of the outcome of $v(x)$. Here we will present the *entropy-based* information measure I_E of Kruglyak et al. (1996).[23] For further discussion

---

[22]Simultaneously considering $|l|$ syntenic loci $\mathbf{x} = (x_1, x_2, \ldots, x_{|l|})$, the complete inheritance picture is summarized by the $m \times |l|$ *inheritance matrix*,

$$v(\mathbf{x}) = \left[v(x_1)^T, v(x_2)^T, \ldots, v(x_{|l|})^T \right],$$

where the ith row and j^{th} column correspond to the equally indexed meiosis and inheritance vector (9) respectively.

[23]Other measures are, for instance, a variance-based measure (cf. e.g. Kruglyak and Lander, 1995) which may be thought of as an estimate of the variance of the NPL score test statistic (see below) and measures based on Fisher information or lod scores (log-odds statistic; see below).

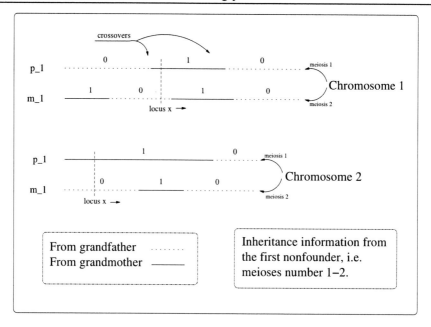

Figure 7. A schematic artificial inheritance vector example of two chromosomes and two meioses.

and suggestions of other information measures, see Teng and Siegmund (1998), Nicolae et al. (1998), Nicolae (1999) and Nicolae and Kong (2004).

Definition 6. *Considering a discrete probability distribution based on $|y|$ distinct outcomes $y_1, y_2, \ldots, y_{|y|}$ with probabilities $p_1, p_2, \ldots, p_{|y|}$, where $p_i = P(y_i)$. The entropy E (Shannon, 1948; Kullback, 1968) with respect to this distribution is,*

$$E = -\sum_{i=1}^{|y|} p_i \log_2 p_i. \tag{10}$$

The minimum value 0 of (10) is attained when the p_y-distribution is one-point and the maximum $\log_2(|y|)$ is attained when the distribution is uniform.

In our case, for a single pedigree, we face $m = 2(n - f)$ meioses and therefore $|y| = 2^m$ different valid outcomes of the corresponding inheritance vector, $w_1, w_2, \ldots, w_{2^m}$, with probabilities,

$$p_i = P(v(x) = w_i | \text{MD}); \quad i = 1, 2, \ldots, 2^m, \tag{11}$$

where MD is the marker data. The *entropy-based information content* is given by,

$$I_E(x) = \left[\frac{E_0 - E(x)}{E_0 - E_{\min}} \right], \tag{12}$$

where $E(x)$ is the observed entropy at locus x, E_0 its maximal possible value m and $E_{\min}$ its minimal value. When the phase of all founders is known we put $E_{\min} = 0$. In general, however, the phase of all founders is unknown and $E_{\min} = f$, since switching of founder

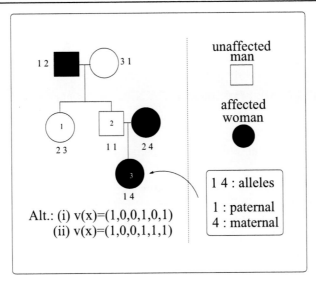

Figure 8. An inheritance vector example of a small three-generational pedigree. In the ideal case when the phase of all founders is known, only two different inheritance vectors are possible given marker genotype data.

alleles results in 2^f equally likely inheritance vectors. In any case, $I_E(x)$ ranges from 0 (no marker information) to 1 (complete marker information).

A monograph on the history, essence and mathematics (including a generous discussion on entropy) of, for instance, information, coding theory and language modeling and approximation is Pierce (1965). Usage of entropy for bioinformatics is examplified in Koski (2001).

1.7.　How to Collect Information

There are two distinct ways of collecting or extracting the available inheritance information from the complete set of genotypes in the pedigree set: (i) *Single-point analysis*, where for each locus x one uses only marker genotypes at this locus when reconstructing the inheritance distribution (11). (ii) *Multipoint analysis*, where for each locus x one uses all marker genotypes from chromosome $c(x)$ when defining algorithms for computing the inheritance distribution (11).

Given a well-defined genetic marker map and assuming no interference, which leads to the Haldane's map function (1), one may use that, for each pedigree, the *inheritance process* $\{v(x);\ x \in \Omega\}$ over the genomic region Ω is a time-homogeneous Markov chain with state-space defined by the 2^m inheritance vectors. Given data, one calculates a multipoint-based inheritance distribution (11) using the theory of *hidden Markov models (HMM)* by interpreting marker genotypes and inheritance vectors as observed and hidden variables respectively. The transition matrices between markers may easily be derived assuming a Poisson distributed number of crossovers between consecutive markers.[24]

---

[24]The relation between hidden and observed variables may be analyzed by checking the consistency between

An original HMM-algorithm performing this task was presented by Lander and Green (1987). A detailed review was recently published in the textbook of Ziegler and Koenig (2006). Later, this algorithm has been updated with several speed-ups. Some extensions are described, for instance, by Kruglyak et al. (1995, 1996), Kruglyak and Lander (1998), Gudbjartsson et al. (2000), Markianos et al. (2001) and Abecasis et al. (2002). General HMM-theory, algorithms, implementations and applications are discussed, for example, by Rabiner (1989) and Cappé et al. (2005); and commonly applied to contexts as speech recognition (Jelinek, 1998) and biological sequence analysis (Durbin et al., 1998).

An alternative algorithmic approach for computing the inheritance distribution (11) was introduced by Elston and Stewart (1971); extended by O'Connell (2001). The complexity of this algorithm increases only linearly with increasing pedigree size but exponentially with the number of markers. The opposite is true for the Lander-Green-Kruglyak algorithm.

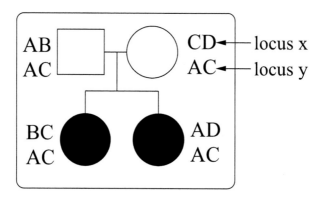

Figure 9. An ASP pedigree, being used in Example 7 for some simple multipoint calculations for loci x and y.

Example 7 (Multipoint versus singlepoint analysis for a small pedigree). *Consider the ASP pedigree of Figure 9 and two loci x and y with recombination fraction θ between them. Assume that the phase of the parental genotypes is known, with the convention that the left haplotype is the paternal one. Now, with probability one $v(x) = (1,0,0,1)$ and the possibilities of consistent inheritance vectors for $v(y)$ are,*

$$(1,0,0,1),\ (1,0,1,0),\ (0,1,0,1),\ (0,1,1,0).$$

Using multipoint analysis the corresponding probabilities are,

$$(1-\theta)^4,\ (1-\theta)^2\theta^2,\ \theta^2(1-\theta)^2,\ \theta^4,$$

where the alternatives correspond to 0, 2, 2 and 4 recombinant meioses between x and y respectively. Note that if using single-point analysis at y all four outcomes above are equally likely.

The multipoint approach increases the information content (12) and hence extracts available information more efficiently at the price of increased computational complexity.

all possible founder genotype configurations and inheritance vectors with actual marker data; see Sobel and Lange (1996) and Kruglyak et al. (1996) for details.

2. Nonparametric Linkage Analysis

Linkage analysis aims at using statistical approaches to find locus (loci) involved in the genetic component of a disease under study. Qualitative-phenotype linkage analysis may be performed in two quite different ways, the *parametric* and the *nonparametric* way, making different assumptions prior to the analysis. In this chapter we mainly adopt the nonparametric approach.

The term nonparametric refers to the fact that no explicit assumptions on the underlying genetic model λ in (4) are made and corresponding nonparametric statistical tests usually depend on the concept of allele-sharing IBD. To perform a test, the actual sharing in the pedigree set is calculated and compared, through a properly chosen *test statistic*, with the expected sharing under the null hypothesis. In nonparametric linkage analysis *genetic linkage* at locus x means that the inheritance of alleles at x is correlated with the phenotype of interest.

This implies, for properly defined disease phenotypes and genetic models, that on average two affected individuals will share more alleles than is expected under the null hypothesis H_0 of no linkage. In the one-locus case one may state the pair of *tested hypotheses* as,

$$\begin{cases} H_0 : \text{Disease locus unlinked to } x. \\ H_1 : \text{A disease locus at } x. \end{cases} \tag{13}$$

when testing a single locus and as,

$$\begin{cases} H_0 : \text{No disease locus linked to } \Omega. \\ H_1 : \text{At least one disease locus along } \Omega. \end{cases} \tag{14}$$

when testing a whole region Ω.

Examples of review articles on linkage analysis are Kruglyak et al. (1996) and Teare and Barrett (2005), and thourough discussions on nonparametric linkage analysis are given in the theses Nicolae (1999), Basu (2005) and Ängquist (2007b). On likelihood-based approaches and simulation, Thompson (2000) is a major work.

In passing, one may note that the future, and survival, of linkage analysis methods has to some extent been debated. Two articles offering somewhat pessimistic and optimistic views respectively are Risch and Merikangas (1996) and Clerget-Darpoux and Elston (2007). In the end, as always, it seems to come down to the premises one accept and the type of evidence one finds convincing. Still, though, many linkage studies are performed and published in well-established scientific journals on a regular basis.

2.1. Parametric Linkage Analysis

Here one assumes a known (or rather estimated) genetic model λ in (4) which must be defined prior to the analysis. The most common test statistic is the likelihood ratio-based *lod-score statistic*, which for the single-point case is defined as,

$$Z(\theta; \lambda) = \log_{10} \left[\frac{P(Y, \text{MD}(x) \mid \theta, \lambda)}{P(Y, \text{MD}(x) \mid 0.5, \lambda)} \right],$$

where Y and $\text{MD}(x)$ refer to phenotypic data and marker data at locus x respectively, θ is the tested recombination fraction between the disease locus and x, and λ is the genetic model. In this case we use $Z(0;\lambda)$ as test statistic for (13) and the maximum lod-score $Z_{\max} = \sup_{0 \leq \theta \leq 0.5} Z(\theta;\lambda)$ as test statistic for (14) with $\Omega = c(x)$, which can be written as,

$$\begin{cases} H_0 : \ \theta = 0.5 \\ H_1 : \ 0 \leq \theta < 0.5 \end{cases}$$

For the multipoint case the lod score statistic is only slightly altered into the form,[25]

$$Z(x;\lambda) = \log_{10}\left[\frac{P(Y, \text{MD} \mid x, \lambda)}{P(Y, \text{MD} \mid \infty, \lambda)}\right], \qquad (15)$$

where MD is the complete set of marker data from Ω and x is the hypothesized position of the disease locus l. We use $Z(x;\lambda)$ as test statistic for (13) and the maximum lod-score $Z_{\max} = \sup_{x \in \Omega} Z(x;\lambda)$ as test statistic for (14).

2.1.1.   Related Notes and References

A great likelihood reference is Edwards (1992); integrating philosophy, statistical history and statistics, and including many genetical examples. The original lod-score reference is Morton (1955) which is based on results and procedures given by Haldane and Smith (1947) and Barnard (1949). Some modern references are Terwilliger and Ott (1994), Kruglyak et al. (1996), Ott (1999), Kurbasic and Hössjer (2004, 2006), Xing and Elston (2006) and the Doctoral Thesis of Kurbasic (2007). Some recent extensions to include for *marker-marker linkage disequilibrium*, caused by using dense set of SNP-markers, are based on marker clustering (Abecasis and Wigginton, 2005) and generalization of the Lander-Green-Kruglyak algorithm (Kurbasic and Hössjer, 2007).

A (single-point) Bayesian perspective is suggested in Vieland (1998) through the concept of *posterior probability of linkage (PPL)*. Here a prior $f(\theta)$ with respect to the recombination fraction between the disease and single marker loci is introduced; see also Smith (1959). A complete Bayesian guide is Gelman et al. (2004).

A conceptually different perspective is offered by Strug and Hodge (2006), who suggest an *evidential framework* as opposed to the frequentist or Bayesian counterparts. They work directly with the likelihood ratio instead of the lod-score and rejects the traditional usage of error probabilities as measuring the amount of evidence; see also Royall (1997). For philosophical discussions on statistical inference and evidence see, for example, Hacking (1965).

2.2.   Score Functions

For the one-locus case, the nonparametric test statistic is based on a score function $S(v)$, which assigns a number to each possible pedigree-wise IBD-sharing structure (or

---

[25]An explanation to the ∞-sign in the denominator of (15) is that, under the null hypothesis, the disease locus is unlinked to all chromosomes constituting Ω. In other words it may be seen as, using (1), being located infinitely far away from Ω.

inheritance vector).[26] As noted above, one is normally interested in increased allele-sharing among affecteds since this indicates presence of genetic linkage between the marker and disease loci.

Next, two commonly used score functions will be briefly introduced and discussed. Firstly, S_{pairs} (Weeks and Lange, 1988) is based on IBD-sharing among *all pairs of affected individuals* in the pedigree,

$$S_{\text{pairs}}(v) = \sum_{(i,j)\in\mathbb{A}} \text{IBD}(i,j), \tag{16}$$

where $i < j$, $\mathbb{A}$ is the set of affecteds and $\text{IBD}(i,j)$ is the number of alleles shared IBD between individuals i and j. Here one basically calculates the number of (pairwisely) shared alleles IBD in the pedigree.

Secondly, S_{all} (Whittemore and Halpern, 1994) is based on the *simultaneous IBD-sharing among all the affecteds* in the pedigree,

$$S_{\text{all}}(v) = \frac{1}{2^{|\mathbb{A}|}} \sum_{h\in\mathbb{H}} \prod_{i=1}^{2f} b_i(h)!, \tag{17}$$

where $|\mathbb{A}|$ is the number of affecteds, $\mathbb{H}$ is a set containing all ways of selecting one allele from each affected, $2f$ is the number of founder alleles in the pedigree and $b_i(h)$ is the number of times the ith founder allele is present in selection h.[27] This quite complex-looking function might be interpreted as, picking one allele from each affected, the mean number of permutations which leaves the ordered selection unaltered with respect to allelic types.[28]

Example 8 (Score function calculations using a small pedigree). *Consider the pedigree in Figure 10. Ordering the nonfounders from left to right the inheritance vector $v=(0,0,1,0,1,0,1,1)$ is fully known if both parents have known phase. Calculating S_{pairs} and S_{all} we get,*

$$S_{pairs}(v)=(1+1+1+2+1+1)=7,$$
$$S_{all}(v)=(24+6+6+6+6+4+2+2+2+2+2+2+2+1+1)/16=35/8.$$

---

[26]One may note that this notion of a score function may be seen as adopting a data-mining perspective where such functions are used for scoring patterns, in this case inheritance patterns (Hand et al., 2001). Note: Therefore it obviously follows that this concept is *not* generally similar to the likelihood first derivative-based score functions met in statistical inference; see Lehmann and Casella (1998).

[27]The selection h consists of $|\mathbb{A}|$ alleles that may be grouped according to their ancestral history, i.e. each allele is a copy of one of the $2f$ founder alleles. The link to the number of members in the ith group is $b_i(h)$.

[28] Both S_{pairs} and S_{all} are calculated, given v, using the group of affecteds $\mathbb{A}$ only. Each such *traditional* score function S corresponds to an *extended* score function S',

$$S'(v) = S'(v|\mathbb{A}\cup\mathbb{UA}) = S(v|\mathbb{A}) + S(v|\mathbb{UA}),$$

where $\mathbb{UA}$ is the set of unaffecteds and $S(\cdot|\mathbb{B})$ is the traditional score function replacing the subgroup of affecteds with the arbitrary subgroup $\mathbb{B}$. This may increase power since we extract more inheritance information, but the computational complexity may be alarmingly increased and the gain small for some genetic disease models and pedigrees. For more information, and additional extended versions, see Ängquist (2006c, 2008).

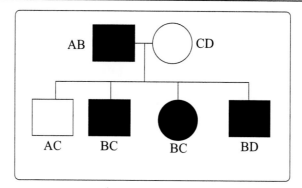

Figure 10. An example pedigree used in Example 8 regarding the comparison of different score functions.

For instance, for S_{all}, if $h = \{B,C,C,B\}$ we get $\prod_{i=1}^{4} b_i(h) = 0!2!2!0! = 4$. Note that the score functions are generally not comparable at this point since we have not yet standardized them with respect to the reference null distribution; see below.

The relative performance of different score functions, in terms of statistical power, depends on the underlying genetic model and the structure, i.e. the relational structure and phenotype setting, of the pedigree set. Given a well-defined genetic model λ in (4) it is possible to derive different kinds of *optimal* score functions, based on different optimality criterions (McPeek, 1999; Hössjer, 2003b, 2005c; Ängquist et al., 2008; Ängquist, 2008), which then implicitly leads to the use of both affecteds and unaffecteds in the sharing-based calculations, in contrast to (16) and (17), through their corresponding definitions. In applications λ is most oftenly not fully known leading to extensive usage of score functions that may be designed to have good robust performance for a (sufficiently) wide range of genetic models. Other references with discussion on choice of and performance with respect to score functions are found in, for example, Davis and Weeks (1997), Feingold et al. (2000), Sengul et al. (2001), Lange and Lange (2004) and Hössjer (2005b).

In Ängquist (2008) classes of score functions are divided into three groups: (i) *Implicitly defined functions*. These are, like in (16) and (17) above, defined through a well-defined high-level function that automatically, given the IBD-sharing or some other information, calculates scores. (ii) *Explicitly defined functions*. That follow no generating rules; allowing for maximum user control. May be formulated based on some vague information or intuition regarding plausible disease model λ. (iii) *Optimality defined funtions*. Creates scores based on an optimality criteria conditioned on an assumed genetic disease model, or several weighted suggested disease models.

For instance, Ängquist et al. (2008) considers optimality based on maximization of the *noncentrality-parameter (NCP)*,

$$NCP = NCP(S,\lambda) = E\left[Z(l)|\lambda\right], \qquad (18)$$

where S is the underlying score function, l the disease locus and $\lambda \in H_1$ the one-locus genetic disease model. It turns out that the score function which maximizes (18) is,

$$S(w) \propto P(v(l) = w|Y,\lambda) - 2^{-m}; \quad \forall w \in \mathbb{V}, \qquad (19)$$

where Y is the phenotype vector, $\mathbb{V}$ the set of possible inheritance vectors and m the number of meioses in the pedigree. One may note that: (i) This score function is proportional to the inheritance distribution difference under the null and (assumed) alternative hypotheses. (ii) It is not a traditional score function in the sense of Footnote 28, since the inheritance distribution generally depends on the inheritance, and hence sharing, related to all nonfounders; affecteds as well as unaffecteds. Similar results are obtained for the unconditional and conditional two-locus cases; see below. See also Sham et al. (1997), Nilsson (1999) and Hössjer (2003a).

2.3. The NPL Score

To ease interpretation and significance calculations we *standardize*[29] the score function under H_0 according to,

$$S(v) \leftarrow \left[\frac{S(v) - \mu}{\sigma}\right], \tag{20}$$

where, for a pedigree with m meioses,

$$\begin{cases} \mu = \sum_i 2^{-m} S(w_i) \\ \sigma^2 = \sum_i 2^{-m} S(w_i)^2 - \mu^2 \end{cases}$$

are the mean and variance of S, prior to standardization, under the null hypothesis H_0 of no linkage. Note that we, after standardization, end up with the properties $E(S|H_0) = 0$ and $V(S|H_0) = 1$. We refer to S prior and post standardization as an *unstandardized* and *standardized* score function respectively. Many unstandardized score functions correspond to the same standardized one, which might be referred to as score function *equivalence*; see Ängquist (2008).

A standardized score function is used to calculate, at locus x, the *pedigree-specific non-parametric linkage (NPL) score*,

$$Z(x) = \sum_i p(w_i) S(w_i), \tag{21}$$

where $p(w_i) = P(v(x) = w_i | \text{MD})$ is the inheritance distribution at x.

For a pedigree set consisting of N pedigrees we combine the pedigree-specific scores (21) into the (total) *NPL score* as,

$$Z(x) = \sum_{k=1}^{N} \gamma_k Z_k(x), \tag{22}$$

where Z_k is the NPL score (21) and γ_k the weight assigned to the kth pedigree. The pedigree *weighting scheme* is chosen as $\sum_{k=1}^{N} \gamma_k^2 = 1$ in order to assure,

$$E(Z(x)|H_0) = 0 \text{ and } V(Z(x)|H_0) \leq 1, \tag{23}$$

---

[29] Also referred to as *normalization*.

with equality for complete marker information.[30] The actual weights may be chosen according to pedigree size, structure, information or inheritance at other loci. We use $Z(x)$ as test statistic for testing a single disease locus through (13). Oftenly, such tests are based on a *perfect-data approximation*, which corresponds to calculating the null distribution assuming that $V(Z(x)|H_0) = 1$ and implies *conservative* p-value estimates when facing imperfect data.[31]

Calculating the NPL score with respect to a set of loci leads to a stochastic process $\{Z(x); x \in \Omega\}$, the *NPL process*. A simulated example is given in Figure 11. According to the blockwise inheritance of chromosomal segments, NPL scores at linked loci are correlated. The multipoint approach increases the extractable information, and hence NPL score correlation, at closely linked loci as an effect of smoothing the observed NPL score process. To test (14) we use the maximum of the NPL process over Ω, referred to as the maximum NPL score,

$$Z_{\max} = \sup_{x \in \Omega} Z(x). \tag{24}$$

2.3.1. Alternative Methods

An alternative, likelihood-based but still nonparametric, definition of total NPL score was proposed by Kong and Cox (1997), which leads to less conservative tests for incomplete marker data. For perfect data it coincides with (22). Leibon et al. (2007) suggested performing linkage analysis using what is called the *shadow method* and Ängquist (2007a) discussed the *maximum selected pedigree subset-score method* as a way to deal with pedigree-wise locus heterogeneity with respect to linkage.

Another option is to use MLS scores, where the likelihood of data is maximized jointly over disease locus and allele sharing probabilities; see Risch (1990) and Holmans (1993). Alternatively, for MOD scores, the LOD score may be maximized jointly over disease locus and genetic model parameters, as suggested by Risch (1984), Clerget-Darpoux et al. (1986) and Strauch et al. (2005).

Quite different approaches are adopted by performing linkage analysis using neural networks (Lucek et al., 1998), and cluster- and latent class analysis (Neuman et al., 2000).

---

[30]The inequality in (23) follows for one pedigree from,

$$1 = V\left(S[v(x)]\right) = V\left[E\left(S[v(x)]|MD\right)\right] + E\left[V\left(S[v(x)]|MD\right)\right]$$
$$\geq V\left[E\left(S[v(x)]|MD\right)\right] = V[Z(x)],$$

assuming expectation and variance is taken under H_0. See also Kruglyak et al. (1996).

[31]Missing genotypes, homozygosity, usage of single-point analysis or multipoint analysis with a sparse marker map leads to loss of inheritance information, i.e. increased inheritance vector ambiguity, implying a decrease of the NPL score variance. The extreme case of totally uninformative marker data for a pedigree leads to a constant score $Z(x) = 0$ in (21). If the underlying combination of pedigree structure, score function and phenotypic configuration is totally uninformative, the unstandardized scores $S(v)$ are independent of v which, using (20), implies a not even well-defined procedure.

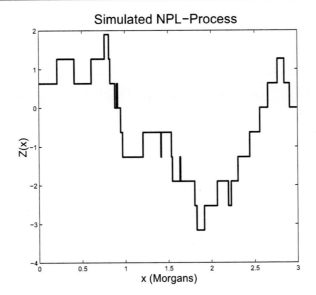

Figure 11. A simulated NPL process along a single chromosome of length 3 Morgans, assuming perfect marker data. The underlying score function is S_{all}, the pedigree set consists of $N = 10$ (homogeneous) pedigrees of the same structure as in Figures 1-2 and 8 with equal weights $\gamma_k = 1/\sqrt{10}$.

2.4. Calculating the Statistical Significance

2.4.1. Basics

Consider a test statistic Z. The *significance level* of a test which rejects H_0 when $Z \geq T$, where T is a given numerical threshold, is

$$\alpha(T) = P(Z \geq T | H_0), \tag{25}$$

and the *power function* is

$$\beta(T, \vartheta) = P(Z \geq T | \vartheta \in H_1). \tag{26}$$

In other words the significance level and power function correspond to the probability of exceeding threshold T under the null and (an instance of) the alternative hypothesis respectively. Given a test result $Z = z$, one may calculate the *p-value* $\alpha(z)$.

In our scenario we will use the NPL score $Z = Z(x)$ from (22) for *pointwise* tests and the maximum NPL score $Z = Z_{\max}$ from (24) for *genomewide* tests.[32] Moreover, our power depends on the genetic model instance ($\vartheta = \lambda$), and T is a NPL score threshold.

---

[32] That is, if Ω is the whole *genome*, i.e. the complete genomic region under study.

2.4.2. Approaches and Common Assumptions

In the pointwise case, when N is large, one may use the approximation,

$$Z(x) \stackrel{H_0}{\in} N(0,1), \tag{27}$$

of (22), based on the *Central Limit Theorem*. This leads to the approximation,

$$\alpha(T) \approx 1 - \Phi(T),$$

where Φ equals the standard normal N(0,1) distribution function. This approximation is usually conservative, due to (23), when marker data is incomplete.

In the genomewide case, the distribution of Z_{max} is less tractable. It may be computed using either *Monte Carlo simulation* or *analytical approximations*. In the latter case one uses *extreme-value theory* of the stochastic process $\{Z(x); x \in \Omega\}$, see e.g. Leadbetter et al. (1983), Siegmund (1985), Aldous (1989), Lander and Botstein (1989), Feingold (1993), Feingold et al. (1993) and Tu and Siegmund (1999). A small-scale example of the genomewide significance level (25) is given in Figure 12.

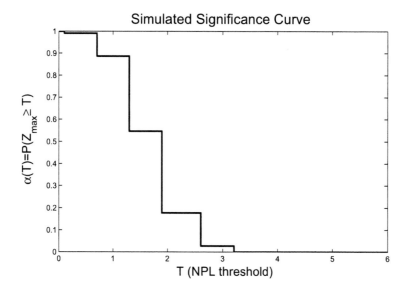

Figure 12. A simulated genomewide significance level curve for a single chromosome of length 3 Morgans, assuming perfect marker data. The score function and pedigree set are as in Figure 11.

The advantage of a closed form analytical approximation (see Section 2.5.) is ease of computation, even for large thresholds. On the other hand, the approximation may be more or less accurate according to the assumptions made. It is common to assume:

One That the NPL score (22) is marginally normally distributed, with expected value equal to zero, for all loci under H_0. Deviations from this assumption will, depending on the type of nonnormality, imply either conservative or anticonservative approximations.

Two That the variance of the NPL score in (22) equals one at all loci, i.e. to use the perfect-data approximation. This implies conservativity according to (23).

Three That Ω is continuous, i.e. an infinitely dense marker map is used. This implies conservativity.

The Monte Carlo approximation (see Section 2.6.) is more accurate provided the number of replicates is large enough. On the other hand, computationally it can be very slow, especially for large thresholds.

2.4.3. *Alternative Standardizations*

In (20) we used the null hypothesis score variance σ^2, based on an assumption of perfect data, in order to standardize our scores. As noted above, this generally leads to conservative analytical significance calculations under imperfect data since the actual variance is lower than one; see (23).

In principle one may for each pedigree calculate all possible scores, i.e. based on all possible inheritance vectors, through for each marker: (i) Conditioning on founder marker genotypes. (ii) Using the inheritance vectors to derive the score from (21). (iii) Calculating the *true* score variance $\overline{\sigma}^2$ based on this set of scores. (iv) Performing score standardization based on $\overline{\sigma}$. The problem is that this procedure in many cases may be refrainingly computer intensive and that some of the justifications of the normal approximations below will be weakened, calling for further approximations with respect to the irregular (not truly stationary) behaviour of the corresponding NPL process.

In Jung et al. (2006) one discusses standardization through using empirical variances $\hat{\sigma}^2$, calculated based on the observed scores squared deviations around either 0 (expected mean under the null hypothesis) or the estimated expected mean $\hat{\mu} = \hat{E}[Z(x)]$.

Using Monte Carlo simulations, sometimes called gene-dropping, adopting the perfect data assumption causes no formal problem since no conservativeness is implied, but some power may be lost according to the unfairness of all markers not having the same score variance. In case of using simulations one may condition on founder marker genotypes or randomly assign such according to observed or assumed marker allele frequencies; see Jung et al. (2006).

2.4.4. *Presentation*

Significance calculations, significance levels and powers, is oftenly suitable to present through standard graphs. The simplest (most obvious) approach is then to plot these as separate curves against the threshold T. An alternative is to plot powers against significance levels which produces so called *receiver operating characteristic (ROC) curves*; see Selin (1965) and Bradley (1996). An example is given in Figure 13.

2.5. Significance Calculations through Theoretical Approximation

The most commonly used theoretical approximation (Lander and Botstein, 1989; Feingold et al., 1993; Lander and Kruglyak, 1995) is based on the assumption that the NPL

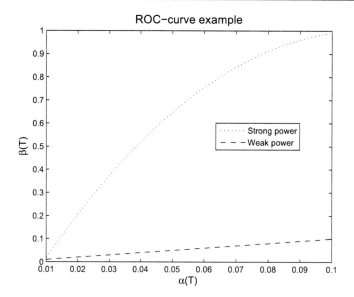

Figure 13. An artificial example of plotting using ROC-curves based on underlying thresholds T. Note that the weak power function corresponds to the straight line $\alpha = \beta$ which means that significance levels and powers equals throughout.

process is a stationary continuous-time Gaussian process with N(0,1) marginals. Moreover, the process is assumed to be *Ornstein-Uhlenbeck*-like (Uhlenbeck and Ornstein, 1930; Hsu and Park, 1985; Blackwell, 2002). By this we mean that the process is non-differentiable and the autocovariance function,

$$r_Z(h) = C\left[Z(x), Z(x+h)\right],$$

has a non-zero right-hand side derivative $r_Z'(0)$ at zero.

Following Lander and Kruglyak (1995), the approximation $\hat{\alpha}(T)$ of the genomewide significance level (25) is defined as

$$\hat{\alpha}(T) = 1 - \exp\left[-\mu(T)\right], \tag{28}$$

where

$$\mu(T) = [C + 2\rho g T^2]\alpha_{\mathrm{pt}}(T). \tag{29}$$

In (28)-(29) we have that $C = C(\Omega)$ is the number of chromosomes in Ω, $g = g(\Omega)$ is the total genome length of Ω in Morgans, $\rho = -r_Z'(0)/2$ is the crossover rate and $\alpha_{\mathrm{pt}}(T) = 1 - \Phi(T)$ the approximative pointwise significance level with respect to the threshold T.

Extensions of (28) are given by Tang and Siegmund (2001) where *correction* for non-normality, in the form of *distributional skewness*, of the NPL score is introduced. Further, based on works of Siegmund (1985) and Feingold et al. (1993) a relaxation of the assumption of a continuous-time process into a finite number of *equidistant markers* is made.

The *crossover rate* ρ in (29) reflects the fluctuation of the NPL process, i.e. how often the score changes and how large these changes are. It may be expressed for arbitrary pedigrees (Hössjer, 2001, 2003a; Ängquist, 2001) as,

$$\rho = \frac{1}{4} 2^{-m} \sum_{w \in \mathbb{V}} \sum_{j=1}^{m} [S(w) - S(w + e_j)]^2, \tag{30}$$

where S is a standardized score function (20), $\mathbb{V}$ is the set of possible inheritance vectors and e_j is an inheritance vector with one in the jth position and zeros elsewhere, corresponding to a crossover of the jth meiosis. In this sense, $\{w + e_j\}_{j=1}^{m}$ are *neighbours* of w.[33] Hence ρ depends on the score function, pedigree structure and pedigree size.

For a pedigree set consisting of N pedigrees one may combine the pedigree-specific crossover rates (30) into an overall crossover rate,

$$\rho = \sum_{k=1}^{N} \gamma_k^2 \rho_k,$$

where γ_k and ρ_k are the kth pedigree weight in (22) and crossover rate respectively.

Example 9 (Crossover rate for a single ASP pedigree). *Consider an ASP pedigree. If using the standardized version (20) of a symmetric score function,[34] $S(w)$ attains the value $-\sqrt{2}$, 0, $\sqrt{2}$ when the ASP shares 0, 1 or 2 alleles IBD.*

In this case the number of meioses $m = 4$, the number of possible inheritance vectors $|\mathbb{V}| = 16$. The IBD-status of the ASP along a chromosome changes at points of crossovers according to a Markov chain with transition matrix,

$$P = [p_{ij}]_{i,j=0}^{2} = \begin{bmatrix} 0 & 1 & 0 \\ 1/2 & 0 & 1/2 \\ 0 & 1 & 0 \end{bmatrix},$$

where p_{ij} refers to the probability of changing from i to j alleles shared IBD. Hence in all cases, $w \in \mathbb{V}$ and $k \in \{1, 2, \ldots, m\}$, $|S(w) - S(w + e_k)| = \sqrt{2}$. Using (30), we obtain $\rho = \rho(ASP) = 2$.

Further, the approximation of the significance level (28)-(29) may be extended to *correct* for *marginal nonnormality*, under an assumption of fully informative inheritance at all markers, of the NPL score (22); see Ängquist and Hössjer (2005). Here, the nonnormality correction is based on introducing a *link function*, h_{link}, which transforms the NPL process to marginal approximate-normality. Formally,

$$Y(x) = h_{\text{link}}^{-1} [Z(x)]; \; x \in \Omega, \tag{31}$$

where $h_{\text{link}} = (F^{-1} \circ \Phi)$, $F(z) = P(Z(x) \le z | H_0)$ is the marginal distribution function of the NPL score process $Z(\cdot)$ under H_0 and $\Phi(z) = \int_{-\infty}^{z} \phi(z) dz$ is the standard normal cumulative distribution function. The transformed process Y in (31) is a stationary process with

---

[33] A neighbour w' to w is an inheritance vector that differs from w at only one vector position, i.e. the corresponding *Hamming distance $H(w', w) = 1$*.

[34] Let s_i be the unstandardized score corresponding to the ASP sharing i alleles IBD. The score function is symmetric if $s_2 - s_1$ equals $s_1 - s_0$. This is true both for S_{pairs} and S_{all}. Since all symmetric score functions in this case leads to the same standardized scores, and score distribution, they are *equivalent* score functions; see Ängquist (2008) for general discussion and conditions.

approximately standard normal marginals, the approximation being due to discreteness of distribution F.

To turn Y into a *continuous-valued* process we use a *linear binning* smoothing procedure to approximate F by a continuous version $\hat{F}$ when defining h_{link}.[35] Hence, basing the transformation (31) on $\hat{F}$ and Φ, one may note that it is then one-to-one and that the method might be seen as a continuous distributional quantile-coupling procedure; an artificial example of the link function h_{link} is given in Figure 14. Further, using that

$$\alpha(T) = P(Z_{\max} \geq T | H_0) = P(Y_{\max} \geq h_{\text{link}}^{-1}(T) | H_0),$$

leads to an improved significance approximation of $\alpha(T)$. It is based on (28), but with a refined upcrossing intensity (29) expressed as,

$$\mu(T) = [C + 2\rho_Y g\, h_{\text{link}}^{-1}(T)^2]\alpha_{\text{pt}}\left[h_{\text{link}}^{-1}(T)\right],$$

where ρ_Y is the updated crossover rate with respect to the transformed process Y and g is the genome length. The calculation of ρ_Y is based on the theory of *subordinated* Gaussian processes (Clark, 1973) and a *Hermite polynomial expansion* of h_{link} (Taqqu, 1975; Hall, 1992).

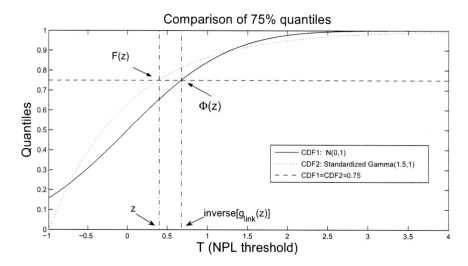

Figure 14. Implicit construction of the link $z \leftrightarrow h_{\text{link}}^{-1}(z)$ in (31) through comparison between the NPL score distribution function F and the standard normal distribution function Φ. Here the unstandardized NPL score is assumed to be distributed as $\Gamma(1.5, 1)$.

Other analytical approaches are, for example, described in Hernández et al. (2005), Bacanu (2005) and Saunders et al. (2007).

2.6. Significance Calculations through Monte Carlo Simulation

Using a simulation algorithm is often the easiest way to capture hard-to-get information, by being well-suited to mimic complicated mechanisms and adapt to complex model-

---

[35]This transforms h_{link} from being a *nondecreasing step-function* to a *strictly increasing* function.

structures. The main drawback usually is the computational complexity.

For complete marker data, the pedigree-specific NPL score (21) simplifies to,

$$Z(x) = S[v(x)], \tag{32}$$

where $v(x) = [v_1(x), v_2(x), \ldots, v_m(x)]$ is the inheritance vector at locus x. When map distance is measured in Morgans and no chiasma interference is assumed, the components $v_j(\cdot)$ can be simulated under H_0 as independent stationary Markov processes[36] with intensity matrix

$$\begin{bmatrix} -1 & 1 \\ 1 & -1 \end{bmatrix} \tag{33}$$

along each chromosome of Ω. From (32) we then get the pedigree-specific NPL score along Ω. Repeating this for all N pedigrees, using (22) and maximizing over Ω, we obtain a maximum NPL score Z_{max} simulated under H_0. Further, repeating this for J simulations, with $Z_{max,j}$ denoting the maximum NPL score for the jth simulation, the crude (standard) Monte Carlo estimate of the significance level is,

$$\hat{\alpha}(T) = \frac{1}{J} \sum_{j=1}^{J} I(Z_{max,j} \geq T); \ T \in \mathbb{T}, \tag{34}$$

where $I(A)$ is the indicator function of the event A and $\mathbb{T}$ is a predefined set of score thresholds.[37]

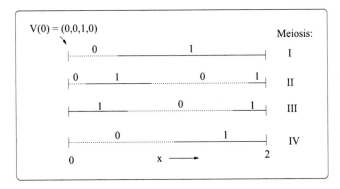

Figure 15. An example illustrating the simulation of the inheritance vector process $v(\cdot)$ of a single ASP along a chromosome of length 2 Morgans.

Example 10 (A small simulation example for a single ASP). *Consider Figure 15-16. The former illustrates a simulated realization of the inheritance process along a chromosome of length 2 Morgans, whereas the latter shows the corresponding pedigree-specific NPL score process. Note that the inheritance vector travels from $v(0) = (0,0,1,0)$, through a total number of seven crossovers, to $v(2) = (1,1,1,1)$.*

---

[36]For an introduction to Markov theory, some good resources are Bremaud (2001) and Häggström (2002).

[37]An indicator function $I(A)$ attains the value 1 if event A is true, otherwise it equals 0.

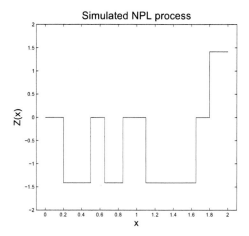

Figure 16. The pedigree-specific NPL score process $Z(x)$ corresponding to the simulated inheritance process of Figure 15.

To use simulation to facilitate calculation of statistical power with respect to the appropriate test statistic, pointwise or genomewide, under a genetic model λ of the alternative hypothesis H_1 and the phenotype vector Y, the H_0-simulation may be modified as follows for the inheritance process of each pedigree: If l is the disease locus, simulate $v(l)$ from $P(v|\lambda, Y)$. Then generate the components $v_j(\cdot)$ of $v(\cdot)$ independently to the right and left of l, starting at $v_j(l)$, according to the same Markov process (33) as in the H_0-simulation.

Some references to simulation procedures for incomplete marker data are Boehnke (1986), Ploughman and Boehnke (1989), Ott (1989) and Terwilliger et al. (1993). The fast, but slightly biased, replicate-pool method is outlined by Song et al. (2004) and Wigginton and Abecasis (2006). A recent reference to (advanced) stochastic simulation is Asmussen and Glynn (2007).

2.6.1. Importance Sampling-Based Simulation

For large thresholds T where $\alpha(T) = P(Z_{\max} \geq T|H_0)$ is very small, we will generally need a very large number of simulations to be able to estimate $\alpha(T)$ using (34) with reasonably small variance. Given a constant time limit the actual number of performed simulations also depends on the computer-time per simulation or, equivalently, the *computational cost*, which is primarily affected by the pedigree sizes. One possible solution to this problem is to use *importance sampling*, or *weighted simulation*, which is a variance reduction technique making it possible to sample from interesting regions of the underlying sample space with higher probability. An early reference is Hammersley and Handscomb (1964) and a modern introduction is given by Ross (2006). See also Hesterberg (1995) and Robert and Casella (2004).

A very brief outline of the general method is as follows. Assume we want to estimate,

$$\alpha = E\left[f(Z)\right] = \int f(Z)dP(Z),$$

where E denotes expectation when Z has distribution P. The following reformulation, using the *change of probability measure* from P to $\tilde{P}$,

$$\alpha = \int f(Z)\frac{dP(Z)}{d\tilde{P}(Z)}d\tilde{P}(Z) = \tilde{E}\left[f(Z)L(Z)\right], \tag{35}$$

is valid if,

$$f(Z)dP(Z) > 0 \Rightarrow d\tilde{P}(Z) > 0.$$

In (35), $\tilde{E}$ denotes expectation when Z has distribution $\tilde{P}$ and the weighting function $L(Z) = dP(Z)/d\tilde{P}(Z)$ is the likelihood ratio with respect to the two measures. Sampling from $\tilde{P}$, the integral in (35) may be estimated using,

$$\tilde{\alpha} = \frac{1}{J}\sum_{j=1}^{J} f(Z_j)L(Z_j), $$

where $\{Z_j\}_{j=1}^{J}$ are independent and identically distributed copies of Z under $\tilde{P}$.

Now, put $\alpha = \alpha(T)$, $Z = \{Z(x);\ x \in \Omega\}$, $f(Z) = I(Z_{\max} \geq T)$ and let the probability measure P be the H_0-distribution of Z. Ängquist and Hössjer (2004) introduced an *exponentially tilted* probability measure $\tilde{P}$, which for complete marker data along one chromosome $c = [0,l]$ has the form,

$$d\tilde{P}(Z) = \left[\frac{\int_0^l \exp\left[\delta Z(x)\right] dx}{lM(\delta)}\right] dP(Z), \tag{36}$$

where $M(\delta) = E\left(\exp\left[\delta Z(X)\right] | H_0\right)$ is the moment generating function of $Z(x)$ under H_0 and δ is a *design* or *tilting* parameter reflecting the amount of change of measure.

When $\delta = 0$, $\tilde{P} = P$ coincides with the H_0-distribution. The larger $\delta > 0$ is, the more likely it is for $Z_{\max}$ to attain large values when $Z \sim \tilde{P}$. To actually simulate such a Z, we proceed similarily as when generating NPL scores under H_1 to estimate power in Section 2.6.. First select an *artificial disease locus* x_0 according to a uniform distribution on $[0,l]$, then generate $Z(x_0)$ by simulating inheritance vectors at x_0 of all pedigrees under a pointwise version of the exponentially tilted distribution and, finally, generate inheritance vectors to the left and right of x_0 for all pedigrees to compute $Z(\cdot)$ along the whole chromosome.

For incomplete marker data, the formula for $\tilde{P}$ is more complicated than (36), but the procedure is analogous.

Example 11 (A simple importance sampling example for a single ASP). *To illustrate the distribution of $Z(x_0)$ under $\tilde{P}$, consider the NPL score (32) of a single pedigree with m meioses. By conditioning on x_0, it follows from (36) that,*

$$\tilde{P}\left(v(x_0) = w\right) = \left[\frac{\exp\left[\delta S(w)\right]}{2^m M(\delta)}\right], \tag{37}$$

where $M(\delta) = 2^{-m}\sum_{w \in \mathbb{V}}\exp\left[\delta S(w)\right]$.

In particular, for an ASP with score function as in Example 9, the $2^m = 16$ inheritance vectors may be divided into three groups of sizes 4, 8 and 4, with $S(w) = -\sqrt{2}$, 0 and

$\sqrt{2}$ *respectively. The three groups correspond to the sib-pair sharing 0, 1 and 2 alleles IBD with allele-sharing probabilities* $z_i = P(IBD = i|H_0)$ *under* H_0, *where all inheritance vectors* $v \in \mathbb{V}$ *are equally likely,*

$$z = (z_0, z_1, z_2) = (0.25, 0.50, 0.25),$$

as discussed in Example 5. On the other hand, using (37) the allele sharing probabilities $\tilde{z}_i = \tilde{P}(IBD = i)$ *under* $\tilde{P}$ *are,*

$$\tilde{z} = (\tilde{z}_0, \tilde{z}_1, \tilde{z}_2) = c_\delta^{-1} \left[\exp(-\delta\sqrt{2}), 2, \exp(\delta\sqrt{2}) \right],$$

where $c_\delta = \left[\exp(\delta\sqrt{2}) + 2 + \exp(-\delta\sqrt{2}) \right]$ *is a normalizing constant.*

In Figure 17 we display the tilted IBD-sharing distribution $\tilde{z}$, *through the* $(\tilde{z}_1, \tilde{z}_2)$*-pair, as function of the tilting parameter* δ.

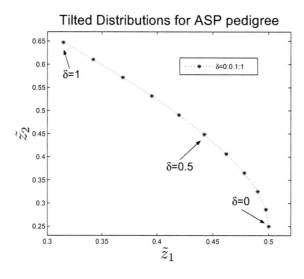

Figure 17. Displaying the IBD-sharing probabilities $(\tilde{z}_1, \tilde{z}_2)$ for an ASP, under $\tilde{P}$, with respect to the tilting parameter $\delta = (0, 0.1, \ldots, 1)$.

As seen from (37), we use $\delta > 0$ at the artificial disease locus x_0 in order to increase the probability $\tilde{P}[v(x_0)] \propto \exp\left(\delta S[v(x_0)]\right)$ of inheritance vectors $v(x_0)$ corresponding to large positive NPL scores. Since the NPL score at x_0 has an expected value depending on δ, the optimal choice of δ clearly depends on the threshold T of interest; see e.g. Naiman and Priebe (2001).[38]

An alternative importance sampling algorithm in the same context is outlined in Malley et al. (2002), where an approximate fast algorithm based on the assumption that Z is a Gaussian process is suggested.[39] This assumption is motivated by the central limit theorem for large N but can be quite inaccurate if N is small and/or there are a few large pedigrees

---

[38]Basically one adjusts $\tilde{P}$ in (35) in order to make $\tilde{V}[f(Z)L(Z)]$ small.

[39]See also Frigessi and Vercellis (1985) and Naiman and Priebe (2001).

in the data set. An advantage of this method is that $\tilde{P}$ puts all it's probability mass on the set $\{Z_{\max} \geq T\}$ where f is positive, but on the other hand it gives, more or less, biased p-value estimates and also requires each marker to be fully polymorphic. An extension in the form of slight assumptional relaxation is outlined in Wu and Naiman (2005) based on, for instance, calculating what is referred to as a *Bonferroni correction factor*.

2.7. Further Topics on Statistical Significance and Evidence

2.7.1. *Fuzzy Significance*

At each locus, with perfect data, all inheritance vectors corresponding to the pedigrees in the pedigree set are known with probability one and the final NPL score is, in this inheritance-information sense, unambiguous. Holding on to this notation, with imperfect data, hence the NPL score is ambiguous since weighting together pedigree-specific scores based on a set of possible inheritance structures. This formation of the NPL score leads to a loss of information through trying to summarize observable data in a more interpretable form.

It is possible to graphically present corresponding results without reducing the information-level through using so called *fuzzy p-values*. Basically, one calculates all possible NPL scores corresponding to all combined possible, consistent with data, pedigree-specific inheritance structures from all instances $S(w_i)$ of (21). Each distinct score is then connected to the corresponding joint pedigree-set probability of outcome. Finally, one may plot the resulting *p-value distribution*.[40]

Fuzzy p-values was discussed in Geyer and Meeden (2005) in the context of treating significance issues for discrete-valued phenomena; more or less a generalization of traditional randomized tests (Lehmann, 1959). Later, this was applied to linkage studies by Thompson (2006), Ängquist (2006b) and Thompson and Geyer (2007).

2.7.2. *Meta-Analyses*

A core concept for strengthening the evidence for true linkage findings is the procedure of replicating positive findings. Another way of enhancing evidence extraction from several (comparable) linkage studies is by performing a combined post-study through a so called *meta-analysis*. The genome-search analysis method (GSMA) is described in Wise et al. (1999) and applied, for example, in Demenais et al. (2003) in the context of type-2 diabetes. This is a ranking-based method applied to the set of linkage results at predefined genome region-bins throughout the observed genome.

---

[40]One consistent outcome may be calculated as in (22) based on the scores $S_1(w_{i_1}), S_2(w_{i_2}), \ldots, S_N(w_{i_N})$ where $S_k(w_{i_k})$ is the score for the kth pedigree corresponding to the consistent inheritance vector $w_{i_k} \in \mathbb{V}_k$, i.e. $p_k(w_{i_k}) > 0$. The probability of this NPL score is then $\prod_{k=1}^{N} p_k(w_{i_k})$. Note that several consistent structures may correspond to the same numeric score leading to merging of probabilities when actually forming the p-value distribution. Moreover, if $|\{w_k\}|$ is the number of consistent inheritance vectors for the kth pedigree, the total number of consistent structures, according to inheritance-independence between pedigrees, are $\prod_{k=1}^{N} |\{w_k\}|$.

2.8.   Some Notes on Related Software

Common analysis programs used when performing nonparametric, and parametric, linkage analysis are GENEHUNTER (Kruglyak et al., 1996),[41] ALLEGRO (Gudbjartsson et al., 2000) and MERLIN (Abecasis et al., 2002). A historically influential program is LINKAGE (Terwilliger and Ott, 1994).

Some programs to perform genetic simulations are SIMLA (Bass et al., 2002) and SIMPED (Leal et al., 2005). Further, the important issue of detecting and correcting genotyping errors may be handled by PEDCHECK (O'Connell and Weeks, 1998; see also Sobel et al., 2002).

An example of a platform for a broad range of genetic analysis programs is commercially offered by BC/GENE (Dahlin, 2005).

3.   Two-Locus NPL Analysis

One may generalize the NPL procedure above in order to simultaneously, or sequentially, search for two distinct disease loci on Ω. We will now briefly outline, and make some comments on, such procedures. The two-locus setting has enriched potential according to its more general approach, but in addition to this comes the increased amount of *multiple testing* inherent in searching for two rather than one disease locus. In spite of this, two-locus linkage analysis can often be worthwile; cf. review articles as Cordell (2002), Strauch et al. (2003) and Hoh and Ott (2003).

3.1.   Unconditional Case

To be able to perform a so called *unconditional two-locus NPL analysis*, we generalize the pedigree-specific NPL score in (21) to,

$$Z(x_1, x_2) = \sum_{i,j} p(w_i, w_j) S(w_i, w_j), \tag{38}$$

where $p(w_i, w_j) = P\big(v(x_1) = w_i, v(x_2) = w_j | \mathrm{MD}\big)$ is the joint inheritance distribution at loci x_1 and x_2, see Strauch et al. (2000). We assume that x_1 and x_2 are unlinked[42] in (38) but otherwise varied independently. This implies allele-sharing independence at the two loci,

$$p(w_i, w_j) = P\big(v(x_1) = w_i | \mathrm{MD}\big) P\big(v(x_2) = w_j | \mathrm{MD}\big),$$

and one-locus multipoint algorithms can be used for calculating (38). The (total) NPL score is then calculated as,

$$Z(x_1, x_2) = \sum_{k=1}^{N} \gamma_k Z_k(x_1, x_2), \tag{39}$$

where $Z_k(x_1, x_2)$ is the pedigree-specific NPL score (38) and γ_k the weight assigned to the kth pedigree.

---

[41]Additional packages exist.   For instance, two-locus analysis is implemented in GENEHUNTER-TWOLOCUS (Strauch et al., 2000).

[42]Formally, $c(x_1) \neq c(x_2)$.

Equation (38) involves a standardized two-locus score function $S(v_1, v_2)$, depending on the pair of inheritance vectors $(v_1, v_2) \in \mathbb{V} \times \mathbb{V}$. Two simple examples of such score functions are: (i) The *additive* two-locus score function,

$$S(v_1, v_2) = \frac{S(v_1) + S(v_2)}{\sqrt{2}}, \tag{40}$$

which essentially is the sum of the corresponding standardized one-locus score functions. (ii) The *multiplicative* two-locus score function,

$$S(v_1, v_2) = S(v_1)S(v_2), \tag{41}$$

which is the product of the corresponding standardized one-locus score functions. Some articles related to such two-locus approaches are Knapp et al. (1994) and Strauch et al. (2000). Consider Li and Reich (2000) for a complete list of distinct two-locus fully penetrant models for binary (biallelic disease loci) traits; an extension to continuous penetrance settings is given in Hallgrimsdottir and Yuster (2006).

Dealing with two disease loci it is possible that the inheritance at these loci interact. We speak of *gene-gene interaction* (also worded as IBD- or allele-sharing correlation); see Lander and Botstein (1986) and Schork et al. (1993). For ASPs:

Definition 7. *Assume two disease loci, l_1 and l_2, with marginal one-locus IBD-sharing vectors (7), given by $z^1 = (z_0^1, z_1^1, z_2^1)$ and $z^2 = (z_0^2, z_1^2, z_2^2)$ respectively, and a joint IBD-sharing matrix (8). Now if,*

$$\exists i, j: \ z_{ij} \neq z_i^1 z_j^2; \ i, j \in \{0, 1, 2\},$$

we say there is gene-gene interaction present.

Since $z = \{z_{ij}\}$ is a function of disease allele frequencies and penetrance values, we may reformulate the genetic model λ in (4) as $\lambda = (z, l)$, where $l = (l_1, l_2)$.

Note that the simple score functions in (40) and (41) are based on one-locus functions $S(v)$. This implies simplicity, but more general two-locus functions might be able to more fully capturing, for example, present IBD-sharing correlation. One example is the two-locus version of the NCP-optimal score function (19),

$$S(w_1, w_2) \propto P(v(l_1) = w_1, v(l_2) = w_2 | Y, \lambda) - 2^{-2m}; \ \forall w_1, w_2 \in \mathbb{V}, \tag{42}$$

which is based on the two-locus disease model λ (Ängquist et al., 2008).

3.1.1. Composite Null Hypotheses

To try to make the search for a second disease loci more efficient one may consider *composite* two-locus null hypotheses; see Ängquist et al. (2005).[43]

---

[43]Generally, defining a simple hypothesis consists of stating an instance $\theta = \theta_0$ of the unknown parameters $\theta \in \Theta$, whereas a composite hypothesis consists of the union of single-instance, or intervals of, parameter values in Θ. In our setting a composite hypothesis refers to allowing for *one* disease locus in the stating of a null hypothesis.

Example 12. *An example of a composite null hypothesis when considering IBD-sharing with respect to a homogeneous set of ASPs is,*

$$H_0 : z \in \mathbb{Z}_0,$$

where

$$\mathbb{Z}_0 = \left\{ z;\ z^1 = (0.25, 0.50, 0.25)\ or\ z^2 = (0.25, 0.50, 0.25) \right\}$$

corresponds to 'At most one disease locus'. In this case the corresponding alternative hypothesis H_1 is 'Two disease loci'.

Ängquist et al. (2005) makes multiple suggestions on how to incorporate a composite null hypothesis into the analysis. Significance calculations may, for instance, be based on: (i) The *least-favourable IBD distribution* to derive theoretical significance level approximations (28). This results in a conservative upper bound,

$$\bar{\alpha}(T) = \max_{\lambda \in H_0} \alpha(T|\lambda),$$

of the significance level. (ii) Estimating a one-locus genetic component, for instance with estimates $\hat{l}_2$ and $\hat{z}^2 = \hat{z}^2(\hat{l}_2)$, constraining the null hypothesis with respect to this component and performing corresponding Monte Carlo simulations to estimate p-values.

3.1.2. Alternative Methods

Unconditional (simultaneous) search can be done for the MLS approach (Cordell et al., 1995; Farrall, 1997; Cordell et al., 2000), by means of regression analysis (Barber et al., 2006) or by specifying the allele-sharing probabilities for affected relative pairs in advance (Dupuis et al., 1995). It may also also be achieved for quantitative traits using variance components methods, either exactly (Almasy and Blangero, 1998) or by means of Markov Chain Monte Carlo (MCMC) approximation (Sung et al., 2007).

3.2. Conditional Case

Restricting the two-locus analysis by letting the locus x_2 in (38) be fixed and conditioning on information at this locus yields a *conditional* two-locus NPL analysis. The *conditioning locus* x_2 may be a verified disease locus, $x_2 = l_2$, a suggested or estimated disease locus, $x_2 = \hat{l}_2$ or a locus considered to be interesting for some other reason. The pedigree-specific type of information to condition on may be one-locus inheritance vectors or NPL scores.

The most well-known example of this kind is the conditional multiplicative two-locus pedigree-specific NPL score introduced by Cox et al. (1999). It may be described through the one-locus pedigree-specific NPL scores in (21) as,

$$Z(x_1, x_2) = Z(x_1) f [Z(x_2)], \tag{43}$$

where $f(\cdot)$ is a given function of the pedigree-specific NPL score at the conditioning locus x_2. When computing the total two-locus NPL score (39) based on pedigree-scores (43), one must replace the unconditional two-locus constraint $\sum_{k=1}^{N} \gamma_k^2 = 1$ on the pedigree weights with,[44]

$$\sum_{k=1}^{N} \gamma_k^2 \, f\,[Z_k(x_2)]^2 = 1.$$

Using (43) is basically a way to utilize gene-gene interaction to increase power. Different choices of the weighting function f is suitable for different types of correlation; for examples see Cox et al. (1999), Ängquist (2001) and Ängquist et al. (2007). We refer to *positive* and *negative* interaction (correlation) as *epistasis* and *heterogeneity* respectively; see Cox et al. (1999) and Holmans (2002).[45] Applications of the approach (43) are described and published, for instance, in Schulze et al. (2004) and Hoffmann et al. (2007).

A quite general conditional approach is outlined in Ängquist et al. (2008) which, for instance, generalizes the methods of Cox et al. (1999) in several ways. The most important extension is to perform *conditioning* on pedigree-specific inheritance vectors rather than NPL scores. This means that one can transfer the general score function-structure of (38) over to the conditional case. For instance, a conditional version of a NCP-optimal score function, see (19 and (42),

$$S(w_1, w_2) \propto P(v(l_1) = w_1 | v(l_2) = w_2, Y, \lambda) - 2^{-m}; \quad \forall w_1 \in \mathbb{V}, \tag{44}$$

may be formulated. Here note that w_2 in (44) is held fixed and in practise, using the NPL score approach, corresponds to the inheritance at a conditioning locus when performing actual calculations based on marker data.

Moreover, the cases of *known* or *unknown* (estimated) conditioning loci are treated separately. The former are defined prior to data analysis, whereas the latter are estimated from an initial one-locus linkage scan. The procedure for unknown loci may be described as follows:

Algorithm 1. *(Unknown conditioning loci)*

1. *Perform chromosome-wise one-locus analysis along Ω.*

2. *Select the (possibly empty) set of conditioning loci $\mathbb{X}_2$ using some predefined selection-criterion.*

3. *Perform conditional two-locus analysis over Ω and $\mathbb{X}_2$. For $x_2 \in \mathbb{X}_2$, we condition on inheritance information at x_2 and calculate scores when x_1 varies along the remaining chromosomes, i.e. $x_1 \in \Omega \setminus c(x_2)$.*

Since possibly (and, if tuned correctly, probably) ending up with several conditioning loci, one needs to account for that in the significance calculations in this multiple testing scenario. A suggestion formulated in a general setting: If performing n different tests and where n may be stochastic, base the global significance test on the p-value,

$$\alpha_{\text{global}} = \min_{1 \le i \le n} \alpha_i(T_i), \tag{45}$$

---

[44]Hence, a conditional two-locus analysis – based on (39) and (43) – may be described within the framework of one-locus analysis by interpreting the product $\gamma_k f\,[Z_k(x_2)]$ as the kth pedigree weight, assigned to $Z_k(x_1)$.

[45]For discussions of the 'correct' usage of these two terms cf. e.g. Vieland and Huang (2003) and its critical replies Cordell (2003) and Farrall (2003). The former view insists on some specific biological interpretations of these terms; I favour the latter one.

where $\alpha_i(T_i)$ is the test-specific p-value corresponding to the ith test, based on test statistic T_i; see Ge et al. (2003). In our case n is the number of conditioning loci and $T_i = Z^i_{\max}$, i.e. the maximum (conditional) two-locus NPL score corresponding to the ith conditioning locus.

3.2.1. Further References and Alternative Methods

The conditional strategy was first proposed by Lander and Botstein (1986) for heterogeneous traits; later to be investigated in, for instance, Dupuis et al. (1995) for nonparametric (allele-sharing) statistics. A related, but somewhat different, methodology is given in Liang et al. (2001) and Chiu and Liang (2004), where conditional linkage analysis are based on generalized estimating equations (GEEs).

It is also possible to proceed sequentially for MLS scores (Pinto et al., 2007) as well as for variance components models (Almasy and Blangero, 1998; Sung et al., 2007).

3.2.2. The Problem with Multiple Testing

Assume a simple situation with a pointwise test for linkage at a single locus x. Choosing a test statistic, calculating your test score and comparing this result to a reference null hypothesis distribution offers no conceptual problem; except perhaps interpretational which is a different story, partially treated in other sections of this chapter.

On the other hand, for instance, when performing a conditional two-locus NPL analysis one scans through, possibly large, genomic regions with many densely positioned markers. In addition one may use a combination of: (i) Several conditioning loci. (ii) Several types of weighting functions. (iii) Several distinct score functions. All this implies a general need for *multiple testing correction*.

A standard approach is the so called *Bonferroni correction*: Assume a combined testing-procedure based on n tests with corresponding test statistics $T_1, T_2, \ldots, T_n$ and critical regions $C_1, C_2, \ldots, C_n$. Now, simple manipulations give

$$P\big(\exists i : \ T_i \in C_i\big) \leq \sum_{j=1}^{n} P(T_j \in C_j).$$

Taking advantage of this inequality, when testing on a global significance level α, the Bonferroni-procedure now perform each individual test on significance level α/n. If the number of tests is large, especially if many test statistics are positively dependent, this is a severe testing problem with respect to power.

In our case, since the individual tests generally are dependent, the standard Bonferroni approach may yield a very conservative upper bound on the *familywise error rate (FWE)*[46] under H_0, i.e. the probability that at least one individual test is declared significant. One possibility to deal with this is to use Monte Carlo simulations to base significance calculations on, for example, (45).

Many refinements of the Bonferroni upper bound of the FWE have been proposed, including the sequential procedures of Holm (1979), Hochberg (1988) and Benjamini and

---

[46]This is also referred to as the *global* multiple testing significance level (Hougaard, 2006).

Hochberg (1995). The latter one controls the *false discovery rate (FDR)*; see Ge et al. (2003) for a up-to-date state-of-the-art review. A very recent monograph is Dudoit and Laan (2008).

3.3. Evidential Interpretation

Let us assume $|l|$ true disease loci constituting a genetic disease model. In practise, these disease loci are fully, or partially, unknown or hidden. Hence, on the basis that the test statistic

$$Z(x_1, x_2, \ldots, x_{|l|}),$$

assumingly correctly rejecting H_0, does not necessarily imply overwhelming evidence for all the correspondingly suggested disease loci

$$\arg\max_{x_1, x_2, \ldots, x_{|l|} \in \Omega} T(x_1, x_2, \ldots, x_{|l|})$$

to be located on true disease regions, or even on true disease chromosomes. The remainder of this section will be dedicated to such evidential and interpretational problems when performing multilocus analysis. We will restrict the discussion to note on differences, in this sense, in interpreting one-locus and, conditional as well as unconditional, two-locus NPL analysis. Related issues are also discussed in Ängquist (2006a).

The general interpretation of significant linkage in the one-locus and unconditional two-locus cases are that there exists 'At least one disease locus on genome Ω'. The latter approach might be restricted by, for instance, introducing composite null hypothesis in the form of allowing for one disease locus in the null hypothesis H_0; taking this into account when performing significance calculations (see Ängquist et al., 2005). In such cases the interpretation is moved towards 'At least two disease loci on Ω'.

The conditional two-locus case is more involved. If using a known conditioning locus x, the interpretation is 'At least one more, except from x, disease locus'. Note that such findings neither automatically strengthen nor weaken the evidence for x as a disease locus. Usually one also assumes that such additional loci is nonsyntenic to x, i.e. not located on chromosome $c(x)$.

If using estimated conditioning loci the interpretation depends on the selection criteria of these loci and the procedure to correct for multiple testing (see Ängquist et al., 2008). Using a somewhat simplified notation, let

$$Z_{\max} = \max_{x_1, x_2} Z(x_1, x_2) \tag{46}$$

be the maximum score found. Here $x_2 \in \mathbb{X}_2$ belongs to the set of selected conditioning loci and x_1 is varied over the corresponding nonsyntenic chromosomes. Now, tightening the selection criteria implies less conditioning loci[47] and hence strengthens the evidence for $x_2 = \hat{l}_2$ corresponding to (46), where $\hat{l}_2$ is an estimate of a second disease locus l_2. At the same time the evidence weakens for the estimate of l_1, a first disease loci, $x_1 = \hat{l}_1$ since a significant finding has increased probability of originating from a conditioning loci being

---

[47]In other words, reduced cardinality of $\mathbb{X}_2$.

a true disease chromosome. The reversed interpretation is valid if weakening the selection criteria.

The core issue here seems to be to clearly separate the interpretation of finding *something*, with some statistical significance, and the evidence for *what* we have found. For instance, assume that $Z_{\max}$ is either attained for one locus or two separate loci; in the latter case nonsyntenic. This surely affects the interpretation of the content of what we have found, but not greatly whether we have an interesting finding or not. If having two peaks of equal height, the evidence for each finding being located in a true disease region is in this sense weakened. This follows since an ambiguity whether only one, or both, of these loci are disease-related. More generally, the complete set of suggested loci

$$\left[\hat{l}_1(x_2), x_2\right], \; x_2 \in \mathbb{X}_2,$$

and their relative distribution of score peak heights, might be interesting for exploratory, descriptive or data mining purposes, even though we have only a signficant finding corresponding to 'At least one disease locus'. See also the discussion of score peaks, and their heights and widths, in Siegmund (2001).

4. Other Statistical Genetics Procedures

In this section we will briefly comment on some related and complementary analysis procedures in the context of gene mapping.

4.1. Segregation Analysis

Prior to linkage analysis, or in its own right, one may perform a *segregation analysis*. This is done in order to determine or estimate genetic model parameters, as well as environmental factors, using phenotype data from pedigrees with many affecteds. For further details see e.g. Khoury et al. (1993) and Haines and Pericak-Vance (2006).

4.2. Association Analysis

Association analysis may be used for fine-mapping regions pinpointed by an initial linkage analysis or directly for genomewide mapping (Risch and Merikangas, 1996). It aims at finding allelic variants associated with disease. Further, one may split association analysis into *population-based* (Clayton, 2001; Balding, 2006)[48] and *family-based* (Terwilliger and Ott, 1992; Spielman et al., 1993; Zhao, 2000) procedures. See also, for instance, Cordell and Clayton (2002, 2005).

Alleles in close physical proximity of the disease locus show association with the disease-causing locus because of linkage disequilibrium. However, association between more distant (and even unlinked) loci may also exist because of *mixture of populations*. The latter source of association is a confounder in gene mapping of disease loci.

---

[48]Including the biostatistical approach of *case-control studies* which often is considered within the field of *epidemiology* (Clayton and Hills, 1993).

In the population-based procedure, one searches for association between phenotypes and allelic variants of single individuals. The method is powerful but sensitive to population admixture. In the family-based procedure, association between phenotypes and transmission of allelic variants from heterozygous parents is of interest. The method is less powerful but robust towards association due to population admixture. These two procedures can also be combined, see e.g. Clayton (1999) and Shih and Whittemore (2002).

Whereas linkage analysis is based on inheritance of alleles within pedigrees, it does not utilize association between allelic variants and phenotypes on the population level. However, the two approaches can be brought together into joint tests for linkage and association, see for example Fulker et al. (1999), Xiong and Jin (2000), Göring and Terwilliger (2000), Sham et al. (2000) and Hössjer (2005a).[49]

4.3. Quantitative Trait Loci Analysis

Using a quantitative phenotype one may perform a *quantitative trait loci (QTL)* analysis, see Lynch and Walsh (1998). Several different subfields of this approach exist. The traditional *Haseman-Elston* approach (Haseman and Elston, 1972) is based on a *regression* of the squared phenotype-difference between sibs with respect to their corresponding allele-sharing and the second, *variance component*, approach is based on splitting up the total phenotype-variation into genetic and environmental factors (Almasy and Blangero, 1998; Cherny et al., 2004). A third approach is based on *IBD-sharing* conditional on observed phenotypes, either by means of a regression model (Sham et al., 2002) or a likelihood score statistic (Tang and Siegmund, 2001; Hössjer, 2005b). For a recent unification of all three types of methods, by means of *generalized estimating equations*, see Chen et al. (2004). A permutation-based procedure related to the visualization of genome-wide significance in the QTL-case is Lystig (2003).

4.4. Further Notes and References

Some general references to linkage analysis, and in a wider sense gene mapping, monographs are Sham (1998), Ott (1999), Almgren et al. (2003), Thomas (2004), Ziegler and Koenig (2006), Haines and Pericak-Vance (2006) and Siegmund and Yakir (2007). Inference on genetic data is outlined in Thompson (2000). A shorter review is given by Ott and Hoh (2000).

A somewhat different field of research is *microarray analysis*. This approach is based on a certain way of measuring *gene expressions* and their relation to a disease of interest. This popularily applied procedure involves a variety of different statistical, as well as technological, aspects and considerations. See e.g. Schena et al. (1995), Schena (2002), Speed (2003) and Bengtsson (2004). Finally, studies of gene-environment interactions are reviewed in Hunter (2005).

---

[49]One may note the following: Assume a disease phenotype caused by a *recent* genetic mutation at locus l and, further, a marker locus m in close vicinity of, but not equivalent to, l. Now, with respect to future generations, m will always be linked to l, but eventually the originally corresponding allelic association will *fade away* according to the recombination process.

Acknowledgements

Many thanks to Professor Ola Hössjer, my former PhD supervisor; now affiliated at Stockholm University. I am most grateful for all inspiration, helpful and thoughtful comments, and numerous suggestions on how to improve this review. For instance, substantial parts of the notes on original research is based on papers which he co-authored; I hope that I referenced this in a proper way, which surely was my intention.

A.   Details on Some Notation

Item	Comments and Examples
$\{a,b,c\}$	A set, i.e. a collection of elements. In this case the elements are a, b and c. In the same sense $\{f(x); x \in [a,b]\}$ may denote a continuous process, i.e. an uncountable set, indexed over interval $[a,b]$.
$f(A\|B)$	Conditioning. The function f applied to A conditioned on, i.e. given, B. [Mainly used with respect to probability functions, i.e. $f = P(\cdot)$ for probability measure P.]
A^T	Transposing. Interchanging rows and columns in matrix A. [Note that A^T is a transposed vector if A is a $(n \times m)$-matrix with $n = 1$ or $m = 1$.]
$a!$	The factorial function, i.e. $a! = a(a-1)\cdots 1$, where a is a positive integer.
$f(x) \propto x^a$	Proportionality symbol. Here $f(x)$ is proportional to the function x^a, i.e. $f(x) = Cx^a$ where C is a constant.
$X \leftarrow Y$	Updating assignment. Declare $X = Y$, where $Y = f(X)$ is a function of X.
$[a(b[c(d)])]$	Our adopted sequential delimiter system. Applied both to composite functions and separators.
$\hat{p}$	Denotes an estimate of a parameter or distribution, i.e. it is a function, depending on the available data, trying to estimate the true p.
$a:b:c$	A vector $(v_1, v_2, \ldots, v_n)$ where: (i) The first value is $v_1 = a$. (ii) For all $k: 2 \le k \le n$, the kth element is $v_k = v_{k-1} + b$. (iii) Moreover, $c - b < v_n \le c$, i.e. the final value v_n is at most c, but larger than $c - b$.
$A \backslash B$	Complementary set to B with respect to A, i.e. $x \in A \backslash B$ if and only if $x \in A \cap x \notin B$. [Also called the *relative complement* of B with respect to A (Khuri, 2003).]

References

Abecasis, G. R., Cherny, S. S., Cookson, W. O. and Cardon, L. R. (2002). MERLIN - rapid analysis of dense genetic maps using sparse gene flow trees. *Nature Genetics*, **30**, 97–101.

Abecasis, G. R. and Wigginton, J. E. (2005). Handling marker-marker linkage disequilibrium: Pedigree analysis with clustered markers. *American Journal of Human Genetics*, **77**, 754–767.

Aldous, D. (1989). *Probability approximations via the Poisson clumping heuristic.* New York: Springer-Verlag.

Almasy, L. and Blangero, J. (1998). Multipoint quantitative-trait linkage analysis in general pedigrees. *American Journal of Human Genetics*, **62**, 1198–1211.

Almgren, P., Bendahl, P. O., Bengtsson, H., Hössjer, O. and Perfekt, R. (2003). *Statistics in genetics.* Department of Mathematical Statistics: Lund University.

Ängquist, L. (2001). *Conditional two-locus NPL-analyses: Theory and applications* (Master's Thesis No. 2001:E22). Lund: Department of Mathematical Statistics, Lund University.

Ängquist, L. (2006a, December). *A discussion on evidential interpretation of two-locus nonparametric linkage analysis.* (Free download from homepage: 'http://www.maths.lth.se/matstat/staff/larsa/'.)

Ängquist, L. (2006b, October). *Interpreting significance in nonparametric linkage analysis: Fuzzy p-values and information levels.* (Free download from homepage: 'http://www.maths.lth.se/matstat/staff/larsa/'.)

Ängquist, L. (2006c, June). *Some notes on the choice of score function in nonparametric linkage analysis.* (Free download from homepage: 'http://www.maths.lth.se/matstat/staff/larsa/'.)

Ängquist, L. (2007a, February). *Improve NPL analysis through pedigree-selection: The maximum selected pedigree subset-score method.* (Free download from homepage: 'http://www.maths.lth.se/matstat/staff/larsa/'.)

Ängquist, L. (2007b). *Pointwise and genomewide significance calculations in gene mapping through nonparametric linkage analysis: Theory, algorithms and applications* (Doctoral Thesis No. 2006:15). Lund: Department of Mathematical Statistics, Lund University.

Ängquist, L. (2008). A unified discussion on the concept of score functions used in nonparametric linkage analysis. *Bioinformatics and Biology Insights*, **2**, 119–132.

Ängquist, L., Anevski, D. and Luthman, H. (2005). *Unconditional two-locus nonparametric linkage analysis: On composite null hypotheses with and without gene-gene interaction* (Tech. Rep. No. 2005:28). Lund: Department of Mathematical Statistics, Lund University.

Ängquist, L. and Hössjer, O. (2004, May). Using importance sampling to improve simulation in linkage analysis. *Statistical Applications in Genetics and Molecular Biology*, **3**(1:5). (Electronic journal, 24 pages)

Ängquist, L. and Hössjer, O. (2005). Improving the calculation of statistical significance in genome-wide scans. *Biostatistics*, **6**(4), 520–538.

Ängquist, L., Hössjer, O. and Groop, L. (2007). *Strategies for conditional two-locus non-*

parametric linkage analysis (Tech. Rep. No. 2007:1). Lund: Department of Mathematical Statistics, Lund University.

Ängquist, L., Hössjer, O. and Groop, L. (2008). *Strategies for conditional two-locus nonparametric linkage analysis* (Tech. Rep. No. 2007:1). Lund: Department of Mathematical Statistics, Lund University. (Accepted for publication in 'Human Heredity'. Most likely published in volume 2008 66:3.)

Asmussen, S. and Glynn, P. W. (2007). *Stochastic simulation: Algorithms and analysis.* New York: Springer-Verlag.

Bacanu, S. A. (2005). Robust estimation of critical values for genome scans to detect linkage. *Genetic Epidemiology*, **28**, 24–32.

Balding, D. J. (2006). A tutorial on statistical methods for population association studies. *Nature Reviews Genetics*, **7**, 781–791.

Barber, M. J., Todd, J. A. and Cordell, H. J. (2006). A multimarker regression-based test of linkage for affected sib-pairs at two linked loci. *Genetic Epidemiology*, **30**, 191–208.

Barnard, G. A. (1949). Statistical inference [Series B (Methodological)]. *Journal of the Royal Statistical Society*, **11**, 115–149.

Bass, M., martin, E. and Hauser, E. (2002). Software for simulation studies of complex traits: SIMLA. *American Journal of Human Genetics*, **71**, 569.

Basu, S. (2005). *Allele-sharing methods for linkage detection using extended pedigrees* (Doctoral Thesis). Washington: Department of Statistics, University of Washington.

Bengtsson, H. (2004). *Low-level analysis of microarray data* (Doctoral Thesis No. 2004:6). Lund: Department of Mathematical Statistics, Lund University.

Bengtsson, O. (2001). *Two-locus affected sib-pair identity by descent probabilities: Constraints, parameterisation and estimation* (Licentiate Thesis). Göteborg: Department of Mathematical Statistics, Chalmers University of Technology, Göteborg University.

Benjamini, Y. and Hochberg, Y. (1995). Controlling the false discovery rate: A practical and powerful approach to multiple testing [Series B (Methodological)]. *Journal of the Royal Statistical Society*, **57**(1), 289–300.

Blackwell, P. (2002). Ornstein-Uhlenbeck process. In R. C. Elston, J. M. Olson and L. Palmer (Eds.), *Biostatistical genetics and genetic epidemiology* (pp. 585–588). New York: John Wiley & Sons.

Boehnke, M. (1986). Estimating the power of a proposed linkage study: A practical computer simulation approach. *American Journal of Human Genetics*, **39**, 513–527.

Bradley, A. P. (1996). ROC curves and the χ^2 test. *Pattern Recognition Letters*, **17**, 287–294.

Bremaud, P. (2001). *Markov chains: Gibbs fields, Monte Carlo simulation, and queues* (Second ed.). New York: Springer.

Cappé, O., Moulines, E. and Rydén, T. (2005). *Inference in hidden Markov models* [Springer Series in Statistics]. New York: Springer.

Chen, W. M., Broman, K. W. and Liang, K. Y. (2004). Quantitative trait linkage analysis by generalized estimating equations: Unification of variance components and Haseman-Elston regression. *Genetic Epidemiology*, **26**, 265–272.

Cherny, S. S., Sham, P. C. and Cardon, L. R. (2004, March). Introduction to the special issue on variance components methods for mapping quantitative trait loci. *Behavior Genetics*, **34**(2), 125–126.

Chiu, Y. F. and Liang, K. Y. (2004). Conditional multipoint linkage analysis using affected sib pairs: An alternative approach. *Genetic Epidemiology*, **26**, 108–115.

Clark, P. K. (1973). A subordinated stochastic process model with finite variance for speculative prices. *Econometrica*, **41**(1), 135–155.

Clayton, D. (1999). A generalization of the transmission/disequilibrium test for uncertian-haplotype transmission. *American Journal of Human Genetics*, **65**, 1170–1177.

Clayton, D. (2001). Population association. In D. J. Balding, M. Bishop and C. Cannings (Eds.), *Handbook of statistical genetics* (pp. 519–540). Chichester: John Wiley & Sons.

Clayton, D. and Hills, M. (1993). *Statistical models in epidemiology*. Oxford: Oxford University Press.

Clerget-Darpoux, F., Bonaiti-Pellie, C. and Hochez, J. (1986). Effects of misspecifying genetic parameters in lod score analysis. *Biometrics*, **42**, 393–399.

Clerget-Darpoux, F. and Elston, R. C. (2007). Are linkage analysis and the collection of family data dead? prospects for family studies in the age of genomewide association. *Human Heredity*, **64**, 91–96.

Collins, A., Frezal, J., Teague, J. and Morton, N. E. (1996). A metric map of humans: 23.500 loci in 850 bands. *Proceedings of the National Academy of Sciences of the United States of America*, **93**, 14771–14775.

Cordell, H. J. (2002). Epistasis: What it means, what it doesn't mean, and statistical methods to detect it in humans. *Human Molecular Genetics*, **11**, 2463–2468.

Cordell, H. J. (2003). Affected sib-pair data can be used to distinguish two-locus hetero-geneity from two-locus epistasis. *American Journal of Human Genetics*, **73**, 1468–1471. (Discussion of article by Vieland and Huang, 2003)

Cordell, H. J. and Clayton, D. G. (2002). A unified stepwise regression procedure for evaluating the relative effects of polymorphisms within a gene using case/control or family data: Applications to HLA in type 1 diabetes. *American Journal of Human Genetics*, **70**, 124–141.

Cordell, H. J. and Clayton, D. G. (2005). Genetic association studies. *Lancet*, **366**, 1121–1131.

Cordell, H. J., Todd, J. A., Bennett, S. T., Kawaguchi, Y. and Farrall, M. (1995). Two-locus maximum lod score analysis of a multifactorial trait: Joint consideration of IDDM2 and IDDM4 with IDDM1 in type 1 diabetes. *American Journal of Human Genetics*, **57**, 920–934.

Cordell, H. J., Wedig, G. C., Jacobs, K. B. and Elston, R. C. (2000). Multilocus linkage tests based on affected relative pairs. *American Journal of Human Genetics*, **66**, 1273–1286.

Cox, N. J., Frigge, M., Nicolae, D. L., Concannon, P., Hanis, C. L., Bell, G. I. and Kong, A. (1999). Loci on chromosomes 2 (NIDDM1) and 15 interact to increase susceptibility to diabetes in Mexican Americans. *Nature Genetics*, **21**, 213–215.

Dahlin, A. (2005, December). *Manual bc/gene: Analysis programs*. (For free download, visit homepage: 'http://www.bcplatforms.com'.)

Davis, S. and Weeks, D. E. (1997). Comparisons of nonparametric statistics for detection of linkage in nuclear families: Single-marker evaluations. *American Journal of Human Genetics*, **61**, 1431–1444.

Demenais, F., Kanninen, T., Lindgren, C. M., Wiltshire, S., Gaget, S., Dandrieux, C., Almgren, P., Sjögren, M., Hattersley, A., Dina, C., Tuomi, T., McCarthy, M. I., Froguel, P. and Groop, L. C. (2003). A meta-analysis of four European genome screens (GIFT Consortium) shows evidence for a novel region on chromosome 17p11.2-q22 linked to type 2 diabetes. *Human Molecular Genetics*, **12**(15), 1865–1873.

Donnelly, K. P. (1983). The probability that related individuals share some section of the genome identical by descent. *Theoretical Population Biology*, **23**, 34–64.

Dudoit, S. and Laan, M. van der. (2008). *Multiple testing procedures with applications in genomics*. New York: Springer.

Dudoit, S. and Speed, T. P. (1998, Jul). *Triangle constraints for sib-pair identity by descent probabilities under a general model for disease susceptibility* (Tech. Rep. No. 527). Department of Statistics, University of California, Berkeley.

Dupuis, J., Brown, P. O. and Siegmund, D. (1995, Jun). Statistical methods for linkage analysis of complex traits from high-resolution maps of identity by descent. *Genetics*, **140**, 843–856.

Durbin, R., Eddy, S., Krogh, A. and Mitchinson, G. (1998). *Biological sequence analysis: Probabilistic models of proteins and nucleic acids*. Cambridge: Cambridge University Press.

Edwards, A. W. F. (1992). *Likelihood: Expanded edition* (Second Edition ed.). New York: John Hopkins University Press.

Elston, R. C. and Stewart, J. (1971). A general model for the analysis of pedigree data. *Human Heredity*, **21**, 523–542.

Farrall, M. (1997). Affected sibpair linkage tests for multiple linked susceptibility genes. *Genetic Epidemiology*, **14**, 103–115.

Farrall, M. (2003). Reports of the death of the epistasis model are greatly exaggerated. *American Journal of Human Genetics*, **73**, 1467–1468. (Discussion of article by Vieland and Huang, 2003)

Feingold, E. (1993). Markov processes for modeling and analyzing a new genetic mapping method. *Journal of Applied Probability*, **30**, 766–779.

Feingold, E., Brown, P. O. and Siegmund, D. (1993). Gaussian models for genetic linkage analysis using complete high-resolution maps of identity by descent. *American Journal of Human Genetics*, **53**, 234–251.

Feingold, E., Song, K. K. and Weeks, D. E. (2000). Comparisons of allele-sharing statistics for general pedigrees. *Genetic Epidemiology*, **19**, 92–98. (Supplement 1)

Frigessi, A. and Vercellis, C. (1985). An analysis of Monte Carlo algorithms for counting problems. *Calcolo*, **22**(4), 413–428.

Fulker, D. W., Cherny, S., Sham, P. C. and Hewitt, J. K. (1999). Combined linkage and association sib-pair analysis for quantitative traits. *American Journal of Human Genetics*, **64**, 259–267.

Ge, Y., Dudoit, S. and Speed, T. P. (2003). Resampling-based multiple testing for microarray data analysis. *Sociedad Española de Estadística e Investigación Operativa Test*, **12**(1), 1–77. (With discussion.)

Gelman, A., Carlin, J. B., Stern, H. S. and Rubin, D. B. (2004). *Bayesian data analysis* (Second Edition ed.) [Texts in Statistical Science]. Boca Raton (Florida): Chapman & Hall/CRC.

Geyer, C. J. and Meeden, G. D. (2005). Fuzzy and randomized confidence intervals and p-values. *Statistical Science*, **20**(4), 358–387. (With discussion.)

Göring, H. H. H. and Terwilliger, J. D. (2000). Linkage analysis in the presence of errors IV: Joint pseudomarker analysis of linkage and/or linkage disequilibrium on a mixture of pedigrees and singletons when the mode of inheritance cannot be accurately specified. *American Journal of Human Genetics*, **66**, 1310–1327.

Gudbjartsson, D. F., Jonasson, K., Frigge, M. and Kong, A. (2000). ALLEGRO, a new computer program for multipoint linkage analysis. *Nature Genetics*, **25**, 12–13.

Gut, A. (1995). *An intermediate course in probability*. New York: Springer-Verlag.

Hacking, I. (1965). *Logic of statistical inference*. New York: Cambridge University Press.

Häggström, O. (2002). *Finite Markov chains and algorithmic applications*. Cambridge: Cambridge University Press.

Haines, J. L. and Pericak-Vance, M. A. (Eds.). (2006). *Genetic analysis of complex disease*. New York: Wiley-Liss.

Haldane, J. B. S. (1919). The combination of linkage values and the calculation of distances between loci of linked factors. *Genetics*, **8**, 299–309.

Haldane, J. B. S. and Smith, C. A. B. (1947). A new estimate of the linkage between the genes for colour-blindness and haemophilia in man. *Annals of Eugenics*, **14**, 10–31.

Hall, P. (1992). *The bootstrap and Edgeworth expansion*. New York: Springer-Verlag.

Hallgrimsdottir, I. B. and Yuster, D. S. (2006). *A complete classification of epistatic two-locus models* (arXiv: q.bio.QM No. 0612044v1). Global: Web-based.

Hammersley, J. M. and Handscomb, D. C. (1964). *Monte Carlo methods*. New York: John Wiley & Sons.

Hand, D. J., Mannila, H. and Smyth, P. (2001). *Principles of data mining*. Cambridge, Massachusetts: The MIT Press.

Haseman, J. K. and Elston, R. C. (1972). The investigation of linkage between a quantitative trait and a marker locus. *Behavior Genetics*, **2**, 3–19.

Hernández, S., Siegmund, D. O. and Gunst, M. D. (2005). On the power for linkage detection using a test based on scan statistics. *Biostatistics*, **6**(2), 259–269.

Hesterberg, R. (1995). Weighted average importance sampling and defensive mixture distributions. *Technometrics*, **37**(2), 185–194.

Hochberg, Y. (1988, December). A sharper Bonferroni procedure for multiple tests of significance. *Biometrika*, **75**(4), 800–802.

Hoffmann, K., Mattheisen, M., Dahm, S., Nürnberg, P., Roe, C., Johnson, J., Cox, N. J., Wichmann, H. E., Wienker, T. F., Schulze, J., Schwarz, P. E. and Lindner, T. H. (2007). A German genomewide linkage scan for type 2 diabetes supports the existence of a metabolic syndrome locus on chromosome 1p36.13 and a type 2 diabetes locus on chromosome 16p12.2. *Diabetologia*, **50**, 1418–1422.

Hoh, J. and Ott, J. (2003, September). Mathematical multi-locus approaches to localizing complex human trait genes. *Nature Reviews Genetics*, **4**, 701–709.

Holm, S. (1979). A simple sequentially rejective multiple test procedure. *Scandinavian Journal of Statistics*, **6**, 65–70.

Holmans, P. (1993). Asymptotic properties of affected-sib-pair linkage analysis. *American Journal of Human Genetics*, **52**, 362–374.

Holmans, P. (2002). Detecting gene-gene interactions using affected sib pair analysis with covariates. *Human Heredity*, **53**, 92–102.

Hössjer, O. (2001). *Asymptotic estimation theory of multipoint linkage analysis under perfect marker information* (Tech. Rep. No. 2001:16). Lund: Department of Mathematical Statistics, Lund University.

Hössjer, O. (2003a). Asymptotic estimation theory of multipoint linkage analysis under perfect marker information. *Annals of Statistics*, **31**(4), 1075–1109.

Hössjer, O. (2003b). Determining inheritance distributions via stochastic penetrances. *Journal of the American Statistical Association*, **98**, 1035–1051.

Hössjer, O. (2005a). Combined association and linkage analysis for general pedigrees and genetic models. *Statistical Applications in Genetics and Molecular Biology*, **4**(1:11). (Electronic journal, 42 pages)

Hössjer, O. (2005b). Conditional likelihood score functions for mixed models in linkage analysis. *Biostatistics*, **6**(2), 313–332.

Hössjer, O. (2005c). Information and effective number of meioses in linkage analysis. *Journal of Mathematical Biology*, **50**(2), 208–232.

Hössjer, O. (2006). Modeling the effect of inbreeding among founders in linkage analysis. *Theoretical Population Biology*, **70**(2), 146–163.

Hougaard, P. (2006). *Multiple testing: A clinical trial perspective* [Draft Notes]. Biostatistics: Lundbeck, Valby, Denmark. (Medicon Valley Academy Course Material: IDEON, Lund, 2006-03-06.)

Hsu, Y. S. and Park, W. J. (1985). Ornstein-Uhlenbeck process [A Wiley-Interscience Publication]. In S. Kotz and N. L. Johnson (Eds.), *Encyclopedia in statistical sciences* (Vol. 6, pp. 518–521). New York: John Wiley & Sons.

Hunter, D. J. (2005, April). Gene-environment interactions in human diseases. *Nature Reviews Genetics*, **6**, 287–298.

International Human Genome Sequencing Consortium. (2001). Initial sequencing and analysis of the human genome. *Nature*, **409**, 860–921.

International Human Genome Sequencing Consortium. (2004, October). Finishing the euchromatic sequence of the human genome. *Nature*, **431**, 931–945.

Jelinek, F. (1998). *Statistical methods for speech recognition*. Cambridge, Massachussets: The MIT Press.

Jung, J., Weeks, D. E. and Feingold, E. (2006). Gene-dropping vs. empirical variance estimation for allele-sharing linkage statistics. *Genetic Epidemiology*, **30**, 652–665.

Khoury, M. J., Beaty, T. H. and Cohen, B. C. (1993). *Fundamentals of genetic epidemiology* [Monographs un Epidemiology and Biostatistics, Volume 22]. New York and Oxford: Oxford University Press.

Khuri, A. I. (2003). *Advanced calculus with applications in statistics* (Second ed.) [Wiley Series in Probability and Statistics]. Hoboken (New Jersey): Wiley-Interscience.

Knapp, M., Seuchter, S. A. and Baur, M. (1994). Two-locus disease models with two marker loci: The power of affected sib-pair tests. *American Journal of Human Genetics*, **55**, 1030–1041.

Kong, A. and Cox, N. (1997). Allele-sharing models: LOD scores and accurate linkage tests. *American Journal of Human Genetics*, **61**, 1179–1188.

Kosambi, D. D. (1944). The estimation of map distances from recombination values. *Annals of Eugenics*, **12**, 172–175.

Koski, T. (2001). *Hidden Markov models for bioinformatics* [Computational Biology, Volume 2]. Dordrecht, Boston and London: Kluwer Academic Publishers.

Kruglyak, L., Daly, M. J. and Lander, E. S. (1995). Rapid multipoint linkage analysis of recessive traits in nuclear families, including homozygosity mapping. *American Journal of Human Genetics*, **56**, 519–527.

Kruglyak, L., Daly, M. J., Reeve-Daly, M. P. and Lander, E. S. (1996). Parametric and nonparametric linkage analysis: A unified multipoint approach. *American Journal of Human Genetics*, **58**, 1347–1363.

Kruglyak, L. and Lander, E. S. (1995). Complete multipoint sib-pair analysis of qualitative and quantitative traits. *American Journal of Human Genetics*, **57**, 439–454.

Kruglyak, L. and Lander, E. S. (1998). Faster multipoint linkage analysis using Fourier transforms. *Journal of Computational Biology*, **5**(1), 1–7.

Kullback, S. (1968). *Information theory and statistics* (Second ed.). New York: Dover.

Kurbasic, A. (2007). *Topics in human gene mapping* (Doctoral Thesis No. 2006:14). Lund: Department of Mathematical Statistics, Lund University.

Kurbasic, A. and Hössjer, O. (2004). On computation of p-values in parametric linkage analysis. *Human Heredity*, **57**, 207–219.

Kurbasic, A. and Hössjer, O. (2006). Relative risks and effective number of meioses: A unified approach for general genetic models and phenotypes. *Annals of Human Genetics*, **70**, 907–922.

Kurbasic, A. and Hössjer, O. (2007). *A general method for linkage disequilibrium correction for multipoint linkage and association.* (Under revision for 'Genetic Epidemiology')

Lander, E. S. and Botstein, D. (1986, October). Strategies for studying heterogeneous genetic traits in humans by using a linkage map of restriction fragment length polymorphisms. *Proceedings of the National Academy of Sciences of the United States of America*, **83**(19), 7353–7357.

Lander, E. S. and Botstein, D. (1989). Mapping Mendelian factors underlying quantitative traits using RFLP linkage maps. *Genetics*, **121**, 185–199.

Lander, E. S. and Green, P. (1987). Construction of multilocus genetic linkage maps in humans. *Proceedings of the National Academy of Sciences of the United States of America*, **85**, 2363–2367.

Lander, E. S. and Kruglyak, L. (1995). Genetic dissection of complex traits: Guidelines for interpreting and reporting linkage results. *Nature Genetics*, **11**, 241–247.

Lange, E. M. and Lange, K. (2004). Powerful allele-sharing statistics for nonparametric analysis. *Human Heredity*, **57**, 49–58.

Leadbetter, R., Lindgren, G. and Rootzén, H. (1983). *Extremes and related properties of random sequences and processes* [Springer Series in Statistics]. Berlin: Springer-Verlag.

Leal, S. M., Yan, K. and Müller-Myhsok, B. (2005). A simulation program to generate haplotype and genotype data for pedigree structures. *Human Heredity*, **60**(2), 119–122.

Lehmann, E. L. (1959). *Testing statistical hypotheses* (First ed.). New York: Wiley.

Lehmann, E. L. and Casella, G. (1998). *Theory of point estimation* (Second ed.). New York: Springer-Verlag.

Leibon, G., Rockmore, D. and Pollak, M. R. (2007). *A simple computational method for the identification of disease-associated loci in complex incomplete pedigrees* (arXiv: q.bio.GN No. 0710.5625v1). Global: Web-based.

Li, W. and Reich, J. (2000). A complete enumeration and classification of two-locus disease models. *Human Heredity*, **50**, 334–349.

Liang, K. Y., Chiu, Y. F., Beaty, T. H. and Wjst, M. (2001). Multipoint analysis using affected sib-pairs: Incorporating linkage evidence from unlinked regions. *Genetic Epidemiology*, **21**, 105–122.

Lucek, P., Hanke, J., Reich, J., Solla, S. A. and Ott, J. (1998). Multi-locus nonparametric linkage analysis of complex trait loci with neural networks. *Human Heredity*, **48**, 275–284.

Lynch, M. and Walsh, B. (1998). *Genetics and analysis of quantitative traits*. Sunderland, Massachusetts: Sinauer Associates , Inc.

Lystig, T. C. (2003, August). Adjusted p-values for genome-wide scans. *Genetics*, **164**, 1683–1687.

Malley, J. D., Naiman, D. and Bailey-Wilson, J. (2002). A comprehensive method for genome scans. *Human Heredity*, **54**, 174–185.

Markianos, K., Daly, M. J. and Kruglyak, L. (2001). Efficient multipoint linkage analysis through reduction of inheritance space. *American Journal of Human Genetics*, **68**, 963–977.

McPeek, M. S. (1999). Optimal allele-sharing statistics for genetic mapping using affected relatives. *Genetic Epidemiology*, **16**, 225–249.

Mendel, J. G. (1866). Versuche über pflanzen-hybriden. *Verhandlungen Naturforschende Vereinigung Brünn*, **4**, 3–47.

Morgan, T. H. (1928). *The theory of genes*. New Haven: Yale University Press.

Morton, N. E. (1955). Sequential tests for the detection of linkage. *American Journal of Human Genetics*, **7**, 277–318.

Naiman, D. Q. and Priebe, C. (2001). Computing scan statistic *p*-values using importance sampling, with applications to genetics and medical image analysis. *Journal of Computational and Graphical Statistics*, **10**(2), 296–328.

Neuman, R. J., Saccone, N. L., Holmans, P., Rice, J. P. and Sun, L. (2000). Clustering methods applied to allele sharing. *Genetic Epidemiology*, **19**, S57–S63. (Supplement 1)

Nicolae, D. L. (1999, Jun). *Allele sharing models in gene mapping: A likelihood approach* (Doctoral Thesis). Chicago: Department of Statistics, University of Chicago.

Nicolae, D. L., Frigge, M. L., Cox, N. J. and Kong, A. (1998). Discussion. *Biometrics*, **54**, 1271–1274. (Discussion of article by Teng and Siegmund, 1998)

Nicolae, D. L. and Kong, A. (2004). Measuring the relative information in allele-sharing linkage studies. *Biometrics*, **60**, 368–375.

Nilsson, S. (1999). *Two contributions to genetic linkage analysis* (Licentiate Thesis). Göteborg: Department of Mathematical Statistics, Chalmers University of Technology and Göteborg University.

O'Connell, J. R. (2001). Rapid multipoint linkage analysis via inheritance vectors in the Elston-Stewart algorithm. *Human Heredity*, **51**, 226–240.

O'Connell, J. R. and Weeks, D. E. (1998). PedCheck: a program for identification of genotype incompatibilities in linkage analysis. *American Journal of Human Genetics*, **63**(1), 259–266.

Ott, J. (1989). Computer-simulation methods in human linkage analysis. *Proceedings of the National Academy of Sciences of the United States of America*, **86**(11), 4175–4178.

Ott, J. (1999). *Analysis of human genetic linkage* (Third ed.). New York: The John Hopkins University Press.

Ott, J. and Hoh, J. (2000). Statistical approaches to gene mapping. *American Journal of Human Genetics*, **67**, 289–294.

Pierce, J. R. (1965). *Symbols, signals and noise: The nature and process of communication*. New York, Evanston and London: Harper Torchbooks.

Pinto, D., Trenité, D. G. A. K.-N., Cordell, H. J., Mattheisen, M., Strauch, K., Lindhout, D. and Koeleman, B. P. C. (2007). Explorative two-locus linkage analysis suggests a multiplicative interaction between the 7q32 and 16p13 myoclonic seizures-related photosensitivity loci. *Genetic Epidemiology*, **31**, 42–50.

Ploughman, L. M. and Boehnke, M. (1989). Estimating the power of a proposed linkage study for a complex genetic trait. *American Journal of Human Genetics*, **44**, 543–551.

Rabiner, L. R. (1989, February). A tutorial on hidden Markov models and selected applications in speech recognition. *Proceedings of the IEEE*, **77**(2), 257–286.

Risch, N. (1984). Segregation analysis incorporating genetic markers. I. single-locus models with an application to type I diabetes. *American Journal of Human Genetics*, **36**, 363–386.

Risch, N. (1990). Linkage strategies for genetically complex traits: Iii. the effect of marker polymorphism on analysis of affected relative pairs. *American Journal of Human Genetics*, **46**(2), 242–253.

Risch, N. and Merikangas, K. (1996, September). The future of genetic studies of complex human diseases. *Science*, **273**(5281), 1516–1517.

Robert, C. P. and Casella, G. (2004). *Monte Carlo statistical methods* (Second ed.) [Springer Texts in Statistics]. New York: Springer.

Ross, S. M. (2006). *Simulation* (Fourth ed.). San Diego: Academic Press.

Royall, R. M. (1997). *Statistical inference: A likelihood paradigm*. London: Chapman & Hall.

Saunders, I. W., Hannan, G. N., Brohede, J., Giles, G. G., Jenkins, M. A. and Southey, J. L. H. M. C. (2007). A range of simple summary genome-wide statistics for detecting genetic linkage using high density marker data. *Genetic Epidemiology*, **31**, 565–576.

Schena, M. (2002). *Microarray analysis*. New Jersey: Wiley-Liss.

Schena, M., Shalon, D., Davis, R. W. and Brown, P. O. (1995, October). Quantitative monitoring of gene expression patterns with a complementary DNA microarray. *Science*, **270**(5235), 467–470.

Schork, N. J., Boehnke, M., Terrwilliger, J. D. and Ott, J. (1993). Two-trait-locus linkage analysis: A powerful strategy for mapping complex genetic traits. *American Journal of Human Genetics*, **53**, 1127–1136.

Schulze, T. G., Buervenich, S., Badner, J. A., Steele, C. J. M., Detera-Wadleigh, S. D., Dick,

D., Foroud, T., Cox, N. J., MacKinnon, D. F., Potash, J. B., Berrettini, W. H., Byerley, W., Coryell, W., Jr, J. R. D., Gershon, E. S., Kelsoe, J. R., McInnis, M. G., Murphy, D. L., Reich, T., Scheftner, W., Jr, J. I. N. and McMahon, F. J. (2004). Loci on chromosomes 6q and 6p interact to increase susceptibility to bipolar affective disorder in the National Institute of Mental Health Genetics Initiative pedigrees. *Biological Psychiatry*, **56**, 18–23.

Selin, I. (1965). *Detection theory* [The RAND Corporation]. Princeton, New Jersey: Princeton University Press.

Sengul, H., Weeks, D. E. and Feingold, E. (2001). A survey of affected-sibship statistics for nonparametric linkage analysis. *American Journal of Human Genetics*, **69**, 179–190.

Sham, P. (1998). *Statistics in human genetics*. London: Arnold Applications of Statistics.

Sham, P., Zhao, J. and Curtis, D. (1997). Optimal weighting scheme for affected sib-pair analysis of sibship data. *Annals of Human Genetics*, **61**, 61–69.

Sham, P. C., Cherny, S., Purcell, S. and Hewitt, J. K. (2000). Power of linkage versus association analysis of quantitative traits, by use of variance-components methods, for sibship data. *American Journal of Human Genetics*, **66**, 1616–1630.

Sham, P. C., Purcell, S., Cherny, S. and Abecasis, G. R. (2002). Powerful regression-based quantitative-trait linkage analysis of general pedigrees. *American Journal of Human Genetics*, **71**, 238–253.

Shannon, C. E. (1948). A mathematical theory of communication. *Bell System Technical Journal*, **27**, 379–423 and 623–656.

Shih, M. C. and Whittemore, A. S. (2002). Tests for genetic association using family data. *Genetic Epidemiology*, **22**, 128–145.

Siegmund, D. (1985). *Sequential analysis: Tests and confidence intervals* [Springer Series in Statistics]. Berlin: Springer-Verlag.

Siegmund, D. (2001). Is peak height sufficient? *Genetic Epidemiology*, **20**, 403–408.

Siegmund, D. and Yakir, B. (2007). *The statistics of gene mapping* [Statistics for Biology and Health]. New York: Springer.

Smith, C. A. B. (1959). Some comments on the statistical methods used in linkage investigations. *American Journal of Human Genetics*, **11**, 289–304.

Sobel, E. and Lange, K. (1996). Descent graphs in pedigree analysis: Applications to haplotyping, location scores, and marker-sharing statistics. *American Journal of Human Genetics*, **58**, 1323–1337.

Sobel, E., Papp, J. C. and Lange, K. (2002). Detection and integration of genotyping errors in statistical genetics. *American Journal of Human Genetics*, **70**, 496–508.

Song, K. K., Weeks, D. E., Sobel, E. and Feingold, E. (2004). Efficient simulation of P values for linkage analysis. *Genetic Epidemiology*, **26**, 88–96.

Speed, T. (Ed.). (2003). *Statistical analysis of gene expression microarray data* [Interdisciplinary Statistics]. Florida: Chapman & Hall.

Spielman, R. S., McGinnis, R. and Ewens, W. J. (1993). Transmission test for linkage disequilibrium: The insulin gene region and insulin-dependent diabetes mellitus (IDDM). *American Journal of Human Genetics*, **52**, 506–516.

Strachan, T. and Read, A. P. (2003). *Human molecular genetics* (Third ed.). London and New York: Garland Science.

Strauch, K., Fimmers, R., Kurz, T., , Baur, M. P. and Wienker, T. F. (2003). How to model a complex trait: 2. analysis with two disease loci. *Human Heredity*, **56**, 200–211.

Strauch, K., Fimmers, R., Kurz, T., Deichmann, K. A., Wienker, T. F. and Baur, M. P. (2000). Parametric and nonparametric multipoint linkage analysis with imprinting and two-locus-trait models: Application to mite sensitization. *American Journal of Human Genetics*, **66**, 1945–1957.

Strauch, K., Fürst, R., Rüschendorf, F., Windemuth, C., Dietter, J., Flaquer, A., Baur, M. P. and Wienker, T. F. (2005). Linkage analysis of alcohol dependence using mod scores. *BMC Genetics*, **6**, S162. (Supplement 1)

Strug, L. J. and Hodge, S. E. (2006). An alternative foundation for the planning and evaluation of linkage analysis: I. decoupling 'error probabilities' from 'measures of evidence'. *Human Heredity*, **61**, 166–188.

Suarez, B. K. (1978). The affected sib-pair IBD distribution for HLA-linked disease susceptibility genes. *Tissue Antigens*, **12**, 87–93.

Sung, Y. J., Thompson, E. A. and Wijsman, E. M. (2007). MCMC-based linkage analysis for complex traits on general pedigrees: Multipoint analysis with a two-locus model and a polygenic component. *Genetic Epidemiology*, **31**(2), 103–114.

Tang, H. K. and Siegmund, D. (2001). Mapping quantitative trait loci in oligogenic models. *Biostatistics*, **2**, 147–162.

Taqqu, M. S. (1975). Weak convergence to fractional Brownian motion and to the Rosenblatt process. *Zeitschrift für Wahrscheinlichkeitstheorie und vervandte Gebiete*, **31**, 287–302.

Teare, M. D. and Barrett, J. H. (2005). Genetic linkage studies. *Lancet*, **366**, 1036–1044. (Series: Genetic Epidemiology 2)

Teng, J. and Siegmund, D. (1998). Multipoint linkage analysis using affected relative pairs and partially informative markers. *Biometrics*, **54**, 1247–1265.

Terwilliger, J. D. and Ott, J. (1992). A haplotype-based 'haplotype relative risk' approach to detecting allelic associations. *Human Heredity*, **42**, 337–346.

Terwilliger, J. D. and Ott, J. (1994). *Handbook of human genetic linkage*. Baltimore and London: The John Hopkins University Press.

Terwilliger, J. D., Speer, M. and Ott, J. (1993). Chromosome-based method for rapid computer simulation in human genetic linkage analysis. *Genetic Epidemiology*, **10**, 217–224.

Thomas, D. C. (2004). *Statistical methods in genetic epidemiology*. New York: Oxford University Press.

Thompson, E. A. (2000). *Statistical inference from genetic data on pedigrees* [NSF-CBMS Regional Conference Series in Probability and Statistics, Volume 6]. Beachwood (Ohio) and Alexandria (Virginia): Institute of Mathematical Statistics and American Statistical Association.

Thompson, E. A. (2006, April). *Uncertainty in inheritance: Assessing evidence for linkage* (Tech. Rep. No. 498). Department of Statistics, University of Washington, Seattle, Washington.

Thompson, E. A. and Geyer, C. J. (2007). Fuzzy p-values in latent variable problems. *Biometrika*, **94**(1), 49–60.

Tu, I. P. and Siegmund, D. (1999). The maximum of a function of a Markov chain and applications to linkage analysis. *Advances in Applied Probability*, **31**, 510–531.

Uhlenbeck, G. E. and Ornstein, L. S. (1930). On the theory of Brownian motion. *Physical Review*, **36**, 823–841.

Vieland, V. J. (1998). Bayesian linkage analysis, or: How I learned to stop worrying and love the posterior probability of linkage. *American Journal of Human Genetics*, **63**, 947–954.

Vieland, V. J. and Huang, J. (2003). Two-locus heterogeneity cannot be distinguished from two-locus epistasis on the basis of affected sib-pair data. *American Journal of Human Genetics*, **73**, 223–232.

Weeks, D. E. and Lange, K. (1988). The affected-pedigree-member method of linkage analysis. *American Journal of Human Genetics*, **42**, 315–326.

Whittemore, A. S. and Halpern, J. (1994). A class of tests for linkage using affected pedigree members. *Biometrics*, **50**, 118–127.

Wigginton, J. E. and Abecasis, G. R. (2006). An evaluation of the replicate-pool method: Quick estimation of genome-wide linkage peak p-values. *Genetic Epidemiology*, **30**, 320–332.

Wise, L. H., Lanchbury, J. S. and Lewis, C. M. (1999). Meta-analysis of genome searches. *Annals of Human Genetics*, **63**, 263–272.

Wu, X. and Naiman, D. Q. (2005). P-value simulation for affected sib pair multiple testing. *Human Heredity*, **59**, 190–200.

Xing, C. and Elston, R. C. (2006, July). Distribution and magnitude of type 1 error of model-based multipoint lod scores: Implications for multipoint mod scores. *Genetic Epidemiology*, **30**(5), 447–458.

Xiong, M. and Jin, L. (2000). Combined linkage and linkage disequilibrium mapping for genome screens. *Genetic Epidemiology*, **19**, 211–234.

Zhao, H. (2000). Family-based association studies. *Statistical Methods in Medical Research*, **9**, 563–587.

Ziegler, A. and Koenig, I. R. (2006). *A statistical approach to genetic epidemiology: Concepts and applications*. Weinheim: Wiley-WCH.

In: Genetic Recombination Research Progress
Editor: Jacob H. Schulz, pp. 139-162

ISBN: 978-1-60456-482-2
© 2008 Nova Science Publishers, Inc.

Chapter 4

The Fate of Phylogenetics in the Face of Lateral Gene Transfers

E. Bapteste[1,], Yan Boucher[2] and W.F. Doolittle[3]*

[1] UPMC UMR 7138, 7 quai Saint-Bernard, Bâtiment A, 4ème étage, 75005, Paris, France.

[2] Department of Civil and Environmental Engineering MIT 48-336A , 77 Massachusetts Avenue , Cambridge, MA 02139

[3] Canadian Institute for Advanced Research and Genome Atlantic, Department of Biochemistry and Molecular Biology, Dalhousie University, Halifax, Nova Scotia, Canada

Abstract

Discovering the impact of biological processes, such as lateral gene transfer, has significantly transformed our understanding of microbial evolution and our views of the natural genetic relationships between diverse life forms. As the diversity of evolutionary processes are revealed, real natural connections appear to be much more complex than initially believed at the time traditional molecular phylogenetics assigned itself the task to reconstruct the universal Tree of life, aka the unique genealogy of species, based on the use of carefully selected genes. Both the notions of species and of a unique inclusive hierarchy of living beings needs in fact to be reevaluated. Importantly, such a reevaluation calls for a substantial renewal of our phylogenetic practices and focuses, opening new perspectives to the whole field of evolutionary biology. Here, these changes are justified by recalling an important recent lesson in the philosophy of biology: how the consideration of evolutionary processes, necessary for species definition, indicates that many incompatible but legitimate definitions of species taxa and taxonomies are more realistic when it comes to represent life's natural relationships. We discuss why, in addition, many other evolutionary units, smaller or larger than the usual species taxa, have also to be considered as real, because they too emerge from the evolutionary process. Thus, even though the trajectories of these multiple evolutionary units may conflict, they deserve equally to be fully investigated by phylogenetics. Instead of the use of a unique tree, we explain how the consideration of multiple databases could help

---

[*] E-mail address: eric.bapteste@snv.jussieu.fr

properly systematize a portion of the real biological diversity. In addition, consideration of networks could partly help investigate the dynamics sustaining the natural genetic connections between all the evolutionary units. Such analyses go beyond the scope of traditional phylogenetics, yet they matter, because, while some evolutionary units are "closed", others are "open", vastly changing, thus presenting fuzzy rather than precisely definable boundaries. We argue why allowing such a distinction at all evolutionary levels has important bearing on our ability and ways to describe life's evolution, which can not be accounted for by a unique model. Finally, after encouraging the development of additional evolutionary metaphors to describe the true course of evolution, we briefly present implications for biotechnologies and conservation biology that makes this pluralistic phylogenetics particularly valuable and promising.

Introduction

1. LGT Forces us to go Beyond the Classical Darwinian Framework in Order to Study Classification and Microbiology with More Realism

To paraphrase Panchen, "in a perfect world, or one made perfect for traditional phylogeneticists; the natural arrangement of organisms would have the following properties: (1) the arrangement would form an inclusive, divergent hierarchy, (2) that arrangement could therefore be represented by an "ordinally stratified hierarchical clustering", (3) all taxa at every rank (including the whole Biota) would be monophyletic, (4) each taxon at every rank could be characterized by at least one apomorph and thus unique character, which, unless that taxon was terminal, would unite the two sister-taxa of immediately lower rank which that taxon included, (5) apormorph characters of every taxon could be distinguished unambiguously by (probably ontogenetic) means that did not depend on any pre-existing classification"[1]. However, acquisition of genetic material by other means than inheritance from a progenitor (lateral gene transfers or LGT) that is observed at all taxonomic levels (i.e. between genera, families, domains, etc.[2, 3]) as well as recombination within groups of closely related organisms[4] make such a view of the world obsolete, especially for microbes. It is important to note that the variety of phenomena introducing modifications in prokaryotic genomes, both internal and external, do not equally affect all types of organisms though. The modifications occurring in the genome of an obligate intracellular symbiont such as *Buchnera* will mostly be gene loss or point mutations[5], while those occurring in a generalist environmental bacteria such as *Vibrio* are more likely to originate from LGT[6]. The fact that the processes bringing modification differ between organisms implicitly means that any naturalistic classification would require some degree of pluralism. This fact leads one to reflect on some conceptual problems, which go beyond more classical concerns linked solely with improving phylogenetic methods searching for the best tree. Acknowledgement of LGT imposes a reversal of the viewpoint the philosopher Elliott Sober applied to traditional phylogeny. This epistemologist wondered «if one wishes to reconstruct the phylogenetic pattern, how much does one need to know about the evolutionary process?"[7] He concluded that ideally "a principle of "less is more" governs this problem. The less we need to know about the evolutionary process to make an inference about pattern, the more confidence we can have in our conclusion»[7]. To the contrary, we state that it is more accurate to claim that we need all available information concerning the process to hope to propose a credible evolutionary scenario. In that regard, a significant proportion of LGT can have a major impact. As Woese asks: «what does it mean, then, to speak of an organismal genealogy when nearly all the genes in the cell – genes that give it its general character – do not share a common history?»[8]. For him, «this question goes beyond the classical Darwinian context»[8].

It thus seems necessary to look for some solutions to the problems raised by LGT within philosophy of sciences, on topics sharing multiple issues with phylogeny but which look at them from a different angle. More precisely, we are interested in concepts developped during the debate on the definition of species in biology and in philosophy of processes to understand what systematics could look like in presence of LGT. It is on the basis of notions borne of this discipline that we will attempt to redefine the fundamental goals of phylogenetics and the future of this discipline outside the traditional Tree of species.

Discussion

2. On the Importance of Processes in Philosophical Definitions of Species and for Taxonomical Purposes in General

a. Brief Description of the Transition from Essentialism to Eliminative Pluralism in the Species Debate

Philosophers have asked themselves many questions on species and their classification and have recognized for a long time that the term "species" can have two different meanings. When talking about species, one must indicate whether he is referring to the general category of species or to a specific taxon of a given species[9]. Definitions of the terms natural/artificial have also been debated by philosophers. Here, we will stand on Splitter's conclusions[10], who considered an object as «real or natural when it is causally efficacious, relative to some explanatory theory»[10]. This way «species concepts may be real in the sense that species classifications are explanatory where other modes of classification are not»[10]. On the opposite, Splitter characterized «as unreal or non-natural, those objects that are identified as the outcomes or artifacts of causal processes in which they do not actually function»[10]. For example, with no causal action in nature, «higher taxa would be artifacts in this sense»[10], conventional constructions. The Proteobacteria for instance might not be considered as a natural group, because there is no such thing as a real causal impact of the Proteobacteria phylum (i.e. there is not a single physiological feature shared by all Proteobacteria that is not a general feature of bacterial cells). We will not retrace the history of the debate on the species concept but limit ourselves to recall some philosophical contributions, which help to identify certain difficulties traditionally associated with the definition and classification of biological species and their possible solutions.

The first preoccupation at the heart of the species problem has been to move beyond an essentialist perspective. This essentialism, according to Karl Popper and David Hull, is «the view held by Plato and many of his followers, that it is the task of pure knowledge or 'science' to discover and to describe the true nature of things; i.e. their hidden reality or essence»[11]. Such an approach assimilates the biological species concept to the identification of a few immutable properties, static and clearly defined, shared by a group of organisms. On the basis of these principles, «from the beginning taxonomists have sought two things – a definition of 'species' which would result in real species and a unifying principle which would result in a natural classification»[11]. The reason for the failure of this essentialist approach – two thousand years of stasis – has nonetheless been established. Such an essentialist view is in direct conflict with the lessons from the theory of evolution. Philosophers like Hull have consequently recommended to give up the aristotelian definition «both for species names and for 'species'»[11] to get out of this paradox and accept descent with modification as a valid unifying principle for natural classification. Nonetheless, in order to avoid an artificial classification, evolutionists must now discover whether it is possible to subdivide a continuous line of descendants in distinct units in a non-arbitrary way. For example, biologists can try to characterize species using «sets of statistically covarying properties arranged in indefinitely long disjunctive definitions»[11] and, armed with such a list, try to decide when a critical (sufficient) combination of properties is shared by a group of organisms that could be identified as belonging to the same species. This formalization uses what is called a cluster concept and underscores that combinations of overlapping properties between members of the same species do not have to be identical. In fact, «several different but overlapping sets of properties are accordingly each sufficient»[11] to define membership in the same natural group. Historically, in the debate surrounding the definition of species, one of the significant steps was thus to try to obtain means to measure the extent of this "family resemblance" (in space) and its persistence (in time).

Hull has suggested that the identification of species through time could rest on the development of a metric, operating partitions in taxonomic space, a «scale» «to delineate evolutionary units»[12]. The length of this scale is not absolute but empirically derived from the observation of the effect of biological processes on phenotype. It then sufficed that the scale simply rests on a real contemporary biological mechanism, sufficiently preserved in the chain of organisms studied, for it to be possible to infer the conservation in space and time of a set of specific properties. However, acknowledgement of the importance of the evolutionary process in the definition of species could also naturally lead to marginalizing the use of such a scale and underscore the lack of relevance for generalizing the characteristic properties for members of a given species[13]. The first version, by Ghiselin, of the thesis claiming that species are individuals, was based on knowledge of evolutionary biology and the essential premise «to try to define the species and other categories in terms of the causes of evolution»[13], demonstrating that species were «systems at various levels of integration», «composite wholes», because «constituent organisms are parts, not members»[13]. In this case, species characterized by this status of individual were real, regardless of the look of their component parts (potentially very divergent): no essential characteristic beside its particular functional integration was giving its reality to the species. Hull nonetheless revisited this thesis[14], which he popularised under a slightly different format, putting forward the notion of an individual essence for specific taxa, where «integration by descent is only a necessary condition for individuality; it is not sufficient. […] A certain cohesiveness is also required»[14] and the fact that genetic exchange was "one means by which such unity [necessary to function as units of evolution] can be promoted"[14].

Hull recognized that in practice, however, the access to this individual essence could be extremely complicated because of interconnection between multiple levels of natural selection. For example, with his concept of the individual, which linked evolution and essentialism, this philosopher left open the debate on the status of species. If some, like Splitter, consequently stated that «the way stands open to develop a strengthened and unified species concept that can restore species to their former role as fundamental units of evolution»[10], the recognition of the importance of evolutionary processes in the definition of the individual species opened the door to proposals of more pragmatic species concepts, to pluralist rather than unified concepts.

There are indeed many versions of the pluralist concept that we will not present here[15, 16]. We will only expose the concept which to our knowledge presents the most complete and pedagogical argument: Ereschefsky's eliminative pluralism. In a seminal paper published in 1992, this author noted that «biologists offer various definitions of the species category»[17], and recalled how the species category now played two closely connected roles in biology. First of all, «species taxa are the basal units»[17] of traditional taxonomy: higher taxa are composed of specific taxa and constitute more inclusive units. Secondly, species taxa are «groups of organisms that evolve as units due to their exposure to common evolutionary forces»[17]. However, Ereschefsky demonstrated that these two roles are inconciliable, because empirical studies show that «interbreeding, ecological and monophyletic lineages do not correspond in nature»[17]. To the contrary, «the Tree of Life on this planet is segmented into a plurality of incompatible but equally legitimate taxonomies»[17], because «different species approaches often classify the same organisms into different lineages»[17].

Mishler and Donoghue also observed that different criteria were leading to different groupings, and that it became hard to decide which one to choose as representative of the species. They ask «why should we necessarily pin species names on sets of organisms delimited by reproductive barriers? Why not choose, for example, to name morphological units instead»[18]. As a result, they «urge explicit recognition and acceptance of a more pluralistic conception of species, one that recognizes the evident variety and complexity of "species situations"»[18], but they thought that «different factors may be "most important" in the evolution of different groups»[18] and that, «through the complex process that is science, the community of involved workers can and will hammer out criteria for making such decisions»[18] to distinguish the best species definition for each case, when presented with conflicting taxonomies for the same organisms. Ereshefsky qualified this view as monist about the species category and as pluralist about the species taxa[17].

As opposed to these pluralists, Ereshesfsky maintained that another position could be embraced if multiple approaches and taxonomies, and not only one, were relevant to classify a given organism. Unlike monists, who «insist that only one correct approach to species exists and consequently only one correct taxonomy of the organic world exists»[17], Ereschefsky argued in favour of a «plurality of equally legitimate though incompatible taxonomies of the organic world»[17]. For him, «the desire for hierarchical classifications does not pose a conceptual roadblock to a realist interpretation of pluralism»[19], as long as «one might be willing to ease up on the requirement that an organism belongs to only one species»[19]. His strong idea was that the fundamental nature of an organism could be something different from its cataloguing in a given species, something richer, because «what makes an individual a single organism is the causal relations that bind its spatial and temporal parts into a single instance of life. In other words, one might give a causal account of an organism's identity»[19], rather than a unique taxonomic label. «Given this suggestion, an organism can maintain its identity even though it belongs to different species»[19], because multiple processes created it. Therefore, to understand evolution, there was no other choice than to «study the various types of theoretically important lineages in the world»[17], because each legitimate taxonomy offers important information. This way, «a taxonomy of monophyletic taxa provides a framework for examining genealogy. A taxonomy of interbreeding units offers a framework for examining the effect of sex on evolution. And a taxonomy of ecological units provides a structure for observing the effect of environmental selection forces»[17]. The practical consequences of this conclusion are numerous, given that «a systematic study that considers just one of these taxonomies provides an overly coarse-grained picture of evolution»[17] and is thus not satisfactory.

b. Two Major Philosophical Lessons: Species Antirealism and the Priority of the Process over the Pattern

Ereshefsky also defended the thesis that, because «various taxa we call 'species' lack a common unifying feature»[19], there is «reason to doubt the existence of the species category»[19] and finally to «be anti-realists when it comes to the species category»[19]. As Dupré clearly demonstrated, «merely being a genealogical entity will not suffice. From an evolutionary perspective, all taxa, whether they be species, genera, or tribes are genealogical entities»[19]. Ereshefsky had also noted that "a popular suggestion among biologists and philosophers is that species taxa share a similar type of cohesion or evolutionary unity. This suggestion has two components: first the commonality of species lies in the similar type of process that renders species taxa cohesive entities. Second, the common nature of species taxa lies in their containing a similar structure. The first claim is about process, the second is about pattern"[19]. On one part, however, «the processes that cause taxa to be species vary»[19], and on the other part, species also fail to show a general unit structure. In the absence of characteristics for the category of species, Ereshefsky proposed to «eliminate the term "species" and replace it with a plurality of more accurate terms»[17] to avoid any ambiguity, for example, the terms bio-species, eco-species and phylo-species could be used to refer to interbreeding units, ecological units or phylogenetic units, respectively.

It is important to note that this profound transition from essentialism to eliminative pluralism has been justified by the primary role attributed to processes in the definition of natural groups. This essential primacy of processes is not rare. Following those lines, the philosophy of processes claims that «natural existence consists in and is best understood in terms of processes rather than things – of modes of change rather than fixed stabilities»[20]. In such a labile context, persistent objects can emerge temporarily but are never «no more than a statistical pattern – a stability wave in a surging sea of process»[20]. Two antagonistic lessons and one fundamental conclusion emerged from this line of thought: (i) in some cases, evolution would naturally lead to some cohesion within a group of organisms, so that definitions of groups and «family resemblances» could help organize such natural groups; (ii) On the other hand, it is entirely possible that natural groups composed of completely disparate parts but evolving in an integrated manner could arise in nature. In such cases, to adequately represent their natural relationships, there is no other means than to re-describe the causes of their origins and how these parts are integrated with one another to operate under a larger unit of selection. These results therefore raise an important question: do species really deserve to be distinguished from other evolutionary entities and to be placed at the center of phylogenetic analyses? If indeed there are no species in general, then phylogenetics must redefine its (principal) objective, which can no longer be to find the Tree of species. We propose that, instead, to become more realist and pragmatic, phylogenetics must focus on describing the evolution of all evolutionary units with a comparable ontology and that a plurality of evolutionary scenarios could account for natural relationships between such real "taxa", including those created by the process of LGT. Such "taxa" could simply be defined using the concept of operational taxonomic units (OTUs). Originally, the term OTU was used by phylogeneticists to describe the organisms being compared when a tree is constructed from morphological data. More recently, however, microbial ecologists have used this term as a synonym for gene or other nucleotide sequences. This movement of the OTU definition towards taxonomic units other than organisms underscores that pluralism is a useful approach that has already made its way in biological thinking.

3. Recent Discoveries about the Evolutionary Process Suggest to Expand the Former Conclusions of Eliminative Pluralism

In the presence of LGT, we not only propose to embrace Ereshefsky's conclusions[9, 17, 19, 21], but also to extend them.

a. Descent with Modification Revisited : Cohesion, LGT and the Unity Relationship

LGT makes it necessary to rethink the general model of evolutionary process used by phylogeneticists. We suggest a general schematic of the evolutionary process that has three parts: (i) creation of original "molecular"/lower-level associations; (ii) selection of "molecular" associations leading to the transformation of "molecular" in "molar"/higher-level associations; (iii) the emergence and maintenance of cohesive "molar" associations. The use of the word "molecular" here should not thus be interpreted literally. In this context, it is simply used to denote a state of organization that is relatively inferior in reference to an organization that is relatively superior, more organized, qualified as "molar". The initial place given here to "molecular" should not be interpreted as an endorsement of reductionism. Indeed, the "molecular"/lower-level does not deserve such a place because it would be a level susceptible to explain all biology, but simply because "molecular"/ lower-level objects are at the foundation of the evolution of life forms, as the creation of natural objects starts when the most elementary data is gathered. For example, we can consider individual genes acquired laterally or new alleles resulting from point mutations as "molecular" elements, and their complex association, integrated and proven (through natural selection), as "molar"/higher-level ensembles. This would match Woese's vision that for biological sciences «the time has come to replace the purely reductionist "eyes-down" molecular perspective with a new and genuinely holistic, "eyes-up," view of the living world, one whose primary focus is on evolution, emergence, and biology's innate complexity»[22].

Under this view, the first step of the evolutionary process consists in the creation of original "molecular" associations within organisms, through internal and external sources of variation (novel gene acquisition by LGT, replication errors creating point mutations, recombination with foreign DNA, etc). This first step of the evolutionary process therefore takes place in a "molecular" transition zone able to ignore organismal boundaries, since contemporary individuals can exchange "molecular" elements with each other. Consequently, sexual relations (in the broad sense), viral transfections, symbiosis, uptake of environmental DNA by transformation, acquisition of extra-chromosomal DNA elements by conjugation, etc., bring together original "molecular" associations. G. Deleuze conceptualized the topology of these natural connections, so much more complex than the hierarchical pattern of a universal tree, by the notion of a «proliferation plan, a populating plan, a contagion plan»[23].

In brief, this «contagion plan» is the space within and between each of the organisms, where contemporary «molecular» fragments are being associated and where they express wildly, with their own temporality. This contagion plan, area of associations of «lower-level» elements, is thus an eminent place of the evolutionary experimentation, "species" desubjectivation, and genetic nomadism[23]. This contagion plan is unfortunately ignored in the rigid classic tree-like phylogenetic framework.

The second step of the evolutionary process is responsible for what happens with these original "lower-level" associations. They are either judged viable and give rise to more organized "higher-level" associations (stabilized complexes of genes for example); or they are judged aberrant and the random associations between these "molecules" are eliminated through the gauntlet of natural selection, because «the total amount of information that could be expressed at any one point in time is highly constrained by the fact that bases (a similar argument holds for genes, tissues, and organisms) are causally linked, so accessing some information will limit (or eliminate) expression of other, [...] therefore putting an upper bound on the amount of information potential/capacity that could be expressed at any one time»[24].

The last step of the evolutionary process is to solidify and stabilize the different "molar"/higher-level ensembles that have not been rejected by the first cycle of natural selection. In this case, a new evolutionary unit (stable and sufficiently cohesive) is created, susceptible to be duplicated and to be (for a certain time) a resilient construct in the evolutionary flux (i.e. "selfish operons"[25, 26]). This last part of the evolutionary process rests on mechanisms that ensure the cohesion between interacting parts of living organisms and the movements of such parts across the contagion plan. The evolutionary pressure judging this last step of the evolutionary process is therefore a positive feedback loop, which ensures a relation of unity between integrated parts.

The philosopher John Collier described this essential logic in the construction of evolutionary units and named it cohesion[27]. We will adopt this general terminology and its implications, namely that «cohesion represents those factors that causally bind the components of something through space and time, so it acts coherently and resists internal and external fluctuations»[27]. Interestingly, causal powers of cohesion can be likened to those playing a role in emerging weak type phenomena and to downward causation[28]. This latter notion has been proposed because «there appear to be robust hierarchies whose higher levels are not merely effects; the higher levels are themselves causally efficacious»[27]. In the evolutionary process, this is a "downward causation", reflexive and diachronic, which plays a capital role in ensuring cohesion, because «it is possible for the [biological] system to be its own source and receiver [...] The source is a genetic system at time t_0, the channel is reproduction and ontogeny, and the receiver is the same genetic system at any given time $t_{1...n}$; thus, the receiver is temporally distinct from the source»[24].

In this case, we expect cohesion to have two linked consequences: the maintenance of a relative coherence at the higher-level selected for, and the possibility for a relative independence of the constituent parts at the lower-level. In short, this evolutionary process would bring the emergence of multiple evolutionary units, susceptible to present a certain stability and integration, therefore real natural genetic relationships, at different levels. In this general model, which underlines the decisive importance of cohesion mechanisms and includes lateral transfers of genetic material, there is no ontological difference between different cohesive evolutionary units, which can rest on elements of various phylogenetic origins. We could qualify as "multiplicities"[23] these different emerging mosaic evolutionary units, which should nonetheless be treated as valid phylogenetic "taxa" (using the term OTUs to identify them and clearly defining its meaning).

b. More «Taxa » in the Eliminative Pluralism Sense

A short (non exhaustive) list of potential OTU's issued from such an evolutionary process would include: genes, cohesive genetic-proteinic associations (fulfilling a biological function that has been subjected to natural selection) and certain ultra-structural features or "organs" (fulfilling functions can be studied as persistant arrangements coded by the cohesive association of multiple genetic markers). Organisms as a whole and natural groups of higher level can also be relevant evolutionary units, according to the doctrin of generalised phenotype which «the phenotypic effects of a gene are the tools by which it levers itself into the next generation, and these tools may "extend" far outside the body in which the gene sits»[29]. The case most familiar to phylogeneticists is a natural and cohesive group that can be characterized and described by «family resemblances», which is composed of organisms interacting with each other under the influence of a mechanism from a superior level (such as a population-level mechanism), leading to the maintenance of their individual cohesion and constraining that of their descendants. This way, a minimal genetic or morphological cohesion is expected between parts of the same taxon. However, cohesion can also happen between phylogenetically unrelated or morphologically dissimilar organisms, constituting a second type of natural group. The latter emerges from cohesion between organisms, themselves cohesive but initially lacking common «family resemblances». Although these mosaic taxa have parts that seem unrelated to each other, cohesion nonetheless operates a group selection, in the meaning defined by Sober, which «subsumes a set of objects under a single selection process»[30], via «some common causal influence acting on the objects which affects their reproductive chances»[30]. For example, syntrophic microbial consortia, composed of multiple organisms with various physiologies, are able to achieve chemical reactions that would be energetically unfavourable if carried out by a single microbe. In our view, one could even be tempted to include bioconstructions amongst the fundamental elements ensuring the cohesion of this kind of natural group. This theory, that we will qualify as a theory of the (very) extended phenotype, rests on the notion that the evolution of natural objects, which are organismal multiplicities, can rely on very diverse environmental structures, thus providing more complex eco-species than the ones identified by the past. This idea echoes Jablonka's proposition, that «in addition to the direct re-production and multiplication of phenotypes through the various inheritance systems, re-production of phenotypes can also be indirect and extended in space and time, mediated by ecological niche construction»[31]. According to this author, interactions mediated by niche construction shape the world and structure its genetic diversity, since «even small amounts of niche construction, or niche construction that only weakly affects resource dynamics, can significantly alter both ecological and evolutionary patterns»[32]. The case of the mound of fungi-cultivating termites could be a good example of such a disparate cohesive natural group built around a bioconstruction. This perennial association of termites (*Macrotermes*), fungi (*Termitomyces*), and the mound confronts the classic evolutionary thinking with a problem. One can wonder if «the mound, therefore, [is] an extended phenotype, that is a product of genes that control the processes whereby termites build mounds?»[33] or if «it is the fungi that are cultivating the termites, using their tendency to build regulated environments as a way to suppress the growth of *Termitomyces*' fungal competitors. Now, precisely whose extended phenotype is the mound? Is it the termites' that build it, or is it the fungi's that perturb the termites' home?»[33] As Turner noted, in such a case, the imprecision is obviously a key element of the (very) extended phenotype theory, where it is futile to point to a particular organism of the association as the true object of natural selection. What is selected rather, is a massive coalition of genes, distributed between two organisms, the *Macrotermes* and the *Termitomyces*»[33]: a "multiplicity". By analogy, this example allows us to consider the case of microbial multiplicities, less well known by biologists, but equally real, disparate natural groups anchored in a specific environment that are qualified of metacommunities by contemporary microbiologists. For instance, the recently characterized acid mine drainage microbial biofilms growing within underground pyrite ore bodies are examples of such metacommunities[34]. These acidophilic biofilms are self-sustaining communities that grow in the deep subsurface and receive no significant inputs of fixed carbon or nitrogen from external sources. They are dominated by two types of microbes from different domains of life, the bacteria *Leptospirillum* and archaea *Ferroplasma*. These organisms form an integrated community and need each other to fulfill all functions required for their survival such as iron oxidation to obtain energy, synthesis of polymer to maintain their biofilm structure (which allows floating at the air-water interface) as well as nitrogen and carbon fixation for the synthesis of biomolecules.

c. More «Taxonomies » and a Plurality of Phylogenetic Drawings

From this diversity of OTU's, multiple legitimate taxonomies could emerge, given that real natural connections between the parts of organisms are complex and all deserve to be studied to describe true evolutionary trajectories of genes. On this topic, Ereshefsky explicitly laid out what he considered to be the «criteria that a taxonomic approach must satisfy to be considered legitimate»[17]. He distinguished two motivations for classification: on one hand the «sorting principles» and on the other hand «motivating principles» "which justify the use of sorting principles"[17]. He explained that these motivating principles «sets out the causal factor responsible for the existence of the lineages in question»; for example, approaches based on interfecondity focused on sex, ecological approaches on environmental selective pressures and phylogenetic approaches on the process of descent with modification from a common ancestor. The important idea was that «a taxonomy (biological or otherwise) consists of entities that are the nodes of causal processes. Those entities are either the results of a common type of causal process, or they are objects that have a similar causally efficacious property»[17]. Among other things, «the motivating principles of a taxonomic approach should be empirically testable»[17] and they «should be consistent with and derivable from the tenets of the theory for which the taxonomy is produced. In particular, a taxonomic approach in biological systematics should be derivable from well-established tenets in evolutionary theory»[17]. Finally, «the sorting principles of a taxonomic approach should produce a single internally consistent taxonomy»[17]. This simply means that the taxa belonging to a category should be «comparable along the appropriate parameters» within a given taxonomy.

Finally, to retrace the origins of these different evolutionary units, in the presence of LGT, we have to recognize different evolutionary models and representations of relationships for various taxa or goals[35]. Two types of complementary phylogenetic representations can be used: phylogenetic pluralism and "post-phylogenetic" (phylogenetic network).

Phylogenetic pluralism describes the composition of evolutionary units and aims to identify «family resemblances» created by cohesion forces, susceptible to lead to the proposition of certain standards to partially characterize biodiversity. Post-phylogenetics aims to retrace the evolutionary dynamic, consistency and genetic extensions of each evolutionary unit. We will briefly present two applications in systematics using such non-traditional representations of relationships, accounting for LGT.

4. Building Databases in a Renewed Phylogenetic Context without a Verticalist *a Priori*

First, we propose to start future phylogenetic analyses by multiplying individual phylogenies without any *a priori* against "bad" gene trees, i.e. against gene phylogenies which do not reflect the Tree of Life, aka without *a priori* regarding the evolutionary pattern or patterns that would result from their analyses. The study of the correlations between the histories of these markers evolving under different ecological and selectives constraints, in different genomic neighborhoods, should then allow to circumscribe coevolving genetic sets for which phylogeneticists could search the cause of their coherence and study their geographic and "taxonomic" distributions. Once such individual phylogenies (molecular ones or ones based on other characters) are reconstructed, there will be indeed enough material to observe strong correlations (when some sets of individual phylogenies show numerous common points), moderate correlations (when some sets of individual phylogenies only show some common points), and even a total lack of correlation (when the different individual phylogenies under study have nothing to do with one another). It is thus possible to identify, for a given set of markers, which ones are shared by which groups of organisms, structural feature, or function. It is then tempting to associate these groupings sustained by perennial "genetic" associations, transmitted either vertically or laterally, to the aforementionned OTU's. Such persisting sets could be very numerous and each of them could provide precious information to elaborate an entry in a taxonomical database, addressing the distribution and the evolution of a specific cluster of features or, in other words, of a specific «family resemblance».

For instance, we identified such strongly correlated sets of gene phylogenies in T4 phages[36], a collection of marine viruses with a large spectrum of hosts, which are particularly sensitive to lateral recompositions of their genomes, to the extent that it does not make sense to talk about an inclusive hierarchy for these viral "taxa". Nonetheless, these T4 have retained two perennial genetic associations, likely during several hundred million years. The nine genes of the replication and the thirteen genes encoding the viral capsid are always present in their genomes, in a syntenic order, absolutely preserved, without a trace of recombination or losses at this evolutionary scale. These sets thus are indicative of a «family resemblance», the "T4 look". Similarly, the "methanogen look" is another «family resemblance». The fifty genes of the hydrogenotrophic methanogenesis and of the biosynthesis of their associated cofactors constitute a persisting genetic association, although the organisms carrying this functional unit belong to more than one clade[37]. Other such evolutionary units could include the genes coding for photosynthetic reaction centers (type I and II), the dissimilatory sulfate reduction pathway, the nitrogenase complex, the deoxyxylulose and mevalonate isoprenoid precursor biosynthesis pathways, AHL quorum sensing system and prokaryotic gas vesicles[38].

Consequently, we propose that a cluster concept definition could be associated to the different grouping of evolutionary units, based on the discoveries of such persisting higher-level associations. At the so called "species" level, in bacteria, for instance, instead of using only a binomial nomenclature which does not indicate anything about the real genetic, structural and functional composition of the prokaryotic organisms[21], a "formula of the multiplicity"[37] could clear up and summarize our knowledge about the real associations of its parts (which are themselves evolutionary units). Naturally, the more "molar" sets identified, the more complex, accurate and characteristic the formula describing each large evolutionary unit will be. We can thus imagine the propositions of several formulas such as: Multiplicity A^{DF} X_{+1} Y_{+2} Z_{+3}, Multiplicity B^{DF} X_0 Y_{-1} Z_{+3} C_{+4}, Multiplicity C^{DF} X_{+1} Y_{+1} Z_{+3}, etc., where Multiplicity I stands for the OTU of interest I ; X,Y and Z correspond to different persisting lower-level systems, and +1, -1, 0 indicate if these lower-level units share different or similar evolutionary states (i.e. how close their components are). Thanks to these cluster definitions, prompted by the phylogenetic correlations between gene phylogenies and for which the biological process responsible for cohesion have been empirically tested, biologists could study the natural connections between the various multiplicities bearing similar higher-level genetic sets and identify "family resemblances", without being dependent on an exclusive *a priori* considering there is only one "true" phylogenetic pattern. The similarity of higher-level systems, thus defined by the biological homology, could support hypotheses of taxic homology (but not necessarily the opposite, as in classic phylogenetics[39]). We would then possess what is needed to elaborate multiple legitimate taxonomies of life forms (sensu Ereshesfky) and a single organism could belong to different taxonomic groups (overlapping or disjoint). For instance, *Thermotoga* could be classified as a sulfur reducer and a thermophile at the same time, in a group that indicate it shares real causal powers with thermophilic *Deinococcus* and sulfur-reducing *Shewanella*, whatever their vertical phylogenetic relationships and real differences for numerous other «family resemblances» are for these three organisms. (*Deinococcus* is a radio-resistant and *Shewanella* is a psychrophile, and for these reasons, they are also unlike *Thermotoga,* but like other organisms, which also deserves to be described from other taxonomical perspectives.) Progresses in informatics, which ease the use of interactive databases, will undoubtledly favor such a dynamic evolution of classificatory practices. It seems that «thus, this is not the end of knowledge, quite the opposite. Databases are the Encyclopedia of tomorrow»[40], in a context of LGT.

5. Explaining Lateral Connections : when Beings Are Thought as «Places and Motions»

One can also connect the different phylogenetic patterns specific of the different evolutionary units in a typical postmodern move[40]. On the basis of these multiple series of phylogenetic data, one would create a phylogenetic network or "post-phylogeny", i.e. a scheme resulting from «the loss of a continuous metanarrative» (the universal Tree) and which «breaks the subject into moments of heterogeneous subjectivities, which do not constitute a coherent identity»[41]. One of the main goal of this second approach would be to «discover the "governing principles"[42] of the evolution of the different evolutionary units, when they are all mapped together.

It would be the task of post-phylogenetics to link all phylogenetic informations by simultaneously representing the diverse influences of all natural relationships (ecological, genealogical, cohesive, etc.) on the different "taxa", by retracing all the genetic connections between the genes, genomes, organs, organisms, species and communities, namely between all cohesive groups below and beyond "species", even if their histories are not all converging in a tree. This approach would aim to enrich our ways to represent the complexity of evolution, to model the contagion plan (or "the intricate network of fields of genetic influence"[43] as phrased by Dawkins) more fairly, when differences in the evolutionary trajectories appear between the parts of the natural objects under study, in agreement with the old notion that «a true reconstruction of the course of evolution is the ideal of every taxonomist»[44].

With current phylogenetic networks, this protocol of comparisons and contrasts of fragments of phylogenetic histories produces a drawing with two significant axes[45-48]. The vertical axis recalls life's memory, i.e. the common descent with modification studied by the classic phylogenetic research. On the other hand, the horizontal axis hosts life's anti-memory, those contemporary events which break the vertical order. In this drawing, lateral and vertical events deserve an equal consideration because none of these phenomena can *a priori* justify some ontological priority. Yet, to account for the differences in the strength of the correlations found in the evolutionary trajectories of different genetic elements, the connections in the phylogenetic network appear more or less strong (i.e. in a phylogenetic network called "synthesis" the thicker the links between the OTU's, the larger the number of genes involved in motions of vertical or lateral inheritance[48, 49]). In addition, as the details of the individual trajectories that built these connections are known, phylogeneticists can see the relative stability of lower-level and higher-level associations on this graph, or their disruption across the evolutionary times. It provides informations about which features coevolve naturally and which ones only present weakly correlated dynamics – punctual associations. Typically, it is possible to follow the movements of key features contributing to the "family ressemblance", to identify natural groups emerging from the association of components whose histories were initially independent (as explained in the extended and (very) extended phenotypes models) and to observe groups which persist with modification since a unique last common ancestor (as described in the classic phylogenetic hypothesis).

This phylogenetic network corresponds to the "map" or the "nomadology" of Deleuze and Guattari[23, 50], which is why some intuitions of these process philosophers could help elaborating the future evolutionary analyses of such drawing. More precisely, the conclusion that beings are not identities in the classical sense of the term but rather "places and motions" can prove useful to discover some traffic rules on this contagion plan. The analysis of its connections can indeed be initiated by assimilating the evolutionary units to places, temporary points of arrival and departure for the genes, and by focusing firstly on the genetic moves that occurr between these cellular locations rather than within each of them. Such a study can fulfill several objectives. Here, we will only comment on three of them: (i) the "deduction" of the ecological/adaptive motives responsible for the genetic associations transgressing the traditional genealogical frontiers; (ii) the search of potential optimisations of the genetic flows maximizing the propagation of the genetic material on the contagion plan;

and (iii) the exploration of the relative potential of contagion of the different evolutionary units.

At the molecular level, the study of the causes of the genetic connections between evolutionary units is related to the "gene ecology"[51], evoked by Ed Delong. This gene ecology develops in the context of comparative metagenomics, which does not bother with the traditional taxonomic categories, such as species or genera, etc. , as it focuses on priority on the metabolic potentials of each environment. For instance, for Edwards, «we have used comparative metagenomics to characterize the metabolic potential of different environments, and identify those genes, pathways, and subsystems that are more common in any particular environment»[52] (or see[53]). This metagenomic conception, which studies evolution while freeing itself from the traditional phylogenetic framework, rests largely on the idea that the biological world is functionally structured (and not only genealogically structured). For metagenomicists of Ed Delong's school, genetic elements may be preferentially diffused between the members of similar metacommunities, which are distinguished from others because they comprise different functional genetic subsets, issued from a broader metagenome. In brief, to emerge and adapt, communities differentially draw on the set of genes evolved on Earth, and these elements evolve subsequently in each community, according to its specific functional constraints. The genomic content of organisms living in a same community obeys then a certain logic and shows some coevolution, coadaptation and complementarity, and preferential lateral exchanges of genetic material are expected, irrespective to the genealogical diversity of the community members. Importantly, such a hypothesis is to the antipodes of classic phylogenetics: belonging to a community could significantly structure the genome of two closely related organisms (in terms of classic genealogy) but they could show very different genetic compositions, if their living conditions expose them to different environmental genetic flows. One of the "nomadologic" projects would then be to test the impact of the environment on the genomic evolution and to decide whether or not it exceeds the historical impact of vertical inheritance over the long run. An hypothetico-deductive approach could thus contribute to explain even some of the thinest lateral connections of the phylogenetic network, by characterizing ecologically significant lateral adaptations. For instance, Slot et al.[54] recently explained the transfer of a nitrate assimilation gene cluster of size three, between basidiomycota and ascomycota fungi, and even possibly from stramenopiles to fungi, to allow a better exploitation of nitrate in aerobic soils. Similarly, Sharma et al.[55] described multiple transfers of the bacteriorhodopsin and two associated markers in Haloarchaea.

Such a post-phylogenetic perspective suggests expanding the traditional concept of phylogeography, which was investigating the distribution of the classic taxonomical categories in the geographical areas of the world, toward an even more dynamic post-phylogeography. This latter idea would investigate the distributions of all the evolutionary units within typical ecological environments, themselves located on special geographical spots of the Earth. "These "gene ecologies" could readily be mapped directly on organismal distributions and interactions, environmental variability, and taxonomic distributions»[51] to understand how and at which pace microbial molecular adaptations are spreading in the world. That way, post-phylogenetics would study the rules of convergences and divergences of the evolutionary trajectories and of the genetic elements, and their frequency of exchange between the different loci of life. Already, «these gene distribution patterns seem more indicative of habitat-specific genetic or physiological trends that have spread through different members of the community»([51, 53] but see [52]).

Most fundamentally, post-phylogeneticists could also evaluate if, at least partly, the shape of the grid of the phylogenetic network would not allow one or more strategies to optimize the distribution of some genetic flows. In the same way that road, railway, maritime, informatic or neuronal networks contribute to maximize the efficiency of transfers of goods or information, post-phylogenetics could look for some particularities of the traffic plans of genetic elements on the contagion plan. This proposition rests on the demonstration by the philosopher A. Parrochia that, in general, a network can play an eminent role, that it would have a capacity to screen, being «used as sieve and filter»[56], and that «the characteristic feature of every network» would be «indeed to realize an optimization under certain constraints»[56]. The different construction steps of a network would thus generally unfold under a positive selection to increase the efficiency of exchange, since «functionally, each time, the networking realizes an economy»[56]. In fact, the fundamental mechanisms to explain this selection would be that «life's possibilities can not be infinitely effectuated» and that they are «limited by space», so that «an optimal spatial occupation becomes necessary»[56]. If some of these intuitions are correct, these analogies are interesting to frame the nomadological study of the network of life.

As this network spreads over time, and because the genetic elements are in competition to occupy the cellular genomic space (which cannot expand to infinity in a given organism) one could test if the post-phylogeny presents some historical phases of organization and optimization of genetic flows and if some genetic elements benefited from it or were penalized (i.e. one could test if some genetic elements had their intensity of diffusion modified over time and if some genetic associations were selected for their ability to laterally invade a maximum of genomes, while other tended to become lost in such a context of active genetic recompositions.) Such trends are very conceivable as we know of many systems of lateral gene transfers that were developped during the course of evolution (like operons[25], integrons[57], conjugative transposons, plasmids, maybe even viruses[58]). Consequently, we had suggested that phylogeneticists try to detect some axes of transmission and some structural patterns in the phylogenetic network, which could correspond to evolutionary highways, rotaries, speed lanes or side issues[59], for which one could explain the distribution and set up. Robert Beiko et al.[60] and Victor Kunin et al[61]. also have seemed motivated by these sorts of problems, and they identified evolutionary "highways"[60] and «hubs»[61], responsible for a higher turn-over of the genetic material across life forms.

In the future, evolutionists could develop more rigourous mathematical models than mentioned in the former studies to compare the trajectories of the genetic flow between different areas of the contagion plan. They could investigate the modes of transmissions of different genetic objects and evaluate their differential capabilities of colonization and diffusion in living beings. They could notably study if some genes or other larger evolutionary units (i.e. plasmids, integrons, operons, mitochondria, chloroplasts, some special "molar" features having significant impacts on human societies, such as the diffusion of antibiotic and antiviral resistances, depolluting properties, etc.) maximize their moves, their spreadings between two regions of the phylogenetic network; if they follow the shortest path or instead if their transmission by contagion is relatively "suboptimal" in a given environment. The different genetic objects could then be classified on a spectrum going from the «mobilons»[59] (those entities which seem equiped to explore the network of life and to ride the contagion plan in all directions) to the "home-bodies" (those elements which can be affected by intergenomic recombination but are rarely mobilized), and even to the "losers" (those endangered genetic elements which are eliminated in the competition for the cellular genomic space).

6. Establishing which Taxonomical Units Are «Open» or «Closed»

Another angle of analysis prompted by the acknowledgement of LGT would concern the genetic movements occuring within a given evolutionary unit (rather than between units) in order to characterize their changing and diffuse dimension, because living bodies are not closed but open, part of a contagion plan. Put in a more philosophical phrasing, phylogeneticists could investigate the dynamic of the BwO (Body without Organs[50]) associated to each mosaic evolutionary unit to try to discern its contours. This concept, developed by G. Deleuze, corresponds to the portion of the contagion plan, centered around a phylogenetically composite OTU. It is the biological matter in motion, a sort of "molecular" soup with vague contours and changing due to the occurrence of particular associations at an indefinite pace, as described in the evolutionary process aforementionned.

At the "species" level, for instance, the study of this ontological blur, would impose itself quite spontaneously, as «the net or web metaphor should remind us that all prokaryotic taxa are in essence imprecisely bounded and ephemeral»[62]. Acknowledging this imprecision is essential, because, while trying to negate it, traditional phylogeneticists often fail to fully study biological phenomena[63]. While the recourse to a terminology (i.e. the classical conventional taxonomical categories) is sometimes only apparently accurate, the use of blurry concepts, which reflects the real diversity of the evolutionary trajectories, should by contrast allow real progresses in systematics. The "bacterial pan genome"[64] is a good example of that. This vague concept should sometimes replace the one of "bacterial species", because «data clearly show that the strategy to sequence one or two genomes per species, which has been used during the first decade of the genomic era, is not sufficient and that multiple strains need to be sequenced to understand the basics of bacterial species»[65]. These two notions are incommensurable, since «given that the number of unique genes is vast, the pan-genome of a bacterial species might be orders of magnitude larger than any single genome»[64]. It even seems that two cases should be distinguished: «species can have an open or a closed pan-genome. An open pan-genome is typical of those species that colonize multiple environments and have multiple ways of exchanging genetic material»[64]. *Streptococci, Meningococci, H. pylori, Salmonellae* and *E. coli* are notable examples of this. In these open-pangenomes, the association of genes changes according to the strains, to the environments and «mathematical modeling predicts that new genes will be discovered even after sequencing hundreds of genomes per species»[64]. For these organisms, «the surprising conclusion from the study is that, in theory, the bacterial species will never be fully described, because new genes will be added to the genome of the species with each new genomic sequence»[64]. Traits such as whether a species pan-genome is open or closed can also vary through time. For example, although *Buchnera* and *E. coli* are both bacteria from the *Enteronacteriaceae* family, the former is an intracellular symbiont experiencing practically no LGT (tightly closed pan-genome) while the latter is a generalist bacterium subject to frequent LGT (wide open pan-genome). Blurry concepts are amenable to change through time, while concepts such as the "bacterial species" are fixed in time and cannot handle the ephemeral nature of bacterial genomes.

A new and ambitious goal in presence of LGT could thus be to investigate the evolution, composition and dynamic of this set of mobile genes within "species" (and within each other evolutionary unit), and to identify which "taxa" deserve to be thought via vague concepts to be correctly understood. These kind of studies will be useful to evaluate some ultra-uniformitarianist evolutionary narratives which too often assimilate the traditional taxonomical categories of the present time (i.e. extant cyanobacteria) to some of their supposed common ancestor (i.e. cyanobacteria of the past), as if the two "taxa" were composed identically. Yet, they are not necessarily so when the qualitative features associated with the extant group do not in fact persist longer than a limited number of generations, as suggested by the high genomic turn-over examplified by the open pan-genomes. For this reason, favoring a vague concept of lineage over the application of a wrongly accurate one, denouncing overly simple definitions of the ancient biodiversity as potentially misleading, will free many biologists from the constraint of a reasoning giving the priority to rigid patterns, and would make them more prone to practice a process-biology. Such a careful use of accurate and vague concepts in time due to the high rate of LGT, should then provide one of the most lively testimonies of lineage evolution.

7. The Appeal for New Metaphors

In the previous sections, we have explained how building databases and reconstructing network would help to reason positively (either by induction or by deduction) about life's complex evolution. However, inevitably there will be results that will simply appear as anarchic events, absolutely unexplainable by the aforementioned approaches. This is also a non negligible fact and reporting the unspeakable dimension to which evolutionary studies confront us could be a third important epistemological project, even if it is not in the traditional spirit of phylogenetic science. We propose nonetheless to phylogeneticists to expose, including less traditional means, the impressive amount of genetic creativity provoked by evolutionary mechanisms in order to convince the largest audience possible (not only the evolutionists) of the fundamental importance of those processes in the study of nature, and of their ontological priority upon the discovery of patterns. This third approach suggests in fact to revolutionize the classical claim about the «best available single proof of the reality of macro-evolution»[66] presented by Zuckerkandl and Pauling, which supported that the congruence of individual genes phylogenies would reflect the species phylogeny and thus would confirm the reality of evolution. Quite the opposite, a better evidence of this phenomenon could be found in our inability to reconstruct the Tree of Life from multiple phylogenetic markers failing to show any significant correlations[49, 67], because evolution is simply more lively than a single model can tell.

Surely, a fairly simple way to stress the priority and diversity of the processes could consist in proposing scenarios insisting on the evolutionary dynamics, carefully expressed in terms of mechanisms between transition states, and integrating metacommunities, horizontal kins[68] and blurry concepts in the analysis, when possible. Defining clades could thus retain an important role as a way to circumscribe a region of the phylogenetic network but not as the ultimate goal of the evolutionary analysis. Clades could serve of reference toward an exhaustive description of the nature in motion, yet the search for synapomorphies should only constitute the starting point of the demonstration. However, the recourse to such dynamic scenarios is not the only option to give all its place to the evolutionary processes. To go further, we suggest rather that phylogeneticists consider the development of new metaphors about evolution, for instance based on some artistic work because «sometimes, showing is more important than demonstrating, suggesting more important than constraining»[69]. This creative approach, in order to present all the complexity of the evolutionary processes, could prove quite effective to bring traditional phylogenetics out of its current state of crisis, because it would help many evolutionists forgetting the nostalgy of the lost tree-like metanarrative. For instance, a short movie retracing the thousands morphogeneses occurring in the elements of a landscape across time, showing multi-paced concresions, sedimentations, erosions and breaks, like the changing aspects of water in a more or less frozen Sea of life, could be more suited to represent the life's biological evolution than the drawing of a Tree of Life.

Conclusion

In the face of LGT, a renewed phylogenetics will have significant epistemological impacts. Three aspects of evolution will be particularly put forward: the complexity of the evolutionary process, the diversity of the natural genetic connections between living beings and the need to adjust our metholodogies to locate ourselves at the appropriate level of analysis[35]. Thus, it will appear more and more legitimate to claim that evolutionists should not build their science only on concepts of classical inspiration, but also partly on blurry and pluralistic concepts. With its diversity of "taxa" (genes, structural features, functions, species, symbionts, metacommunities, etc.), its diversity of analyses ("family resemblance" searches, nomadologic studies even metaphoric explorations), its diversity of representations (pluralistic, reticulated, tree-like for certain OTU's) and its taste for blurry concepts (open pan-genomes, contagion plan), a renewed phylogenetics will find its place onto a general path towards pluralism in evolutionary biology, a path which is not unprecedented in science[70]. One of the lessons of such a renewed phylogenetics will then be that, undoubtledly, in evolution as elsewhere, multiplying the questions independently of an *a priori* unique metanarrative is to be encouraged[49]. In fact, a context which discourages theoretical pluralism "tends to promote both dogmatism and a self-fulfilling confirmation of received views»[71], like the continuous and quite meaningless reconstruction of the tree of prokaryotes[72], largely due to the fact that lateral gene transfers were discovered relatively lately by phylogeneticists.

The refusal of monism in phylogenetic science will logically be accompanied by the refusal to «believe that there is only one scientific method», and will prevent «hiding behind the "fetishism" of the method»[73], which has already been denounced several times[74]. Phylogeneticists should then start supporting the idea that «(1) there are many different systems of representation for scientific use in understanding nature; (2) there is no coherent ideal of a complete account of nature»[75]. Instead of getting lost for the quest of big names and big myths, aka a big unique species Tree at life level, this pluralism will acknowledge that «true questions must have a precise context and a precise stake»[73], and will promote the «research of an intermediate way between a reactionary metaphysics and an irresponsible relativism»[73]. Interestingly, this choice in favor of a methodological pluralism will be consistent with an actual conception entertained by philosophers regarding the internal structure of the evolutionary theory, characterised by its "modular or reticular" dimension[76]. It will notably make clear the «necessary complementarity of its constituent sub-theories»[73]. The main consequence of this claim will be to acknowledge that specialists of different biological disciplines deserve equal consideration in an analysis of renewed phylogenetics, which means that, necessarily, a good phylogenetic analysis will require multidisciplinary collaborations and can not be left to a single phylogeneticist as has happened in the past. In a postmodern phrasing, it will confirm that the end of the tree metanarrative «means the Death of the Professor»[40]. More and more in evolutionary biology, the motto will thus have to become «interdisciplinarity»[40], the collaboration of Professors, as already preached by W. F. Doolittle[62] on similar considerations.

Rather than looking for a single global explanation, little narratives will likely become the norm in evolutionary biology, as this pluralistic approach will probably lead to the proposition of multiple middle-range evolutionary "laws", based on the discovery of more or less perennial, more or less stable, "molar" sets. These biological laws will resemble the description of «family resemblances», such as the composition and the order of conserved genes in the T4 phages modules described before[36]. At a certain evolutionary scale, some robust generalizations will indeed be possible and Wake's dream to «make evolutionary biology predictive»[77] won't vanish. By adjusting themselves to what's out there, the discovery of lower-level evolutionary units – not necessarily dependent of the genealogy- will in fact increase the predictive and deductive power of biology, and they will potentially lead to some very concrete applications.

That's why there will also be important ethical consequences, particularly regarding the usage of genetically modified organisms and in the management of biodiversity. Applied ecology and bioremediation will benefit directly from a renewed phylogenetics. The lateral mobility of some «molar» and «molecular» systems, in some metacommunities, will be a useful source of inspiration for the biotechnologies. It will become easier to consider the creation of what one could call GEMOs (Genetically and Ethically Modified Organisms), that the knowledge of «molar» systems and of their rules of dynamic associations would give the means to generate. Even if it is obvious that social issues do not have to be solved by scientific applications (politics and education seem more powerful and legitimate to us), it is nonetheless certain that a renewed phylogenetics could contribute significantly in the debates that will certainly arise around these likely inevitable GEMOs (whether we like them or not ([78]and reference therein)). For instance, phylogeneticists will have to test the potential of propagation in nature of engineered genetic associations, their likely survival duration, etc.

Furthermore, a renewed phylogenetics could lead to a reinforced involvement of the evolutionist community in the management of biodiversity, because of a deep understanding of the diversity of evolutionary units. Evolutionists works will strongly support one of Popper's opinions, who, commenting what he had learned from science, concluded that «a strange picture of the world emerges which opens up many problems. It is a dualistic world: a world of structures in chaotically distributed motion. The small structures (such as the so-called elementary particles) build up larger structures; and this is brought about mainly by chaotic or random motion of the small structures, under special conditions of pressure and temperature. The larger structures may be atoms, molecules, crystals galaxies and galactic clusters. Many of these structures have a seeding effect, like drops of water in a cloud, or crystals in a solution; that is to say, they can grow and multiply by instruction; and they may persist, or disappear by selection. Some of them, such as the DNA crystals which constitute the gene structure of organisms and, with it, their building instructions, are almost infinitely rare and, we may perhaps say, very precious»[79]. However, a renewed phylogenetics will also show that these precious structures are sensitive to the influence of contemporary processes, so that their origins are multiplied, and a certain anti-historicism becomes relevant to describe them. Then, the preciosity and the rarity of life forms will be raised even more, since «with the understanding of the origin, the insignificance of the origin raises: meanwhile what's close to us, what's in us and around us begin little by little to appear rich in colors, in beauties, in enigmas and in meanings, that the ancient mankind was not even suspecting in its dreams»[80]. For all these reasons, in the long run, a renewed phylogenetics shall propose to the evolutionists more than a simple archeaological mission but in fact to work to acknowledge these real "molecular", "molar", organic, communitary associations and the bioconstructions, which result from the evolutionary processes, and that their predecessors did not suspect even in their dreams. The renewed phylogenetic era would thus be characterised as a period of constant ethical challenge, which would allow these scientists to practice an ethics in the noble sense of the word, that is «not a ban and prohibition function, but a function of invention and protection»[81], as they will realize that they live in a much richer biodiversity than previously thought.

Glossary

Essentialism: Essentialism is the view that some permanent, unalterable, and eternal properties of objects are essential to them, so that, for any specific kind of entity, it is at least theoretically possible to specify a finite list of characteristics —all of which any entity must have to belong to the group defined. (i.e. for an essentialist, all species remain unchanging throughout time.)

Ontology: Ontology is the science of being and existence. It seeks to describe or posit the basic categories and relationships of being (or existence) to define entities and types of entities within its framework, and to assert which one can be said to be real.

Nomadology: This cartographic perspective stems from Gilles Deleuze's philosophical work, which assumes that reality is an ongoing process. During this process, new connections and new opportunities are created, establishing new relations, so that systems can be radically open, and are always carrying the potential of connecting with other systems. As a result, seemingly established systems can be reversed by small, initial events and nomadology, as opposed to traditional history, can help thinking this situation of transformation, by celebrating the discontinuities, motions and changes that can transgress the apparent boundaries of the historically stratified reality.

Plan of Contagion: This notion describes the complex and constantly changing network of natural connections existing in nature between the different evolutionary units within which all beings are inserted. For instance, "if we consider that there are 10^{31} bacteriophages on earth, which infect 10^{24} bacteria per second, we can imagine that a continuous flow of genetic material occurs between bacteria sharing the same environments"[65]. Therefore one can argue that evolutionary units are not only nested in a hierarchical genealogical structure but also embedded within an intricate genetic tapestry.

Pluralism: Pluralism opposes to monism, by endorsing the view that several methods and theories are legitimate in an evolutionary study, because no single explanatory system can account for all the diverse phenomena of life. Eliminative pluralism is a special form of pluralism, that recommands to abandon a single large concept, which has proven not useful, in favor of several more precise concepts that denote natural categories.

Acknowledgments

We thank Dr. Christopher Lane and Dr. Adrian K. Sharma for careful reading of this chapter, and Pr. Jean Gayon, Pr. Michel Morange, Pr. Armand de Ricqlès, and Pr. Pascal Tassy for stimulating critical discussions on many of these complex issues.

References

[1] Panchen, A.L., *Classification, Evolution, and the Nature of Biology*. 1992: Cambridge University Press. 415.

[2] Koonin, E.V., K.S. Makarova, and L. Aravind, *Horizontal gene transfer in prokaryotes: quantification and classification. Annu Rev Microbiol*, 2001. 55: p. 709-42.

[3] Doolittle, W.F., et al., *Lateral Gene Transfer*, in *Evolutionary Genomics and Proteomics*. 2007, Sinauer.

[4] Hanage, W.P., C. Fraser, and B.G. Spratt, The impact of homologous recombination on the generation of diversity in bacteria. *J Theor Biol,* 2006. 239(2): p. 210-9.

[5] Tamas, I., et al., 50 million years of genomic stasis in endosymbiotic bacteria. *Science,* 2002. 296(5577): p. 2376-9.

[6] Polz, M.F., et al., Patterns and mechanisms of genetic and phenotypic differentiation in marine microbes. *Philos Trans R Soc Lond B Biol Sci,* 2006. 361(1475): p. 2009-21.

[7] Sober, E., Reconstructing the Past: Parsimony, *Evolution, and Inference.* 1991*: The MIT Press*; Reprint edition. 288.

[8] Woese, C.R., Bacterial evolution. *Microbiol Rev,* 1987. 51(2): p. 221-71.

[9] Ereshefsky, M. *Species.* 2006 [cited; Edward N. Zalta:[Available from: http://plato.stanford.edu/entries/species/.

[10] Splitter, L.J., *Species and Identity.* Philosophy of Science, 1988. 55(3): p. 323-348.

[11] Hull, D.L., The Effect of Essentialism on Taxonomy --Two Thousands Years of Stasis (I). *The British Journal for the Philosophy of Science*, 1965. 15(60): p. 314-326.

[12] Hull, D.L., The Effect of Essentialism on Taxonomy--Two Thousand Years of Stasis (II). *The British Journal for the Philosophy of Science*, 1965. 16(61): p. 1-18.

[13] Ghiselin, M.T., A Radical Solution to the Species Problem. *Systematic Zoology,* 1974. 23(4): p. 536-544.

[14] Hull, D.L., Are Species Really Individuals? *Systematic Zoology,* 1976. 25(2): p.174-191.

[15] Dupre, J., Natural Kinds and Biological Taxa. *The Philosophical Review*, 1981. 90(1): p. 66-90.

[16] Kitcher, P., Species. *Philosophy of Science, 1984.* 51(2): p. 308-333.

[17] Ereshefsky, M., Eliminative Pluralism. *Philosophy of Science,* 1992. 59(4): p. 671-690.

[18] Mishler, B.D. and M.J. Donoghue, *Species Concepts: A Case for Pluralism.* Systematic Zoology, 1982. 31(4): p. 491-503.

[19] Ereshefsky, M., Species Pluralism and Anti-Realism. *Philosophy of Science*, 1998. 65(1): p. 103-120.

[20] Rescher, N. *Process Philosophy.* 2002 [cited; Edward N. Zalta:[Available from: http://plato.stanford.edu/entries/process-philosophy/.

[21] Ereshefsky, M., *The poverty of the linnaean Hierarchy. A philosophical study of biological taxonomy.* 2001: Cambridges studies in Philosophy and Biology.

[22] Woese, C.R., A new biology for a new century. *Microbiol Mol Biol Rev*, 2004. 68(2): p. 173-86.

[23] Deleuze, G. and F. Guattari, *Capitalisme et Schizophrénie, tome 2: Mille Plateaux.* Critique. 1980: Editions de Minuit. 645.

[24] Brooks, D.R., Evolution in the Information Age: Rediscovering the Nature of the Organism. *Semiosis, Evolution, Energy, Development*, 2001. 1(1): p. 1-29.

[25] Lawrence, J., Selfish operons: the evolutionary impact of gene clustering in prokaryotes and eukaryotes. *Curr Opin Genet Dev*, 1999. 9(6): p. 642-8.

[26] Lawrence, J.G., Selfish operons and speciation by gene transfer. *Trends Microbiol*, 1997. 5(9): p. 355-9.

[27] Collier, J.D. and S.J. Muller. The dynamical basis of emergence in natural hierarchies. in Emergence, Complexity, Hierarchy and Organization. 1998: ECHO III Conference, *Acta Polytechnica Scandinavica*, MA91.

[28] Bedau, M.A., Downward causation and autonomy in weak emergence. *Principia*, 2003. 6: p. 5-50.

[29] Dawkins, R., *Extended Phenotype* - But Not Too Extended. A Reply to Laland, Turner and Jablonka. *Biology and Philosophy*, 2004. 19: p. 377-396.

[30] Sober, E., *Holism, Individualism, and the Units of Selection.* PSA: Proceedings of the Biennal Meeting of the Philosophy of Science Association, 1980. 2(Symposia and Invited Papers): p. 93-121.

[31] Jablonka, E., *From Replicators to Heritably Varying Phenotypic Traits: The Extended Phenotyper Revisited.* Biology and Phylosophy, 2004. 19: p. 353-375.

[32] Laland, K., *Extending the* Extended Phenotype. *Biology and Phylosophy*, 2004. 19: p. 313-325.

[33] Turner, J.S., Extended Phenotypes and Extended Organisms. *Biology and Phylosophy*, 2004. 19: p. 327-352.

[34] Tyson, G.W., et al., Community structure and metabolism through reconstruction of microbial genomes from the environment. *Nature*, 2004. 428(6978): p. 37-43.

[35] Doolittle, W.F. and E. Bapteste, Pattern Pluralism and the Tree of Life hypothesis. *Proc Natl Acad Sci U S A*, 2007. 104(7): p. 2043-9.

[36] Filee, J., et al., A Selective Barrier to Horizontal Gene Transfer in the T4-Type Bacteriophages That Has Preserved a Core Genome with the Viral Replication and Structural Genes. *Mol Biol Evol*, 2006. 23(9): p. 1688-1696.

[37] Bapteste, E., C. Brochier, and Y. Boucher, *Higher-level classification of the Archaea: evolution of methanogenesis and methanogens.* Archaea, 2005. 1(5): p. 353-63.

[38] Boucher, Y., et al., Lateral gene transfer and the origins of prokaryotic groups. *Annu Rev Genet*, 2003. 37: p. 283-328.

[39] Rieppel, O., Modules, Kinds, and Homology. *Journal of Experimental Zoology*, 2005. 304B: p. 18-27.

[40] Lyotard, J.-F., La Condition postmoderne: rapport sur le savoir. *Critique*, ed. E.d. Minuit. 1979, Paris. 109.

[41] Aylesworth, G. *Postmodernism*. 2005 [cited; Edward N. Zalta:[Available from: http://plato.stanford.edu/entries/postmodernism/.

[42] Martin, W., Mosaic bacterial chromosomes: a challenge en route to a tree of genomes. *Bioessays*, 1999. 21(2): p. 99-104.

[43] Dawkins, R., *The Extended Phenotype*. 1999: Oxford University Press. 233.

[44] Lyons, S., Thomas Kuhn Is Alive and Well, the evolutionary relationships of simple life forms-a paradigm under siege? *Perspectives in Biology and Medicine*, 2002. 45(3): p. 359-76.

[45] Huson, D.H. and D. Bryant, Application of Phylogenetic Networks in Evolutionary Studies. *Mol Biol Evol*, 2006. 2: p. 254-67.

[46] Nakhleh, L., et al., Reconstructing reticulate evolution in species-theory and practice. *J Comput Biol*, 2005. 12(6): p. 796-811.

[47] Makarenkov, V. and P. Legendre, From a phylogenetic tree to a reticulated network. J *Comput Biol*, 2004. 11(1): p. 195-212.

[48] MacLeod, D., et al., Deduction of probable events of lateral gene transfer through comparison of phylogenetic trees by recursive consolidation and rearrangement. *BMC Evol Biol*, 2005. 5(1): p. 27.

[49] Susko, E., et al., Visualizing and assessing phylogenetic congruence of core gene sets: a case study of the gamma-proteobacteria. Mol Biol Evol, 2006. 23(5): p. 1019-30.

[50] Deleuze, G., *Différence et Répétition*. Epimethee. 2000: Presses Universitaires de France. 416.

[51] DeLong, E.F., et al., *Community genomics among stratified microbial assemblages in the ocean's interior. Science*, 2006. 311(5760): p. 496-503.

[52] Edwards, R.A., et al., Using pyrosequencing to shed light on deep mine microbial ecology. *BMC Genomics*, 2006. 7: p. 57.

[53] Tringe, S.G., et al., *Comparative metagenomics of microbial communities. Science*, 2005. 308(5721): p. 554-7.

[54] Slot, J.C. and D.S. Hibbett, Horizontal Transfer of a Nitrate Assimilation Gene Cluster and Ecological Transitions in Fungi: A Phylogenetic Study. *PLOS One*, 2007. in press.

[55] Sharma, A.K., et al., Evolution of rhodopsin ion pumps in haloarchaea. *BMC Evol Biol*, 2007. 7: p. 79.

[56] Parrochia, D., Philosophie des réseaux. *La Politique éclatée*. 1993: Presses Universitaires de France. 304.

[57] Boucher, Y., et al., Integrons: mobilizable platforms that promote genetic diversity in bacteria. *Trends Microbiol*, 2007. 15(7): p. 301-9.

[58] Osborn, A.M. and D. Boltner, When phage, plasmids, and transposons collide: genomic islands, and conjugative- and mobilizable-transposons as a mosaic continuum. *Plasmid*, 2002. 48(3): p. 202-12.

[59] Bapteste, E., D. Macleod, and W.F. Doolittle. heuristic of the Synthesis of Life: the case study of gamma-proteobacteria. in *Microbial Genomes Conference, TIGR Meeting*. 2005. Halifax, Canada.

[60] Beiko, R.G., T.J. Harlow, and M.A. Ragan, *Highways of gene sharing in prokaryotes.* *Proc Natl Acad Sci U S A*, 2005. 102(40): p. 14332-7.

[61] Kunin, V., et al., The net of life: reconstructing the microbial phylogenetic network. *Genome Res*, 2005. 15(7): p. 954-9.

[62] Doolittle, W.F., Lateral genomics. *Trends Cell Biol*, 1999. 9(12): p. M5-8.

[63] Staley, J.T. and A. Konopka, Measurement of in situ activities of nonphotosynthetic microorganisms in aquatic and terrestrial habitats. *Annu Rev Microbiol*, 1985. 39: p. 321-46.

[64] Medini, D., et al., The microbial pan-genome. *Curr Opin Genet Dev*, 2005. 15(6): p. 589-94.

[65] Tettelin, H., V. Masignani, and e. al., Genome analysis of multiple pathogenic isolates of Streptococcus agalactiae: implications for the microbial "pan-genome". *Proc Natl Acad Sci U S A.*, 2005. 102(39): p. 13950-5.

[66] Zuckerkandl, E. and L. Pauling, *Evolutionary divergence and convergence in proteins*, in *Evolving Genes and Proteins*, V. Bryson and H.J. Vogel, Editors. 1965, Academic Press: New York. p. 97-166.

[67] Bapteste, E., et al., Do orthologous gene phylogenies really support tree-thinking? *BMC Evol Biol*, 2005. 5(1): p. 33.

[68] Bapteste, E., et al., Phylogenetic reconstruction and lateral gene transfer. *Trends Microbiol*, 2004. 12(9): p. 406-11.

[69] Moles, A. and E. Rohmer-Moles, *Les sciences de l'imprécis*. Points Sciences, ed. Seuil. 1998. 359.

[70] Suppes, P., The Plurality of Science. PSA: *Proceedings of the Biennal Meeting of the Philosophy of Science Association*, 1978. 2(Symposia and Invited Papers): p. 3-16.

[71] Tsou, J.Y., *Reconsidering Feyerabend's "Anarchism".* Perspectives on Science, 2003. 11(2): p. 208-235.

[72] Dagan, T. and W. Martin, The tree of one percent. *Genome Biology*, 2006. 7: p. 118.

[73] Tiercelin, C., *Hilary Putnam, l'héritage pragmatiste.* 2002: Presses Universitaires de France. 126.

[74] Bucknam, J., Y. Boucher, and E. Bapteste, Refuting phylogenetic relationships. *Biol Direct*, 2006. 1: p. 26.

[75] Steel, D., Can a reductionist be a pluralist? *Biology and Philosophy*, 2004. 19: p. 55-73.

[76] Duchesneau, F., Philosophie de la biologie. 1997, Paris: Presses Universitaires de France. 437.

[77] Wake, D.B., *A Tree Grows in Manhattan*, in *Assembling the Tree of Life*, M.J.D. Joel Cracraft, Editor. 2004, Oxford University Press. p. 543-544.

[78] Lartigue, C., et al., *Genome transplantation in bacteria: changing one species to another. Science*, 2007. 317(5838): p. 632-8.

[79] Popper, K., *The rationality of scientific revolutions*, in *Scientific Revolutions*, I. Hacking, Editor. 1981, Oxford University Press. p. 80-106.

[80] Nietzsche, F., *The Dawn of Day.* 2007: Dover Publications. 395.

[81] Kahn, A. and D. Lecourt, *Bioéthique et liberté.* 2004: Presses Universitaires de France. 128.

In: Genetic Recombination Research Progress
Editor: Jacob H. Schulz, pp. 163-187

ISBN: 978-1-60456-482-2
© 2008 Nova Science Publishers, Inc.

Chapter 5

Intricacies of Integration

René Daniel and Johanna A. Smith

Division of Infectious Diseases - Center for Human Virology, Kimmel Cancer Center,
Thomas Jefferson University, Philadelphia, PA, U.S.A.

Abstract

This article describes the molecular mechanisms which underly joining of human and animal retroviral DNA to host cell DNA. We discuss contributions of the viral protein integrase, as well as cellular co-factors, to the efficiency of integration and selection of integration sites in the host cell genome. We also address the questions related to the treatment of HIV (human immunodeficiency virus) infection, since the HIV-1 integrase is an attractive target for the development of anti-HIV-1 therapeutics. Finally, we present opportunities as well as the broblems associated with the integration of retroviral vectors, which are used in gene therapy applications.

Introduction

Early Steps of the Retroviral Life-Cycle

The life-cycle of a retrovirus can be divided into early and late steps. The early steps (Figure 1) include entry, where the retroviral particle binds to a receptor by means of the envelope proteins that are embedded in the envelope lipid bilayer. This binding is followed by fusion of the envelope with the cell membrane and injection of the innards of the particle into the cellular cytoplasm. In the cytoplasm, the viral reverse transcriptase proteins transcribe the viral RNA genome into double-stranded DNA. Once reverse transcription has occurred, the viral complex is called the preintegration complex. This complex consists of viral DNA along with other viral proteins, including the above mentioned reverse transcriptase, integrase, matrix, and possibly capsid [1]. These complexes also contain certain cellular proteins (discussed below) which appear to participate in the retroviral life-cycle. The preintegration complex then proceeds towards the cellular DNA. Through a process called integration, the viral DNA is incorporated into the host cell genome. In order to have access to host cell DNA however, the preintegration complex must traverse the nuclear membrane. Preintegration complexes of a classic model retrovirus, MLV (murine leukemia virus), can not pass the membrane and must remain in the cytoplasm until the membrane is dissolved during mitosis [2]. As a consequence, MLV integration can not occur in nondividing cells, which thus limits MLV replication to take place only in actively dividing cells. In contrast, the preintegration complexes of some other retroviruses, including the human immunodeficiency virus (HIV) and avian sarcoma virus (ASV), can pass through the nuclear pore and enter the nucleus [3]. These viruses are thus capable of efficiently infecting nondividing cells. This property is greatly utilized in retroviral vectors derived from these viruses (see below). Following integration, the late steps of retroviral replication transpire. First, retroviral DNA is efficiently transcribed by cellular enzymes. Viral RNAs are then exported to the cytoplasm, where translation and processing of viral proteins occurs. The viral proteins, as well as two copies of the viral RNA genome assemble into viral particles at the cell membrane, where the viral envelope is acquired during budding of the virus particle from the cellular membrane. The newly created retrovirus can now infected another cell.

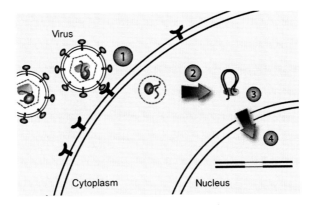

Figure 1. *Early steps of a retroviral life-cycle.* Infection is initiated upon viral binding to receptors of a host cell (1). Subsequently, the viral envelope fuses with the cell membrane, and viral components enter the cytoplasm. Here, the viral genetic material, which is in the form of RNA, is reversely transcribed to DNA by the viral protein reverse transcriptase (2). Following reverse transcription, the preintegration complex traverses the nuclear envelope through the nuclear pore complex (3). Through the process of integration, the viral DNA is inserted into the genome of the infected cell (4) completing the early steps of the retroviral life-cycle. Integration is catalyzed by the viral protein integrase.

Mechanism of Integration

An essential step or feature of the life-cycle of retroviruses, as well as retrotransposons, is the insertion of viral or transposon DNA into host cell genomic DNA, termed integration. This enables efficient transcription of viral (or retrotransposon) genes, which is performed by cellular enzymes [4, 5]. The integration step itself, however, is performed by a retroviral enzyme called integrase (or, in the case of transposons, transposase, [5-7]). A typical example of an integrase (IN) enzyme is that which is encoded by the human immunodeficiency type 1 (HIV-1) virus. This is a relatively small protein, consisting of 288 amino acids and encoded by the viral pol gene [8]. IN consists of three functional domains: N-terminus, core domain, and the C-terminus, which are all required for the first two steps of integration [9-11]. The catalytic domain is within the core domain and has a typical catalytic motif consisting of three amino acids (DDE motif, [12]). Mutation of any of these amino acids abolishes IN activity. The N-terminus consists of the N-terminal fifty amino acids and contains a zinc finger motif [13]. The C-terminus contains a DNA-binding domain and multimerization determinants [14-17]. It should be noted that IN functions as a multimer, probably a tetramer or even an octamer [18-24]. However, it appears that a dimer is also catalytically active [25-27].

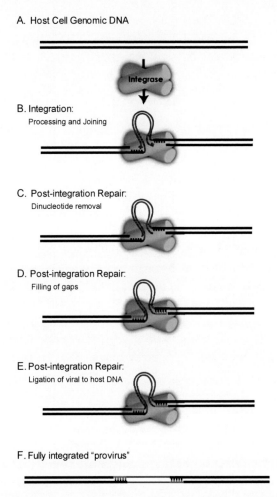

Figure 2. *Integration and Post-integration Repair.* Integrase is the retroviral protein responsible for the catalysis of the integration reaction. First, 3' viral DNA ends are processed to expose hydroxyl groups (Processing) which concomitantly attack host cell DNA (Joining, B). This essentially produces a staggered double stranded break in host DNA usually 4-6 base pairs apart (depending on the retrovirus). During the events which follow in the process of Post-integration Repair, DNA repair enzymes are employed. Post-integration Repair consists of cleavage of the 5' viral DNA ends (dinucleotide removal, C), filling of the gaps created by the staggered cut (D), and ligation of viral 5' ends to the host cell DNA (E). An appropriate chromatin structure must also be constructed around the viral DNA, and vicinity of the integration event. Upon the completion of Integration and Post-integration Repair, the viral genetic material is now referred to as a "stably transduced provirus". It should be noted that the order of the steps that take place in Post-integration Repair, is not yet known.

IN is a very unusual enzyme, which performs the unique integration reaction. Integration consists of two steps, called Processing and Joining. During Processing, which can take place in the cytoplasm of the host (infected) cell, IN (now a part of the preintegration complex, mentioned above) removes two nucleotides from the 3'-ends of the viral DNA. The Joining step occurs in the nucleus and consists of concerted cutting of host cell DNA and joining of viral DNA 3'-ends to the phosphates at the ends of the cut host cell DNA (Figure 2). In this reaction, IN produces a staggered cut on both strands of host cell DNA (Figure 2). The result of the integration reaction is thus an intermediate, where the viral DNA is flanked by single-stranded gaps in host cell DNA. These gaps are repaired in a step of the retroviral life-cycle, which is usually called Post-integration Repair (Figure 2). This step is discussed here since it is necessary for stable insertion of the viral DNA into host cell DNA and as such is often called the final, third step of integration. However, in this article, the term integration refers specifically to the Processing and Joining steps.

Post-integration Repair (PIR) includes processing of the 5'-ends of viral DNA, filling of the single-strand DNA gaps flanking the provirus, ligation of the 5'-ends of viral DNA to the newly synthesized DNA, and finally reconstitution of appropriate chromatin structure at the integration site. Upon completion of these requisite steps, viral DNA is referred to as a stably transduced provirus. Unlike integration, PIR is performed by cellular enzymes. A large body of evidence suggests that the enzymes involved are those of the cellular response pathway to double-strand DNA breaks (for reviews see [28-30]). It has been demonstrated that efficient PIR involves proteins of the cellular non-homologous end joining pathway (NHEJ, [31-35]). The NHEJ proteins shown to be involved in PIR include the DNA-dependent protein kinase (DNA-PK), ligase IV, and XRCC4. Additionally, certain other proteins that are involved in double-strand DNA break repair are also required for efficient PIR. These include the ATM (ataxia telangiectasia mutated) and ATR (ATM and Rad3 related) kinases [36, 37]. Interestingly, one of the ATM and ATR substrates, histone H2AX, is not required for efficient PIR, although it is phosphorylated at sites of retroviral DNA integration, and is required for efficient double-strand DNA break repair [38]. These latter data indicate that although PIR resembles the double-strand DNA break repair, there are significant differences between these two repair systems. Cellular proteins that are involved in PIR are potential targets for anti-retroviral therapy, as discussed below.

Filling of the single-stranded gaps by PIR leads to duplications of the host DNA sequences flanking the integrated viral DNA (referred to as "proviral" DNA). This duplication originates as a result of the staggered cut of host cell DNA, created by IN, as described above (see Figure 2). The length of the duplication depends on the species of retrovirus that performed the cut. In the case of HIV-1, the duplication consists of 5 base pairs. On the other hand, the IN enzyme of another model retrovirus, MLV (murine leukemia virus), creates the staggered double stranded break 4 base pairs apart, which generates a duplication of 4 base pairs. Finally, another extensively studies retrovirus, ASV (avian sarcoma virus), produces 6 base pair duplications [5].

As noted above, PIR is concluded following the reconstitution of an appropriate chromatin structure at the site of integration. This involves the incorporation of histones along proviral DNA, as well as in its vicinity, in order to assemble the proper nucleosomal structure. This process may include modifications made to histone tails in order to maintain/create the appropriate chromatin environment in the newly altered host cell DNA. At present, it is not known to what extent the chromatin structure at the integration site is determined by proviral DNA, as opposed to the flanking host cell DNA.

Integration Studies

In vitro Reconstitution of the Integration Reaction

The detailed biochemical studies of the integrase-mediated reaction have been exposed subsequent to the development of *in vitro* systems. These systems usually include: purified IN protein, model viral DNA, and cellular DNA. These systems led to the discoveries that IN is necessary and sufficient for both the Processing and Joining steps of integration, requires no source of energy, but necessitates a divalent cation (Mg^{2+} or Mn^{2+}) for its catalysis [39-42]. However, it has been clear early on that the utilized *in vitro* systems only imperfectly mimic the situation *in vivo*. The model viral DNA is usually represented by an oligonucleotide mimicking one end of viral DNA and the observed "integrations" were thus only half-reactions, since *in vivo* both ends of viral DNA are joined to host cell DNA in a concerted process. Thus, the oligonucleotides representing one end of viral DNA were replaced by "miniviral" DNAs, that allowed the study of the concerted joining of both ends [7, 43]. Ultimately, viral DNA models and purified IN can be replaced by extracting preintegration complexes (PICs) from infected cells instead (see above, [44]). Interestingly, it had been shown that the recombinant IN enzyme can catalyze *in vitro* integration at an efficiency comparable to that of PICs [27]. This finding can be explained in two ways. Either integration does not involve co-factors that could be present in PICs, or these putative co-factors were stripped or inactivated during preparation of PICs. Recent findings, discussed below, appear to support the latter hypothesis.

Similar to model viral DNAs, attempts were made to improve the model host cell DNA substrates. Most studies have been conducted using naked DNA substrates. However, DNA is tightly wrapped in a high-order chromatin structure *in vivo*. Initially, these substrates were modified by addition of a single nucleosome [45] or a minichromosome consisting of nucleosomes and circular DNA [46, 47]. Finally, an extended 13-nucleosome array was used, which was modified with the addition of histone H1, which compacted the array into a higher-order chromatin-like structure [48]. What has been learned from these innovative DNA substrates? It has been shown that nucleosomes affect the integration site selection by IN. With a naked DNA substrate, integration can occur anywhere in DNA, but usually occurs as a ladder of integration sites [49, 50]. Different IN proteins exhibit distinct integration patterns [49, 50]. Certain DNA-binding proteins can prevent integration by blocking target DNA or distorting the DNA structure [51-55]. Addition of nucleosomes showed that integration occurs preferentially into DNA that is wrapped around the nucleosome, rather then into a naked DNA substrate [45-47]. Further, chromatin compaction affects IN proteins of different species, in opposing ways. Efficiency of integration is decreased when HIV-1 IN is used in the system containing compacted chromatin with histone H1, while compaction increases integration efficiency mediated by ASV IN [48]. Taken together, chromatin proteins and structures influence both integration efficiency and integration site selection. In the following sections, we discuss how these *in vitro* results correlate with the *in vivo* findings, as well as new discoveries obtained from *in vivo* systems.

To study PIR, *in vitro* systems are less developed than those used to study integration. Two labs attempted to set up a rudimentary system based on an oligonucleotide with a single-strand gap in the DNA sequence [56, 57]. As expected, it has been shown that this gap can be filled by addition of a purified polymerase and a ligase. However, these data do not show what proteins are actually involved in integration *in vivo*. These systems could be made eventually more useful if they included cell extracts in the place of purified proteins.

In vivo Studies

In contrast to the *in vitro* systems described above, *in vivo* systems allow the study of integration within cells. In general, these systems rely on model retroviruses, such as HIV-1, MLV and ASV, as noted above. These are replication-competent viruses, performing other steps of the retroviral life-cycle, in addition to integration. A major advance in integration studies was made subsequent to the development of retroviral vectors. These are replication-deficient viruses that can perform the early steps of the life-cycle, up to PIR, but can not perform the late steps. This is due to a lack of retroviral genes in their genome, which usually carries only a marker gene, such as GFP (green fluorescent protein) or a therapeutic gene, in gene therapy applications (see below). The viral proteins, which, in addition to the genome, constitute the vector particle, are provided *in trans* by transfection of DNA encoding these genes into vector-producing cells. These vectors could then be used to study integration in laboratories that do not provide the safety level required for work with replication-competent HIV-1. Additionally, it is possible to pseudotype the vectors, as well as viruses. This means that the retroviral envelope proteins are replaced by envelope proteins from another virus. As a consequence, the pseudotyped virus, or vector, can infect a wider ranger of cells than the original virus, such as common lab cell lines (for example, HeLa). This method also simplifies integration studies.

Several methods were developed to study efficiency of integration within cells. These include Southern blotting, *Alu*-PCR and real-time *Alu*-PCR. Three species of viral DNA can be found in retrovirus-infected cells. These are unintegrated viral DNA, circularized viral DNA (unintegrated dead-end product formed in the cell nucleus) and viral DNA that is joined to host cell DNA. If DNA is extracted from infected cells, and analyzed by Southern blotting with a probe for viral DNA after electrophoresis in a low percentage agarose gel, 3 bands can be detected, each representing one of these species of viral DNA. The fastest species is the circularized viral DNA, then linear DNA. The viral DNA that was joined to host DNA runs with other host DNA in the high molecular weight fraction. Alternatively, the *Alu*-PCR method can be also used to determine the efficiency of integration. The *Alu*-PCR method employs two PCR primers. One primer targets cellular *Alu* sequences and the other primer binds to viral LTR sequence. *Alu*-PCR thus detects covalent joining of viral DNA to host DNA. This method can be improved as real-time *Alu*-PCR, which improves the quantitation of integrated viral DNA [58, 59]. However, these methods (Southern and *Alu*-PCR) can not determine if PIR occurred, they detect only the initial IN-mediated joining.

In contrast to the above-described methods to measure integration efficiency, methods to measure PIR are not as well developed. These methods usually rely on the finding that a failed PIR induces apoptosis, and cells in which integration (joining of 3'ends of viral to host DNA) occurred are thus lost from the population [35]. Alternatively, a strategy was developed by Dr. P. Brown and coworkers for MLV [60], which uses a mutant that carries a *Bpm*I restriction endonuclease recognition site near the end of its LTR, in U5. *Bpm*I cleaves at a fixed distance away from its recognition site. In this case, the cleavage site for the right LTR is always in host DNA. *Bpm*I digestion of cellular DNA from infected cells thus always yields an LTR-containing fragment of a distinct size, regardless of the integration site. The fragments are then analyzed on a denaturing gel, which can distinguish cleavages in single-strand from double-strand flanks by hybridization with strand-specific probes. However, this technically demanding method is not in widespread use.

In contrast to the *in vitro* methods, infection of cells also allows an analysis of PICs which are present in these cells. Identification of cellular proteins present in PICs led to identification of several integration co-factors (see below). Additionally, yeast two-hybrid systems were employed to identify IN-interacting proteins, with a successful outcome (also see below). Finally, *in vivo* studies permit comparison of integration efficiency of cells that carry known mutations in cellular genes and we can thus evaluate the effect of these genes on integration or PIR.

In contrast to the methods that are used to study integration efficiency, investigation of integration site selection (i.e. where in the cellular DNA integration occurs) relies on cloning and sequencing of integration sites. This demanding method was simplified to a degree by sequencing of the human genome, which lets us quickly identify the location of integration sites [61]. Integration preferences of various viruses and vectors will be further discussed in the later sections of this article.

What have these methods showed? Just like *in vitro*, the integration reaction *in vivo* is dependent on IN and mutations in the IN DDE motif block integration [5, 7, 12, 62]. The major difference between integration *in vivo* and *in vitro*, as described above, appears to be the role of cellular proteins, which act as co-factors in integration. In this review, we will focus on several of these proteins, which have been extensively studied.

Two of the potential integration co-factors were identified by yeast two-hybrid screens. The first of these was the human immunodeficiency type 1 (HIV-1) IN-binding protein termed Ini-1 (IN-interacting protein 1, [63]). Ini-1 increased integration efficiency *in vitro* [63]. The remaining question was, of course, if Ini-1 performs the same function *in vivo* and extensive studies were performed to address this question. Experiments with siRNA targeting Ini-1 [64] reported that HIV-1 replication was significantly reduced in cells with knocked down Ini-1. However, Boese *et al.* reported that in Ini-1-deficient cultured cells, integration proceeds just as efficiently as when Ini-1 is present [65]. Thus, Ini-1, although required for efficient HIV-1 replication, does not seems to be necessary for efficient integration *in vivo*. However, Ini-1 demonstrated to be involved in other steps of the retroviral life-cycle [64, 66, 67].

On the other hand, the Debyser laboratory employed the yeast two-hybrid system to isolate another HIV-1 IN-binding protein [19]). This protein was already known and termed the lens epithelium-derived growth factor (LEDGF/p75) [68, 69]. Ironically, in recent experiments where the mouse homolog of LEDGF/p75, Psip1, was knocked out, it was demonstrated that LEDGF/p75 is not a lens growth factor [70]. Instead, skeletal abnormalities were observed in animals lacking Psip1. These data indicate that this protein is involved in bone development [70]. SiRNA-mediated knockdown of LEDGF/p75 and experiments with primary LEDGF/p75 null cells showed that integration of HIV-1-based vectors is reduced 89-96% in the absence of LEDGF/p75 [71, 72]. Taken together, LEDGF/p75 is required for efficient integration of HIV-1 DNA. How about other retroviruses? It was shown that LEDGF/p75 does not bind to the MLV IN, nor is required for MLV integration [71, 73, 74].

In vitro, LEDGF/p75 enhances integration, which could explain one aspect of the LEDGF/p75 function in integration [19, 75, 76]. However, LEDGF/p75 also plays a major role in integration site selection of HIV-1 and HIV-1-based vectors, as discussed below ([71, 77-79].

Biochemical analysis in collusion with *in vitro* systems (as described above) identified another set of potential integration co-factors these are high mobility group proteins (HMGs) and the barrier-to-autointegration factor (BAF). Similar to Ini-1 a cellular high-mobility group protein 1 (HMG-1) was found to enhance integration *in vitro* [80]. Like other HMGs, HMG-1 is a non-histone chromatin protein and its effect could be due to their DNA-bending ability [7, 80].

An HMG-1-related protein, HMG I(Y), was found by analysis of HIV-1 preintegration complexes, which were isolated from HIV-1-infected cells [81]. Similar to HMG-1, HMG I(Y), as wells as HMG-2, increase the integration efficiency *in vitro* [80-82]. However, integration efficiency in cells lacking HMG I(Y) is just as high as in cells that have normal levels of this protein [83]. Thus, HMG proteins may not be necessary for efficient integration. A thorough biochemical analysis of murine leukemia virus (MLV) preintegration complexes identified a presence of another cellular protein, called the barrier-to-autointegration factor (BAF) [84]. BAF is a very small protein (89 amino acids). BAF prevents autointegration (integration into viral DNA itself), which otherwise leads to an aborted retroviral live-cycle. BAF performs the same function in HIV-1 preintegration complexes [85].

In vivo systems were also used to identify proteins that participate in PIR. PIR is not performed by IN, but exclusively by cellular proteins. As noted above in the second section of this article, these are proteins that are involved in the cellular double-strand DNA break repair [28-30].

In summary, *in vivo* systems show that IN plays critical role in integration. However, *in vivo*, IN is assisted by a number of cellular co-factors, of which apparently only some were identified.

Practical Implications

Since integration is an essential step of the retroviral life-cycle, efforts were made to exploit integration in both HIV-1 treatment and gene therapy. Obviously, inhibition of integration should result in a block of the retroviral life-cycle, and thus benefit HIV-1 positive individuals. But in the context of gene thereapy, the understanding and investigation of integration is also of major importance. A vast number of gene therapy trials have employed retroviral-based vectors to facilitate gene transfer. The efficiency of stable retroviral transduction is of critical concern, and may determine whether gene therapy approaches to treatment of human diseases result in success or failure. In many cases, the efficiency of retroviral transduction is quite low, and insufficient to achieve the therapeutic objectives [11]. Therefore, to increase the efficiency of integration, which is again an essential step of transduction, would be highly beneficial for the field of gene therapy.

In addition to integration efficiency, integration site selection plays an increasing role in both HIV-1 treatment and gene therapy applications. This process affects the likelihood that

HIV-1 enters latency, a major complication in HIV-1 treatment. Finally, integration site selection has assumed a large significance recently due to the propensity of retroviruses to integrate in the vicinity of active genes. These issues are discussed in the next two sections of this article.

Methods to Manipulate the Efficiency of Retroviral DNA Integration

Integrase Inhibitors

As noted above, IN is an attractive target for HIV-1 inhibitors, and as such, a large number of potential IN inhibitors have been developed. Their inhibitory activity was usually explored first *in vitro*, with processing and joining assays, and later in tissue culture experiments with replication-competent HIV-1 and HIV-1-based vectors. Some of these inhibitors showed promising inhibition of IN activity *in vitro* but failed to inhibit viral replication *in vivo* [86, 87]. Other types of inhibitors inhibited the IN activity *in vitro*, and also inhibited HIV-1 replication. However, the mechanism of inhibition *in vivo* was dissimilar from that *in vitro*. For example, the well-known IN inhibitor, L-chicoric acid, inhibits HIV-1 replication by targeting the envelope protein gp120 in tissue culture, instead of IN [88]. Similarly, our laboratory developed a naphatalene-derived inhibitor, 3,8-dibromo-7-amino-4-hydroxyl-2-naphtalene sulfonic acid, which was effective in the *in vitro* IN assays [89]. This compound also inhibited transduction by HIV-1-based vectors and HIV-1 replication in cultured cells, but it appeared to primarily block reverse transcription. Interestingly, it did not block the reverse transcriptase per se, but rather appeared to disrupt certain protein-protein interactions in the reverse transcriptase complex [89]. Therefore, development of IN inhibitors has turned out to be more complicated than originally envisaged.

A breakthrough in the inhibitor development came with the appearance of diketo acid (DKA) inhibitors. These were developed by researchers at Merck (Darmstadt, Germany) and at Shionogi Inc (Osaka, Japan) [90, 91]. These compounds block the joining reaction (also termed strand transfer) as well as inhibit HIV-1 replication in cultured cells primarily by the same mechanism. One of these compounds was reported to inhibit simian immunodeficiency virus (SIV) replication in infected macaques [92]. However, some of these early DKA compounds showed toxicity in canine trials [93]. Fortunately, latter DKA derivatives lacked the toxicity and entered clinical trials, culminating in the 2007 approval of the first IN inhibitor by FDA for anti-HIV-1 therapy [94, 95].

Another class of inhibitors was developed by the utilization of assays based on TN5 transposase inhibition [96]. As noted above, transposases are closely related to retroviral integrases. This approach revealed several compounds which block IN processing activity. It remains to be seen if these inhibitors will prove to be clinically useful.

Another class of IN inhibitors include nucleotide analogs and other related structures, which block IN-DNA interactions [97, 98]. Finally, styrylquinolines, which are being developed by BioAlliance (Paris, France), were reported to inhibit the formation of the IN-DNA complex, as well as IN interactions with host factors [99-103].

Taken together, it is likely that IN inhibitors will prove to be valuable drugs in HIV-1 treatment. However, as is the case of any viral protein, it is probable that HIV-1 strains will eventually emerge that carry IN resistant to inhibition by these compounds. Indeed, tissue culture experiments have already demonstrated the ability of HIV-1 resistance to these compounds [90]. One possible way to address the resistance problem plaguing inhibitors of viral proteins, is to develop inhibitors of cellular proteins, which are required for HIV-1 replication. Given the low mutation rate of these proteins, development of drug resistance is very unlikely. Indeed this approach was shown to work when an inhibitor of cellular deoxyhypusine synthase (DHS) was used to block activation of the eukaryotic initiation factor 5A (eIF-5A), which is a co-factor of the HIV-1 protein Rev. Even long-term treatment of HIV-1-infected cells with this inhibitor did not lead to selection of HIV-1 mutants resistant to this compound [104]. However, this kind of inhibitor has not yet entered the clinical practice. Similarly, in the case of IN, no such effective inhibitor has yet been developed to target a relevant co-factor, although proof-of-principle experiments have shown potential. HIV-1 replication was observed to be suppressed to a degree with a dominant negative mutant of the IN co-factor LEDGF/p75, as well as with siRNA targeting the protein [72]. It remains to be seen if the development of these inhibitors leads to clinically practical compounds.

PIR Inhibitors

In contrast to integration, inhibitors of cellular proteins have been extensively developed to inhibit PIR, since PIR is performed exclusively by cellular enzymes (see above). Our laboratory was the first to show that HIV-1 PIR, as well as transduction with HIV-1-based vectors can be suppressed with wortmannin, an inhibitor of DNA-PK and ATM [32]. These results indicated that PIR inhibition may be a feasible approach to anti-HIV therapy. We later built on these experiments and showed that replication of diverse HIV-1 strains can be inhibited by caffeine and the caffeine-related methylxanthines: theobromine, theophylline and paraxanthine [36, 105, 106]. These compounds are known or presumed to inhibit the ATM and ATR kinases [107-109]. Interestingly, one of these compounds, theophylline, is currently used for treatment of diseases other than AIDS, at plasma concentrations that were close to the IC_{50} obtained with one of the HIV-1 strains used in our experiments. We also presented evidence that this inhibition occurs at the PIR step in HIV-1-infected cells [36, 106]. Finally, we showed that HIV-1 transduction and PIR can be inhibited by pentoxifylline, which is a synthesized caffeine-related compound clinically used for the treatment of vasooclusive diseases [110]. An alternative approach was used by the Tang group, which showed that HIV-1 replication could be blocked by a ribozyme targeting the Ku80 protein [34], a component of DNA-PK, which is a trimer consisting of Ku80, Ku70 and the DNA-PK catalytic subunit [111]. Similarly, it has been shown by the Mouscadet group that an antisense probe against Ku80 suppresses HIV-1 replication [112]. The O'Connor group at KuDos Pharmaceuticals very elegantly demonstrated that HIV-1 replication can be inhibited by a newly developed inhibitor of the ATM kinase, Ku-55933 [37]. These investigators also demonstrated that the inhibition occurs at the PIR step of the HIV-1 life-cycle. In addition, they showed that Ku-55933 can even inhibit replication of HIV-1 strains that carry resistance to abacavir and lamivudine, which are otherwise used for HIV-1 treatment.

In summary, tissue culture experiments showed PIR inhibitors have a potential for therapeutic use. These inhibitors, if further developed, may prove to be valuable for the treatment of HIV-1 infection, particularly of infections with strains that are resistant to currently used inhibitors of HIV-1 proteins. However, as with any inhibitor, it is essential that we be concerned with the possible negative side effects. This is especially true for inhibitors of HIV-1 cellular co-factors, since these proteins, which unlike viral proteins, may participate in important cellular processes. However, it should be pointed out that cells carrying mutations in a number of these proteins, such as DNA-PK, ATM and LEDGF/p75, are viable [32, 35, 71] so their inhibition may not have such negative consequences. Also, inhibitors of these proteins do not seem to be cytotoxic at concentrations that are necessary for HIV-1 inhibition [72, 113-115]. Thus, it is possible that treatment with these inhibitors will *not* lead to insurmountable problems, and they may eventually prove to be clinically valuable.

Techniques to Increase the Efficiency of Integration

The efficiency of integration plays an important role in gene therapy, since a retroviral vector has to integrate in order to complete gene transfer. Efficiency of retroviral transduction can be raised by several methods which include: increasing the multiplicity of infection (number of infectious viral particles per cell), using agents such as polybrene to increase binding of viral particles to cell surfaces, and spinoculation, where cells are centrifuged with infectious, viral supernatant in order to increase binding of particles to cells [116]. However, little has been done to increase the efficiency of integration per se. One reason is that integration is a highly concerted and complex reaction. An increase of IN activity may lead to unspecific cutting of DNA, rather than an increase in integration efficiency. Hypothetically, one could stimulate integration by suppressing cellular co-factors which are known to inhibit this process, such as RAD52 [117]. Another possibility is to increase the efficiency of PIR by overexpressing such proteins involved in this process. However, these approaches have yet to be fully explored. It is hoped that future experiments will address some of these questions and lead to more efficient genetic transfer.

Integration Site Selection

Integration Site Preferences by Retroviruses and Retroviral Vectors

As noted above, *in vitro* systems have revealed certain integration site preferences. However, it is not clear if similar preferences can be found *in vivo*. Inside cells, genomic DNA is found in a higher-order chromatin structure, which is much more complex than most *in vitro* systems, which usually employ naked DNA and purified IN. Only one *in vitro* system has utilized a higher-order chromatin structure to study integration [48]. It showed that IN proteins from different species may exhibit unique preferences when it comes to the influence of chromatin structure on integration site selection (see above).

It has been shown recently that the human genome possesses approximately 25.000 genes [118]. Early studies, prior to sequencing of the human genome, suggested that retroviruses may integrate in or around the vicinity of transcription units [119, 120]. Due to the laborious sequencing methods at the time, prior to the automated sequencing we utilize today, these studies relied on a relatively low number of cloned integration sites [121]. Consequently, most investigators found it difficult to conclude if these integration site preferences were "real".

The existence of the completely sequenced human genome enabled a truly statistical analysis, which at first determined that HIV-1 in human T cell lines preferentially integrates in genes [121, 122]. The expected rate of integration in genes, if integration were random, would be 30%. However, it was found that the actual rate is close to 70%. Interestingly, inside the genes, HIV-1 did not show any integration preferences and appeared just as likely to integrate at the 5'-end of the gene as at the 3'-end. One related question is whether similar preferences can be found in the case of HIV-1-based vectors. This was addressed early on, and indeed, these vectors show integration site preferences similar to that of the replication-competent HIV-1 [122]. The HIV-1 study triggered considerable interest and led to examination of integration site preferences by other retroviruses. Retroviral species that are closely related to HIV-1, HIV-2, SIV, and FIV (feline immunodeficiency virus), exhibit integration site preferences similar to that of HIV-1 [123-126].

In contrast to HIV-1 and other related species, MLV, and MLV-based vectors, target other areas of the genome. MLV has shown a tendency to integrate close to 5'-ends of transcription units (about 20% of total integration events). In addition, approximately 17% of MLV integration events occur in the vicinity of CpG islands [127]. No clear preferences were found for the rest of MLV integration sites [127].

Finally, integration preferences of ASV and ASV-based vectors differ from both HIV-1 and MLV. ASV has been detected to have only a weak preference for genes (around 40%), and no MLV-like preference for 5'-ends of transcription units [127, 128]. Interestingly, very high levels of transcription may even inhibit ASV integration in the vicinity of genes [129, 130]. These preferences seem to be consistent with and validate the above described *in vitro* system, which utilized a nucleosomal array [48]. Lastly, the human T-leukemia virus type 1 (HTLV-1), similar to ASV, does not seem to prefer genes or transcription start sites [131].

Integration sites were also examined in more detail by two groups [132, 133]. Their results revealed symmetrical base preferences surrounding integration sites. The symmetry likely reflects the topology of the integration reaction. These weak consensus sequences appear to be preferred integration sites and are virus-specific. [132].

In summary, integration site selection is a nonrandom process, and is specific for each retroviral species. In the next section, we examine the molecular mechanism underlying this process.

Mechanism of Integration Site Selection

As described above, the cellular protein LEDGF/p75 is required for efficient integration of HIV-1 DNA. LEDGF/p75 is a transcription factor and has a C-terminal IN-binding domain and N-terminal chromatin-binding domain [76, 79, 134-136]. The chromatin-binding domain has PWWP and AT-hook motifs, which mediate chromatin binding [72, 135]. Interestingly, the IN-binding domain is missing in the LEDGF isoform, p52 [134]. LEDGF/p75 was found associated with preintegration complexes of HIV-1 and FIV in cells infected with these viruses [73]. A detailed analysis of the integration sites in LEDGF/p75 knockout cells determined that the residual HIV-1 DNA integration which occurs in these cells, does not show the typical HIV-1-associated preference for genes [71]. Instead, preferences were found to occur in the vicinity of promoter and CpG islands, resembling MLV [71]. However, the symmetrical base preferences surrounding the integration site were still observed [71, 132]. Thus, it appears that LEDGF/p75 targets HIV-1 (and other lentiviruses) integration into active genes. This presumably occurs by means of the C-terminal end of LEDGF/p75 binding to IN, and the LEDGF/p75 N-terminus binding to chromatin in the vicinity of active genes.

It is unlikely that LEDGF/p75 would be the single factor influencing integration site selection by lentiviruses, although it does apparently play a major role. An analysis of a large number of integration sites suggested that integration occurs in the vicinity of certain epigenetic marks, such as histone H4 K4 methylation, H4 acetylation or H3 acetylation [137]. These latter data suggest that the chromatin structure itself could be involved in the targeting of integration sites. Other recently identified factors that may influence integration site selection include the SATB1 protein. This is a T-cell lineage-specific chromatin organizer, and its knockdown results in reduction of HIV-1 integration near SATB1 binding sites. SATB1 thus appears to be involved in integration site selection, by a so far unknown mechanism [138]. Finally, it has been suggested that the DNA repair protein Ku80, which is in the preintegration complex, drives integration toward chromatin domains prone to silencing [112, 139].

LEDGF/p75 does not bind to MLV IN, nor, likely, ASV IN (see above). Thus, integration site selection by these viruses is likely determined by other mechanisms. At this point, we do not know what determines ASV integration preferences. A study of HIV chimeras with MLV genes showed that MLV IN likely determines integration site preferences by MLV [140]. An auxiliary role in integration site selection was shown for the Gag-derived proteins, since an HIV chimera containing MLV gag has targeting preferences different from both HIV and MLV [140]. In addition to gag proteins, other virus proteins could possibly be involved in integration site selection. However, *vif, vpr, vpx, nef, env* and promoter and enhancer regions of SIV do not appear to be required for preferential SIV integration into genes [141].

Taken together, although we do not yet fully understand the process of integration site selection, cellular co-factors of retroviral DNA integration, together with IN, are intimately involved in this process.

Practical Implications of Integration Site Preferences

Integration site selection plays a role in both HIV-1 infection and gene therapy applications. Integration in certain regions of the human genome, such as heterochromatin, can lead to suppression (silencing) of the retroviral DNA by epigenetic mechanisms. Indeed, some studies suggested that the site of HIV-1 integration determines the basal transcriptional activity and response to Tat transactivation [142, 143]. Therefore, integration in the transcriptionally inactive chromatin may lead to establishment of HIV-1 latency, which is a major complication of HIV-1 treatment [144]. Integration site selection thus plays an important role in HIV-1 infection. However, this process assumed recently truly dramatic proportions in the field of gene therapy.

Gene therapy trials, as well as gene therapy experiments in animal systems, usually employ MLV-based or HIV-1-based vectors. It has been long speculated that predilection of these vectors to integrate in the vicinity of promoters (MLV) or in any portion of transcription units (HIV-1), could lead to an undesirable activation or disruption of cellular gene expression. Alternatively, it has also been speculated that integration around undesirable sites occurs with such a low frequency, that it may not affect a therapeutic outcome. Finally, one could also imagine that activation of a gene due to integration in its vicinity may be even a positive event in the gene therapy trial, since it may be a beneficial scenario for the transduced cells.

The first piece of evidence suggesting that an adverse event may arise due to integration at a "wrong spot" of the genome came from a murine model, where retroviral transduction led to the development of leukemia, which was associated with integration and activation of the murine gene Evi1 (ecotropic viral integration site-1, [145]). A second piece of evidence was the unfortunate outcome of a gene therapy trial conducted with X-linked SCID (SCID-X1) patients [146-148]. Four out of 11 patients involved in this trial, which used an MLV-based vector, eventually developed T cell leukemia. A sequencing analysis of the T cells from these patients showed an expansion of T cell clones, which contained an insertion around the Lim domain only 2 (LMO2) protooncogene, and subsequently induced its activation. Third, leukemias were also reported after high-copy retroviral gene transfer of the multidrug resistance 1 (MDR1) cDNA [149]. These results shed light on the consequences of integration site selection and integration site preferences on the outcome of gene therapy trials.

In contrast to these events, no cancer development was described in another recent SCID-X1 trial [150], nor in a recent ADA-SCID gene therapy trial [151]. In these instances, no adverse effects associated with vector integration were observed [151]. Likewise, there were no clinically significant consequences found in leukemic patients treated with allogeneic stem cell transplantation and donor lymphocytes genetically modified with a suicide gene [152].

Finally, the speculation that an integration event itself could be beneficial for the outcome of a trial was somewhat vindicated in a successful gene therapy trial for the X-linked chronic granulomatous disease, where activating insertions in MDS1-EVI1, PRDM16 and SETBP1 genes apparently influenced regulation of long-term hematopoiesis by expanding gene-corrected myelopoiesis three- to four-fold in treated individuals [153]. Therefore, site-specific integration events may at times enhance the likelihood of a positive outcome of a gene therapy trial.

Gene therapy trials usually employ MLV-based vectors. The safety of HIV-1-based vectors was evaluated very recently in a tumor-prone mouse model [154]. Here, as in other cases, MLV-based vectors were found to accelerate oncogenesis in this model. In contrast, HIV-1-based vectors did not appear to accelerate tumor formation. Therefore, it is possible that HIV-1-based vectors may be a safer gene transfer tool than MLV-based vectors. This hypothesis, however, may require further testing.

In summary, insertion of retroviral DNA into chromosomes was associated with development of leukemias in both animal models and human gene therapy trials. These cases highlight a necessity for further improvements in design of retroviral vectors.

Efforts to Direct Integration of Retroviral DNA to Predetermined Chromosomal Regions

Retroviral vectors should ideally integrate in predetermined regions of the human genome, where the likelihood of activation of an oncogene is minimal. The initial attempts to target retroviral DNA integration date to the mid '90s. Since integration is catalyzed by IN, this protein became the focus of early efforts. Three laboratories (Skalka, Chow and Bushman) have used a strikingly similar approach to the problem. All three groups constructed fusion proteins, which consisted of the integrase protein, either from HIV-1 or ASV, and a DNA binding sequence from a eukaryotic or bacterial protein. In two cases, the DNA-binding domain (DBD) was from the *E. coli* lexA repressor [54, 155], while another case utilized the zinc finger protein, E2C [156], and finally the zinc finger zif268 [157]. DBDs were fused to either ends of the IN protein. In all cases, the fusion protein(s) was shown to target integration to a predetermined sequence in the test tube [54, 155, 157, 158]. However, attempts to extend these experiments to the cultured cells brought a disappointment. The IN-lexA fusion proved to be a target for the viral protease protein, which deleted the heterologous DBD [54]. We then attempted to mutate the protease recognition site between DBD and the rest of the protein. The resulting mutant indeed proved to be resistant to the viral protease, however, it failed to incorporate into the virion, probably due to its size (Daniel and Skalka, unpublished data). Similar problems, i.e. incorporation into the virion, were encountered when lexA DBD was replaced with DBD of the cellular GATA6 protein (Daniel and Skalka, unpublished data). As for HIV-1 IN, the fusion proteins proved in one case to be unfunctional *in vivo* [157]. In the other case, the Chow laboratory demonstrated that integrase-E2C fusion proteins do increase integration into pre-determined chromosomal regions *in vivo* (about 10 fold, [159]). Unfortunately, the overall efficiency of retroviral DNA integration in this case decreased up to 100 fold, which is a very undesirable side effect. Therefore, an efficient method that combines high rates of integration and targeting to a predetermined chromosomal region, remains to be developed.

Perspectives

Recent advances in the field of retroviral DNA integration led to development of a first clinically used integrase inhibitor. Additionally, our understanding of the process of integration site selection has dramatically increased over the recent years. Given the proven danger of cancer development associated with integration in undesirable regions of the human genome, this knowledge will be of utmost importance in the advancement of new vectors for gene therapy that have an increased safety margin. It is hoped that more laboratory investigators will join in this effort, at this exciting time of dramatic developments in the field of integration.

References

[1] Lewinski, M.K. and F.D. Bushman, Retroviral DNA integration--mechanism and consequences. *Advances in Genetics*, 2005. 55: p. 147-81.

[2] Roe, T., et al., Integration of murine leukemia virus DNA depends on mitosis. *EMBO Journal,* 1993. 12(5): p. 2099-108.

[3] Katz, R.A., J.G. Greger, and A.M. Skalka, Effects of cell cycle status on early events in retroviral replication. *Journal of Cellular Biochemistry*, 2005. 94(5): p. 880-9.

[4] Pereira, L.A., et al., A compilation of cellular transcription factor interactions with the HIV-1 LTR promoter. *Nucleic Acids Research*, 2000. 28(3): p. 663-8.

[5] Coffin, J.M., S.H. Hughes, and H.E. Varmus, *Retroviruses.* 1997, Plainview, NY: Cold Spring Harbor Laboratory Press.

[6] Craigie, R., HIV integrase, a brief overview from chemistry to therapeutics. *Journal of Biological Chemistry*, 2001. 276(26): p. 23213-6.

[7] Flint, S.E., LW; Racaniello, VR; Skalka, AM, *Principles of Virology.* second edition ed. 2004, Washington, DC: ASM Press.

[8] Engelman, A., et al., Multiple effects of mutations in human immunodeficiency virus type 1 integrase on viral replication. *Journal of Virology*, 1995. 69(5): p. 2729-36.

[9] Drelich, M., R. Wilhelm, and J. Mous, Identification of amino acid residues critical for endonuclease and integration activities of HIV-1 IN protein in vitro. *Virology*, 1992. 188(2): p. 459-68.

[10] Schauer, M. and A. Billich, The N-terminal region of HIV-1 integrase is required for integration activity, but not for DNA-binding *Biochemical & Biophysical Research Communications,.* 1992. 185(3): p. 874-80.

[11] Vink, C., A.M. Oude Groeneger, and R.H. Plasterk, Identification of the catalytic and DNA-binding region of the human immunodeficiency virus type I integrase protein. *Nucleic Acids Research*, 1993. 21(6): p. 1419-25.

[12] Kulkosky, J., et al., Residues critical for retroviral integrative recombination in a region that is highly conserved among retroviral/retrotransposon integrases and bacterial insertion sequence transposases. *Molecular & Cellular Biology,* 1992. 12(5): p. 2331-8.

[13] Bushman, F.D., et al., Domains of the integrase protein of human immunodeficiency virus type 1 responsible for polynucleotidyl transfer and zinc binding. *Proceedings of the National Academy of Sciences of the United States of America*, 1993. 90(8): p. 3428-32.

[14] Andrake, M.D. and A.M. Skalka, Multimerization determinants reside in both the catalytic core and C terminus of avian sarcoma virus integrase. *Journal of Biological Chemistry*, 1995. 270(49): p. 29299-306.

[15] Esposito, D. and R. Craigie, Sequence specificity of viral end DNA binding by HIV-1 integrase reveals critical regions for protein-DNA interaction. *EMBO Journal*, 1998. 17(19): p. 5832-43.

[16] Jenkins, T.M., et al., A soluble active mutant of HIV-1 integrase: involvement of both the core and carboxyl-terminal domains in multimerization. *Journal of Biological Chemistry,* 1996. 271(13): p. 7712-8.

[17] Lutzke, R.A. and R.H. Plasterk, Structure-based mutational analysis of the C-terminal DNA-binding domain of human immunodeficiency virus type 1 integrase: critical

residues for protein oligomerization and DNA binding. *Journal of Virology*, 1998. 72(6): p. 4841-8.

[18] Barsov, E.V., et al., Inhibition of human immunodeficiency virus type 1 integrase by the Fab fragment of a specific monoclonal antibody suggests that different multimerization states are required for different enzymatic functions. *Journal of Virology*, 1996. 70(7): p. 4484-94.

[19] Cherepanov, P., et al., HIV-1 integrase forms stable tetramers and associates with LEDGF/p75 protein in human cells. *Journal of Biological Chemistry*, 2003. 278(1): p. 372-81.

[20] Ellison, V., et al., An essential interaction between distinct domains of HIV-1 integrase mediates assembly of the active multimer. *Journal of Biological Chemistry*, 1995. 270(7): p. 3320-6.

[21] Petit, C., O. Schwartz, and F. Mammano, Oligomerization within virions and subcellular localization of human immunodeficiency virus type 1 integrase. *Journal of Virology*, 1999. 73(6): p. 5079-88.

[22] Wang, L.D., et al., Constructing HIV-1 integrase tetramer and exploring influences of metal ions on forming integrase-DNA complex. *Biochemical & Biophysical Research Communications*, 2005. 337(1): p. 313-9.

[23] Yang, Z.N., et al., *Crystal structure of an active two-domain derivative of Rous sarcoma virus integrase. Journal of Molecular Biology*, 2000. 296(2): p. 535-48.

[24] Zheng, R., T.M. Jenkins, and R. Craigie, Zinc folds the N-terminal domain of HIV-1 integrase, promotes multimerization, and enhances catalytic activity. Proceedings *of the National Academy of Sciences of the United States of America*, 1996. 93(24): p. 13659-64.

[25] Faure, A., et al., HIV-1 integrase crosslinked oligomers are active in vitro. *Nucleic Acids Research*, 2005. 33(3): p. 977-86.

[26] Guiot, E., et al., Relationship between the oligomeric status of HIV-1 integrase on DNA and enzymatic activity. *Journal of Biological Chemistry*, 2006. 281(32): p. 22707-19.

[27] Sinha, S. and D.P. Grandgenett, Recombinant human immunodeficiency virus type 1 integrase exhibits a capacity for full-site integration in vitro that is comparable to that of purified preintegration complexes from virus-infected cells. *Journal of Virology*, 2005. 79(13): p. 8208-16.

[28] Daniel, R., DNA repair in HIV-1 infection: a case for inhibitors of cellular co-factors? *Current HIV Research*, 2006. 4(4): p. 411-21.

[29] Skalka, A.M. and R.A. Katz, Retroviral DNA integration and the DNA damage response. *Cell Death & Differentiation*, 2005. 12 Suppl 1: p. 971-8.

[30] Smith, J.A. and R. Daniel, Following the path of the virus: the exploitation of host DNA repair mechanisms by retroviruses. *ACS Chemical Biology [Electronic Resource]*, 2006. 1(4): p. 217-26.

[31] Daniel, R., et al., Evidence that stable retroviral transduction and cell survival following DNA integration depend on components of the nonhomologous end joining repair pathway *Journal of Virology.*, 2004. 78(16): p. 8573-81.

[32] Daniel, R., et al., Wortmannin potentiates integrase-mediated killing of lymphocytes and reduces the efficiency of stable transduction by retroviruses.[erratum appears in Mol Cell Biol 2001 Apr;21(7):2617]. *Molecular & Cellular Biology*, 2001. 21(4): p. 1164-72.

[33] Jeanson, L., et al., Effect of Ku80 depletion on the preintegrative steps of HIV-1 replication in human cells. *Virology*, 2002. 300(1): p. 100-8.

[34] Waninger, S., et al., Identification of cellular co-factors for human immunodeficiency virus replication via a ribozyme-based genomics approach. *J Virol*, 2004. 78(23): p. 12829-37.

[35] Daniel, R., R.A. Katz, and A.M. Skalka, A role for DNA-PK in retroviral DNA integration. *Science*, 1999. 284(5414): p. 644-7.

[36] Daniel, R., et al., Evidence that the retroviral DNA integration process triggers an ATR-dependent DNA damage response. *Proc Natl Acad Sci U S A*, 2003. 100(8): p. 4778-83.

[37] Lau, A., et al., Suppression of HIV-1 infection by a small molecule inhibitor of the ATM kinase. *Nat Cell Biol*, 2005. 7(5): p. 493-500.

[38] Daniel, R., et al., Histone H2AX is phosphorylated at sites of retroviral DNA integration but is dispensable for postintegration repair. *J Biol Chem*, 2004. 279(44): p. 45810-4.

[39] Ellison, V. and P.O. Brown, A stable complex between integrase and viral DNA ends mediates human immunodeficiency virus integration in vitro. *Proceedings of the National Academy of Sciences of the United States of America*, 1994. 91(15): p. 7316-20.

[40] Vink, C., R.A. Lutzke, and R.H. Plasterk, Formation of a stable complex between the human immunodeficiency virus integrase protein and viral DNA. *Nucleic Acids Research*, 1994. 22(20): p. 4103-10.

[41] Wolfe, A.L., et al., The role of manganese in promoting multimerization and assembly of human immunodeficiency virus type 1 integrase as a catalytically active complex on immobilized long terminal repeat substrates. *Journal of Virology*, 1996. 70(3): p. 1424-32.

[42] Yi, J., E. Asante-Appiah, and A.M. Skalka, Divalent cations stimulate preferential recognition of a viral DNA end by HIV-1 integrase. *Biochemistry*, 1999. 38(26): p. 8458-68.

[43] Bischerour, J., et al., The (52-96) C-terminal domain of Vpr stimulates HIV-1 IN-mediated homologous strand transfer of mini-viral DNA. *Nucleic Acids Research*, 2003. 31(10): p. 2694-702.

[44] Chen, H. and A. Engelman, Characterization of a replication-defective human immunodeficiency virus type 1 att site mutant that is blocked after the 3' processing step of retroviral integration. *Journal of Virology*, 2000. 74(17): p. 8188-93.

[45] Pruss, D., F.D. Bushman, and A.P. Wolffe, Human immunodeficiency virus integrase directs integration to sites of severe DNA distortion within the nucleosome core. *Proceedings of the National Academy of Sciences of the United States of America*, 1994. 91(13): p. 5913-7.

[46] Pryciak, P.M., A. Sil, and H.E. Varmus, *Retroviral integration into minichromosomes in vitro. EMBO Journal*, 1992. 11(1): p. 291-303.

[47] Pryciak, P.M. and H.E. Varmus, Nucleosomes, DNA-binding proteins, and DNA sequence modulate retroviral integration target site selection. *Cell*, 1992. 69(5): p. 769-80.

[48] Taganov, K.D., et al., Integrase-specific enhancement and suppression of retroviral DNA integration by compacted chromatin structure in vitro. *Journal of Virology*, 2004. 78(11): p. 5848-55.

[49] Katzman, M. and M. Sudol, Mapping domains of retroviral integrase responsible for viral DNA specificity and target site selection by analysis of chimeras between human immunodeficiency virus type 1 and visna virus integrases. *Journal of Virology*, 1995. 69(9): p. 5687-96.

[50] Shibagaki, Y. and S.A. Chow, Central core domain of retroviral integrase is responsible for target site selection. *Journal of Biological Chemistry*, 1997. 272(13): p. 8361-9.

[51] Bor, Y.C., F.D. Bushman, and L.E. Orgel, In vitro integration of human immunodeficiency virus type 1 cDNA into targets containing protein-induced bends. *Proceedings of the National Academy of Sciences of the United States of America*, 1995. 92(22): p. 10334-8.

[52] Goulaouic, H. and S.A. Chow, Directed integration of viral DNA mediated by fusion proteins consisting of human immunodeficiency virus type 1 integrase and Escherichia coli LexA protein. *Journal of Virology*, 1996. 70(1): p. 37-46.

[53] Katz, R.A., K. Gravuer, and A.M. Skalka, A preferred target DNA structure for retroviral integrase in vitro. *Journal of Biological Chemistry*, 1998. 273(37): p.24190-5.

[54] Katz, R.A., G. Merkel, and A.M. Skalka, Targeting of retroviral integrase by fusion to a heterologous DNA binding domain: in vitro activities and incorporation of a fusion protein into viral particles. *Virology*, 1996. 217(1): p. 178-90.

[55] Muller, H.P. and H.E. Varmus, DNA bending creates favored sites for retroviral integration: an explanation for preferred insertion sites in nucleosomes. *EMBO Journal*, 1994. 13(19): p. 4704-14.

[56] Brin, E., et al., Modeling the late steps in HIV-1 retroviral integrase-catalyzed DNA integration. *Journal of Biological Chemistry*, 2000. 275(50): p. 39287-95.

[57] Yoder, K.E. and F.D. Bushman, *Repair of gaps in retroviral DNA integration intermediates. Journal of Virology*, 2000. 74(23): p. 11191-200.

[58] Brussel, A., O. Delelis, and P. Sonigo, Alu-LTR real-time nested PCR assay for quantifying integrated HIV-1 DNA. *Methods in Molecular Biology*, 2005. 304: p. 139-54.

[59] O'Doherty, U., et al., A sensitive, quantitative assay for human immunodeficiency virus type 1 integration. *Journal of Virology*, 2002. 76(21): p. 10942-50.

[60] Roe, T., S.A. Chow, and P.O. Brown, 3'-end processing and kinetics of 5'-end joining during retroviral integration in vivo. *Journal of Virology*, 1997. 71(2): p. 1334-40.

[61] Lander, E.S., et al., Initial sequencing and analysis of the human genome.[see comment][erratum appears in Nature 2001 Aug 2;412(6846):565 Note: Szustakowki, J [corrected to Szustakowski, J]]. *Nature*, 2001. 409(6822): p. 860-921.

[62] Naldini, L., et al., In vivo gene delivery and stable transduction of nondividing cells by a lentiviral vector.[see comment]. *Science*, 1996. 272(5259): p. 263-7.

[63] Kalpana, G.V., et al., Binding and stimulation of HIV-1 integrase by a human homolog of yeast transcription factor SNF5.[see comment]. *Science*, 1994. 266(5193): p. 2002-6.

[64] Ariumi, Y., et al., The integrase interactor 1 (INI1) proteins facilitate Tat-mediated human immunodeficiency virus type 1 transcription. *Retrovirology*, 2006. 3: p. 47.

[65] Boese, A., et al., Ini1/hSNF5 is dispensable for retrovirus-induced cytoplasmic accumulation of PML and does not interfere with integration. *FEBS Letters*, 2004. 578(3): p. 291-6.

[66] Treand, C., et al., Requirement for SWI/SNF chromatin-remodeling complex in Tat-mediated activation of the HIV-1 promoter. *EMBO Journal*, 2006. 25(8): p. 1690-9.

[67] Mahmoudi, T., et al., The SWI/SNF chromatin-remodeling complex is a cofactor for Tat transactivation of the HIV promoter.[erratum appears in J Biol Chem. 2006 Sep 8;281(36):26768]. Journal *of Biological Chemistry*, 2006. 281(29): p. 19960-8.

[68] Ge, H., Y. Si, and R.G. Roeder, Isolation of cDNAs encoding novel transcription coactivators p52 and p75 reveals an alternate regulatory mechanism of transcriptional activation. *EMBO Journal*, 1998. 17(22): p. 6723-9.

[69] Singh, D.P., et al., Lens epithelium-derived growth factor (LEDGF/p75) and p52 are derived from a single gene by alternative splicing. *Gene*, 2000. 242(1-2): p. 265-73.

[70] Sutherland, H.G., et al., Disruption of Ledgf/Psip1 results in perinatal mortality and homeotic skeletal transformations. *Molecular & Cellular Biology*, 2006. 26(19): p. 7201-10.

[71] Shun, M.C., et al., LEDGF/p75 functions downstream from preintegration complex formation to effect gene-specific HIV-1 integration. *Genes & Development*, 2007. 21(14): p. 1767-78.

[72] Llano, M., et al., An essential role for LEDGF/p75 in HIV integration. *Science*, 2006. 314(5798): p. 461-4.

[73] Llano, M., et al., LEDGF/p75 determines cellular trafficking of diverse lentiviral but not murine oncoretroviral integrase proteins and is a component of functional lentiviral preintegration complexes. *Journal of Virology*, 2004. 78(17): p. 9524-37.

[74] Busschots, K., et al., The interaction of LEDGF/p75 with integrase is lentivirus-specific and promotes DNA binding. *Journal of Biological Chemistry*, 2005. 280(18): p. 17841-7.

[75] Cherepanov, P., LEDGF/p75 interacts with divergent lentiviral integrases and modulates their enzymatic activity in vitro. *Nucleic Acids Research*, 2007. 35(1): p. 113-24.

[76] Cherepanov, P., et al., Identification of an evolutionarily conserved domain in human lens epithelium-derived growth factor/transcriptional co-activator p75 (LEDGF/p75) that binds HIV-1 integrase. *Journal of Biological Chemistry*, 2004. 279(47): p. 48883-92.

[77] Ciuffi, A., et al., Modulating target site selection during human immunodeficiency virus DNA integration in vitro with an engineered tethering factor. *Human Gene Therapy*, 2006. 17(9): p. 960-7.

[78] Ciuffi, A., et al., A role for LEDGF/p75 in targeting HIV DNA integration. *Nature Medicine*, 2005. 11(12): p. 1287-9.

[79] Llano, M., et al., Identification and characterization of the chromatin-binding domains of the HIV-1 integrase interactor LEDGF/p75. *Journal of Molecular Biology*, 2006. 360(4): p. 760-73.

[80] Aiyar, A., et al., Concerted integration of linear retroviral DNA by the avian sarcoma virus integrase in vitro: dependence on both long terminal repeat termini. *Journal of Virology*, 1996. 70(6): p. 3571-80.

[81] Farnet, C.M. and F.D. Bushman, HIV-1 cDNA integration: requirement of HMG I(Y) protein for function of preintegration complexes in vitro. *Cell*, 1997. 88(4): p. 483-92.

[82] Hindmarsh, P., et al., HMG protein family members stimulate human immunodeficiency virus type 1 and avian sarcoma virus concerted DNA integration in vitro. *Journal of Virology*, 1999. 73(4): p. 2994-3003.

[83] Beitzel, B. and F. Bushman, Construction and analysis of cells lacking the HMGA gene family. Nucleic *Acids Research*, 2003. 31(17): p. 5025-32.

[84] Lee, M.S. and R. Craigie, *A previously unidentified host protein protects retroviral DNA from autointegration. Proceedings of the National Academy of Sciences of the United States of America*, 1998. 95(4): p. 1528-33.

[85] Lin, C.W. and A. Engelman, The barrier-to-autointegration factor is a component of functional human immunodeficiency virus type 1 preintegration complexes. *Journal of Virology*, 2003. 77(8): p. 5030-6.

[86] Johnson, A.A., C. Marchand, and Y. Pommier, HIV-1 integrase inhibitors: a decade of research and two drugs in clinical trial. *Current Topics in Medicinal Chemistry*, 2004. 4(10): p. 1059-77.

[87] Witvrouw, M., et al., Novel inhibitors of HIV-1 integration. *Current Drug Metabolism*, 2004. 5(4): p. 291-304.

[88] Pluymers, W., et al., Viral entry as the primary target for the anti-HIV activity of chicoric acid and its tetra-acetyl esters *Molecular Pharmacology.*, 2000. 58(3): p.641-8.

[89] Daniel, R., et al., *Characterization of a naphthalene derivative inhibitor of retroviral integrases. AIDS Research & Human Retroviruses*, 2004. 20(2): p. 135-44.

[90] Hazuda, D.J., et al., Inhibitors of strand transfer that prevent integration and inhibit HIV-1 replication in cells. *Science*, 2000. 287(5453): p. 646-50.

[91] Goldgur, Y., et al., Structure of the HIV-1 integrase catalytic domain complexed with an inhibitor: a platform for antiviral drug design. *Proceedings of the National Academy of Sciences of the United States of America*, 1999. 96(23): p. 13040-3.

[92] Hazuda, D.J., et al., Integrase inhibitors and cellular immunity suppress retroviral replication in rhesus macaques. *Science*, 2004. 305(5683): p. 528-32.

[93] Ramcharan, J. and A.M. Skalka, Strategies for identification of HIV-1 integrase inhibitors. Future *Virology*, 2006. 1(6): p. 717-731.

[94] anonymous, FDA expert panel to review Merck's new HIV treatment Isentress in September. *AIDS Reader*, 2007. 17(8): p. 421.

[95] Morales-Ramirez, J.O., H. Teppler, and S. Kovacs. Antiviral effect of MK-0518, a novel HIV-1 integrase inhibitor, in ATR-naive HIV-1 infected patients. in *Proceedings of the 10th European AIDS Conference*. 2005. Dublin, Ireland.

[96] Ason, B., et al., Targeting Tn5 transposase identifies human immunodeficiency virus type 1 inhibitors. *Antimicrobial Agents & Chemotherapy*, 2005. 49(5): p. 2035-43.

[97] Este, J.A., et al., Human immunodeficiency virus glycoprotein gp120 as the primary target for the antiviral action of AR177 (Zintevir). *Molecular Pharmacology*, 1998. 53(2): p. 340-5.

[98] John, S., T.M. Fletcher, 3rd, and C.B. Jonsson, Development and application of a high-throughput screening assay for HIV-1 integrase enzyme activities. *Journal of Biomolecular Screening*, 2005. 10(6): p. 606-14.

[99] Bonnenfant, S., et al., Styrylquinolines, integrase inhibitors acting prior to integration: a new mechanism of action for anti-integrase agents. *Journal of Virology*, 2004. 78(11): p. 5728-36.

[100] d'Angelo, J., et al., HIV-1 integrase: the next target for AIDS therapy? *Pathologie Biologie*, 2001. 49(3): p. 237-46.

[101] Deprez, E., et al., Mechanism of HIV-1 integrase inhibition by styrylquinoline derivatives in vitro. *Molecular Pharmacology*, 2004. 65(1): p. 85-98.

[102] Mousnier, A., et al., Nuclear import of HIV-1 integrase is inhibited in vitro by styrylquinoline derivatives. *Molecular Pharmacology*, 2004. 66(4): p. 783-8.

[103] Zouhiri, F., et al., Structure-activity relationships and binding mode of styrylquinolines as potent inhibitors of HIV-1 integrase and replication of HIV-1 in cell culture. *Journal of Medicinal Chemistry*, 2000. 43(8): p. 1533-40.

[104] Hauber, I., et al., Identification of cellular deoxyhypusine synthase as a novel target for antiretroviral therapy. *Journal of Clinical Investigation*, 2005. 115(1): p. 76-85.

[105] Daniel, R., et al., Caffeine inhibits human immunodeficiency virus type 1 transduction of nondividing cells. *Journal of Virology*, 2005. 79(4): p. 2058-65.

[106] Nunnari, G., et al., Inhibition of HIV-1 replication by caffeine and caffeine-related methylxanthines. *Virology*, 2005. 335(2): p. 177-84.

[107] Blasina, A., et al., Caffeine inhibits the checkpoint kinase ATM. *Current Biology*, 1999. 9(19): p. 1135-8.

[108] Hall-Jackson, C.A., et al., ATR is a caffeine-sensitive, DNA-activated protein kinase with a substrate specificity distinct from DNA-PK. *Oncogene*, 1999. 18(48): p.6707-13.

[109] Sarkaria, J.N., et al., Inhibition of ATM and ATR kinase activities by the radiosensitizing agent, caffeine. *Cancer Research*, 1999. 59(17): p. 4375-82.

[110] Smith, J.A., et al., Pentoxifylline suppresses transduction by HIV-1-based vectors. *Intervirology*, 2007. 50(5): p. 377-86.

[111] Collis, S.J., et al., The life and death of DNA-PK. *Oncogene*, 2005. 24(6): p. 949-61.

[112] Masson, C., et al., Ku80 participates in the targeting of retroviral transgenes to the chromatin of CHO cells. *Journal of Virology*, 2007. 81(15): p. 7924-32.

[113] Daniel, R., et al., Evidence that stable retroviral transduction and cell survival following DNA integration depend on components of the nonhomologous end joining repair pathway. *J Virol*, 2004. 78(16): p. 8573-81.

[114] Daniel, R., et al., Evidence that the retroviral DNA integration process triggers an ATR-dependent DNA damage response. *Proceedings of the National Academy of Sciences of the United States of America*, 2003. 100(8): p. 4778-83.

[115] Daniel, R., et al., Wortmannin potentiates integrase-mediated killing of lymphocytes and reduces the efficiency of stable transduction by retroviruses. *Mol Cell Biol*, 2001. 21(4): p. 1164-72.

[116] O'Doherty, U., W.J. Swiggard, and M.H. Malim, Human immunodeficiency virus type 1 spinoculation enhances infection through virus binding. *Journal of Virology*, 2000. 74(21): p. 10074-80.

[117] Lau, A., et al., Suppression of retroviral infection by the RAD52 DNA repair protein. *Embo J*, 2004. 23(16): p. 3421-9.

[118] International Human Genome Sequencing, C., Finishing the euchromatic sequence of the human genome.[see comment]. *Nature*, 2004. 431(7011): p. 931-45.

[119] Mooslehner, K., U. Karls, and K. Harbers, Retroviral integration sites in transgenic Mov mice frequently map in the vicinity of transcribed DNA regions. *Journal of Virology*, 1990. 64(6): p. 3056-8.

[120] Scherdin, U., K. Rhodes, and M. Breindl, Transcriptionally active genome regions are preferred targets for retrovirus integration. *Journal of Virology*, 1990. 64(2): p. 907-12.

[121] Bushman, F., et al., Genome-wide analysis of retroviral DNA integration. *Nature Reviews. Microbiology*, 2005. 3(11): p. 848-58.

[122] Schroder, A.R., et al., HIV-1 integration in the human genome favors active genes and local hotspots. *Cell*, 2002. 110(4): p. 521-9.

[123] Crise, B., et al., Simian immunodeficiency virus integration preference is similar to that of human immunodeficiency virus type 1. *Journal of Virology*, 2005. 79(19): p. 12199-204.

[124] Hematti, P., et al., Distinct genomic integration of MLV and SIV vectors in primate hematopoietic stem and progenitor cells. *Plos Biology*, 2004. 2(12): p. e423.

[125] MacNeil, A., et al., Genomic sites of human immunodeficiency virus type 2 (HIV-2) integration: similarities to HIV-1 in vitro and possible differences in vivo *Journal of Virology.*, 2006. 80(15): p. 7316-21.

[126] Kang, Y., et al., Integration site choice of a feline immunodeficiency virus vector. *Journal of Virology*, 2006. 80(17): p. 8820-3.

[127] Mitchell, R.S., et al., Retroviral DNA integration: ASLV, HIV, and MLV show distinct target site preferences *Plos Biology.*, 2004. 2(8): p. E234.

[128] Narezkina, A., et al., *Genome-wide analyses of avian sarcoma virus integration sites.* *Journal of Virology*, 2004. 78(21): p. 11656-63.

[129] Maxfield, L.F., C.D. Fraize, and J.M. Coffin, Relationship between retroviral DNA-integration-site selection and host cell transcription.[see comment]. *Proceedings of the National Academy of Sciences of the United States of America*, 2005. 102(5): p. 1436-41.

[130] Weidhaas, J.B., et al., *Relationship between retroviral DNA integration and gene expression. Journal of Virology*, 2000. 74(18): p. 8382-9.

[131] Derse, D., et al., Human T-cell leukemia virus type 1 integration target sites in the human genome: comparison with those of other retroviruses. *Journal of Virology*, 2007. 81(12): p. 6731-41.

[132] Holman, A.G. and J.M. Coffin, Symmetrical base preferences surrounding HIV-1, avian sarcoma/leukosis virus, and murine leukemia virus integration sites.[see comment][erratum appears in Proc Natl Acad Sci U S A. 2005 Apr 26;102(17):6238]. *Proceedings of the National Academy of Sciences of the United States of America*, 2005. 102(17): p. 6103-7.

[133] Wu, X., et al., Transcription start regions in the human genome are favored targets for MLV integration.[see comment]. *Science*, 2003. 300(5626): p. 1749-51.

[134] Maertens, G., et al., LEDGF/p75 is essential for nuclear and chromosomal targeting of HIV-1 integrase in human cells. *Journal of Biological Chemistry*, 2003. 278(35): p. 33528-39.

[135] Turlure, F., et al., A tripartite DNA-binding element, comprised of the nuclear localization signal and two AT-hook motifs, mediates the association of LEDGF/p75 with chromatin in vivo *Nucleic Acids Research.*, 2006. 34(5): p. 1653-75.

[136] Vanegas, M., et al., Identification of the LEDGF/p75 HIV-1 integrase-interaction domain and NLS reveals NLS-independent chromatin tethering. *Journal of Cell Science*, 2005. 118(Pt 8): p. 1733-43.

[137] Wang, G.C., A; Leipzig, J; Berry, CC; Bushman, FD, HIV integration site selection: analysis by massively parallel pyrosequencing reveals association with epigenetic modifications. *Genome Research*, 2007. 17: p. 1186-94.

[138] Kumar, P.P., et al., SATB1-binding sequences and Alu-like motifs define a unique chromatin context in the vicinity of human immunodeficiency virus type 1 integration sites. *Journal of Virology*, 2007. 81(11): p. 5617-27.

[139] Li, L., et al., Role of the non-homologous DNA end joining pathway in the early steps of retroviral infection. *EMBO Journal*, 2001. 20(12): p. 3272-81.

[140] Lewinski, M.K., et al., Retroviral DNA integration: viral and cellular determinants of target-site selection. *PLoS Pathogens*, 2006. 2(6): p. e60.

[141] Monse, H., et al., Viral determinants of integration site preferences of simian immunodeficiency virus-based vectors. *Journal of Virology*, 2006. 80(16): p. 8145-50.

[142] Jordan, A., D. Bisgrove, and E. Verdin, HIV reproducibly establishes a latent infection after acute infection of T cells in vitro. *EMBO Journal*, 2003. 22(8): p. 1868-77.

[143] Jordan, A., P. Defechereux, and E. Verdin, The site of HIV-1 integration in the human genome determines basal transcriptional activity and response to Tat transactivation. *EMBO Journal*, 2001. 20(7): p. 1726-38.

[144] Finzi, D., et al., Latent infection of CD4+ T cells provides a mechanism for lifelong persistence of HIV-1, even in patients on effective combination therapy.[see comment]. *Nature Medicine,* 1999. 5(5): p. 512-7.

[145] Li, Z., et al., Murine leukemia induced by retroviral gene marking. *Science*, 2002. 296(5567): p. 497.

[146] Bushman, F.D., Retroviral integration and human gene therapy.[comment]. *Journal of Clinical Investigation*, 2007. 117(8): p. 2083-6.

[147] Deichmann, A., et al., Vector integration is nonrandom and clustered and influences the fate of lymphopoiesis in SCID-X1 gene therapy.[see comment]. *Journal of Clinical Investigation*, 2007. 117(8): p. 2225-32.

[148] Hacein-Bey-Abina, S., et al., LMO2-associated clonal T cell proliferation in two patients after gene therapy for SCID-X1.[see comment][erratum appears in Science. 2003 Oct 24;302(5645):568]. *Science*, 2003. 302(5644): p. 415-9.

[149] Modlich, U., et al., Leukemias following retroviral transfer of multidrug resistance 1 (MDR1) are driven by combinatorial insertional mutagenesis. *Blood*, 2005. 105(11): p. 4235-46.

[150] Schwarzwaelder, K., et al., Gammaretrovirus-mediated correction of SCID-X1 is associated with skewed vector integration site distribution in vivo.[see comment]. *Journal of Clinical Investigation*, 2007. 117(8): p. 2241-9.

[151] Aiuti, A., et al., Multilineage hematopoietic reconstitution without clonal selection in ADA-SCID patients treated with stem cell gene therapy.[see comment]. *Journal of Clinical Investigation*, 2007. 117(8): p. 2233-40.

[152] Recchia, A., et al., Retroviral vector integration deregulates gene expression but has no consequence on the biology and function of transplanted T cells. *Proceedings of the National Academy of Sciences of the United States of America,* 2006. 103(5): p. 1457-62.

[153] Ott, M.G., et al., Correction of X-linked chronic granulomatous disease by gene therapy, augmented by insertional activation of MDS1-EVI1, PRDM16 or SETBP1.[see comment]. *Nature Medicine,* 2006. 12(4): p. 401-9.

[154] Montini, E., et al., Hematopoietic stem cell gene transfer in a tumor-prone mouse model uncovers low genotoxicity of lentiviral vector integration. *Nature Biotechnology*, 2006. 24(6): p. 687-96.

[155] Holmes-Son, M.L. and S.A. Chow, Integrase-lexA fusion proteins incorporated into human immunodeficiency virus type 1 that contains a catalytically inactive integrase

gene are functional to mediate integration. *Journal of Virology*, 2000. 74(24): p. 11548-56.

[156] Bushman, F.D., Tethering human immunodeficiency virus 1 integrase to a DNA site directs integration to nearby sequences. *Proceedings of the National Academy of Sciences of the United States of America*, 1994. 91(20): p. 9233-7.

[157] Tan, W., et al., Fusion proteins consisting of human immunodeficiency virus type 1 integrase and the designed polydactyl zinc finger protein E2C direct integration of viral DNA into specific sites. *Journal of Virology*, 2004. 78(3): p. 1301-13.

[158] Bushman, F.D. and M.D. Miller, Tethering human immunodeficiency virus type 1 preintegration complexes to target DNA promotes integration at nearby sites. *Journal of Virology*, 1997. 71(1): p. 458-64.

[159] Tan, W., et al., Human immunodeficiency virus type 1 incorporated with fusion proteins consisting of integrase and the designed polydactyl zinc finger protein E2C can bias integration of viral DNA into a predetermined chromosomal region in human cells. *Journal of Virology*, 2006. 80(4): p. 1939-48.

In: Genetic Recombination Research Progress
Editor: Jacob H. Schulz, pp. 189-203

ISBN: 978-1-60456-482-2
© 2008 Nova Science Publishers, Inc.

Chapter 6

Phage Integrases for Mediating Genomic Integration and DNA Recombination

Christof Maucksch [1, 2], *Manish K. Aneja* [1] *and Carsten Rudolph* [1, 2,*]

[1]Division of Molecular Pulmonology, Department of Pediatrics,
Ludwig-Maximilians University, Lindwurmstrasse 2A
80337 Munich, Germany.
[2]Department of Pharmacy, Free University of Berlin
Takustrasse 3, 14166 Berlin, Germany.

Abstract

φC31 integrase is a site specific recombinase derived from the *Streptomyces* phage. In the phage lifecycle, the enzyme mediates lysogeny by mediating recombination between specific sequences termed *attB* (present in the bacterial DNA) and *attP* (present in the phage genome). Screening the enzyme activity in mammalian cells provided positive results and also showed that the enzyme retained its property of site specific recombination into mammalian genomes. Mammalian genomes have been shown to contain sequences that are similar to the wild type *attP* sequence of the *Streptomyces* phage genome and experiments with the integrase in mammalian cells showed that it could mediate recombination and subsequent integration of any DNA bearing an *attB* site into these pseudo*attP* sites. These properties have made φC31 integrase emerge as a promising tool for nonviral gene therapy to achieve long-term gene expression in different tissues *in vitro* and *in vivo*. The present chapter introduces the different recombination mediating enzymes but focuses primarily on the activity of φC31 integrase in different tissues. Results of this enzyme system in the lung tissue and hematopoietic cells are also discussed. In murine and human lung cell lines, it could be shown that the φC31 integrase mediates specific recombination between *attB* and *attP* and subsequently long-term gene expression could be achieved without any selection pressure. The results were further validated *in vivo* in mice. However, when integration into a specific site "mpsL1" was

---

* E-mail address: Carsten.Rudolph@med.uni-muenchen.de, Telephone: (49) 89 5160 7711, Fax: (49) 89 5160 4421
(Corresponding author)

investigated, it was revealed that this site was targeted only in 50% of the mice. This was in contrast to the reports from other tissues where all the mice showed integration at mpsL1. These results indicated that the activity of this enzyme may actually be tissue dependent.

Tissue specific efficiency of the ϕC31 integrase was confirmed in the human hematopoietic system. The activity of the ϕC31 integrase was extremely reduced in CD34$^+$ hematopoietic stem cells, primary T lymphocytes and T cell derived cell lines in comparison with mesenchymal stem cells and cell lines derived from lung -, liver - and cervix tissue. No enhanced long-term expression mediated by the ϕC31 integrase could be observed in T cell lines. A direct comparison of hematopoietic Jurkat T and A549 alveolar type-II lung cell lines indicated up to a 100-fold higher activity of the ϕC31 integrase in lung tissue compared to hematopoietic cells. Looking for possible mechanisms responsible for this discrepancy of ϕC31 integrase activity in different tissues, revealed that Jurkat cells contain significantly higher amounts of DAXX protein compared to A549 cells. As DAXX has been reported to interact with the ϕC31 integrase and inhibit recombination, higher levels in hematopoietic cells may be one reason for low activity of ϕC31 integrase in these cells.

These studies in lungs and the hematopoietic cells provide evidence for the tissue specificity of the ϕC31 integrase system and also raise the need for development of novel strategies for achieving recombination and long-term expression.

1. Introduction

Genetic engineering strategies often require permanent modification of the target genome. The ability to manipulate eukaryotic genomes has expanded simultaneously in several distinct yet interrelated directions. Most methods available for stable transgene integration are useful in one or more of the following three broadly defined applications: manipulations of cells in culture, construction of transgenic organisms, and gene therapies. Great sophistication typically is applied in the design of the introduced genetic material. However, cruder methods prevail for placement of the introduced gene into the genome, random integration often being state of the art. Lack of control over the position of introduced DNA results in unpredictable gene expression and potentially undesirable mutagenesis of important genes. Viral gene delivery vectors, especially those of the lentivirus subfamily, have recently become popular for the stable modification of cells in culture. Lentiviral vectors are able to infect even non dividing cells with high efficiency and integrate their payloads into unmodified chromosomes. However, lentiviral integration is highly promiscuous and tends to favour sites proximal to actively expressing genes [1, 2]. Little control can be exercised over lentiviral integration multiplicity and specificity. Similarly, the Sleeping Beauty transposon also mediates random integratrion in TA dinucleotides [3]. Both Sleeping Beauty and the viral systems additionally have size limits in transgene capacity as well. A better solution would be a method that produces efficient site-specific integration into safe locations in the target genome.

Homologous recombination can provide great specificity in integration sites, but it occurs at too low frequency to be optimal for genetic engineering in multicellular organisms [4]. Moreover, few non viral techniques exist for stable and efficient eukaryotic gene delivery and fewer still are broadly useful in both cell culture and whole-organism applications. Therefore, the need of the hour is a system for mediating DNA recombination and integration in a highly efficient and specific manner.

2. DNA Recombinases

Enzymes of the site-specific recombinase family combine features of higher specificity and greater efficiency comparable to viral methods. The use of these site-specific recombinases (SSRs) both *in vitro* and *in vivo*, have proven to be useful tools in the analysis of gene function. Based on evolutionary and mechanistic relationships, most site-specific recombinases (SSRs) can be classified into the tyrosine or serine recombinase families. The site-specific recombinases from both families are structurally and functionally diverse [5]. The first widely used SSR in mammalian cultured cells and animals was the P1 bacteriophage-derived *Cre*, a member of the λ integrase family that recognizes homotypic 34 bp loxP recognition sites [6, 7]. To date, *Cre* recombinase remains the enzyme of choice to efficiently mediate DNA recombination both *in vitro* and *in vivo*. A second SSR from the λ integrase family, FLP from *Saccharomyces cerevisiae*, has also been used in mammals [8] and recognizes distinct 34 bp FRT sites [9]. Initial use of FLP in mammalian cells revealed inefficient recombinase activity due to thermo-instability of the protein [10]. Subsequent screening for thermo-stable mutants resulted in the identification of FLPe, with a 4-fold increase in recombination efficiency [11]. Despite this improvement, the recombination efficiency of FLPe in cells remains quite low, at most a 6%, with mosaic recombination found in almost all ES clones [12]. Both these recombinases use a catalytic tyrosine in their strand exchange mechanism [13, 14] Cre and FLP require no host-specific co-factors and have been successfully used in mammalian cells, providing important and widely used tools for genome manipulation [7, 15, 16].

Other site-specific recombinases, such as γδ, use a catalytic serine and are members of the evolutionarily unrelated serine recombinase family [17]. Cre, FLP, and γδ are resolvases and catalyze bi-directional recombination between identical sites. Upon binding to their target recognition sequences, SSRs can induce the deletion, insertion, or inversion of DNA sequences leading to conditional gene inactivation or expression [18].

Recombinases such as *Cre*, *FLP*, and β-recombinase [19] perform both integration and excision with the same target sites. Therefore, although these recombinases efficiently perform excision in mammalian cells, the net integration frequency that they mediate is low (~ 0.03% for *Cre*) [19] because of the excisive back reaction. These enzymes are therefore most useful for efficiently carrying out site-specific deletions [7, 15, 19]. Additionally, use of *Cre* and *FLP* enzymes requires that target sequences (*loxP* and FRT sequences respectively) are artificially placed in the genome before manipulation, limiting their applicability in the de novo generation of transgenic cells [7, 20]. Thygarajan et al. 2000 [21] have documented the occurrence of pseudo *loxP* sites for *Cre* in mammalian cells, and it appears that these sites can be used in vivo in mice, at least under conditions of continuous high-level expression of *Cre*. The use of such sites resulted in *Cre*-dependent chromosome rearrangements in transgenic mice spermatids [22]. Use of special *loxP* cassettes designed to limit the reverse reaction has resulted in higher integration frequencies [23, 24]. However, this strategy is not applicable when using endogenous sequences as targets. These two recombinases are prototypical of techniques used in manipulations of cells in culture, construction of transgenic organisms.

3. Phage Integrases

Perhaps more ideal, when integration alone is the goal, are recombinases that perform only the integration reaction and require accessory factors for the reverse reaction. In this case, once integrated, the transgene cannot be excised by the recombinase. Such a property would lend the integration reaction a considerable degree of efficiency. In contrast to DNA recombinases like *Cre* and FLP, phage integrases mediate unidirectional recombination between two dissimilar attachment (*att*) sequences, the bacterial *attB* and phage *attP* sites, resulting in hybrid sites, *attL,* and *attR* [25]. Because these integrases cannot carry out the reverse excision reaction without additional co-factors, these enzymes are especially helpful for catalyzing integration reactions.

Integrases, from phages R4 (469 amino acids) [26] and TP901-1 (485 amino acids) [27], were evaluated in mammalian cells. Both enzymes could perform intra-molecular recombination between their *attB* and *attP* sites in the human cell environment [28, 29], but were less useful for recombination into chromosomal pseudo *att* sites. The R4 integrase could integrate efficiently at endogenous sites, but frequently appeared to produce aberrant chromosomal events, while the TP901-1 integrase did not perform integration at native sequences at a detectable frequency.

Integrases have been isolated from variety of other phages of different Gram-positive bacteria. The integrase from *Listeria monocytogenes* phage A118 [30] is a 452 amino acid protein that is 50% similar to the TP901-1 integrase from *Lactococcus lactis* subsp *cremoris* [31]. U153 is also a bacteriophage of *L. monocytogenes* [32] and possesses a closely related 452 amino acid serine integrase [33]. The U153 and A118 integrases share 89% DNA sequence identity and 93% identity at the protein level. These two similar integrases also share a common *attB* site, and their *attP* sites are 94% identical [33].

Mycobacteriophage Bxb1 is a temperate phage of *Mycobacterium smegmatis* [34, 35]. Integration of the Bxb1 genome into the mycobacterial genome is mediated by the 500 amino acid Bxb1 integrase that catalyzes site-specific recombination between its *attP* and *attB* sites [35 – 37].

Bacteriophage φFC1 infects *Enterococcus faecalis* KBL 703 and integrates into the host chromosome by a site-specific mechanism. The φFC1 integrase is 465 amino acids long [38] and shares 57% overall homology with the integrases from phages A118 [30] and TP901-1 [31]. The φFC1 integrase requires only its *attB* and *attP* sites to carry out integration in *E. coli* as well as in *E. faecalis* [38].

The prophage element φRV1 resides within the *Mycobacterium tuberculosis* genome. A "resurrected" non-replicative form of the phage was shown to efficiently integrate within the non-host strain *Mycobacterium bovis* BCG [39]. φRV1 contains a 469 amino acid serine integrase that shares 27% identity with the Bxb1 integrase [40, 41].

In a recent study, Keravala *et al.* [42] evaluated the ability of these five serine phage integrases (from phages A118, U153, Bxb1, φFC1, and φRV1), to mediate recombination in mammalian cells. Two types of recombination were investigated, including the ability of an integrase to mediate recombination between its own phage *att* sites in the context of a mammalian cell and the ability of an integrase to perform genomic integration pairing a phage *att* site with an endogenous mammalian sequence. They demonstrated that the A118 integrase mediated precise intra-molecular recombination of a plasmid containing its *attB* and *attP* sites at a frequency of ~ 50% in human cells. The closely related U153 integrase also performed efficient recombination in human cells on a plasmid containing the *attB* and *attP* sites of A118. The integrases from phages Bxb1, φFC1, and φRV1 carried out such recombination at their *attB* and *attP* sites at frequencies ranging from 11 to 75%. Furthermore, the A118 integrase mediated recombination between its *attP* site on a plasmid and pseudo *attB* sites in the human genome, i.e. native sequences with partial identity to *attB*. Fifteen such A118 pseudo *att* sites were analyzed, and a consensus recognition site was identified. The other integrases did not mediate integration at genomic sequences at a frequency above background.

The activity of Bxb1 integrase in mammalian cells was also reported by Russell *et al.* [43]. These site-specific integrases represent valuable new tools for manipulating eukaryotic genomes.

3.1. Phage φC31 Integrase

Like other phage integrases as mentioned in the earlier section, φC31 integrase, is isolated from a *Streptomyces* phage [44, 45] and was shown to mediate intra-molecular recombination of plasmids in *Escherichia coli* and in vitro, requiring no host-specific co-factors [46]. φC31 integrase, is a member of the serine recombinase protein family that has been developed as a system for genetic manipulation [47 - 49]. It has been shown that the enzyme is able to catalyze chromosomal transgene insertion under a diverse range of experimental and therapeutic conditions.

Like other phage integrases, φC31 integrase recognises bacterial chromosome attachment sequences (*attB* sites) and phage genome attachment sequences (*attP* sites) of 30-40 bp and catalyze a recombination reaction between the two (Figure 1A) resulting in the integration of the phage genome. The process of phage integration typically cleaves the *attB* and *attP* sites and joins them to each other, generating two hybrid sequences (*attL* and *attR*) that flank the integrated phage genome.

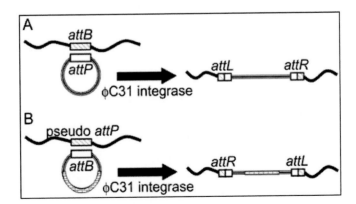

Figure 1. Normal and adapted reactions of φC31 integrase. (A) In nature, φC31 integrase catalyses the integration of phage genome into the *Streptomyces* host genome via recombination between *attP* and *attB* sites. (B) φC31 integrase also mediates integration of an episomal plasmid with *attB* sites into pseudo*attP* sites in mammalian genomes. Taken from Chalberg *et al.* [59]

Groth *et al.* [47] reported the activity of the 605 amino acid φC31 integrase system in mammalian cells. A plasmid assay system was constructed that measured intramolecular integration of *attP* into *attB*. This assay was used to demonstrate that in the presence of the φC31 integrase, precise unidirectional integration occured with an efficiency of 100% in *Escherichia coli* and >50% in human cells. This assay system was also used to define the minimal sizes of *attB* and *attP* at 34 bp and 39 bp, respectively. Furthermore, precise and efficient intermolecular integration of an incoming plasmid bearing *attP* into an established Epstein-Barr virus plasmid bearing *attB* was documented in human cells. This work was an elegant demonstration of efficient, site-specific, unidirectional integration in mammalian cells. These observations then formed the basis for site-specific integration strategies potentially useful in a broad range of genetic engineering applications.

Thyagarajan *et al.* [48] subsequently showed that phage *attP* sites inserted at various locations in human and mouse chromosomes served as efficient targets for precise site-specific integration. In such cell lines, frequencies detected for integration into the inserted *attP* sites were approximately 10- to 20-fold above the spontaneous background frequency of random integration. Furthermore, in the absence of an inserted *attP* site, the authors detected integrase-mediated recombination at endogenous sites in the genome at frequencies ~5- to 10-fold above the background of random integration. They showed that in the presence of the φC31 integrase, a plasmid bearing *attB* was efficiently integrated into mammalian genomes and in human cells, nearly 90% of these integration events were integrase mediated. These integration events were shown to occur at sets of native sequences having partial sequence identity to *attP*, which were termed pseudo *attP* sites (Figure 1B). Integration of the transgene bearing plasmid also resulted in the long-term expression of the transgene. This study thus presented the phage φC31 integrase as an effective site-specific integration system for higher cells and a valuable non-viral integrating mechanism for gene therapy and other chromosome engineering strategies.

Since the reports by Groth *et al.* [47] and Thyagarajan *et al.* [48], the φC31 integrase system has achieved considerable success in an analogous application as a tool for gene therapy and the construction of transgenic organisms. It was shown by Olivares *et al.* [50] that site-specific genomic integration produces therapeutic factor IX levels in mice. Other reports demonstrated genetic correction of inherited human skin disease [51] and *in vivo* correction of murine hereditary tyrosinemia type I [52]. Moreover, φC31 integrase was used in enhanced gene therapy approaches for muscular dystrophy [53] and for genomic integration of the common cytokine receptor γ chain in human T cell lines [54]. The enzyme also catalyzes efficient deletion and cassette exchange at pre-integrated *att* sites [48, 49, 55, 56]. In addition, φC31 integrase has been used in the construction and manipulation of transgenic *Drosophila* [49], *Xenopus* [57], and mice [56]. The integration frequency mediated by the φC31 integrase is approximately 10- to 100-fold higher than *Cre*-mediated integration at an inserted wild-type *loxP* site [20] or integration mediated by FLP at an inserted FRT site [58].

The advantages of this system thus include its relative site specificity and use in a wide variety of applications, as well as its relative simplicity. Integration catalyzed by the enzyme is efficient due to its ability to function in the eukaryotic environment and the absence of any reverse excision reaction. Efficient delivery of φC31 integrase and *attB* donor plasmid to the tissue or cells of interest remains the most challenging aspect of the system. Unlike viral methods of genome manipulation, use of ϕC31 integrase almost always requires an additional method of stimulating cellular DNA uptake. However the relative simplicity of the plasmid based system means that nearly any proven method of introducing exogenous DNA into cells can be used with ϕC31 integrase. Indeed, most methods of transfection technology have at some time been shown to be suitable for this application.

3.2. Integrations Mediated by ϕC31 Integrase

Thayagarajan *et al.* [48] identified sequences present in the mammalian genomes which are similar to wild type sequence of *attP* site of ϕC31 and termed as "pseudo*attP*" site. By sequencing the junctions at 67 rescued integration events, they identified a hierarchy of pseudo *attP* sites, some of which were used repeatedly. The pattern of single and recurring sites in this collection of different pseudo *attP* sites suggested that the total number of pseudo *attP* sites may be between 10^2 and 10^3 in mammalian cells, with some sites significantly preferred over others. In another study, Olivares *et al.* [50] identified a specific site in murine liver (mpsL1) that was frequently used after ϕC31-mediated integration. A number of these pseudo*attP* sites have been identified in both human and mouse cells, and published primer pairs exist to screen clones for the most common integration events [48, 50, 59].

In addition to the numerous reports about the successful application of the ϕC31 integrase, there has been an accumulation of evidence that ϕC31-mediated integration may have unwanted effects. A study which showed the successful application of this integrase system in the liver, also suggested integrase-related toxicity as one of the probable causes for the transient abnormal hepatocytes (as observed on histology images), which were observed in mice that received a ϕC31 integrase-encoding plasmid [52]. Ehrhardt *et al.* [60] demonstrated that ϕC31-mediated integration results in frequent micro-deletions and micro-insertions at sites of insertion. More importantly, Liu *et al.* [61] found that primary human fibroblasts show chromosomal abnormalities after transduction with ϕC31 integrase. Furthermore, it was discussed by Chalberg *et al.* [59] that aberrant chromosomal events may occur after ϕC31-mediated integration. More recently, Ehrhardt *et al.* [62] provided additional molecular evidence that ϕC31 integrase leads to chromosomal rearrangements in different cell lines and also proposed that the integration specificity of the integrase may be cell line dependent. For example, a predominant hot spot of integration in HCT116 cell line (derived from a colon carcinoma) was found on chromosome 12 (Chr 12q22) and the hot spot of integration in human cell lines Huh7 and 293 was identified on chromosome 19 (Chr 19q13.31). A hot spot of integration on chromosome 19 was also identified by Chalberg *et al.* [59], but the hotspot on chromosome 12 was specific for HCT116 cells, indicating that this hot spot is specific for this cell line.

Ehrhardt *et al.* [62] noted that after ϕC31 mediated integration, 15% of all integrated transgenes were flanked by chromosomal sequences originating from different chromosomes. Careful analysis of these chromosomal DNA sequences revealed that hot spots of integration may also be involved in the formation of chromosomal rearrangements. Performing a detailed characterization of insertion sites revealed that ~70% of all integration sites were in genes and the remaining ~30% were in intergenic regions. Within genes, more than 90% of the integrations were observed in introns. This is in contrast to the study performed by Chalberg *et al.* [59], which reported that only 38.7% of all sites of insertion were in genes. One reason for these discrepancies may be the vector that was used as a substrate for φC31-mediated integration and subsequent identification of insertion sites. The size of the plasmid may play a role, as may the enzymes used to cut genomic DNA when performing the plasmid rescue protocol. A smaller plasmid may be less biased toward the smaller chromosomal DNA fragments that are rescued when performing the plasmid rescue protocol.

At present, one can only speculate about the mechanism involved in the induction of chromosomal rearrangements by bacteriophage integrase φC31. After integration of the transgene, the integrated foreign DNA is flanked by *attR* and *attL*, which are hybrid sites consisting of part of the pseudo-*attP* site in the host genome and part of the φC31 recognition site *attB* contained in the substrate. It is speculated that after integration of the transgene, integrase φC31 may recombine either *attR* or *attL* with other pseudo *attP* sites present on other chromosomes. It is estimated that up to 1000 pseudo*attP* sites are present in the mammalian genome, and in addition it is known that for integration to occur, the required sequence homology between the wild-type *attP* site and pseudo*attP* sites is less than 30%. However, it remains to be shown whether bacteriophage integrase φC31 is capable of recombining pseudo*attP* sites on different chromosomes even without any substrate being present.

It however needs to be mentioned that none of the chromosomal translocations observed in cell culture have been verified in in-vivo experiments. Moreover, transgenics created using this integrase [49, 56, 57] do not show any abnormalities. In the following section, we discuss about the activity of φC31 integrase in lungs. Moreover, we also present our results in the hematopoietic system which further support the assumption that the activity of φC31 integrase may be cell type dependent.

3.3. Activity of φC31 Integrase in Lungs and Hematopoeitic System

Recent investigations into the activity of the φC31 integrase in lungs and in hematopoietic cells have given new insights on the feasibility and adaptability of the φC31 integrase as a tool for safe and stable nonviral gene therapy for these tissues.

We have demonstrated for the first time the activity of φC31 integrase in lung cells [63]. In murine (MLE12) as well as in human (A549) alveolar type II cell lines, long-term reporter gene expression could be observed in the presence of integrase without selection pressure. However, initial expression values at day two post transfection, arising from mostly episomal plasmid DNA, were substantially higher than values after 2-3 weeks, which represent steady state expression from integrated plasmid DNA. This extensive drop of expression has been reported in other tissues also [48, 50, 54]. This trend is in accordance with the shift in expression from mainly episomal plasmid DNA to that from stably integrated plasmid DNA. Interestingly however, under comparable transfection efficiences in both cell lines, the final long-term expression levels in MLE 12 cells were ~ 1.000-fold lower compared to the initial expression (day 2 post transfection), whereas for A549 cells, a difference ~ 50.000-fold was observed. This result provided the initial clue that the integrase activity may be cell line dependent.

We could further show that the φC31 integrase mediates integration of pDNA and long-term transgene (luciferase) expression in lungs of BALB/c mice. A stable level of luciferase expression was observed in the mice which were co-injected with the φC31 integrase expression plasmid, whereas transgene expression the control group dropped to background values by day 7 post gene delivery. As episomal pDNA persists in nondividing tissues like lungs for longer than two weeks, the transient expression of luciferase could be assigned to gene silencing, which has been investigated as the main reason for the decrease of transgene expression in eukaryotic cells [64]. Nevertheless, *in vivo*, the same extensive decrease of transgene expression was observed, as has been detected before *in vitro*. This decline in expression values could be due to post integration silencing of the transgene either due to the CpG motifs present in the plasmid backbone [65] or promoter silencing [66, 67] or both. Moreover, the loss of stably transfected cells (due to routine turnover of lung epithelial cells) would also contribute to the steady decline in gene expression. These experiments however provided the first *in vivo* results about the activity of φC31 integrase in lungs.

For any clinical pulmonary gene therapy application of the φC31 integrase system, a certain threshold of transgene expression is critical to observe a physiological effect. We have undertaken different efforts to achieve higher long-term transgene expression levels. In one experiment we administered higher doses of integrase plasmid and in another, a second dose was administered to the mice. However, in both these experiments, there was no increase in the final stable expression levels observed [63]. Detection of φC31 integrase mediated integration of the plasmid DNA in the lungs gave surprising results. Whereas in previous studies in liver [50] and muscle [53] a integration hot spot mpsL1 was found in all animals treated with the integrase plasmid, only 7 out of 15 mice showed integration at this particular hotspot. This again pointed towards the tissue dependent integration specificity of φC31 integrase. Based on results in cell lines, similar hypothesis has been raised by Ehrhardt et al. [62].

As we discussed above, the φC31 integrase has been shown to function in many tissues and cell types. However, it is not yet predictable to which extend the φC31 integrase could be used for different monogenetic diseases in clinical applications. Based on our own results and also from others which proposed that the acitivty of φC31 integrase may be cell type dependent [62], we made experiments using this enzyme in hematopoietic cells. We could show that the activity of the φC31 integrase is strongly reduced in hematopoietic cell types (Maucksch et al., unpublished). In analogy to out experiments in MLE12 and A549 lung cells, we performed co-transfections of different T cell derived cell lines with pEGFPLuc*attB* (a plasmid coding for the fusion protein of EGFP and luciferase and also carried the attB site recognised by φC31 integrase) and pCMV-Int (φC31 integrase expression plasmid). Interestingly, no enhanced luciferase long-term expression above control could be observed in any of the hematopoietic cell lines. In an episomal φC31 recombination assay we compared the activity of the φC31 integrase in T cell derived cell lines, primary T cells and human CD34[+] hematopoietic stem cells (huHSC) with cell lines and primary cells of different non-hematopoietic tissues. We confirmed the reduced efficiency of the φC31 integrase in all hematopoietic cell types, whereas we observed varying but distinct activity in other tissues. Relatively high φC31 integrase activity could be observed in human bronchial epithelial (BEAS-2B), liver (HepG2) and cervix carcinoma (HeLa) cell lines. Moderate activity was observed in primary human mesenchymal stem cells (huMSC) and A549 cells.

In a direct comparison of A549 and Jurkat cells, we observed a 265-fold higher recombination activity (as measured by the expression following recombination) in the former cell line. Also quantification of recombination products (direct quantification by quantitative Real time PCR) generated by the φC31 integrase resulted in significant higher values for A549 (~ 100 fold higher) than for Jurkat cells. In a recent report, Chen et al., [68] published that DAXX (death domain-associated protein) interacts with the φC31 integrase and inhibits its recombination efficiency. DAXX has originally been found as a Fas-interacting protein and modulator of Fas-induced cell death [69]. DAXX has been reported to interact with various proteins involved in cell death regulation. To look if DAXX levels varied between lung cells and hematopoietic cells and if this could be one of the reasons for the negligible activity of φC31 integrase in the latter, we quantified DAXX on mRNA levels (by quantative reverse transcriptase PCR) and observed a significant higher amount of DAXX in Jurkats (5.4 fold higher) than in A549 cells. This might cause the reduced activity of φC31 integrase in Jurkat cells. Also, the possibility of other hematopoeitic cell specific proteins interacting with φC31 integrase and inhibiting it cannot be ruled out. Therefore, investigations need to be carried out to identify the interacting partners of φC31 integrase in hematopoeitic cells before this system can reach the clinic.

4. Conclusion

To summarize, phage integrases offer themselves as valuable tools in the construction of transgenics and also are an arsenal for gene therapy. The advantages of phage integrase based system includes its relative site specificity and capability to work in diverse species of cells and tissues. Especially with φC31 integrase, integration catalyzed by this enzyme is efficient due to its ability to function in the eukaryotic environment and the absence of any reverse excision reaction. Efficient delivery of φC31 integrase and *attB* donor plasmid to the tissue or cells of interest remains the most challenging aspect of the system. Also, in addition to better delivery systems, optimization of the plasmid backbone (depending upon the target tissue), φC31 integrase enzyme with enhanced activity (generation of highly active mutants) and better understanding of the integration profile in different tissues are major prerequisites to provide this system with a relevant clinical potential.

References

[1] Shroder, A. R., Shinn, P., Chen, H., Berry, C., Ecker, J. R., and Bushman, F. (2002) HIV-1 integration in the human genome favors active genes and local hotspots. *Cell* **110**: 521-529.

[2] Mitchell, R. S., Beitzel, B. F., Schroder, A. R., Shinn, P., Chen, H., Berry, C. C., Ecker, J. R., and Bushman, F. D. (2004) Retroviral DNA integration: ASLV, HIV, and MLV show distinct target site preferences. *PLos Biol.* **2**: E234.

[3] Vigdal, T. J., Kaufman, C. D., Izsvak, Z., Voytas, D. F., and Ivics, Z. (2002) Common physical properties of DNA affecting target site selection of *Sleeping Beauty* and other Tc1/mariner transposable elements. *J. Miol. Biol.* **323**: 441-452.

[4] Vega, M. A. (1991) Prospects for homologous recombination in human gene therapy. *Hum. Genet.* **87**: 245-253.

[5] Smith, M. C., and Thorpe, H. M. (2002) Diversity in serine recombinases. *Mol. Microbiol.* **44**: 299–307

[6] Sauer, B., and Henderson, N. (1988) Site-specific DNA recombination in mammalian cells by the Cre recombinase of bacteriophage P1. *Proc. Natl. Acad. Sci. U S A.* **85**: 5166–5170.

[7] O'Gorman, S., Fox, D. T., and Wahl, G. M. (1991) Recombinase-mediated gene activation and site-specific integration in mammalian cells. *Science* **251**:1351–1355.

[8] Dymecki, S. M. (1996) *Flp* recombinase promotes site-specific DNA recombination in embryonic stem cells and transgenic mice. *Proc. Natl. Acad. Sci. U S A.* **93**: 6191–6196.

[9] McLeod, M., Craft, S., and Broach, J. R. (1986) Identification of the crossover site during FLP-mediated recombination in the *Saccharomyces cerevisiae* plasmid 2 microns circle. *Mol. Cell Biol.* **6**: 3357–3367.

[10] Buchholz, F., Ringrose, L., Angrand, P. O., Rossi, F., and Stewart, A. F. (1996) Different thermostabilities of FLP and Cre recombinases: implications for applied site-specific recombination. *Nucleic Acids Res.* **24**: 4256–4262.

[11] Buchholz, F., Angrand, P. O., and Stewart, A. F. (1998) Improved properties of FLP recombinase evolved by cycling mutagenesis. *Nat. Biotechnol.* **16**: 657–662.

[12] Schaft, J., Ashery-Padan, R., van der Hoeven, F., Gruss, P., and Stewart, A. F. (2001) Efficient FLP recombination in mouse ES cells and oocytes. *Genesis* **31**: 6–10.

[13] Esposito, D., and Scocca, J. J. (1997) The integrase family of tyrosine recombinases: evolution of a conserved active site domain. *Nucleic Acids Res.* **25**: 3605-3614.

[14] Nunes-Duby, S. E., Kwon, H. J., Tirumalai, R. S., Ellenberger, T., and Landy, A. (1998) Similarities and differences among 105 members of the Int family of site-specific recombinases. *Nucleic Acids Res.* **26**: 391-406.

[15] Sauer, B. (1994) Site-specific recombination: developments and applications. *Curr. Opin. Biotechnol.* **5**:521-527.

[16] Sorrell, D. A., and Kolb, A. F. (2005) Targeted modification of mammalian genomes. *Biotechnol. Adv.* **23**: 431–469

[17] Stark, W. M., Boocock, M. R., and Sherratt, D. J. (1992) Catalysis by site-specific recombinases. *Trends Genet.* **8**: 432-439

[18] Branda, C. S., and Dymecki, S. M. (2004) Talking about a revolution: The impact of site-specific recombinases on genetic analyses in mice. *Dev. Cell* **6**: 7–28.

[19] Diaz, V., Rojo, F., Martinez-A, C., Alonso, J. C., and Bernard, A. (1999) The prokaryotic beta-recombinase catalyzes site-specific recombination in mammalian cells. *J. Biol. Chem.* **274**: 6634-6640.

[20] Sauer, B., and Henderson, N. (1990), Targeted insertion of exogenous DNA into the eukaryotic genome by the *Cre* recombinase, *New Biol.* **5**: 441-449.

[21] Thyagarajan, B., Guimaraes, M. J., Groth, A. C., and Calos, M. P. (2000) Mammalian genomes contain active recombinase recognition sites. *Gene* **244**: 47-54

[22] Schmidt, E. E., Taylor, D. S., Prigge, J. R., Barnett, S., and Capecchi, M. R. (2000) Illegitimate Cre-dependent chromosome rearrangements in transgenic mouse spermatids. *Proc. Natl. Acad. Sci. USA* **97**:13702-13707

[23] Sauer, B. (1996) Multiplex Cre/lox recombination permits selective site-specific DNA targeting to both a natural and an engineered site in the yeast genome. *Nucleic Acids Res.* **24**: 4608-4613.

[24] Feng, Y. Q., Seibler, J., Alami, R., Eisen, A., Westerman, K. A., Leblouch, P., Fiering, S., and Bouhassira, E. (1999) Site-specific chromosomal integration in mammalian cells: highly efficient CRE recombinase-mediated cassette exchange. *J. Mol. Biol.* **292**: 779-785.

[25] Landy, A. (1989) Dynamic, structural, and regulatory aspects of lambda site-specific recombination. *Annu. Rev. Biochem.* **58**: 913-949

[26] Matsuura, M., Noguchi, T., Yamaguchi, D., Aida, T., Asayama, M., Takahashi, H., and Shirai, M. (1996) The *sre* gene (ORF469) encodes a site-specific recombinase responsible for integration of the R4 phage genome. *J. Bacteriol.* **178**: 3374–3376

[27] Christiansen, B., Brondsted, L., Vogensen, F. K., and Hammer, K. (1996) A resolvase-like protein is required for the site-specific integration of the temperate lactococcal bacteriophage TP901-1. *J. Bacteriol.* **178**: 5164–5173

[28] Olivares, E. C., Hollis, R. P., and Calos, M. P. (2001) Phage R4 integrase mediates site-specific integration in human cells. *Gene* **278**: 167–176

[29] Stoll, S. M., Ginsburg, D. S., and Calos, M. P. (2002) Phage TP901-1 site-specific integrase functions in human cells. *J. Bacteriol.* **184**: 3657–3663

[30] Loessner, M. J., Inman, R. B., Lauer, P., and Calendar, R. (2000) Complete nucleotide sequence, molecular analysis and genome structure of bacteriophage A118 of *Listeria monocytogenes*: implications for phage evolution. *Mol. Microbiol.* **35**: 324–340

[31] Christiansen, B., Johnsen, M. G., Stenby, A., Vogensen, F. K., and Hammer, K. (1994) Characterization of the lactococcal temperate phage TP901-1 and its site-specific integration. *J. Bacteriol.* **176**:1069–1076

[32] Hodgson, D. A. (2000) Generalized transduction of serotype 1/2 and serotype 4b strains of *Listeria monocytogenes*. *Mol. Microbiol.* **35**: 312–323

[33] Lauer, P., Chow, M. Y., Loessner, M. J., Portnoy, D. A., and Calendar, R. (2002) Construction, characterization, and use of two *Listeria monocytogenes* site-specific phage integration vectors. *J. Bacteriol.* **184**: 4177–4186

[34] Barletta, R. G., Kim, D. D., Snapper, S. B., Bloom, B. R., and Jacobs, W. R. Jr. (1992) Identification of expression signals of the mycobacteriophages Bxb1, L1 and TM4 using the *Escherichia–Mycobacterium* shuttle plasmids pYUB75 and pYUB76 designed to create translational fusions to the *lacZ* gene. *J. Gen. Microbiol.* **138**: 23–30

[35] Mediavilla, J., Jain, S., Kriakov, J., Ford, M. E., Duda, R. L., Jacobs, W. R. Jr., Hendrix, R. W., and Hatfull, G. F. (2000) Genome organization and characterization of mycobacteriophage Bxb1. *Mol. Microbiol.* **38**: 955–970

[36] Kim, A. I., Ghosh, P., Aaron, M. A., Bibb, L. A., Jain, S., and Hatfull, G. F. (2003) Mycobacteriophage Bxb1 integrates into the *Mycobacterium smegmatis* groEL1 gene. *Mol. Microbiol.* **50**: 463–473

[37] Ghosh, P., Kim, A. I., and Hatfull, G. F. (2003) The orientation of mycobacteriophage Bxb1 integration is solely dependent on the central dinucleotide of *attP* and *attB*. Mol. *Cell* **12**: 1101–1111

[38] Yang, H. Y., Kim, Y. W., and Chang, H. I. (2002) Construction of an integration-proficient vector based on the site-specific recombination mechanism of enterococcal temperate phage φFC1. *J. Bacteriol.* **184**: 1859–1864

[39]Bibb, L. A., and Hatfull, G. F. (2002) Integration and excision of the *Mycobacterium tuberculosis* prophage-like element, φRV1. *Mol. Microbiol.* **45**: 1515–1526

[40]Cole, S. T., Brosch, R., Parkhill, J., Garnier, T., Churcher, C., Harris, D., Gordon, S. V., Eiglmeier, K., Gas, S., Barry, C. E. 3rd, Tekaia, F., Badcock, K., Basham, D., Brown, D., Chillingworth, T., Connor, R., Davies, R., Devlin, K., Feltwell, T., Gentles, S., Hamlin, N., Holroyd, S., Hornsby, T., Jagels, K., Krogh, A., McLean, J., Moule, S., Murphy, L., Oliver, K., Osborne, J., Quail, M. A., Rajandream, M. A., Rogers, J., Rutter, S., Seeger, K., Skelton, J., Squares, R., Squares, S., Sulston, J. E., Taylor, K., Whitehead, S., and Barrell, B. G. (1998) Deciphering the biology of *Mycobacterium tuberculosis* from the complete genome sequence. *Nature* **393**: 537–544

[41]Hendrix, R. W., Smith, M. C., Burns, R. N., Ford, M. E., and Hatfull, G. F. (1999) Evolutionary relationships among diverse bacteriophages and prophages: all the world's a phage. *Proc. Natl. Acad. Sci. USA* **96**: 2192–2197

[42]Keravala, A., Groth A. C., Jarrahian, S., Thayagarajan, B., Hoyt, J. J., Kirby, P. J., and Calos, M. P. (2006) A diversity of serine phage integrases mediate site-specific recombination in mammalian cells. *Mol. Genet. Genomics* **276**: 135-146

[43]Russell, J. P., Chang, D. W., Tretiakova, A., and Padidam, M. (2006) Phage Bxb1 integrase mediates highly efficient site-specific recombination in mammalian cells. *Biotechniques* **40**: 460–464.

[44]Kuhstoss, S., and Rao, R. N. (1991) Analysis of the integration function of the Streptomycete bacteriophage φC31. *J. Mol. Biol.* **222**: 897–908.

[45]Rausch, H., and Lehmann, M. (1991) Structural analysis of the actinophage φC31 attachment site. *Nucleic Acids Res.* **19**: 5187–5189

[46]Thorpe, H. M., and Smith, M. C. (1998) In vitro site-specific integration of bacteriophage DNA catalyzed by a recombinase of the resolvase/invertase family. *Proc. Natl. Acad. Sci. USA* **95**: 5505–5510

[47]Groth, A. C., Olivares, E. C., Thyagarajan, B., and Calos, M. P. (2000) A phage integrase directs efficient site-specific integration in human cells. *Proc. Natl. Acad. Sci. USA* **97**: 5995–6000

[48]Thyagarajan, B., Olivares, E. C., Hollis, R. P., Ginsburg, D. S., and Calos, M. P. (2001) Site-specific genomic integration in mammalian cells mediated by phage φC31 integrase. *Mol. Cell Biol.* **21**: 3926–3934

[49]Groth, A. C., Fish, M., Nusse, R., and Calos, M. P. (2004) Construction of transgenic *Drosophila* by using the site-specific integrase from phage φC31. *Genetics* **166**:1775–1782

[50]Olivares, E. C., Hollis, R. P., Chalberg, T. W., Meuse, L., Kay, M. A., and Calos, M. P. (2002) Site-specific genomic integration produces therapeutic Factor IX levels in mice. *Nat. Biotechnol.* **20**: 1124–1128

[51]Ortiz-Urda, S., Thyagarajan, B., Keene, D. R., Lin, Q., Fang, M., Calos, M. P., and Khavari, P. A. (2002) Stable nonviral genetic correction of inherited human skin disease. *Nat. Med.* **8**: 1166–1170

[52]Held, P. K., Olivares, E. C., Aguilar, C. P., Finegold, M., Calos, M. P., and Grompe, M. (2005) In vivo correction of murine hereditary tyrosinemia type I by φC31 integrase-mediated gene delivery. *Mol. Ther.* **11**: 399–408

[53] Bertoni, C., Jarrahian, S., Wheeler, T. M., Li, Y., Olivares, E. C., Calos, M. P., and Rando, T. A. (2006) Enhancement of plasmid-mediated gene therapy for muscular dystrophy by directed plasmid integration. *Proc. Natl. Acad. Sci. USA* **103**: 419–424

[54] Ishikawa, Y., Tanaka, N., Murakami, K., Uchiyama, T., Kumaki, S., Tsuchiya, S., Kugoh, H., Oshimura, M., Calos, M. P., and Sugamura, K. (2006) Phage phiC31 integrase-mediated genomic integration of the common cytokine receptor gamma chain in human T-cell lines. J. Gene Med. **8**: 646–653.

[55] Thomason, L. C., Calendar, R., and Ow, D. W. (2001) Gene insertion and replacement in *Schizosaccharomyces pombe* mediated by the *Streptomyces* bacteriophage φC31 site-specific recombination system. Mol. Genet. *Genomics* **265**: 1031–1038

[56] Belteki, G., Gertsenstein, M., Ow, D. W., and Nagy, A. (2003) Site-specific cassette exchange and germline transmission with mouse ES cells expressing φC31 integrase. *Nat. Biotechnol.* **21**:321–324

[57] Allen, B. G., Weeks, D. L. (2005) Transgenic *Xenopus laevis* embryos can be generated using φC31 integrase. *Nat. Methods* **2**: 975–979

[58] Merrihew, R. V., Sargent, R. G., and Wilson, J. H. (1995) Efficient modification of the APRT gene by FLP/FRT site-specific targeting. Somat. *Cell Mol. Genet.* **21**: 299-307

[59] Chalberg, T. W., Portlock, J. L., Olivares, E. C., Thyagarajan, B., Kirby, P. J., Hillman, R. T., Hoelters, J., and Calos, M. P. (2006) Integration specificity of phage φC31 integrase in the human genome. *J. Mol. Biol.* **357**:28–48

[60] Ehrhardt, A., Xu, H., Huang, Z., Engler, J. A., and Kay, M. A. (2005) A direct comparison of two non viral gene therapy vectors for somatic integration: in vivo evaluation of the bacteriophage integrase phiC31 and the Sleeping Beauty transposase. *Mol. Ther.* **11**: 695-706.

[61] Liu, J., Jeppesen, I., Nielsen, K., and Jensen T. G. (2006) PhiC31 integrase induces chromosomal aberrations in primary human fibroblasts. *Gene Ther.* **13**: 1188-1190.

[62] Ehrhardt, A., Engler, J. A., Xu, H., Cherry, A. M., and Kay, M. A. (2006) Molecular analysis of chromosomal rearrangements in mammalian cells after phiC31 mediated integration. *Human Gene Ther.* **17**: 1077-1094.

[63] Aneja, M. K., Imker, R. and Rudolph, C. (2007) Phage phiC31 integrase-mediated genomic integration and long-term gene expression in the lung after nonviral gene delivery. *J. Gene Med.* **9**: 967-975.

[64] Chen, Z.Y., He, C. Y., Meuse, L. and Kay, M. A. (2004) Silencing of episomal transgene expression by plasmid bacterial DNA elements in vivo. *Gene Ther.* **11**: 856-864.

[65] Chen, Z. Y., He, C.Y., Ehrhardt, A. and Kay, M. A. (2003) Minicircle DNA vectors devoid of bacterial DNA result in persistent and high-level transgene expression in vivo. *Mol. Ther.* **8**: 495-500.

[66] Gill, D. R., Smyth, S. E., Goddard, C. A., Pringle, I. A., Higgins, C. F., Colledge, W. H. and Hyde, S. C. (2001) Increased persistence of lung gene expression using plasmids containing the ubiquitin C or elongation factor 1alpha promoter. *Gene Ther.* **8**: 1539-1546.

[67] Yew, N. S., Przybylska, M., Ziegler, R. J., Liu, D. and Cheng, S. H. (2001) High and sustained transgene expression in vivo from plasmid vectors containing a hybrid ubiquitin promoter. *Mol Ther.* **4**: 75-82.

[68]Chen, J. Z., Ji, C. N., Xu, G. L., Pang, R. Y., Yao, J. H., Zhu, H. Z., Xue, J. L. and Jia, W. (2006) DAXX interacts with phage PhiC31 integrase and inhibits recombination. *Nucleic Acids Res*. **34**: 6298-6304.

[69]Yang, X., Khosravi-Far, R., Chang, H. Y. and Baltimore, D. (1997) Daxx, a novel Fas-binding protein that activates JNK and apoptosis. *Cell* **89**: 1067-1076.

In: Genetic Recombination Research Progress
Editor: Jacob H. Schulz, pp. 205-227

ISBN: 978-1-60456-482-2
© 2008 Nova Science Publishers, Inc.

Chapter 7

The Role of Recombination in the Post-genomic Era: Challenges and Perspectives

Antonio Carvajal-Rodríguez[*]

Departamento de Bioquímica, Genética e Inmunología. Universidad de Vigo
36310 Vigo, Spain

Abstract

Recombination is a key evolutionary mechanism that should not be ignored. It is directly related with the amount of linkage disequilibrium, which is important in order to characterize populations from an evolutionary point of view and to localize genes in humans and other organisms. Recombination effects are not independent of other evolutionary forces, such as natural selection and genetic drift. Recently, new methods became available to infer positive selection in the presence of recombination and *viceversa*. These new methods are computationally very intensive and as data sets become larger, their analysis becomes unavoidable. Therefore, faster and efficient algorithms are needed to estimate recombination at the fine-scale and genomic level. Such approaches need also to disentangle recombination and natural selection signals on DNA. Additionally, it is usually difficult to evaluate the adequacy of the methods since there are few simulation tools capable to produce data under complex evolutionary models. Consequently, developing programs that allow simulating both, natural selection and recombination, is also necessary.

In what follows we will review recently developed algorithms to estimate recombination at the fine scale. In addition, we will also consider approaches that allow the simultaneous estimation of recombination and selection. We will explore possible future directions for recombination and selection estimation research. We will stress the importance of incorporating recombination jointly with other genetic factors both in evolutionary and epidemiological contexts, e.g. to model drug resistance emergence. In this framework, we will present a new result concerning the faster evolution of resistance favoured by the minimum co-infection rate combined with the higher recombination value. Finally, we will consider simulation tools. We will briefly point out how to simulate DNA sequences under a specific

---

[*] E-mail address: acraaj@uvigo.es

nonsynonymous/synonymous (dN/dS) ratio and recombination rate (inter and intracodon level). In doing this we will introduce a new method of forward simulation and an efficient way of simulating different substitution models forward in time.

Introduction

From the beginning of population genetics theory, the assumption of random mating and independent segregation of loci allowed us to completely describe a population from a genetic point of view just by considering the product of the gene frequency at individual loci without any concern for the joint distribution of frequencies. Such a description will not be adequate if the loci are not independent, i.e. linkage disequilibrium (LD) exists. In this case, a complete description requires the specification of all haplotype frequencies, i.e. the distinct gene combination in the gamete population pool. However, in the absence of other evolutionary forces, linkage disequilibrium will approach zero over time due to recombination [1]. Therefore, from the long time perspective of most selective changes (thousands of generations) linkage should not play an important role in the presence of recombination. This reasoning will not be true if both linkage and selection interact favouring the coupling of beneficial alleles and the elimination of the deleterious ones by selection [1]. This is what produces the so-called Hill-Robertson effect in populations of low census size under stochastic conditions, because genetic drift causes deviations from linkage equilibrium, both positive and negative, and positive deviations are purged by selection faster than negative ones [2]. Indeed with large population sizes, where the Hill-Robertson effect does not apply, if epistasis is strong enough or recombination low enough, we expect large departures from independence even when gene frequencies are at equilibrium [1, 3]. Summarizing, we should expect to find linkage disequilibrium in genome regions with low recombination and epistatic selection. Or in populations where founder events could provide the context for the Hill-Robertson effect.

Knowledge about patterns of LD in humans will be very important from a genomic point of view. Importantly, recombination is usually not uniform through genomes with hotspots and coldspots observed in many organisms [4-6].The existence of linkage or haplotype blocks [7], or at least, networks of SNPs at high LD [8-10], will facilitate the assembly of human genome haplotype maps [11, 12] that will enormously improve, among others, the efficiency of disease gene mapping. It seems that these blocks are mainly defined by recombination hotspots [13-17]. However, haplotype blocks can also be generated by genetic drift in regions of uniform recombination if this is low enough [18-20]. Consequently, long linkage blocks could exist as a consequence of low or inexistent recombination. On the other hand, many shorter blocks could also exist with a mainly stochastic origin [21, 22]. Additionally, population variation in haplotype blocks could also be due to population-specific recombination hotspots [23].

Besides its influence on linkage disequilibrium patterns, recombination, jointly with positive selection, allows very high rates of evolution [24]. This has important consequences for a better understanding of the evolutionary process, and also in the impact on human health and economy through very different organisms; for example, regarding pathogenicity, [25-28] resistance to viral drugs [29-35], animal or plant pathogens [36-39], radiation [40, 41], and so on.

Recombination has also important consequences in the reliability of estimation of important evolutionary parameters. Ignoring its presence can have a misleading effect when estimating other population genetic parameters [42, 43] or phylogenetic relationships [42, 44]. Hence, it will be appropriate to have methods to estimate recombination in the fine genomic scale, mapping regions with low and high recombination rates [45-47]. It will be also very important that such estimations can be improved taking into account other evolutionary forces, as variable population size, selection, genetic drift, migration and so on [48-50].

In what follows, we will review recently developed methods to estimate recombination at the fine scale, jointly, with some other ones allowing its simultaneous estimation with selection. We will explore possible future directions for recombination and selection estimation research. In addition we will stress the importance of incorporating recombination jointly with other genetic factors in evolutionary and epidemiological models, e.g. to model drug resistance emergence. Finally, a new algorithm for efficient forward simulation of Markovian substitution models taking into account recombination and different evolutionary scenarios will be also outlined.

Recombination Estimation

Recombination can be defined in general as the exchange of genetic information between two nucleotide sequences [51]. As a consequence, previously unlinked DNA fragments will be jointed together. We can consider several ways in which this new DNA (intragenic) or allelic combination (intergenic) can emerge. For example, in higher organisms, this may happen by crossing over. In RNA viruses, recombination may occur by template switching provided that a double infection event has occurred in the host cell. Recombination can also occur by gene conversion, which means that one allele directs the conversion of a partner allele to its own form. In bacteria recombination occurs via the transformation, conjugation and transduction processes. Non homologous recombination may also produce the result of a new gene combination. Consequently, we will also refer to all these processes under the general term of recombination provided that exchange of genetic information occurs.

The distribution of recombination rate across genomic regions is essential for association studies of genetic diseases [52] and to understand the evolution of pathogens and drug resistance [26, 53-55]. Recombination can also generate new allelic variants within a population, or improve the estimation of population genetic parameters like gene flow [56, 57]. Conversely, ignoring its presence can have confounding effects in the estimation of population genetic parameters or phylogenetic relationships [42, 44].

Recombination can be estimated with high accuracy via the single-sperm typing technique [58, 59], that allows the analysis of semen samples with hundreds of millions of meiotic crossover events. Or by the total-sperm DNA with allele-specific PCR technique to selectively amplified crossover molecules [13, 45, 60]. However, sperm-typing methods require molecular markers at the extremes of the DNA region of interest and a large number of sperm samples to ensure informative individuals for any desired chromosomal segment. In addition, due to variation in individual recombination rates, multiple individuals need to be studied for each target DNA interval to compare with population average recombination rate [45]. These methods only measure male recombination as well. A final drawback is that sperm-typing methods are expensive and not yet practical for large genomic regions.

Upcoming improvements in the techniques used in sperm analysis will come via hybrid approaches between single and allelic-specific PCR sperm detection and also with gains in the efficiency of allelic-specific amplification [45].

A less direct, but complementary and powerful strategy, is measuring the average population recombination rate using coalescent approaches to infer historical recombination rates from DNA diversity or LD patterns [61]. From population genetic data, we can obtain estimates of the recombination rate averaged over many individuals of both sexes and over a long period of time [61]. Potential problems with this approach concern to factors as genetic drift, selection and migration which can influence the population recombination rates. The population recombination rate is a parameter difficult to estimate [62]. There are a plethora of methods that use different strategies to estimate it from population genetic data. The simplest methods are counting recombination events and moment-based estimators that use statistics of the data. Full-likelihood approaches use all information of the data assuming a coalescent-based model and the pseudo-likelihood ones use only the most significant information [55]. The efficiency of the different strategies depends on the amount of recombination, the genetic diversity represented in the data, and the degree of rate variation among sites. However, in general, coalescent based algorithms perform better [55, 63].

Methods based on full-likelihood estimates [64-66] are very computationally intensive, becoming impractical for many real data sets. To avoid this difficulty, the pseudo-likelihood methodologies were developed to approach the likelihood surface [55]. Both kinds of strategies assume a coalescent model with neutral evolution, random mating and constant population size. However, the pseudo-likelihood methods make some extra assumptions in order to reduce computation time [62, 67]. For example, the composite likelihood estimator [62] combines the coalescent likelihoods of all pairwise comparisons for segregating sites. This estimator has been adapted and improved in several ways [47, 49, 68-70]. McVean et al. [68] using the importance sampling method of Fearnhead and Donnely [64] to estimate the pairwise likelihoods, extended Hudson's method to allow for a finite-sites mutation model, introducing also a likelihood permutation test (LPT). They assume a two-allele model with reversible, symmetric mutation, with a mutation rate per generation homogeneous across sites. Carvajal-Rodríguez et al. [49] extended this method to allow for more complex models and rate variation among sites. They also studied the power of the LPT with small recombination rates and low diversity values. Indicating that this test is better than other 14 methods previously studied [63]. Nevertheless, approximate likelihood estimators seem to underestimate the recombination rate when the neutral coalescent model assumptions are violated as in the presence of rapid exponential growth, bottlenecks, positive selection, population structure, or non-contemporaneous sampling [48, 49]. Such situations are common in the case of rapid evolving pathogens but also in the history of human populations.

An interesting and recent new approach relays on phylogenetic detection using a genetic algorithm to search the parameter space of multiple recombination break points positions [71, 72]. Genetic algorithms have the advantage of being easy parallelizable. They are, however, computationally intensive making the study of long sequence fragments or genomes difficult. Finally, as was said, recombination is usually not uniform with hot and coldspots distributed through the genome [4-6]. In humans much of the recombination appears to be concentrated into 1–2kb hotspots [13, 45, 60, 73-75]. Therefore new improved recombination estimators are being developed and new methods are been applied to infer patterns of fine scale recombination rate.

Fine-Scale Recombination Variation Estimation

Statistical methods to infer recombination rates at a fine scale are actively being developed. One of the main research directions tries to deal with regions with a high variation in the recombination rates, the so called recombination hot and cold spots. In the next paragraph we will briefly comment some of such methods. The reader is referred to each method reference for a deeper glace on the algorithms.

The approximate likelihood based on the region-wise likelihood of Fearnhead [67] used the importance sampling (IS) to compute the likelihoods, allowing to reveal rate variation and recombination hotspots [76]. This approach was extended to locate hotspots more efficiently by considering overlapping sub-regions [77]. Fearnhead [78] has improved the above methods using a pre-specified grid of hotspot locations and allowing more efficient computation by stopping the simulations in the importance sampling step if there was little or overwhelming evidence of hotspot at a given position. Other model used by McVean and co-workers [47, 75] extends the composite likelihood to allow different recombination rates between each pair of segregating sites and uses a Bayesian reversible-jump Markov Chain Monte Carlo (rjMCMC) to calculate the posterior distribution. This model was recently extended to account for specific recombination hot spot models [70]. Another kind of method is the truncated, weighted pairwise log-likelihood (TWPLL) [79] that uses weighted pairwise likelihoods [80] adapted to estimate recombination with hotspot models. One more approach is the product of approximate conditionals (PAC) model [81] defined in a bayesian context and using MCMC to approximate the joint posterior distribution [46]. Recently, the wavelet-based analysis (WBA) [82] has been used to describe the nature of the variation in the recombination rate measuring relative variation at different physical scales [83]. Some of the above methods have been implemented in programs that are usually freely available at the authors webpage (see Table 1).

Table 1. Recombination hotspot estimation programs

Program	Method	Reference
HotspotFisher	TWPLL	[79]
Hotspotter	PAC	[81]
Interval (LDhat)	rjMCMC	[47]
Rhomap (LDhat)	rjMCMC	[70]
SequenceLDhot	IS	[78]

TWPLL: truncated,weighted pairwise log-likelihood. PAC: product of approximate conditionals. rjMCMC: Reversible-jump Markov Chain Monte Carlo. IS: Importance Sampling to compute approximate likelihood

Joint Estimation of Recombination, Selection and Other Evolutionary Forces

The above mentioned estimation methods consider recombination, based on LD patterns, under a standard neutral model. However, it is well known that different evolutionary forces can have a profound effect on LD. For example, natural selection can alter the LD between neutral markers and hence will affect hotspot estimation [50, 84]. Simulations have also shown that positive selection acting on a few sites can diminish the power to detect recombination [49]. Moreover, long standing balancing selection maintaining multiple alleles can generate LD pattern that could lead to the misinterpretation of recombination hotspots [85]. Therefore, complex patterns of variation emerge due to the interaction of selection and recombination [86-88]. Likewise, there are different demographic situations that will also influence the recombination estimation. For example, bottlenecks will augment LD levels, the same as population structure, but growing populations will reduce it [89]. The kind of mating system will also affect the patterns of diversity and LD [90]. Sometimes, a more profound glace at the population variability can disentangle between different causes, e.g. bottlenecks produce a greater effect on LD levels than on polymorphism levels [91, 92]. Conversely, recombination will have important effects on the estimation of other selective forces such as selection [43, 93, 94] and migration [95].

Given the complex nature of current and future data sets, methods, that can take into account the impact of other forces, are needed to estimate the evolutionary parameters of interest [96]. For example, the so called, complex selection graph (CSG) [97], which extends the ancestral selection graph [98] allows modelling selection and unlinked loci in the context of complex diseases. Recent results using this CSG provided a simple relationship between the joint stationary distribution under selection with the neutral one [99].

To infer the trace of natural selection in the DNA sequences, the ratio of nonsynonymous to synonymous substitutions (dN/dS) is useful and widely applied. Commonly, this ratio is estimated via a maximum likelihood phylogenetic approach using codon based mutation models [100, 101]. However, it is known that recombination have large effect on selection estimation because, in the presence of recombination, there is no unique tree topology to describe the evolutionary history of the sequence [43, 94]. Recently, new methods become available to infer positive selection in the presence of recombination [102, 103]. The method of Scheffler et al. uses a maximum likelihood to infer a different tree topology linked to each detected recombination break point avoiding in this way the distortion effect of recombination over the selection estimates. Under this method recombination is, however, estimated without considering the effect of selection. The approach of Wilson and McVean [103] is, as far as we know, the only one allowing the joint estimation of selection and recombination. It is a coalescent model-based approach using Bayesian inference, via a reversible-jump MCMC algorithm [104]. Nevertheless, the joint estimation of selection and recombination is computationally very intensive, being difficult to apply to large datasets. Therefore, as data sets become larger, faster measures of recombination and/or selection are needed. New future direction in the joint estimation of recombination, selection and other evolutionary parameters of interest could come from the field of genetic algorithms (GAs). A genetic algorithm is a heuristic search technique or solving problem method based on the ideas of evolutionary biology [105, 106]. Briefly, this means that the candidate solutions of a problem are encoded as a population of subjects. Each subject or solution has different "genes" which represent different parameters of the problem and can undergo mutation and recombination. Each subject has a fitness value that depends on how well it performs with respect to the problem to solve. Depending on its fitness, individuals will leave more or less progeny to the next generation. Through the process of finding the optimal solution, the population will evolve during a number of generations or until a maximum or average fitness is reached. GAs are well suited for searches in large spaces of potential solutions [107]. They have been previously used for maximum-likelihood phylogenetic inference [108] and have been recently applied to infer recombination [71, 72], selection [109] and model choice [110]. They could also be used to explore the complex parameter space under models with recombination, selection and demography issues, such as variable population size and population structure. We would define the adequate model parameterization and the corresponding fitness or objective function that we want to maximize. For example, given a sequence sample S and regarding recombination (R) and selection (W) we want to define an objective function proportional to the likelihood $L(R, W|S)$. The subjects will be the 2-tuplas (r, w) where r and w are particular realizations of the parameters belonging to the domain of R and W, respectively. Finally, we should define the genetic operators of the GA to explore the parameter space by evolving some individuals from others. The underlying idea is to explore heuristically, climbing, the likelihood surface instead to compute or to approach it. Of course, how to compute an efficient and reliable objective function is the main problem to apply the GA machinery to evolutionary parameter estimation under complex models. One possibility, for example, would be to approximate the likelihood values with a set of summary statistics as in rejection sampling methods [111] or approximate bayesian computation [112]. Importantly, GAs, though computationally intensive, have the advantage of being easily parallelizable. In any case, whatever the method to estimate recombination from DNA sequence data, the future will likely come from the smart combination of sperm typing and data analysis [85]. This will allow the straightforward confirmation of recombination hotspots previously estimated from, for example, LD patterns.

Recombination and HIV Drug Resistance

Recombination is an important evolutionary process to understand how genetic diversity is generated and maintained in populations. Jointly with positive selection, allows very high rates of evolution in virus as the RNA ones [113, 114], bacteria [27, 28, 115, 116] and eukaryote [24, 117, 118]. Recent work has given some clues on how this fast evolution can occur in the context of protein evolution. Intragenic recombination allows a larger exploration of the sequence space compared to random mutation with a relatively low cost in the loss of function [119]. The impact of recombination in the functionality of proteins has been modelled concerning the protein fragments or schemas, that can be recombined without disturbing the three dimensional structure [120]. The schema concept, borrowed from the genetic algorithms theory, was equated to the building blocks from which novel proteins can be assembled by recombination. Voigt et al [120] computed the interaction between aminoacid residues and determined the number of interactions disrupted by recombination. They concluded that crossovers leading to functional hybrid proteins coincide with positions that minimize the number of disrupted interactions. It seems that evolutionary analysis of the recombination process should be improved by considering nucleotide sequences and protein products jointly.

Given its ability to produce genetic diversity and faster evolution, recombination should be a key factor in evolutionary and epidemiological models. From herein we will focus in one important feature as is the improved pathogen ability to get drug resistance in the presence of recombination. We will consider the concrete case of HIV viruses. Recombination has been coupled with the development of HIV-1 resistance to drug therapy both, by empirical studies [29, 30, 32, 36, 121] and by theoretical predictions [31, 33, 122, 123], though the generality of this result has been disputed [124-126]. Certainly, population genetics theory predicts that recombination can favour adaptation in finite populations via the avoidance of the Hill-Robertson effect [2]. In finite populations, positive or negative random deviations from linkage equilibrium will be generated by genetic drift. However, in the presence of selection, positive deviations from linkage equilibrium will be rapidly fixed or eliminated. On the other hand, negative deviations are not so quickly solved by selection, because of the intermediate fitness of single mutants (a fitter allele in a less fit background). This is known as the Hill-Robertson effect. The accumulation of negative disequilibrium can be significant even in fairly large populations as long as the allele frequencies are initially low and subject to drift [127]. Thus, the Hill-Robertson effect predicts that genetic drift will interact with directional selection to produce negative linkage disequilibrium. Negative linkage disequilibrium is all that recombination needs to favour the apparition of beneficial mutations by breaking up the unpaired combinations of resistant and susceptible alleles [128, 129]. Consequently, fully resistant genotypes and fully susceptible ones will be generated. The resistant ones will spread rapidly within the population while the others will be eliminated fairly quickly [33].

One criticism to the current models for the emergence of resistance in HIV rests in the somewhat poor description of the multiple infection dynamics [130]. Multiple infection of a cell is a necessary condition for the subsequent formation of recombinant virions. The frequency of multiple infection depends on the viral load and on CD4 receptor down-regulation [131-133]. Therefore, a variable frequency of multiple infection depending on the cell type will imply a variable proportion of recombination. Given a recombination frequency, lower co-infection rates will imply lower overall recombination rates and vice versa. In any case, recent studies have provided a growing evidence of both, predominance of multiple infected cells in HIV [134-137] and high intrapatient recombination [135, 138, 139].

The question arises if we can consider the overall recombination rate as a compound parameter $R = f \times r$ and, therefore not to be worried about different f (co-infection rate) or r (recombination rate) values but about different R values between different cells. We expect that the answer to this question is yes since the recombination effect is linked to the frequency of co-infection as is logical, and also has been formally shown at least for a kind of models [33]. However, it is also true that due to the phenotype mixing (when two proviruses are transcribed simultaneously within the same infected cell and the newly assembled virions contain mixed proteins from both parental proviruses), the co-infection rate will have its own impact onto the fitness evolution of the population. We can further study this with an *in silico* experiment. Consider a population of haploid organisms, e.g. proviruses, consisting of RNA sequences, with a given population size, constant selection and discrete generations. Let these organisms to reproduce inside host cells from which they are released as virions (diploids) that survive depending on their fitnesses values. These virions will enter the cells as proviruses (haploids) and will mutate and, if more than one provirus infected the cell, with probability f, they will recombine, with probability r, to produce the next generation of released virions. This is a typical genetic population model of HIV life cycle [124] and can be managed in several ways, e.g. to allow for finite population size [33, 125] or to generate different kinds of cells with different infection rates [126].

We will implement the same model as in Carvajal-Rodriguez [33] to study the effect of different co-infection rates given an overall composite recombination rate $R = f \times r = 0.08$ (see Table 2). Briefly, this model consider four types of proviruses, *ab*, *aB*, *Ab* and *AB*, where lowercase denotes drug-sensitive wild-type alleles and uppercase indicates drug-resistant mutant alleles. Under drug therapy, provirus type *ab* is fully sensitive, *aB* and *Ab* are partially drug resistant, and *AB* is fully resistant. The change in provirus frequencies during the replication cycle is best described by dividing this cycle in three steps: cell infection, virion release, and provirus transcription, see Carvajal-Rodriguez [33] for details on the model. Provirus fitness during therapy is computed under a multiplicative model in which the wild type *ab* has fitness $1 - s$. Genotypes carrying one mutant allele have fitness $(1 - s)^{1/2}$ and the double mutant has fitness 1. We set a selection coefficient of $s = 0.5$, then the fitness at the beginning of the drug treatment will be 0.25. When a double infection occurred the newly assembled virions can contain mixed proteins from both parental proviruses, i.e. the fitness of the new virion will be due to a phenotype mixing. The particular implementation of the phenotypic mixing is as the geometric mean of the fitness of both parental proviruses. That is, in the absence of epistasis, the fitness of virions emerging from double infected cells with provirus *AB* and *ab* will be the same as the fitness of virions emerging from double infected cells with provirus *Ab* and *aB*. Of course the fixation of resistance will vary depending on the viral population size [33, 125]. We choose a particular value of population size of $N = 10^4$. The system will evolve with a mutation rate (μ) of 3×10^{-5} [140] and a given overall recombination rate (R) of 0.08. In consequence we can use distinct, f and r values to get the same given R (see Table 2). Then, we can study if the average fixation of resistance is the same or varies through the different cases. One thousand replicates were performed for each combination of $f \times r$.

Table 2. Different parameter combination assayed for an overall recombination rate of 0.08 (see text for details) and 0.0

R	*f* × *r*
0.0	0.0×0.0
0.0	1.0×0.0
0.0	0.0×1.0
0.08	1.0×0.08
0.08	0.8×0.1
0.08	0.4×0.2
0.08	0.2×0.4
0.08	0.1×0.8

In Figure 1 we can appreciate the percentage of time reduction for fixation of resistance under the corresponding $f \times r$ rate with respect to the worst case 1×0, i.e. always co-infection but no recombination. We can conclude two main things from the figure. First, recombination favours the fixation of resistance. Second, given an overall recombination rate $R = 0.08$, the lower the co-infection rate the fastest the fixation time. That is, if $R = 0.08$ the best was the case with the lowest co-infection rate ($f = 0.1$). Indeed with $R = 0$ any case with $f = 0$ was better than the case with $f = 1$. The reason could be the phenotype mixing, which provokes that the fitness of emerging virions depends on both the parental provirus fitness and the cellular type [141, 142]. In real life, phenotypic mixing occurs when two proviruses that are being transcribed simultaneously within the same infected cell mix their corresponding viral proteins in the assembly of new virions. It is possible then that the fitness of a released virion does not necessarily reflect the genomic RNA that it carries, but that it is rather somewhere in between that of both parental proviruses. In a previous work it was shown that different implementations of the phenotype mixing have a similar effect, at least on a kind of model as the implemented here [33]. Therefore, what happens is a reduction in the selection intensity due to the phenotype mixing. The reason is that if we have two genotypes with very different fitnesses then will produce a phenotype with intermediate fitness. This will slow the evolution of resistance because proviruses with the highest fitness will not be selected so rapidly and conversely, proviruses with a reduced fitness will be not eliminated so quickly. Then, the phenotype mixing effect is opposite to the recombination one. Thus, given a fixed $R = f \times r$ the highest the co-infection rate f, the higher the non-recombinant fraction and the slowest the evolution of resistance because of the opposite effects of phenotype mixing and recombination. On the contrary, the lowest the co-infection rate, f, the higher the recombinant fraction and then the fixation of resistance will be achieved faster. As a result, for a given value of overall recombination rate the lower co-infection rate and the higher recombination rate combination will be preferred. It seems, somewhat unexpectedly, that co-infection and recombination rates are very important per se. Surprisingly, the inhibition of multiple infection by CD4 down-modulation effect given a high recombination rate r could be favourable for the virus.

Thus, given the recent evidences for predominance of multiple infected cells in HIV, the high intrapatient recombination and also the recent modelling results [31, 33], it seems clear that recombination should have a key role on generating of HIV resistance phenotypes.

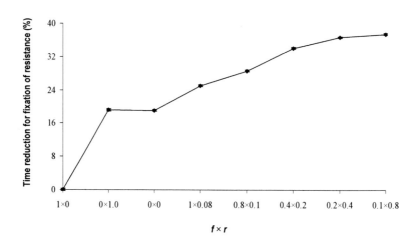

Figure 1. Effect of recombination on the evolution of multidrug resistance with constant population size $N = 10^4$. The y-axis indicates the % of time reduction for fixation of resistance under the corresponding $f \times r$ rate with respect to the worst case of 1×0. f: double infection rate. r: recombination rate.

If resistance mutations pre-exist as it seems to be the case [143, 144] the effect of recombination favouring the achievement of resistance will be magnified [33, 123]. As a conclusion we should say that a combination of drugs that target close regions of the HIV-1 genome will be less susceptible to failure due to the emergence of drug resistance because of a reduced recombination rate between target loci.

Simulation of Recombination under Complex Evolutionary Scenarios

There are many different situations in current population biology research in which simulating populations of genomes undergoing recombination is very useful. As we have shown above, some interesting examples include exploring complex situations such as the evolution of drug-resistance in HIV-1 [33, 123-126, 145].

When simulating biological populations under different evolutionary genetic models, backward or forward strategies can be followed. Backward simulations, also called coalescent-based simulations, are computationally very efficient because they are backward based on the history of lineages with survived offspring in the current population, ignoring all those lineages whose offspring did not arrive to the present [146]. Hence, coalescence is a sample-based theory relevant to the study of population samples and DNA sequence data. Due to its efficiency, it has also been used to derive several algorithms to estimate parameter values that maximize the probability of the given data [147]. From its beginnings, the basic coalescence has been extended in several useful ways to include structured population models [148-152], changes in population size [153-155], recombination [156, 157] and selection [98, 158-162]. By contrast, forward simulations are less efficient because the whole history of the sample is followed from past to present. However, the coalescent framework imposes some limitations not present in the forward ones. First, as it has been indicated before, the coalescence does not keep track of the complete ancestral information. Thus, if the interest is focused on the evolutionary process itself, rather than on its outcome, forward simulations should be preferred [163]. Second, coalescent simulations are complicated by simple genetic forces such as natural selection. Although different evolutionary scenarios have been incorporated (see above), it is still difficult to implement models incorporating complex evolutionary situations with selection, variable population size, recombination, complex mating systems, and so on. Similarly, coalescent methods cannot yet simulate realistic samples of complex human diseases [164]. Moreover, the coalescent model is based on specific limiting values and relationships between some important parameters [165].

Table 3. Different simulators for complex evolutionary models.

Method	Name	Sel	Rec	VRec	VarN	M	Reference
Forward	EasyPop	No	Yes	No	No	Yes	B01
Coalescent	ms	No	Yes	No	Yes	Yes	H02
Coalescent	SelSim	Yes	Yes	No	No	No	S04
Forward	SimuPop	Yes	Yes	Yes	Yes	Yes	P05
Coalescent	Cosi	No	Yes	Yes	Yes	Yes	S05
Coalescent	hap	Yes	No	No	Yes	Yes	F06
Coalescent	msHot	No	Yes	Yes	Yes	Yes	He07
Coalescent	GENOME	No	Yes	Yes	Yes	Yes	L07
Coalescent	ReCodon	No	Yes	No	Yes	Yes	A07
Forward	FREEGENE	Yes	Yes	Yes	Yes	Yes	H07
Forward	GenomePop	Yes	Yes	Yes	Yes	Yes	C08

Sel: Selection. Rec: recombination. VRec: Variable recombination rates. Var N: Variable population size. M: Migration. B01:[166]. H02: [167]. S04: [168]. P05: [169]. S05: [170]. F06: [162]. He07: [171]. L07: [172]. A07: [173]. H07: [174]. C08: [175]

Consequently, both, forward and backward simulation strategies complement each other. To simulate recombination in a complex evolutionary context, several new recent simulation tools, both backward and forward, has being developed (Table 3). Some of them allow for a variable recombination rate and therefore, for the simulation of recombination hotspots. Here, we will not review in detail each one. But we will better consider a kind of evolutionary framework that would be interesting to simulate and will suggest the algorithm to implement it.

We have now growing empirical knowledge about haplotype block and tagSNPS diversity. However, less is known about the effect of population demographic history. We have no clear ideas on how the combined effect of genetic drift, mutation, recombination and migration affect LD patterns though it is known they do [22]. For that reason, it will be of great interest to develop software tools that could efficiently simulate this kind of situation. Additionally, we have seen the importance of recombination on the emergence of viral drug resistance. However, from an epidemiological point of view we have to consider viral populations at intra and inter-patient levels. The inter-patient level could be seen as a metapopulation with complex migration schemes and complex fitness relationships at the intra-patient level.

Developing tools that allow at the same time simulating, natural selection, recombination, complex demography and migration patterns will be of great importance. Nevertheless, this is not a simple task since coalescence does not allow to easily simulate dN/dS, which implies codon units, and recombination, which implies nucleotide units. Indeed, one of the major impacts of recombination on dN/dS estimation should be due to intracodon variation [43, 94] but we cannot simulate recombination at intracodon level under the coalescent framework. Currently, there are few simulators that allow efficient simulation of complex models. One of these is SimuPop [169], which is powerful and flexible object oriented software for forward-time simulation of human populations with complex diseases [164]. Another important tool is the software Cosi [170], which performs coalescent simulation with demography and recombination to simulate human genetic data. A new recently developed coalescent tool is the software GENOME [172], which provides with a new approach to simulate multiple coalescent and recombination events in the same generation. It also allows varying the recombination rate through the genome. Other recent and powerful tool is FREEGENE [174], which introduces a technique that makes large-scale forward-in-time simulation feasible. FREEGENE allows for a complete battery of evolutionary forces (see Table 3) assuming however a bi-allelic model and simplistic migration and demography schemes.

An additional important use of simulators is to study how different situations affect the estimation of a given parameter. For example, how is the recombination rate estimation affected by the joint effect of positive selection and migration? Suppose that a population evolves under a complex nucleotide model, e.g. GTR [176] and with positive selection and recombination. Will be the recombinant signal recovered from that sequences?. And the dN/dS?. As said above it is difficult to simulate intracodon recombinaton under the coalescent. Nevertheless, this can be easily done under a forward framework. We will briefly point out how to efficiently simulate DNA sequences under specific dN/dS ratio and recombination rate (inter and intracodon level).

In molecular evolution, Markov processes are used to describe the change between nucleotides, aminoacids or codons over evolutionary time. Time is measured in number of substitutions because molecular sequence data do not allow separate estimation of the rate and the time but only as their product [177]. Then, on developing a forward simulator for codon models, we first have to introduce the question of simulating Markov processes forward in time to describe the change between nucleotides or codons over time. The question arises if forward simulation could be used to generate sequence samples under different substitution models. That is, if a conditional transition mutation matrix can generate the corresponding substitution matrix. To be more explicit, does a specific transition mutation matrix, which operates forward in time, with generation as the unit of time, lead to the corresponding transition substitution matrix, which concerns only to present and works in a different time scale (measured in number of substitutions)? We expect this to occur, given long time enough, under strictly neutral conditions, because then, the rate per generation of mutant substitution in the population coincides with the mutation rate per gamete per generation [178]. Hence, the Markov chain theory tell us that given a continuous Markov chain model we can expect, under certain conditions, that the stationary distribution be the same when considering the corresponding discrete chain. The opposite, however, not need to be true. Therefore, at equilibrium, we should be able to produce the exact desired nucleotide or codon substitution model using a forward simulator.

Consider a given instantaneous substitution rate matrix Q, the matrix $M = qQ - I$ is the conditional transition matrix to go from substitution i to substitution j given that substitution occurs, where q = diagonal $(1/q_i)$ and I is the identity matrix [179]. The matrix M, considered in a time scale of single instead of $2N$ generations, provide us with the mutation transition matrix. Then, such matrix can be obtained from the instantaneous rate matrix Q. Therefore, given an instantaneous substitution matrix Q which allows for a complete definition of any Markovian substitution model [177], we can obtain the corresponding transition matrix M that can be used to produce the adequate mutation process in a forward in time evolutionary process. To allow for explicit codon models the transition between codons should be implemented typically i.e. the matrix Q should be defined to deal with the codon model of interest [100, 101] [177].

Now, we briefly mention an algorithm to simulate genomes, under a given nucleotide or codon model, forward in time, which allows for an efficient use of computer memory. The basic algorithm can be complicated with complex demography patterns. We call our algorithm SMS (Simulating in the Mutation Space). The basic idea of SMS considers an individual as the differences (mutations) between this individual in reference to an original or consensus genotype (the master sequence). Thus, SMS provides for a forward simulation framework to represent individuals just as the mutations they carry with respect to the wild genotype. Therefore, the dimensionality of the problem of representing genomes is reduced by several–fold factor. By using the SMS representation, efficiency is gained in both computation space and time. However, there is also a necessity to redefine the implementation of some processes such as mutation, migration, recombination and fitness evaluation to adjust to the new way of storing genomes in this less-redundant manner (Carvajal-Rodríguez 2007, Simulating Genomes and Populations in the Mutation Space: An example with the evolution of HIV drug resistance, unpublished). The SMS algorithm is very similar to that used in FREEGENE [174] though was derived independently. This SMS methodology is implemented in the software GenomePop [175], which is a flexible program that allows the forward simulation of DNA sequences, user provided or generated by the program, in a metapopulation context under a variety of conditions. GenomePop have some distinctive features respect other similar programs. We can use different nucleotide or codon models of DNA mutation (not substitution) that, as we have seen, should generate the corresponding substitution models given the adequate conditions. We can simulate any number of populations with any migration model. There is no limit on the complexity and hierarchy level in the migration relationships. We can study SNPs segregating independently or linked in the same chromosome. In addition, selective nucleotide sites can be defined the same as specific positions and amino acids undergoing directional selection. The generated sequence sample can be obtained with serial sampling at different generations or just in the last one. Therefore we can use GenomePop to, for example, study the impact of different migration schemes on the recombination signal detected in populations evolving under different evolutionary models.

It seems that simulation software is already a key part of the current research in evolutionary biology and it seems that it will be a primary tool in the future research of genome and post-genomic evolutionary biology.

Conclusion

We have shown the importance that the occurrence of recombination within a locus or at very tight loci have at the evolutionary, epidemiological, medical and health levels. The impact that recombination, or its absence, may have under different evolutionary scenarios could make difficult the interpretation of the trace in the genetic data of complex evolutionary processes. The feasible understanding of evolutionary processes will provide humans with the tools to meliorate human health and fitness. The future should find us in the effort of combining the insight provided by complex stochastic models with the thoughtful use of simulation methods for both, inference and modelling of complex evolutionary scenarios. Currently, all points to this will dominate the field from now on [96].

Acknowledgments

I am grateful to Silvia Rodríguez-Ramilo, Armando Caballero and Humberto Quesada for useful comments on the manuscript. I am currently funded by an Isidro Parga Pondal research fellowship from Xunta de Galicia (Spain).

References

[1]   Lewontin RC. The genetic basis of evolutionary change. *New York: Columbia University Press*; 1974.

[2]   Hill WG, Robertson A. The effect of linkage on limits to artificial selection. *Genet Res* 1966;8(3):269-94.

[3]   Lewontin RC. The Interaction of Selection and Linkage. I. General Considerations; *Heterotic Models. Genetics* 1964;49(1):49-67.

[4]   Lichten M, Goldman AS. Meiotic recombination hotspots. *Annu Rev Genet* 1995;29: 423-44.

[5]   Petes TD. Meiotic recombination hot spots and cold spots. *Nat Rev Genet* 2001;2(5): 360-9.

[6]   Nachman MW. Variation in recombination rate across the genome: evidence and implications. *Curr Opin Genet Dev* 2002;12(6):657-63.

[7]   Goldstein DB. Islands of linkage disequilibrium. *Nature Genetics* 2001;29:109-111.

[8]   Gabriel SB, Schaffner SF, Nguyen H, Moore JM, Roy J, Blumenstiel B, et al. The structure of haplotype blocks in the human genome. *Science* 2002;296(5576):2225-9.

[9]   Carlson CS, Eberle MA, Rieder MJ, Yi Q, Kruglyak L, Nickerson DA. Selecting a maximally informative set of single-nucleotide polymorphisms for association analyses using linkage disequilibrium. *American Journal of Human Genetics* 2004;74(1): 106-120.

[10]   Nothnagel M, Rohde K. The effect of single-nucleotide polymorphism marker selection on patterns of haplotype blocks and haplotype frequency estimates. *American Journal of Human Genetics* 2005;77(6):988-998.

[11]   International-HapMap-Consortium. A haplotype map of the human genome. *Nature* 2005;437(7063):1299-320.

[12]   International-HapMap-Consortium. The International HapMap Project. *Nature* 2003; 426(6968):789-96.

[13]   Jeffreys A, Kauppi L, Neummann R. Intensely punctate meiotic recombination in the class II region of the major histocompatibilty complex. *Nature Genetics* 2001;29: 217-222.

[14]   Jeffreys AJ, Holloway JK, Kauppi L, May CA, Neumann R, Slingsby MT, et al. Meiotic recombination hot spots and human DNA diversity. *Philos Trans R Soc Lond B Biol Sci* 2004;359(1441):141-52.

[15]   Kauppi L, Sajantila A, Jeffreys AJ. Recombination hotspots rather than population history dominate linkage disequilibrium in the MHC class II region. *Hum Mol Genet* 2003;12(1):33-40.

[16]   Greenwood TA, Rana BK, Schork NJ. Human haplotype block sizes are negatively correlated with recombination rates. *Genome Res* 2004;14(7):1358-61.

[17]   Greenawalt DM, Cui X, Wu Y, Lin Y, Wang HY, Luo M, et al. Strong correlation between meiotic crossovers and haplotype structure in a 2.5-Mb region on the long arm of chromosome 21. *Genome Res* 2006;16(2):208-14.

[18]   Phillips MS, Lawrence R, Sachidanandam R, Morris AP, Balding DJ, Donaldson MA, et al. Chromosome-wide distribution of haplotype blocks and the role of recombination hot spots. *Nat Genet* 2003;33(3):382-7.

[19]   Zhang K, Akey JM, Wang N, Xiong M, Chakraborty R, Jin L. Randomly distributed crossovers may generate block-like patterns of linkage disequilibrium: an act of genetic drift. *Hum Genet* 2003;113(1):51-9.

[20]   Liu N, Sawyer SL, Mukherjee N, Pakstis AJ, Kidd JR, Kidd KK, et al. Haplotype block structures show significant variation among populations. *Genet Epidemiol* 2004;27(4):385-400.

[21]   Sawyer SL, Mukherjee N, Pakstis AJ, Feuk L, Kidd JR, Brookes AJ, et al. Linkage disequilibrium patterns vary substantially among populations. *Eur J Hum Genet* 2005; 13(5):677-86.

[22]   Gu S, Pakstis AJ, Li H, Speed WC, Kidd JR, Kidd KK. Significant variation in haplotype block structure but conservation in tagSNP patterns among global populations. *European Journal of Human Genetics* 2007;15(3):302-312.

[23]   Graffelman J, Balding DJ, Gonzalez-Neira A, Bertranpetit J. Variation in estimated recombination rates across human populations. *Hum Genet* 2007;122(3-4):301-10.

[24]   Marais G, Charlesworth B. Genome evolution: recombination speeds up adaptive evolution. *Curr Biol* 2003;13(2):R68-70.

[25]   Urwin R, Holmes EC, Fox AJ, Derrick JP, Maiden MC. Phylogenetic evidence for frequent positive selection and recombination in the meningococcal surface antigen PorB. *Mol Biol Evol* 2002;19(10):1686-94.

[26]   Awadalla P. The evolutionary genomics of pathogen recombination. *Nat Rev Genet* 2003;4(1):50-60.

[27]   Andrews TD, Gojobori T. Strong positive selection and recombination drive the antigenic variation of the PilE protein of the human pathogen Neisseria meningitidis. *Genetics* 2004;166(1):25-32.

[28]   Orsi RH, Ripoll DR, Yeung M, Nightingale KK, Wiedmann M. Recombination and positive selection contribute to evolution of Listeria monocytogenes inlA. *Microbiology* 2007; 153(Pt 8):2666-78.

[29]   Kellam P, Larder BA. Retroviral recombination can lead to linkage of reverse transcriptase mutations that confer increased zidovudine resistance. *Journal of Virology* 1995;69(2):669-674.

[30]   Moutouh L, Corbeil J, Richman DD. Recombination leads to the rapid emergence of HIV-1 dually resistant mutants under selective drug pressure. *Proc Natl Acad Sci U S A* 1996;93(12):6106-11.

[31]   Suryavanshi GW, Dixit NM. Emergence of recombinant forms of HIV: dynamics and scaling. *PLoS Comput Biol* 2007;3(10):2003-18.

[32]   Nora T, Charpentier C, Tenaillon O, Hoede C, Clavel F, Hance AJ. Contribution of recombination to the evolution of human immunodeficiency viruses expressing resistance to antiretroviral treatment. *J Virol* 2007;81(14):7620-8.

[33]   Carvajal-Rodriguez A, Crandall KA, Posada D. Recombination favors the evolution of drug resistance in HIV-1 during antiretroviral therapy. *Infect Genet Evol* 2007;7(4): 476-83.

[34]   Suzuki Y, Gojobori T, Nakagomi O. Intragenic recombination in rotaviruses. *FEBS Letters* 1998;427:183-187.

[35]   Phan TG, Okitsu S, Maneekarn N, Ushijima H. Evidence of intragenic recombination in G1 rotavirus VP7 genes. *J Virol* 2007;81(18):10188-94.

[36]   McDowell JM, Dhandaydham M, Long TA, Aarts MG, Goff S, Holub EB, et al. Intragenic recombination and diversifying selection contribute to the evolution of downy mildew resistance at the RPP8 locus of Arabidopsis. *Plant Cell* 1998; 10(11):1861-74.

[37]   Aaziz R, Tepfer M. Recombination in RNA viruses and in virus-resistant transgenic plants. *J Gen Virol* 1999; 80(Pt 6):1339-46.

[38]   Kover PX, Caicedo AL. The genetic architecture of disease resistance in plants and the maintenance of recombination by parasites. *Mol Ecol* 2001;10(1):1-16.

[39]   Wicker T, Yahiaoui N, Keller B. Illegitimate recombination is a major evolutionary mechanism for initiating size variation in plant resistance genes. *Plant J* 2007;51(4):631-41.

[40] Sprong D, Janssen HL, Vens C, Begg AC. Resistance of hypoxic cells to ionizing radiation is influenced by homologous recombination status. *Int J Radiat Oncol Biol Phys* 2006;64(2):562-72.

[41] Sale JE. Radiation resistance: resurrection by recombination. *Curr Biol* 2007;17(1):R12-4.

[42] Schierup MH, Hein J. Consequences of recombination on traditional phylogenetic analysis. *Genetics* 2000;156:879-891.

[43] Anisimova M, Nielsen R, Yang Z. Effect of Recombination on the Accuracy of the Likelihood Method for Detecting Positive Selection at Amino Acid Sites. *Genetics* 2003;164(3):1229-1236.

[44] Posada D, Crandall KA. The effect of recombination on the accuracy of phylogeny estimation. *J Mol Evol* 2002;54(3):396-402.

[45] Arnheim N, Calabrese P, Nordborg M. Hot and cold spots of recombination in the human genome: the reason we should find them and how this can be achieved. *Am J Hum Genet* 2003;73(1):5-16.

[46] Crawford DC, Bhangale T, Li N, Hellenthal G, Rieder MJ, Nickerson DA, et al. Evidence for substantial fine-scale variation in recombination rates across the human genome. *Nat Genet* 2004;36(7):700-6.

[47] McVean GA, Myers SR, Hunt S, Deloukas P, Bentley DR, Donnelly P. The fine-scale structure of recombination rate variation in the human genome. *Science* 2004; 304(5670):581-4.

[48] Smith NGC, Fearnhead P. A comparison of three estimators of the population-scaled recombination rate: Accuracy and robustness. *Genetics* 2005;171(4):2051-2062.

[49] Carvajal-Rodriguez A, Crandall KA, Posada D. Recombination Estimation under Complex Evolutionary Models with the Coalescent Composite Likelihood Method. *Mol Biol Evol* 2006;23(4):817-827.

[50] Reed FA, Tishkoff SA. Positive selection can create false hotspots of recombination. *Genetics* 2006;172(3):2011-4.

[51] Perez-Losada M, Porter ML, Tazi L, Crandall KA. New methods for inferring population dynamics from microbial sequences. *Infect Genet Evol* 2007;7(1):24-43.

[52] Weiss KM, Clark AG. Linkage disequilibrium and the mapping of complex human traits. *Trends Genet* 2002;18(1):19-24.

[53] Posada D, Crandall KA, Holmes EC. Recombination in evolutionary genomics. *Annu Rev Genet* 2002;36:75-97.

[54] Rambaut A, Posada D, Crandall KA, Holmes EC. The causes and consequences of HIV evolution. *Nature Review Genetics* 2004;5:52-61.

[55] Stumpf MPH, McVean GAT. Estimating recombination rates from population-genetic data. *Nature Reviews Genetics* 2003;4:959-968.

[56] Hudson RR, Boos DD, Kaplan NL. A statistical test for detecting geographic subdivision. *Molecular Biology and Evolution* 1992;9:138-151.

[57] Hudson RR, Slatkin M, Maddison WP. Estimation of levels of gene flow from DNA sequence data. *Genetics* 1992;132:583-589.

[58] Li HH, Gyllensten UB, Cui XF, Saiki RK, Erlich HA, Arnheim N. Amplification and analysis of DNA sequences in single human sperm and diploid cells. *Nature* 1988; 335(6189):414-7.

[59]   Cui XF, Li HH, Goradia TM, Lange K, Kazazian HH, Jr., Galas D, et al. Single-sperm typing: determination of genetic distance between the G gamma-globin and parathyroid hormone loci by using the polymerase chain reaction and allele-specific oligomers. *Proc Natl Acad Sci U S A* 1989;86(23):9389-93.

[60]   Kauppi L, Jeffreys AJ, Keeney S. Where the crossovers are: recombination distributions in mammals. *Nat Rev Genet* 2004;5(6):413-24.

[61]   Hellenthal G, Stephens M. Insights into recombination from population genetic variation. *Curr Opin Genet Dev* 2006;16(6):565-72.

[62]   Hudson RR. Two-locus sampling distributions and their application. *Genetics* 2001; 159(4):1805-17.

[63]   Posada D, Crandall KA. Evaluation of methods for detecting recombination from DNA sequences: computer simulations. *Proc Natl Acad Sci U S A* 2001;98(24):13757-62.

[64]   Fearnhead P, Donnelly P. Estimating recombination rates from population genetic data. *Genetics* 2001;159(3):1299-318.

[65]   Griffiths RC, Marjoram P. Ancestral inference from samples of DNA sequences with recombination. *Journal of Computational Biology* 1996;3(4):479-502.

[66]   Kuhner MK, Yamato J, Felsenstein J. Maximum likelihood estimation of recombination rates from population data. *Genetics* 2000;156:1393-1401.

[67]   Fearnhead P, Donnelly P. Approximate likelihood methods for estimating local recombination rates. Journal of the Royal Statistical *Society Series B-Statistical Methodology* 2002;64:657-680.

[68]   McVean GAT, Awadalla P, Fearnhead P. A coalescent based-method for detecting and estimating recombination from gene sequences. *Genetics* 2002;160:1231-1241.

[69]   Ptak SE, Voelpel K, Przeworski M. Insights into recombination from patterns of linkage disequilibrium in humans. *Genetics* 2004;167(1):387-97.

[70]   Auton A, McVean G. Recombination rate estimation in the presence of hotspots. *Genome Res* 2007;17(8):1219-27.

[71]   Kosakovsky Pond SL, Posada D, Gravenor MB, Woelk CH, Frost SD. Automated phylogenetic detection of recombination using a genetic algorithm. *Mol Biol Evol* 2006;23(10):1891-901.

[72]   Kosakovsky Pond SL, Posada D, Gravenor MB, Woelk CH, Frost SD. GARD: a genetic algorithm for recombination detection. *Bioinformatics* 2006;22(24):3096-8.

[73]   May CA, Shone AC, Kalaydjieva L, Sajantila A, Jeffreys AJ. Crossover clustering and rapid decay of linkage disequilibrium in the Xp/Yp pseudoautosomal gene SHOX. *Nat Genet* 2002;31(3):272-5.

[74]   Jeffreys AJ, May CA. Intense and highly localized gene conversion activity in human meiotic crossover hot spots. *Nat Genet* 2004;36(2):151-6.

[75]   Myers S, Bottolo L, Freeman C, McVean G, Donnelly P. A fine-scale map of recombination rates and hotspots across the human genome. *Science* 2005; 310(5746):321-4.

[76]   Fearnhead P, Harding RM, Schneider JA, Myers S, Donnelly P. Application of coalescent mediods to reveal fine-scale rate vairiation and recombination hotspots. *Genetics* 2004;167(4):2067-2081.

[77]   Fearnhead P, Smith NGC. A novel method with improved power to detect recombination hotspots from polymorphism data reveals multiple hotspots in human genes. *American Journal of Human Genetics* 2005;77(5):781-794.

[78]   Fearnhead P. SequenceLDhot: detecting recombination hotspots. *Bioinformatics* 2006; 22(24):3061-3066.

[79]   Li J, Zhang MQ, Zhang X. A new method for detecting human recombination hotspots and its applications to the HapMap ENCODE data. *Am J Hum Genet* 2006; 79(4): 628-39.

[80]   Fearnhead P. Consistency of estimators of the population-scaled recombination rate. *Theoretical Population Biology* 2003;64(1):67-79.

[81]   Li N, Stephens M. Modeling linkage disequilibrium and identifying recombination hotspots using single-nucleotide polymorphism data. *Genetics* 2003;165(4):2213-33.

[82]   Percival DB, Walden AT. Wavelet methods for time series analysis. Cambridge: *Cambridge University Press*; 2000.

[83]   Myers S, Spencer CC, Auton A, Bottolo L, Freeman C, Donnelly P, et al. The distribution and causes of meiotic recombination in the human genome. *Biochem Soc Trans* 2006;34(Pt 4):526-30.

[84]   McVean G. The structure of linkage disequilibrium around a selective sweep. *Genetics* 2007;175(3):1395-406.

[85]   Clark VJ, Ptak SE, Tiemann I, Qian Y, Coop G, Stone AC, et al. Combining sperm typing and linkage disequilibrium analyses reveals differences in selective pressures or recombination rates across human populations. *Genetics* 2007;175(2):795-804.

[86]   Vander Molen J, Frisse LM, Fullerton SM, Qian Y, Del Bosque-Plata L, Hudson RR, et al. Population genetics of CAPN10 and GPR35: implications for the evolution of type 2 diabetes variants. *Am J Hum Genet* 2005;76(4):548-60.

[87]   Charlesworth D. Balancing selection and its effects on sequences in nearby genome regions. *PLoS Genet* 2006;2(4):e64.

[88]   Takuno S, Fujimoto R, Sugimura T, Sato K, Okamoto S, Zhang SL, et al. Effects of Recombination on Hitchhiking Diversity in the Brassica Self-incompatibility *Locus Complex. Genetics* 2007;177(2):949-58.

[89]   McVean GAT. A genealogical interpretation of linkage disequilibrium. *Genetics* 2002;162:987-991.

[90]   Glemin S, Bazin E, Charlesworth D. Impact of mating systems on patterns of sequence polymorphism in flowering plants. *Proc Biol Sci* 2006;273(1604):3011-9.

[91]   Ardlie KG, Kruglyak L, Seielstad M. Patterns of linkage disequilibrium in the human genome. *Nat Rev Genet* 2002;3(4):299-309.

[92]   Wall JD, Andolfatto P, Przeworski M. Testing models of selection and demography in Drosophila simulans. *Genetics* 2002;162(1):203-16.

[93]   Nordborg M, Charlesworth B, Charlesworht D. The effect of recombination on background selection. Genet. *Res. Camb.* 1996;67:159-174.

[94]   Shriner D, Nickle DC, Jensen MA, Mullins JI. Potential impact of recombination on sitewise approaches for detecting positive natural selection. *Genetical Research* 2003;81:115-121.

[95]   Chen Y, Marsh BJ, Stephan W. Joint effects of natural selection and recombination on gene flow between Drosophila ananassae populations. *Genetics* 2000;155(3):1185-94.

[96]   Marjoram P, Tavare S. Modern computational approaches for analysing molecular genetic variation data. *Nat Rev Genet* 2006;7(10):759-70.

[97]   Fearnhead P. Ancestral processes for non-neutral models of complex diseases. *Theoretical Population Biology* 2003;63(2):115-130.

[98]   Krone SM, Neuhauser C. Ancestral processes with selection. *Theoretical Population Biology* 1997;51(3):210-237.

[99]   Fearnhead P. The stationary distribution of allele frequencies when selection acts at unlinked loci. *Theoretical Population Biology* 2006;70(3):376-386.

[100] Goldman N, Yang Z. A codon-based model of nucleotide substitution for protein-coding DNA sequences. *Molecular Biology and Evolution* 1994;11:725-736.

[101] Muse SV, Gaut BS. A likelihood approach for comparing synonymous and nonsynonymous nucleotide substitution rates, with application to the chloroplast genome. *Molecular Biology and Evolution* 1994;11(5):715-724.

[102] Scheffler K, Martin DP, Seoighe C. Robust inference of positive selection from recombining coding sequences. *Bioinformatics* 2006;22(20):2493-9.

[103] Wilson DJ, McVean G. Estimating diversifying selection and functional constraint in the presence of recombination. *Genetics* 2006;172(3):1411-25.

[104] Green PJ. Reversible Jump MCMC Computation and Bayesian Model Determination. *Biometrika* 1995;92(4):711-732.

[105] Holland JH. Adaptation in Natural and Artificial Systems: *An Introductory Analysis with Applications to Biology, Control and Artificial Intelligence*; 1992.

[106] Mitchell AM. An *Introduction to Genetic Algorithms*; 1998.

[107] Michalewicz Z. Genetic algorithms + data structures = evolution programs. 3rd, rev. and extended ed. Berlin: *Springer*; 1996.

[108] Lewis PO. A genetic algorithm for maximum-likelihood phylogeny inference using nucleotide sequence data. *Molecular Biology and Evolution* 1998;15(3):277-283.

[109] Pond SLK, Frost SDW. A genetic algorithm approach to detecting lineage-specific variation in selection pressure. *Molecular Biology and Evolution* 2005;22(3):478-485.

[110] Kosakovsky Pond SL, Mannino FV, Gravenor MB, Muse SV, Frost SD. Evolutionary model selection with a genetic algorithm: a case study using stem RNA. *Mol Biol Evol* 2007;24(1):159-70.

[111] Weiss G, von Haeseler A. Inference of population history using a likelihood approach. *Genetics* 1998;149(3):1539-46.

[112] Beaumont MA, Zhang W, Balding DJ. Approximate Bayesian computation in population genetics. *Genetics* 2002;162(4):2025-35.

[113] Worobey M, Holmes EC. Evolutionary aspects of recombination in RNA viruses. *Journal of General Virology* 1999;80:2535-2543.

[114] Simmonds P. Recombination and selection in the evolution of picornaviruses and other Mammalian positive-stranded RNA viruses. *J Virol* 2006;80(22):11124-40.

[115] Cohen E, Kessler DA, Levine H. Recombination dramatically speeds up evolution of finite populations. *Phys Rev Lett* 2005;94(9):098102.

[116] Cohen E, Kessler DA, Levine H. Analytic approach to the evolutionary effects of genetic exchange. *Phys Rev E Stat Nonlin Soft Matter Phys* 2006;73(1 Pt 2):016113.

[117] Marais G, Mouchiroud D, Duret L. Does recombination improve selection on codon usage? Lessons from nematode and fly complete genomes. *PNAS* 2001;98(10):5688-5692.

[118] de Silva E, Kelley LA, Stumpf MP. The extent and importance of intragenic recombination. *Hum Genomics* 2004;1(6):410-20.

[119] Drummond DA, Silberg JJ, Meyer MM, Wilke CO, Arnold FH. On the conservative nature of intragenic recombination. *Proc Natl Acad Sci U S A* 2005;102(15):5380-5.

[120] Voigt CA, Martinez C, Wang ZG, Mayo SL, Arnold FH. Protein building blocks preserved by recombination. *Nat Struct Biol* 2002;9(7):553-8.

[121] Gu Z, Gao Q, Faust EA, Wainberg MA. Possible involvement of cell fusion and viral recombination in generation of human immunodeficiency virus variants that display dual resistance to AZT and 3TC. *Journal of General Virology* 1995;76(10):2601-2605.

[122] Bocharov G, Ford NJ, Edwards J, Breinig T, Wain-Hobson S, Meyerhans A. A genetic-algorithm approach to simulating human immunodeficiency virus evolution reveals the strong impact of multiply infected cells and recombination. *Journal of General Virology* 2005;86:3109-3118.

[123] Rouzine IM, Coffin JM. Evolution of human immunodeficiency virus under selection and weak recombination. *Genetics* 2005;170(1):7-18.

[124] Bretscher MT, Althaus CL, Muller V, Bonhoeffer S. Recombination in HIV and the evolution of drug resistance: for better or for worse? *Bioessays* 2004;26(2):180-8.

[125] Althaus CL, Bonhoeffer S. Stochastic interplay between mutation and recombination during the acquisition of drug resistance mutations in human immunodeficiency virus type 1. *Journal of Virology* 2005;79(21):13572-13578.

[126] Fraser C. HIV recombination: what is the impact on antiretroviral therapy? *J R Soc Interface* 2005;2(5):489-503.

[127] Barton NH, Otto SP. Evolution of recombination due to random dirift. *Genetics* 2005;169(4):2353-2370.

[128] Roze D, Barton NH. The Hill-Robertson effect and the evolution of recombination. *Genetics* 2006;173(3):1793-811.

[129] McVean GA, Charlesworth B. The effects of Hill-Robertson interference between weakly selected mutations on patterns of molecular evolution and variation. *Genetics* 2000;155(2):929-44.

[130] Dixit NM, Perelson AS. HIV dynamics with multiple infections of target cells. *Proc Natl Acad Sci U S A* 2005;102(23):8198-203.

[131] Chen BK, Gandhi RT, Baltimore D. CD4 down-modulation during infection of human T cells with human immunodeficiency virus type 1 involves independent activities of vpu, env, and nef. *J Virol* 1996;70(9):6044-53.

[132] Lama J. The physiological relevance of CD4 receptor down-modulation during HIV infection. *Curr HIV Res* 2003;1(2):167-84.

[133] Piguet V, Schwartz O, Le Gall S, Trono D. The downregulation of CD4 and MHC-I by primate lentiviruses: a paradigm for the modulation of cell surface receptors. *Immunol Rev* 1999;168:51-63.

[134] Jung A, Maier R, Vartanian JP, Bocharov G, Jung V, Fischer U, et al. Multiply infected spleen cells in HIV patients. *Nature* 2002;418(6894):144.

[135] Levy DN, Aldrovandi GM, Kutsch O, Shaw GM. Dynamics of HIV-1 recombination in its natural target cells. *Proc Natl Acad Sci U S A* 2004;101(12):4204-9.

[136] Dang Q, Chen J, Unutmaz D, Coffin JM, Pathak VK, Powell D, et al. Nonrandom HIV-1 infection and double infection via direct and cell-mediated pathways. *Proc Natl Acad Sci U S A* 2004;101(2):632-7.

[137] Chen J, Dang Q, Unutmaz D, Pathak VK, Maldarelli F, Powell D, et al. Mechanisms of nonrandom human immunodeficiency virus type 1 infection and double infection: preference in virus entry is important but is not the sole factor. *J Virol* 2005;79(7): 4140-9.

[138] Charpentier C, Nora T, Tenaillon O, Clavel F, Hance AJ. Extensive recombination among human immunodeficiency virus type 1 quasispecies makes an important contribution to viral diversity in individual patients. *J Virol* 2006;80(5):2472-82.

[139] Mild M, Esbjornsson J, Fenyo EM, Medstrand P. Frequent intrapatient recombination between human immunodeficiency virus type 1 R5 and X4 envelopes: implications for coreceptor switch. *J Virol* 2007;81(7):3369-76.

[140] Mansky LM, Temin HM. Lower in vivo mutation rate of human immunodeficiency virus type 1 than that predicted from the fidelity of purified reverse transcriptase. *J Virol* 1995;69(8):5087-94.

[141] Novick A, Szilard L. Virus Strains of Identical Phenotype but Different Genotype. *Science* 1951;113(2924):34-35.

[142] Brenner S. Genetic Control and Phenotypic Mixing of the Adsorption Cofactor Requirement in Bacteriophage-T2 and Bacteriophage-T4. *Virology* 1957;3(3):560-574.

[143] Clavel F, Hance AJ. HIV drug resistance. N Engl *J Med* 2004;350(10):1023-35.

[144] Palma AC, Araujo F, Duque V, Borges F, Paixao MT, Camacho R. Molecular epidemiology and prevalence of drug resistance-associated mutations in newly diagnosed HIV-1 patients in Portugal. *Infect Genet Evol* 2007;7(3):391-8.

[145] Liu Y, Mullins JI, Mittler JE. Waiting times for the appearance of cytotoxic T-lymphocyte escape mutants in chronic HIV-1 infection. *Virology* 2006;347(1):140-6.

[146] Rosenberg NA, Nordborg M. Genealogical trees, coalescent theory and the analysis of genetic polymorphisms. *Nat Rev Genet* 2002;3(5):380-90.

[147] Fu Y-X, Li W-H. Coalescing into the 21st century: An overview and prospects of coalescent theory. *Theoretical Population Biology* 1999;56:1-10.

[148] Bahlo M, Griffiths RC. Coalescence time for two genes from a subdivided population. *Journal of Mathematical Biology* 2001;43(5):397-410.

[149] Bahlo M, Griffiths RC. Inference from gene trees in a subdivided population. *Theoretical Population Biology* 2000;57(2):79-95.

[150] Beerli P, Felsenstein J. Maximum likelihood estimation of a migration matrix and efective population sizes in *n* subpopulations by using a coalescent approach. *Proceedings of the National Academy of Sciences, U.S.*A. 2001;98(8):4563-4568.

[151] Notohara M. The coalescent and the genealogical process in geographically structured population. *J. Math. Biol.* 1990;29:59-75.

[152] Wilkinson-Herbots HM. Genealogy and subpopulation differentiation under various models of population structure. Journal of Mathematical Biology 1998;37(6):535-585.

[153] Griffiths RC, Tavare S. Sampling theory for neutral alleles in a varying environment. *Philosophical Transactions of the Royal Society of London, Series B* 1994;344:403-410.

[154] Mohle M, Sagitov S. A classification of coalescent processes for haploid exchangeable population models. *Annals of Probability* 2001;29(4):1547-1562.

[155] Tajima F. The effect of change in population size on DNA polymorphism. *Genetics* 1989;123:597-601.

[156] Hey J, Wakeley J. A coalescent estimator of the population recombination rate. *Genetics* 1997;145:833-846.

[157] Hudson RR, Kaplan NL. The coalescent process in models with selection and recombination. *Genetics* 1988;120:831-840.

[158] Kaplan NL, Darden T, Hudson RR. The coalescent process in models with selection. *Genetics* 1988;120:819-829.

[159] Neuhauser C, Krone SM. The genealogy of samples in models with selection. *Genetics* 1997;145:519-534.

[160] Donnelly P, Nordborg M, Joyce P. Likelihoods and simulation methods for a class of nonneutral population genetics models. *Genetics* 2001;159(2):853-867.

[161] Barton NH, Etheridge AM, Sturm AK. Coalescence in a random background. *Annals of Applied Probability* 2004;14(2):754-785.

[162] Fearnhead P. Perfect simulation from nonneutral population genetic models: Variable population size and population subdivision. *Genetics* 2006;174(3):1397-1406.

[163] Calafell F, Grigorenko EL, Chikanian AA, Kidd KK. Haplotype evolution and linkage disequilibrium: A simulation study. Hum Hered 2001;51(1-2):85-96.

[164] Peng B, Amos CI, Kimmel M. Forward-Time Simulations of Human Populations with Complex Diseases. *PLoS Genet* 2007;3(3):e47.

[165] Wakeley J. The limits of theoretical population genetics. Genetics 2005;169(1):1-7.

[166] Balloux F. EASYPOP (version 1.7): a computer program for population genetics simulations. *J Hered* 2001;92(3):301-2.

[167] Hudson RR. Generating samples under a Wright-Fisher neutral model of genetic variation. *Bioinformatics* 2002;18(2):337-338.

[168] Spencer CC, Coop G. SelSim: a program to simulate population genetic data with natural selection and recombination. *Bioinformatics* 2004;20(18):3673-5.

[169] Peng B, Kimmel M. simuPOP: a forward-time population genetics simulation environment. *Bioinformatics* 2005;21(18):3686-7.

[170] Schaffner SF, Foo C, Gabriel S, Reich D, Daly MJ, Altshuler D. Calibrating a coalescent simulation of human genome sequence variation. *Genome Res* 2005;15(11): 1576-83.

[171] Hellenthal G, Stephens M. msHOT: modifying Hudson's ms simulator to incorporate crossover and gene conversion hotspots. *Bioinformatics* 2007;23(4):520-1.

[172] Liang L, Zollner S, Abecasis GR. GENOME: a rapid coalescent-based whole genome simulator. *Bioinformatics* 2007;23(12):1565-7.

[173] Arenas M, Posada D. Recodon: Coalescent simulation of coding DNA sequences with recombination, migration and demography. *BMC Bioinformatics* 2007;8(1):458.

[174] Hoggart CJ, Chadeau-Hyam M, Clark TG, Lampariello R, Whittaker JC, De Iorio M, et al. Sequence-level population simulations over large genomic regions. *Genetics* 2007; 177(3): 1725-31.

[175] Carvajal-Rodríguez A. GenomePop: A software to simulate the evolution of genomes and populations. 2008 [cited; Available from: http://webs.uvigo.es/acraaj/ GenomePop.htm

[176] Rodríguez F, Oliver JF, Marín A, Medina JR. The general stochastic model of nucleotide substitution. *Journal of Theoretical Biology* 1990;142:485-501.

[177] Yang Z, Balding D, Bishop M, Cannings. Adaptive Molecular Evolution. *In: Handbook of Statistical Genetics:* Wiley J. and Sons Ltd.; 2003.

[178] Kimura M. *The Neutral Theory of Molecular Evolution.* Cambridge: Cambridge University Press; 1983.

[179] Karlin S, Taylor HM. *A second course in stochastic processes.* New York: Academic Press; 1981.

In: Genetic Recombination Research Progress
Editor: Jacob H. Schulz, pp. 229-246

ISBN: 978-1-60456-482-2
© 2008 Nova Science Publishers, Inc.

Chapter 8

A Model for the Chiasma Process Allowing Detection of Differing Location Distributions for Single and Multiple Chiasmata

Friedrich Teuscher, Gerd Nürnberg and Volker Guiard†

Research Unit Genetics and Biometry, Research Institute for the Biology of Farm
Animals FBN, W.-Stahl-Allee 2, 18196 Dummerstorf, Germany

Abstract

Direct observations of chiasmata showed that there were chromosome intervals where only single chiasmata appeared and other intervals where rather chiasmata appeared which were accompanied by other chiasmata elsewhere. Particularly, double chiasmata were involved in both chromosome ends, but single chiasmata only in one end. There is no model of a chiasma process developed so far which is appropriate to such events. Therefore, a new model was developed here to infer the phenomenon. It takes into account suppression interference and allows the determination of the chiasma number distribution and the distribution of the chiasma locations for each number of appearing chiasmata. It was shown with an example that the novel model explains the chiasma formation process much better than other models.

1. Introduction

In improving models for the chiasma formation process, the aims are to obtain more precise estimates of genetic distances and to gain a better understanding of the recombination process. The first model was created by Haldane (1919) and has been improved by several authors (reviewed in Karlin and Liberman, 1994 and McPeek and Speed, 1995). Ideally, the sequence of models of the chiasma formation process should converge to the real chiasma formation process.

Before the generation of the new model is motivated and derived, the general task and principle of models of chiasma processes is described. In diploids, chiasmata occur at the four-strand stage of meiosis. The four-strand bundle consists of two identical paternal and two identical maternal chromatids. Identical chromatids are called sister strands and different chromatids are called non-sister strands. A chiasma represents an event where two non-sister strands break and reunite. Afterwards, both strands consist of a maternal and a paternal segment. After meiosis, where several chiasmata may appear, one of the four resulting gametes is chosen for inheritance. If the gamete was involved in one or several chiasmata, then the points reflecting these events are called crossovers. Although the terms chiasma and crossover describe essentially the same event, the term chiasma is used for the four-strand stage of meiosis, while the term crossover is used for gametes, here.

It has been proven that chiasma events are not independent. Genetic interference describes how a chiasma is affected by its neighboring chiasma(ta). Goldgar and Fain (1988) described the three components of interference in this way:

i.   Non-(complete)-randomness in the appearing number of chiasmata. The non interference model applies if the chiasma numbers are Poisson distributed. All other count distributions yield deviations from non interference.

ii.   Non-(complete)-randomness in the chiasma locations on the four-strand bundle. The suppression of nearby chiasmata has been modeled by non-uniformly distributed locations and by non-exponential distributed inter-chiasma distances with renewal point processes.

iii. Chromatid interference: The strands actually involved in a chiasma depend in some way on those strands being involved in a neighboring chiasma.

The creation of a model of a chiasma process involves defining assumptions for these three components. For the first and second component, there are two classes of models of particular interest; the count location models and the stationary renewal models.

The count-location models of Karlin and Liberman (1979) and Risch and Lange (1979) focus on the first component of genetic interference by defining certain chiasma number distributions. For the second component, a uniform distribution of the chiasma locations was assumed. For the third component, no chromatid interference (NCI) was assumed. Such count-location models are abbreviated with CL1 here.

Stationary renewal models (cf. McPeek and Speed, 1995), like the $\chi^2(m)$-model of recombination of Foss et al. (1993) and Zhao et al. (1995), are models for the first two components of interference, while for the third component NCI was assumed. They reflect the amount of interference by the appropriate choice of the distribution of the inter-chiasma distance. While the expected inter-chiasma distance on the genetic scale is consistently 0.5, the variance of it may vary. If the variance is zero, there is complete interference, and with increasing variance interference decreases. In many applications the renewal models behaved better than alternative models. Foss et al. (1993) and Zhao et al. (1995) assumed that the $\chi^2(m)$-model of recombination has a biological background: adjacent crossovers should be separated by a certain number (depending on the degree of interference) of non-crossover events. However recent investigations of Martini et al. (2006) showed that this background does not always applies. Nevertheless the $\chi^2(m)$-model can be considered as a well behaving and elegant statistical model. It was of considerable value that Armstrong et al. (2006) incorporated it into widely used mapping software.

For the third component of interference, i.e. for chromatid interference, Zhao and Speed (1998) and Teuscher et al. (2000) developed models. Principally, these models can be combined with each model for the first two components. However, commonly NCI is assumed and therefore the same was done in this article.

Browning (2000) investigated the relationship between CL1 models and stationary renewal models and concluded that a model like that of Goldgar and Fain (1988) could be ideal, if the drawbacks discussed by Zhao et al. (1995) were removed. Such a model, put on the four-strand bundle instead on the gametes, is a count-location model which considers the first and second component of interference separately by allowing appropriate distributions for chiasma numbers and chiasma locations. Such a count-location model is abbreviated with CL2, here.

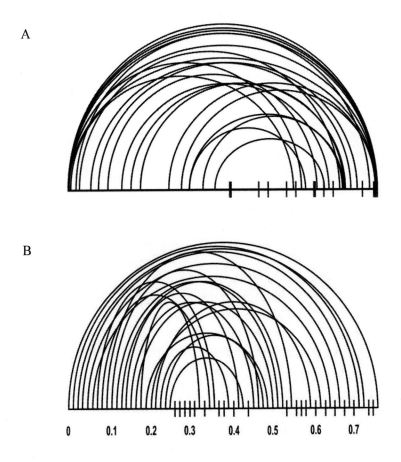

Figure 1. Typical chiasma location distribution of a mouse chromosome. Single chiasmata are represented by vertical bars crossing the axis (A: physical scale of the bivalent, B: genetic scale in Morgan). Chiasmata in dichiasmate bivalents are joined by an arc.

Such a new CL2 model is not only desirable from theoretical reasons, but also from biological: There are spacial properties of the chiasma formation process which have not been modeled so far, but which are obvious in chiasma distributions observed by Povey et al. (1992), Lawrie, Tease, and Hultén (1995), and Hultén, Tease, and Lawrie (1995). Figure 1A shows a typical distribution of single and double chiasmata on the physical scale.

On the physical scale there may be hot spots, e.g. chiasmata often appeared at the right end, and intervals, where no chiasma appeared at all. From the 43 observed meioses of the example, there were 21 with a single chiasma and 22 with double chiasmata, yielding a genetic length of 0.756 Morgans (M), since the expected number of chiasmata is $1 \times 21/43 + 2 \times 22/43 = 1.512$ and the expected number of crossovers is half of it. A rough transformation of the chiasma locations onto the genetic scale, where an interval of length x M means that x is the expected number of crossovers, led to Figure 1B. The mentioned unmodeled properties of the chiasma formation process can be seen: There are intervals where only single chiasmata appeared, and other intervals where chiasmata appeared, which were accompanied by another chiasma elsewhere. Particularly, only double chiasmata were involved in the outermost left interval of length 0.1 M, while single and double chiasmata were equally involved in the outermost right interval of the same length. With common recombination models such findings cannot be handled. For example, with CL1 models single chiasmata were assumed to be uniformly distributed on the genetic scale. Also, it is known for stationary renewal chiasma formation models that it does not make any difference from which side the chiasma formation process starts, i.e. the locations of single and double chiasmata are assumed to be symmetric. If, e.g, the complete interference process would apply (i.e. the distance between two chiasmata is exactly 0.5 M), the single chiasmata would all appear in the interval (0.256, 0.5). From the double chiasmata, the left one would appear in the interval (0, 0.256) and the right one in the interval (0.5, 0.756). The hypothesis is, that this is not the case in the example given with Figure 1 and that a stationary renewal process with less interference would also not fit. So the aim is to establish a model which allows to test, whether the phenomenon appears by chance or if it is significant.

The problem of differing location distributions depending on the number of appearing chiasmata demands an explicit model for the chiasma count distribution. Therefore it can be concluded, that a count-location model would be appropriate, but a stationary renewal model would be not. In order to regard the proposal of Browning (2000), we did not simply use a CL1 model, but rather we established a novel CL2 model, picking up ideas of Goldgar and Fain (1988).

The study is accompanied by the analysis of the $\chi^2(m)$-model of recombination and the method of Yu and Feingold (2002) to determine the spatial distribution of chiasmata from recombination data. The latter method is not a model of high abstraction: it contains a large number of parameters, just the number of equations to be fitted, and allows no general inferences on the chiasma formation process itself. It bases on a simplification of the theoretically possible chiasma formation processes because it assumes at most one chiasma per marker interval and NCI.

2. The Method

2.1. The CL2 Model – a Count-Location Model of a New Kind

We consider r intervals, spanned by $r+1$ markers on a chromosome. The distribution of the chiasma numbers is defined to be $\{c_n\}$, $0 \le n \le N$ (i.e. c_n is the probability that n chiasmata appear between the bounding markers). N is the maximum number of chiasmata, i.e. it is an unobserved number underlying the particular data set. Let $f_n(d_1, \cdots, d_n)$ be the probability density function of the locations d_1, d_2, ..., d_n, $0 \le d_1 \le \cdots \le d_n \le 1$, of n chiasmata on the model scale D. The relation of the scale D to the genetic scale x will be considered later.

For the case of one appearing chiasma we assume its location to be uniformly distributed (i.e. $f_1(d_1) = 1$ for $0 \le d_1 \le 1$). For $n \ge 2$ appearing chiasmata, the joint density function of the locations is defined by

$$f_n(d_1, \cdots, d_n) = (k+1) \cdots (k+n) \left\{ (d_n - d_{n-1})^k + \cdots + (d_2 - d_1)^k \right\} / (n-1), \tag{2.1}$$

generalizing an idea of Goldgar and Fain (1988). The interference parameter $k \ge 0$ describes the suppression interference. If k equals zero, there is no suppression interference since (2.1) is the density of the multivariate uniform distribution. The larger the value of k, the larger the suppression of nearby chiasmata. The locations of the ordered markers must then be defined on the model scale. The normal way would be to assume a certain position m_κ for the κ-th marker with $0 = m_0 \le \cdots \le m_r = 1$. We wish to reflect, however, that multiple chiasmata may have a distribution other than single chiasmata. In an extreme example imagine a marker interval, the κ-th interval, where a single chiasma never appears, but where a chiasma may appear accompanied with another chiasma elsewhere. This would demand $m_{\kappa-1} = m_\kappa$ for $n = 1$ chiasmata, since otherwise the probability of a chiasma in that interval would be $\int_{m_{\kappa-1}}^{m_\kappa} f_1(d_1)\,dd_1 = m_\kappa - m_{\kappa-1} > 0$. On the other hand, $m_{\kappa-1} < m_\kappa$ must apply for $n = 2$ chiasmata, since the probability that at least one of the two chiasmata is appearing in interval κ, which is

$$\int_0^{m_{\kappa-1}} \int_{m_{\kappa-1}}^{m_\kappa} f_2(d_1, d_2)\,dd_1 dd_2 + \int_{m_{\kappa-1}}^{m_\kappa} \int_{m_{\kappa-1}}^{d_2} f_2(d_1, d_2)\,dd_1 dd_2 + \int_{m_\kappa}^{1} \int_{m_{\kappa-1}}^{m_\kappa} f_2(d_1, d_2)\,dd_1 dd_2 ,$$

is zero for $m_{\kappa-1} = m_\kappa$. Generally, the probability density functions (2.1) are not intended to assign a zero probability to a non-empty marker interval (in a sense that the markers do not agree on the genetic or physical scale). Therefore, we have to allow for vanishing marker intervals for a certain n. We indicate that the modelled marker locations depend on the actual appearing chiasma number n by defining the locations on scale D to be $m_{\kappa, n}$ with

$0 = m_{0,n} \leq \cdots \leq m_{r,n} = 1$. Strictly speaking we then have N separate scales D_n, dependent on the actual number n ($n \leq N$) of appearing chiasmata. The probability density functions (2.1) are assigned to these scales.

2.2. The Probability of Recombination Patterns

Chiasmata and crossovers are difficult to observe directly but it can be observed whether or not a recombination appeared between markers. Thus, the aim of models for the chiasma and crossover formation process is to fit an observed distribution of recombination patterns. In order to do this, one has to determine the theoretical probabilities for all possible recombination patterns. Recombination pattern $j = (j_1, \cdots, j_r)$ describes the appearance or absence of recombinations on a randomly chosen gamete, where $j_\kappa = 1$ indicates a recombination and $j_\kappa = 0$ indicates absence of recombination in the κ-th interval. A certain recombination pattern may result from an appropriate class of crossover patterns. Specifically, a recombination is the result of an odd number of crossovers. Crossover patterns may result from certain classes of chiasma patterns. Chiasma pattern $i = (i_1, \cdots, i_r)$ describes the distribution of the chiasmata on the r intervals, i_κ defining the number of chiasmata in the κ–th interval. A recombination pattern $j = (j_1, \cdots, j_r)$ may be generated by a chiasma pattern i, if $i \geq j$ holds (i.e. if $i_\kappa \geq j_\kappa$ obeys for all κ). Weinstein (1936) found the relation

$$\gamma(j) = \sum_{i \geq j} (1/2)^{\#\kappa : i_\kappa > 0} \tau(i) \tag{2.2}$$

between the probabilities $\tau(i)$ of chiasma patterns i, and the probability $\gamma(j)$ of recombination pattern j, valid under NCI. Since a certain chiasma number pattern i arises if $n = \sum_{\kappa=1}^{r} i_\kappa$ chiasmata appear and if these n chiasmata have the demanded distribution, $\tau(i)$ equals $c_n J_i$, with $J_i = P(i) = P\left(i_1 = \#\text{chiasma} \in (0, m_{1,n}) \wedge \cdots \wedge i_r = \#\text{chiasma} \in (m_{r-1,n}, 1)\right)$.

Clearly, for $n = 0$, i.e. for $i = (0, \cdots, 0)$, one obtains $J_i = 1$. For $n \geq 1$, the probability J_i can be obtained by integrating expression (2.1) over all possible locations of the n chiasmata, defined with vector i. It can be evaluated interval-wise. In the first interval, there are i_1 chiasmata. For $i_1 > 1$, the bounds of the chiasma locations result from $0 = m_{0,n} \leq d_1 \leq \cdots \leq d_{i_1} \leq m_{1,n}$. For $i_1 = 1$, the bounds of the location result from $0 = m_{0,n} \leq d_1 \leq m_{1,n}$, and for $i_1 = 0$, integration is not necessary. Thus for the first interval,

$$
J_{i,1} = \begin{cases}
f_n(d_1, \cdots, d_n) & \text{for } i_1 = 0 \\[2mm]
\displaystyle\int_0^{m_{1,n}} f_n(d_1, \cdots, d_n)\,\mathrm{d}d_1 & \text{for } i_1 = 1 \\[2mm]
\displaystyle\int_0^{m_{1,n}}\int_0^{d_{i_1}}\cdots\int_0^{d_2} f_n(d_1, \cdots, d_n)\,\mathrm{d}d_1 \cdots \mathrm{d}d_{i_1} & \text{for } i_1 > 1
\end{cases}.
$$

is obtained. Integration for the κ-th interval, $\kappa = 2, \cdots, r$, yields

$$
J_{i,\kappa} = \begin{cases}
J_{i,\kappa-1} & \text{for } i_\kappa = 0 \\[2mm]
\displaystyle\int_{m_{\kappa-1,n}}^{m_{\kappa,n}} J_{i,\kappa-1}\,\mathrm{d}d_{n(\kappa)} & \text{for } i_\kappa = 1 \\[2mm]
\displaystyle\int_{m_{\kappa-1,n}}^{m_{\kappa,n}}\int_{m_{\kappa-1,n}}^{d_{n(\kappa)}}\cdots\int_{m_{\kappa-1,n}}^{d_{2+n(\kappa-1)}} J_{i,\kappa-1}\,\mathrm{d}d_{1+n(\kappa-1)} \cdots \mathrm{d}d_{n(\kappa)} & \text{for } i_\kappa > 1
\end{cases}. \qquad (2.3)
$$

where $n(\kappa) = \sum_{v=1}^{\kappa} i_v$ denotes the number of chiasmata left from marker κ. $J_{i,r}$ equals the demanded probability J_i. The probability of recombination pattern j can then be determined by

$$
\gamma(j) = \sum_{i \geq j} (1/2)^{\# \kappa : i_\kappa > 0}\, c_{\sum_{\kappa=1}^{r} i_\kappa}\, J_i, \qquad (2.4)
$$

due to formulas (2.2) and (2.3).

2.3. The Relation Between Model and Genetic Scale

The locations of the markers were $m_{\kappa,n}$, $\kappa = 0, 1, \cdots, r$, $n = 1, \cdots, N$, on the model scales D_n. Hence there are N model distances between two markers. These model distances have a so far unknown correspondence to the genetic distance between the markers. This correspondence will now be derived.

Let the genetic distances between each consecutive marker pair be $x_1, x_2, \cdots, x_r$. The relation between the expected number y of chiasmata on the four-strand bundle and the genetic scale x, defining the expected number of crossovers on a gamete, is $y = 2x$, since from one chiasma, a crossover on a randomly selected gamete results with the probability of 0.5. The genetic distance x_κ between markers $\kappa - 1$ and κ therefore equals half the expected number of chiasmata appearing between these markers. Therefore

$$x_\kappa = \tfrac{1}{2} \sum_{n=1}^{N} \sum_{\mu=1}^{n} \mu\, P\Big(\mu \text{ chiasmata in } \big(m_{\kappa-1,n}, m_{\kappa,n}\big) \wedge (n-\mu) \text{ chiasmata outside}\Big) \qquad \text{applies,}$$

where n is the number of chiasmata appearing between the bounding markers and μ is the number of chiasmata appearing in the κ–th interval. First, let κ define a non-bounding interval. Then, the n chiasmata may be distributed over the three intervals $\big(0, m_{\kappa-1,n}\big)$, $\big(m_{\kappa-1,n}, m_{\kappa,n}\big)$, and $\big(m_{\kappa,n}, 1\big)$. Thus, the r interval case can be considered as a $r = 3$ interval case with marker locations 0, $m_{\kappa-1,n}$, $m_{\kappa,n}$, and 1. With the notations from above,

$$P\Big(\mu \text{ chiasmata in } \big(m_{\kappa-1,n}, m_{\kappa,n}\big) \wedge (n-\mu) \text{ chiasmata outside}\Big) \quad \text{equals} \quad \sum_{v=0}^{n-\mu} c_n\, J_{(v,\mu,n-\mu-v)}\,,$$

where v is the number of chiasmata located left from the considered marker interval. Therefore,

$$x_\kappa = \tfrac{1}{2} \sum_{n=1}^{N} c_n \sum_{\mu=1}^{n} \mu \sum_{v=0}^{n-\mu} J_{(v,\mu,n-\mu-v)}\,, \tag{2.5}$$

is valid for the genetic distance. For the bounding intervals, one obtains

$$x_1 = \tfrac{1}{2} \sum_{n=1}^{N} c_n \sum_{\mu=1}^{n} \mu J_{(\mu,n-\mu)} \quad \text{and} \quad x_r = \tfrac{1}{2} \sum_{n=1}^{N} c_n \sum_{\mu=1}^{n} \mu J_{(n-\mu,\mu)} \quad \text{analogously, with the marker}$$

locations 0, $m_{\kappa,n}$, and 1 and 0, $m_{\kappa-1,n}$, and 1, respectively.

3. Hierarchical Sequence of Models and Testing Tools

From Haldane (1919) it is known that the no interference case applies if the crossover numbers follows a Poisson distribution and if the chiasma locations are independently distributed. This is a special case of the CL1-model obtained by putting $c_n = e^{-2x}(2x)^n/n!$. Here, x is the genetic length of the considered chromosome part.

Generally, a CL1-model with chiasma number distribution $\{c_n\}$ is a special case of a CL2 model, since both coincide if the CL2 model has the same chiasma number distribution $\{c_n\}$ like the CL1 model, and if the chiasma locations are independently distributed. In principle all count distributions are allowed for $\{c_n\}$. The most general case is to define c_n as arbitrary constants and to estimate them under the constraint $\sum_{n=0}^{N} c_n = 1$.

Additional interference from suppression of nearby chiasmata (i.e. non-(complete)-randomness in the chiasma locations), is considered by parameter k of the CL2 model $CL2(\{c_n\}, k)$.

The heterogeneity of the distribution of chiasma locations was incorporated by varying the locations $m_{\kappa,n}$ of marker κ with respect to the different scales D_n depending on the actual number n of chiasmata. This model was named CL2($\{c_n\}$, k, $\{m_{\kappa,n}\}$). Often, the observed number of gametes with the maximum number N of crossovers is small. In this case it cannot be assumed to obtain reliable estimates for the locations $m_{\kappa,N}$. Therefore the CL2($\{c_n\}$, k, $\{m_{\kappa,n}\}$) model was first applied with the constraint $m_{\kappa,N-1} = m_{\kappa,N}$, i.e. the marker location distributions were assumed to agree for N and $N-1$ chiasmata. Afterwards the most general model with $m_{\kappa,N-1} \neq m_{\kappa,N}$ was used.

Thus, there is a hierarchical sequence of models. The most parsimonious model is Haldane's, which is equivalent to CL1($\{c_n\}$=Poisson). The next model is CL1($\{c_n\}$) = CL2($\{c_n\}$, $k = 0$), where the chiasma number distribution $\{c_n\}$ is arbitrary. It is followed by CL2($\{c_n\}$, k), regarding suppression of nearby chiasmata, and CL2($\{c_n\}$, k, $\{m_{\kappa,n}\}$) with $m_{\kappa,N-1} = m_{\kappa,N}$. The least parsimonious model is the general CL2-model CL2($\{c_n\}$, k, $\{m_{\kappa,n}\}$), the model allowing different chiasma location distributions for all numbers n of appearing chiasmata.

For the models with an arbitrary chiasma number distribution $\{c_n\}$, it was useful to know the maximum number N of chiasmata. An estimate $\hat{N}$ for N was obtained with the method of Yu and Feingold (2001), which allows to determine estimates and confidence intervals for chiasma frequencies from recombination data. Since we applied the models with arbitrary $\{c_n\}$ under the constraint $c_n = 0$ for $n > \hat{N}$, each model was nested in the next less parsimonious one from the CL1-model on. Therefore, the asymptotic theory of the likelihood ratio test was applied.

For directly observable multilocus recombination patterns, the criterion of fit was the log-likelihood

$$\ln L(\text{model}) = \max\left[\sum_j n(j) \ln \gamma(j)\right], \tag{3.1}$$

where $n(j)$ is the observed number and $\gamma(j)$ is the theoretical probability of recombination pattern j. The maximum was reached say, for $\{\hat{c}_n\}$, $\hat{k}$, and $\hat{m}_{1,n}, \cdots, \hat{m}_{r,n}$ for $n = 1, \cdots, \hat{N}$.

Following the asymptotic theory of the likelihood ratio test, a model M1 with p_1 parameters was significantly better than a nested model M0 with $p_0 < p_1$ parameters, if $2(\ln L(\text{M1}) - \ln L(\text{M0})) > \chi^2(1-\alpha, p_1 - p_0)$ held.

There might arise the question whether the CL2($\{c_n\}$, k, $\{m_{\kappa,n}\}$)-model is better or worse than the $\chi^2(m)$- model of recombination or the method of Yu and Feingold (2002).

Since these models are not nested they could not be compared by the likelihood ratio test. Therefore, Akaike's Information Criterion $AIC = -2\ln(L) + 2$ no. parameter and the Bayesian Information Criterion $BIC = -2\ln(L) + $ no. parameters $* \ln($no. observations$)$ were applied.

The fit of a model to the data is often analyzed with the χ^2 – test of goodness of fit. However this test preserves the desired type-one error probability only for class sizes greater than five. This demands a union of the rare classes which are the most interesting ones in view of interference. Furthermore it would not be stringent to apply the χ^2 – test of goodness of fit by using maximum likelihood estimates. Therefore, the parametric bootstrap method was used for the goodness of fit test by simulating the fitted models with 1,000 replications. The differences Δ_i between the log-likelihoods of the simulated data and the log-likelihoods of the fitted models were stored. The log-likelihood of the data $\ln L(\text{data})$ was calculated via equation (3.1) by substituting the theoretical gamete probability by the observed fraction, i.e. $\gamma(j) = n(j)/\sum n(j)$. The percentiles of the distribution of Δ_i gave the thresholds for rejecting the null hypothesis that the model is correct. The P-value indicating the goodness of fit of a model was than obtained by determining the fraction of Δ_i which were larger than Δ_d, the difference between the log-likelihood of the real data and the log-likelihood of the fitted model.

The simulation of the models was simultaneously used to determine the 95% confidence intervals of the estimates. A specific feature appeared with the simulation of the general CL2-model. When the number of crossover was smaller than four, the model often did not converge, obviously because of the missing information on $m_{K,4}$. Therefore, the model was fitted under the constraint $m_{K,3} = m_{K,4}$ in these cases, which appeared with 23 percent of the simulated meioses.

4. Example: Drosophila Data

The seven marker recombination data of Weinstein (1936) were used repeatedly for investigations on the phenomenon of interference (e.g. Pascoe and Morton, 1987, Zhao et al., 1995, and Teuscher et al., 2000). Although significant gains in fit had been obtained, the deviation from the data was still considerable. Hence, the method derived here was applied. By applying the approach of Yu and Feingold (2001), the confidence intervals for the chiasma number probabilities were (0.05, 0.07) for c_0, (0.62, 0.66) for c_1, (0.27, 0.30) for c_2, (0.01, 0.02) for c_3, $(10^{-9}, 0.003)$ for c_4, and $(10^{-9}, 0.0004)$ for c_5. With the original interpretation we had to conclude that $\hat{N} = 3$ was the maximum number of chiasmata. However there appeared two quadruples of crossovers, hence we had to admit $\hat{N} = 4$. The expected numbers of the recombination patterns for the sequence of CL1 and CL2-models are shown in Table 1.

All refinements of the models proved to be significant. The significance levels in most cases were less than 10^{-14}. The most general CL2-model was still better than the model under the constraint $m_{\kappa, 3} = m_{\kappa, 4}$.

For the general CL2-model, the estimated lengths $\hat{m}_{\kappa, n} - \hat{m}_{\kappa-1, n}$ of the marker intervals are given in Table 2.

Single chiasmata most often appear in interval 4, followed by intervals 5, 3, 2, 6, and 1. Double chiasmata were most often involved in interval 4, followed by intervals 5, 6, 2, 3, and 1. Triples of chiasmata were most often involved in interval 4, followed by intervals 6, 1, 2, and 5. They rarely appeared in interval 3. Quadruples of chiasmata were most often involved in interval 3, followed by intervals 6, 4, 5, 1 and 2. With exception of intervals 2 and 1, the confidence intervals were large for quadruples.

In the first interval, mainly single, triples, or quadruples of chiasmata were involved, relatively to their absolute appearance. In the second interval, the same fraction of single and double chiasmata was observed, exceeding the fraction of triples and quadruples. In the third interval, single and quadruples of chiasmata dominated and triples were observed least frequently. In the fourth interval, single, double, and quadruples of chiasmata occurred with the same frequency, which was smaller than that of triples. In the fifth interval, double chiasmata primarily appeared, followed by singles and triples. In the sixth interval, the fraction of doubles was larger than the fraction of singles. The fraction of triples and quadruples did not deviate significantly from those of singles or doubles.

AIC and BIC values of the CL2($\{c_n\}$, k, $\{m_{\kappa, n}\}$)- models, the $\chi^2(m)$- model, and the method of Yu and Feingold (2002) are shown in Table 3.

With regard to the AIC-values, the general CL2-model was best, followed by the CL2-model with $m_{\kappa, 3} = m_{\kappa, 4}$, Yu and Feingold's method and the $\chi^2(m)$- model. When judged by BIC-values, which impose a higher penalty to the number of parameters for large sample sizes, CL2 with $m_{\kappa, 3} = m_{\kappa, 4}$ was the best model, followed by the $\chi^2(m)$- model, the general CL2-model and Yu and Feingold's method.

The results of the simulations of the CL2-models, the $\chi^2(m)$- model, and the method of Yu and Feingold are shown in Table 4.

The difference between the log-likelihoods of the simulated data and the fitted models was largest for the $\chi^2(m)$- model, followed by the general CL2-model, the CL2-model with $m_{\kappa, 3} = m_{\kappa, 4}$, and the method of Yu and Feingold. The estimated percentiles had the same order. The differences obtained with the fit to the data of Weinstein (1936) were 100.1, 18.1, 28.3, and 2.3, respectively (the log-likelihood of the Weinstein data was −54,850.1). Comparison with the percentiles showed, that the general CL2-model and the method of Yu and Feingold fitted the data well (P- values larger than 0.05), while the $\chi^2(m)$- model and the CL2-model with $m_{\kappa, 3} = m_{\kappa, 4}$ did not. The P values for the tests on deviations of the model from the data were zero for the $\chi^2(m)$- model, 0.001 for the CL2-model with $m_{\kappa, 3} = m_{\kappa, 4}$, 0.09 for the method of Yu and Feingold, and 0.20 for the general CL2-model. Hence, the general CL2-model yielded the best fit.

Table 1. Observed gamete counts of the data of Weinstein (1936) and counts expected for different models of genetic interference, fitted to the data.

Gamete	Observed	Expected				
		Haldane	CL1($\{c_n\}$)	CL2($\{c_n\}$, k)	CL2($\{c_n\}$, k, $\{m_{\kappa,n}\}$) $m_{\kappa,3} = m_{\kappa,4}$	CL2($\{c_n\}$, k, $\{m_{\kappa,n}\}$)
			$\hat{c}_1 = 0.576$	$\hat{c}_1 = 0.638$	$\hat{c}_1 = 0.630$	$\hat{c}_1 = 0.632$
			$\hat{c}_2 = 0.333$	$\hat{c}_2 = 0.275$	$\hat{c}_2 = 0.274$	$\hat{c}_2 = 0.260$
			$\hat{c}_3 = 0.028$	$\hat{c}_3 = 0.024$	$\hat{c}_3 = 0.031$	$\hat{c}_3 = 0.042$
			$\hat{c}_4 = 0.003$	$\hat{c}_4 = 0.003$	$\hat{c}_4 = 0.004$	$\hat{c}_4 = 0.008$
				$\hat{k} = 1.90$	$\hat{k} = 1.88$	$\hat{k} = 1.89$
(0,0,0,0,0,0)	12,776	14,432.3	12,776.8	12,785.4	12,782.6	12,769.9
(1,0,0,0,0,0)	1,407	1,110.8	1,465.7	1,341.4	1,407.2	1,406.5
(0,1,0,0,0,0)	2,018	1,573.0	2,044.9	1,944.6	2,009.3	2,019.2
(0,0,1,0,0,0)	1,976	1,354.8	1,776.2	1,815.4	1,973.6	1,982.9
(0,0,0,1,0,0)	3,378	2,508.0	3,170.4	3,406.3	3,374.5	3,388.1
(0,0,0,1,1,0)	2,356	1,826.4	2,350.1	2,453.4	2,354.0	2,363.2
(0,0,0,0,0,1)	2,067	1,855.9	2,394.2	2,255.3	2,065.0	2,069.2
(1,1,0,0,0,0)	9	121.1	91.7	11.5	10.3	9.3
(1,0,1,0,0,0)	16	104.3	79.7	26.7	16.3	18.1
(1,0,0,1,0,0)	142	193.0	142.2	152.6	136.4	144.3
(1,0,0,0,1,0)	198	140.6	105.4	202.9	182.4	181.2
(1,0,0,0,0,1)	206	142.8	107.4	227.5	227.1	217.3
(0,1,1,0,0,0)	11	147.7	111.1	21.1	11.2	11.1
(0,1,0,1,0,0)	136	273.4	198.4	160.4	125.9	131.2
(0,1,0,0,1,0)	261	199.1	147.1	247.9	264.2	259.1
(0,1,0,0,0,1)	318	202.3	149.8	295.4	330.7	316.1
(0,0,1,1,0,0)	42	235.4	172.3	78.6	53.1	40.0

Table 1. Continued

Gamete	Observed	Expected				
		Haldane	CL1({c_n})	CL2({c_n}, k)	CL2({c_n}, k, {$m_{k,n}$}) $m_{k,3} = m_{k,4}$	CL2({c_n}, k, {$m_{k,n}$})
(0,0,1,0,1,0)	148	171.5	127.7	163.8	146.4	147.3
(0,0,1,0,0,1)	212	174.2	130.1	218.4	208.3	209.7
(0,0,0,1,1,0)	123	317.4	228.0	118.3	141.8	135.3
(0,0,0,1,0,1)	315	322.5	232.3	208.6	293.0	297.9
(0,0,0,0,1,1)	59	234.9	172.2	37.6	57.5	58.2
(1,1,0,1,0,0)	3	21.0	2.9	1.7	3.2	3.3
(1,1,0,0,1,0)	1	15.3	2.2	2.5	1.1	0.6
(1,1,0,0,0,1)	2	15.6	2.2	2.9	3.1	2.1
(1,0,1,0,1,0)	3	13.2	1.9	2.1	0.7	1.3
(1,0,1,0,0,1)	3	13.4	1.9	2.5	2.1	3.3
(1,0,0,1,1,0)	10	24.4	3.4	4.4	6.3	4.8
(1,0,0,1,0,1)	15	24.8	3.4	4.5	16.5	15.9
(1,0,0,0,1,1)	1	18.1	2.5	3.6	3.1	2.0
(0,1,1,1,0,0)	1	25.7	3.5	1.8	1.0	0.4
(0,1,0,1,1,0)	2	34.6	4.7	5.2	2.8	2.9
(0,1,0,1,0,1)	10	35.2	4.8	5.8	7.7	9.9
(0,1,0,0,1,1)	1	25.6	3.5	4.6	1.4	1.0
(0,0,1,1,0,1)	5	30.3	4.1	4.7	5.2	3.4
(0,0,1,0,1,1)	5	22.0	3.1	3.2	0.9	3.4
(0,0,0,1,1,1)	1	40.8	5.5	3.0	3.3	3.1
(1,1,1,1,0,0)	1	2.0	0.10	0.03	0.03	0.02
(1,1,1,0,0,1)	1	1.5	0.08	0.06	0.03	0.08
ln$L$	-54850.1[1]	-56394.3	-55483.4	-54957.3	-54878.4	-54868.2
P² (degrees of freedom)			0 (3)	0 (1)	0 (10)	0.001 (5)

[1] Ideal case: theoretical and observed distribution agree. ² Significance level for a model to differ from the model left from it, P=0 means P<10^{-14} .

Table 2. Estimated lengths[1] of marker intervals on the model scales D_m in dependence on the number n of chiasmata appeared, together with 95% confidence intervals. Data of Weinstein (1936), analyzed with the CL2($\{c_n\}$, k, $\{m_{\kappa,n}\}$) model.

Number of chiasmata n	Marker interval					
	1	2	3	4	5	6
1	0.097^A	0.147^A	0.176^A	0.289^A	0.178^A	0.113^A
	(0.088, 0.108)	(0.136, 0.158)	(0.165, 0.188)	(0.267, 0.305)	(0.165, 0.190)	(0.102, 0.125)
2	0.069^B	0.140^A	0.124^B	0.252^A	0.249^B	0.165^B
	(0.061, 0.078)	(0.127, 0.155)	(0.106, 0.145)	(0.211, 0.289)	(0.224, 0.275)	(0.152, 0.180)
3	0.088^{AB}	0.064^B	0.001^C	0.685^B	0.041^C	0.120^{AB}
	(0.051, 0.165)	(0.034, 0.112)	(10^{-9}, 0.076)	(0.420, 0.791)	(0.014, 0.096)	(0.073, 0.220)
4	0.056^{AB}	0.022^B	0.524^{ABC}	0.106^{AB}	0.101^{ABC}	0.190^{AB}
	(0.019, 0.194)	(10^{-6}, 0.106)	(10^{-5}, 0.630)	(0.026, 0.734)	(0.063, 0.328)	(0.085, 0.506)

ABC Different capitals in a column indicate significant differences between the lengths. [1] The lengths add to one for each chiasma number n.

Table 3. Comparison of models fitted to the data of Weinstein (1936).

	Data	Model			
	CL2($\{c_n\}$, k, $\{m_{\kappa,n}\}$)	CL2($\{c_n\}$, k, $\{m_{\kappa,n}\}$)	CL2($\{c_n\}$, k, $\{m_{\kappa,n}\}$) $m_{\kappa,3} = m_{\kappa,4}$	$\chi^2(m)$	Yu and Feingold (2002)
Log-likelihood	-54,850.1	-54,868.2	-54,878.4	-54,950.2	-54,852.3
Number of parameters	38	25	20	7	63
AIC	109,776.2	109,786.4	109,796.8	109,914.4	109,830.6
BIC	110,089.6	109,992.6	109,961.8	109,972.1	110,350.2

Table 4. Goodness of fit of three models to the data of Weinstein (1936) determined via parametric bootstrap simulation.

		Model			
		CL2($\{c_n\}$, k, $\{m_{\kappa,n}\}$)	CL2($\{c_n\}$, k, $\{m_{\kappa,n}\}$) $m_{\kappa,3} = m_{\kappa,4}$	$\chi^2(m)$	Yu and Feingold (2002)
Simulation of the model	average of $\ln L(\text{data}) - \ln L(\text{model})$	14.3	13.8	19.1	0.76
	0.95 (0.99) percentile	23.7 (28.6)	20.4 (23.1)	27.0 (30.9)	2.4 (3.4)
Fit to data of Weinstein	$\ln L(\text{model})$	-54,868.2	-54,878.4	-54,950.2	-54,852.3
	$\ln L(\text{data}) - \ln L(\text{model})$	18.1	28.3	100.1	2.2
	P value for deviation from the data	0.20	0.001	0	0.09

5. Discussion

The count-location model derived here is a modification and generalization of the model of Goldgar and Fain (1988) in four respects. (i) It was applied to the four-strand bundle instead of the gametes, (ii) two parameters were saved by considering only the range of the markers and by not extrapolating on the whole chromosome (arm), (iii) the number of chiasmata is now arbitrary, and (iv) the location distribution may differ depending on the number of chiasmata involved in the meiosis.

With the data of Weinstein (1936) we have shown that genetic interference was present. The $CL2(\{c_n\}, k)$ model was better than the $CL1(\{c_n\})$ model, i.e. suppression interference acted. The advantage of the CL2-model is that it enables the consideration of findings from cytological observations on chiasmata. With the $CL2(\{c_n\}, k, \{m_{\kappa, n}\})$ model the particular assumption that chiasmata may prefer different intervals, depending on the actual number of chiasmata appearing at a meiosis, proved to be relevant. This finding may stimulate researchers to figure out the reasons on the level of energetic mechanisms.

Although the tests of the sequence of CL2-models were sufficient to prove the relevance of the different sources of interference, we compared the best CL2-models (the most general CL2-model and that under the constraint $m_{\kappa, 3} = m_{\kappa, 4}$) with the $\chi^2(m)$ – model and the method of Yu and Feingold (2002). The three criterions AIC, BIC, and goodness of fit gave different answers on the question which model was best. Summarizing the different criterions, the most general CL2-model behaved best. Thus, the newly developed CL2-model turned out to be a useful tool to unravel another mystery of meioses.

The method of Yu and Feingold (2002) is based on the estimation of $2^m - 1$ parameters, just the number of gamete types to be fitted. From this point of view one would expect a perfect fit of the data. This was not the case in the example. Since the only assumptions of the method are NCI and the formation of at most one chiasma per marker interval, the imperfect fit was caused by no complete interference within the intervals or by chromatid interference.

The CL2-model allows the estimation of the distribution of chiasma numbers on the four-strand bundle and is thus an alternative to the methods of Yu and Feingold (2001, 2002). The frequency of achiasmatic meioses can be estimated with $\hat{c}_0 = 1 - \sum_{i=1}^{\hat{N}} \hat{c}_i$. The interpretation of $\hat{c}_0$ is particularly useful if the bounding markers are located at the ends of the chromosome, what was not the case in the example. From the model assumptions it is clear that for tight marker distances, the methods of Yu and Feingold (2001, 2002) are preferable, while for medium or large marker distances, the CL2-model is the method of choice. With the general CL2-model applied to the example we obtained $\hat{c}_0 = 0.058$, $\hat{c}_1 = 0.632$, $\hat{c}_2 = 0.260$, $\hat{c}_3 = 0.042$, and $\hat{c}_4 = 0.008$, with the method of Yu and Feingold (2001) we obtained $\hat{c}_0 = 0.060$, $\hat{c}_1 = 0.637$, $\hat{c}_2 = 0.286$, $\hat{c}_3 = 0.016$, and $\hat{c}_4 = 0.001$, and with the method of Yu and Feingold (2002) $\hat{c}_0 = 0.060$, $\hat{c}_1 = 0.638$, $\hat{c}_2 = 0.284$, $\hat{c}_3 = 0.017$, and $\hat{c}_4 = 0.0004$. Note that the differences were small.

It must been noticed that the $CL2(\{c_n\}, k, \{m_{\kappa, n}\})$-model is rather sophisticated and contains a lot of parameters, although with biological meaning. Therefore we suggest to use it only in such cases, where more elegant models like the $\chi^2(m)$ – model of recombination failed to fit the data. This was the case with the chosen example, but also with the data of Morgan et al. (1935), Blank et al. (1988), and Lien et al. (2000). We mention that the genetic distances estimated with the CL2($\{c_n\}$, k, $\{m_{\kappa,n}\}$) model were very similar to those estimated with the $\chi^2(m)$ – model. Thus for mapping purposes, the use of the $\chi^2(m)$ – model seems to be sufficient.

In this study, it was assumed that suppression interference between chiasmata acted constantly, independently of the actual number of chiasmata and of the site of the chiasmata. In general this may not necessarily be the case. To generalize the model, one can assign interference parameter $k_{\kappa,n}$ to markers κ and $\kappa+1$, $0 \le \kappa < r-1$, given n chiasmata. The generalized model can be handled like above by defining

$$f_n(d_1,\cdots,d_n) = \sum_{v=1}^{n-1}\left\{(d_{v+1}-d_v)^{k_{\kappa,n}}\prod_{\mu=1}^{n}(k_{v,n}+\mu)\right\}\bigg/(n-1) \quad \text{for equation (2.1). Such an}$$

extensive model, however, seems applicable only if chiasma counts show evidence for certain events and if a large data set is available.

A common property of map functions $\theta(x)$ is that each interval of genetic length x is assigned to one and the same recombination fraction θ. In this sense the CL2 model does not have a map function and two intervals of the same genetic length may be assigned different recombination fractions, depending on their site. Since the CL2 model turned out to be well fitting, the existence of a unique map function seems questionable. While it was clear that the recombination process is not homogenous with respect to the physical scale, it became clear now, that it is also not homogenous with respect to the genetic scale.

Acknowledgement

The authors are indebted to Karsten Schlettwein, Norbert Reinsch, Klaus Wimmers, and to Brent Sørensen for having participated in the generation and improvement of the manuscript.

References

Armstrong, N., McPeek, M., and Speed, T. (2006), Incorporating interference into linkage analysis for experimental crosses. *Biostatistics* **7**, 374-386.

Blank, R., Campbell, G., Calabro, A., and D'Eustachio, P. (1988) A linkage map of mouse chromosome 12: location of *Igh* and effects of sex and interference on recombination. *Genetics*; **120**, 1073-1083.

Browning, S. (2000) The relationship between count-location and stationary renewal models for the chiasma process. *Genetics*; **155**, 1955-1960.

Foss, E., Lande, R, Stahl, F., and Steinberg, C. (1993) Chiasma interference as a function of genetic distance. *Genetics*; **133**, 681-691.

Goldgar, D. and Fain, P. (1988) Models of multilocus recombination: nonrandomness in chiasma number and crossover positions. *American Journal of Human Genetics*; **43**, 38-45.

Haldane, J. (1919) The combination of linkage values and the calculation of distance between the loci of linkage factors. *J Genet,* **8**, 299-309.

Hultén, M., Tease, C., and Lawrie, N. (1995) Chiasma–based genetic map of the mouse X chromosome. *Chromosoma*; **140**, 223-227.

Karlin, S. and Liberman, U. (1979) A natural class of multilocus recombination processes and related measures of crossover interference. *Adv Appl Prob*; **11**, 479-501.

Karlin, S. and Liberman, U. (1994) Theoretical recombination processes incorporating interference effects. *Theoretical Population Biology*; **46**, 198-231.

Lawrie, N., Tease, C., and Hultén, M. (1995) Chiasma frequencies, distributions and interference maps of mouse autosomes. *Chromosoma*; **140**, 308-314.

Lien, S., Szyda, J., Schechinger, B., Rappold, G., and Arnheim, N. (2000) Evidence for heterogeneity in recombination in the human pseudoautosomal region: high resolution analysis by sperm typing and radiation-hybrid mapping. *American Journal of Human Genetics*; **66**, 557-566.

Martini, E., Diaz, R., Hunter, N., and Keeney, S. (2006) Crossover homeostasis in yeast meiosis. *Cell*; **126**: 285-295.

McPeek, M. and Speed, T. (1995) Modeling interference in genetic recombination. *Genetics*; **139**, 1031-1044.

Morgan, T., Bridges, C., and Schultz, J. (1935) Report of investigations on the constitution of the germinal material in relation to heredity. *Carnegie Institute Washington*; **34**, 284-291.

Pascoe, L. and Morton, N. (1987) The use of map functions in multipoint mapping. *American Journal of Human Genetics*; **40**, 174-183.

Povey, S., Smith, M., Haines, J., Kwiatkowski, D. et al. (1992) Report on the first international workshop on chromosome 9. *Annals of Human Genetics*; **56**, 167-221.

Risch, N. and Lange, K. (1979) An alternative model of recombination and interference. *Ann Hum Genet*; **43**, 61-70.

Teuscher, F., Brockmann, G., Rudolph, P., Swalve, H., and Guiard, V. (2000) Models for chromatid interference with application to recombination data. *Genetics*; **156**, 1449-1460.

Weinstein, A. (1936) The theory of multiple-strand crossing over. *Genetics*; **21**, 155-199.

Yu, K. and Feingold, E. (2001) Estimating the frequency distribution of crossovers during meiosis from recombination data. *Biometrics*; **57**, 427-434.

Yu, K. and Feingold, E. (2002) Methods for analyzing the spatial distribution of chiamata during meiosis based on recombination data. *Biometrics*; **58**, 369-377.

Zhao, H. and Speed, T.. (1998) Stochastic modeling of the crossover process during meiosis. *Communications in Statistics-Theory and Methods*; **26**, 1557-1580.

Zhao, H., Speed, T., and McPeek, M. (1995) Statistical analysis of crossover interference using the Chi-square model. *Genetics*; **139**, 1045-1056.

In: Genetic Recombination Research Progress
Editor: Jacob H. Schulz, pp. 247-264

ISBN: 978-1-60456-482-2
© 2008 Nova Science Publishers, Inc.

Chapter 9

The Fanconi Anemia Pathway: Directing DNA Replication-Associated Repair

John M. Hinz

School of Molecular Biosciences, Washington State University
Pullman, WA 99164-4660

Abstract

The disease Fanconi anemia (FA) is a cancer predisposition disorder involving progressive anemia, caused by deficiency in any of 13 known (FANC) genes. At the cellular level FA is a classical chromosomal instability disease associated with sensitivity to DNA damage, particularly interstrand crosslinks (ICLs). Unlike classical DNA repair pathways, it is not understood how most of the FANC proteins function to maintain chromosome stability, but many of them are necessary for the monoubiquitylation of two of the FANC proteins, FANCD2 and FANCI, both posttranslational modifications that are considered necessary for a functional FA "pathway". Through genetic and biochemical studies, this pathway is variously implicated in a number of key S-phase events such prevention of replication fork breaks by promotion of translesion synthesis (TLS), and the repair of broken chromatids through both of the double-strand break (DSB) repair pathways: homologous recombination repair (HRR) and non-homologous end joining (NHEJ). However, many published results concerning the functional links between the FA pathway and the two mutually exclusive DSB repair processes are, not surprisingly, contradictory. Recent use of the mutagenesis model system of hamster CHO cells, with knock-out clones defective in one of the FANC proteins (FANCG), or an HRR protein (Rad51D), have been highly informative in distinguishing functional differences between the FA and HRR pathways. Importantly, the experimental precision of CHO cells has clarified seemingly conflicting mutagenesis measurements in human FA cells, and implies contribution of the FA pathway in promoting all three replication-associated damage-recovery processes: TLS, HRR, and NHEJ. FA cells, therefore, are not completely deficient in any one of these processes, but are less efficient in all of them, suggesting a role for the FA proteins in co-ordination or optimization of these genome stabilizing processes, rather than direct participation in any one of them.

Introduction

Replication of the genome is a difficult task, and must be performed with high fidelity to prevent accumulation of mutations with each cell division. During replication, the replicative polymerases, Pol δ and Pol ε, must manage complex chromatin and DNA structures that can interfere with normal synthesis and increase the chances of error. These hurdles include regions of tightly compacted chromatin, complex DNA sequences, such as polynucleotide repeats (microsatellites), and damaged DNA, including chemical modifications to nucleobases or the phosphate-sugar backbone, as well as breaks and gaps in the DNA strands. The cell is equipped with a number of coordinated protein networks, or "pathways", that help assure the complete and accurate replication of the chromosomes in the context of the specific difficulties posed by the DNA.

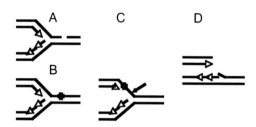

Figure 1. DNA breaks and lesions can cause DSBs during replication. When a replication fork reaches a gap or break in one of the parental DNA strands (A), the fork can run off the end, leading to a one-sided DSB (D). Replication forks reaching a polymerase-blocking DNA lesion (B) can arrest and break, due to nuclease activity (C, arrow), also leading to a one-sided DSB (D).

Of greatest priority is the prevention of DNA double-strand breaks (DSB). A DSB is the most lethal DNA lesion, signaling cell cycle arrest prior to mitosis, or should the cell continue through mitosis, leading to loss of an entire arm of a chromosome [1]. Though the repair of DSBs can be a source of chromosomal rearrangements and other genetic aberrations, it is biologically the more prudent choice over an almost certain death for the cell.

DSBs primarily occur during replication when a replication fork encounters a break or gap in the backbone of a DNA strand (**Fig. 1A**), or a DNA modification, or lesion, through which the replicative polymerase cannot synthesize (**Fig.1B**). Such lesions can occur spontaneously, by interactions with the byproducts of cellular metabolism such as reactive oxygen species, or can be the result of exposure to DNA-reactive radiation or chemicals in the environment, such as ultraviolet light, industrial chemicals, or chemotherapeutic agents. Polymerase-blocking lesions cause an arrest in the progression of the replication fork, leaving it vulnerable to collapse (loss and destabilization of replication-associated proteins from the fork), and eventual breakage (due to activity of nucleases, such as Mus81, that cleave one of the parental DNA strands [2]; **Fig. 1C**). In either case, a DSB, specifically called a one-sided DSB due to the lack of a complimentary end, will be created (**Fig. 1D**).

The process of DSB mitigation during replication takes the form of two stages, DSB prevention and DBS repair. In order to prevent these replication-associated DSBs when a replication fork reaches a polymerase-blocking lesion, the cell can perform the process of translesion synthesis (TLS), swapping out the replicative polymerase and exchanging it with an error-prone, translesion polymerase. Eukaryotic genomes encode several DNA polymerases with variable processivity and fidelity. For example, members of the Y family of translesion polymerases are characterized by a flexible catalytic site that allows them to bypass bulky DNA lesions, but impairs their accuracy while copying undamaged DNA [3]. Use of these polymerases allows the replication fork to maintain its integrity and continue synthesis through DNA lesions, thus preventing fork collapse and breakage, but risking the base substitution mutations associated with these low-fidelity bypass events.

Repair is required if the replication fork does collapse and break. The resulting DSB can then be resolved by one of two major DSB repair pathways, homologous recombination repair (HRR), or non-homologous end joining (NHEJ). HRR is an error-free method of repair that utilizes the recently synthesized sister chromatid as a template for repair of a DSB. Whereas NHEJ repair of DSBs can occur throughout the cell cycle, HRR is restricted to the S and G2 phases of the cell cycle [4-6]. HRR can also be used to repair direct DSBs, such as those induced by ionizing radiation, but appears to have its primary role during DNA synthesis [7, 8]. When a replication fork breaks, HRR can re-establish or restart the fork [7, 9]. **(Fig. 2)** The primary enzyme responsible for catalyzing the search for homologous DNA and subsequent strand exchange in HRR is the Rad51 recombinase, which requires the activity of mediator proteins including BRCA2 (encoded by the breast cancer susceptibility gene 2) and five paralogs of Rad51 (XRCC2, XRCC3, RAD51B, RAD51C, RAD51D) [10]. The BRCA2 protein is intimately involved in HRR where it is thought to participate in the loading of Rad51 onto long 3′ single-stranded DNA ends resulting from nucleolytic processing of the DSB [11-14]. Cancer associated mutations in BRCA2 that disrupt the interaction with Rad51 prevent formation of these Rad51 nucleoprotein filaments [11, 15]. The Rad51 paralogs are also necessary for Rad51 nucleoprotein filament formation, but their precise roles in HRR are not clearly understood [10].

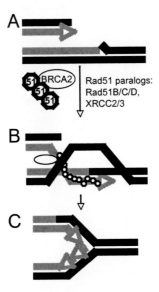

Figure 2. HHR-directed replication fork restart. From a broken replication fork (A) invasion of the 3' strand of the broken chromatid into the sister chromatid for synthesis (B). Nucleolytic cleavage of the resultant Holliday Junction restores the replication fork (C). Rad51and accessory proteins are required for strand invasion.

The alternate process of DSB repair, NHEJ, is simply the ligation of the two broken ends of DNA pieced together, sometimes after limited processing of the ends, thus resulting in an efficient, but error-prone, repair of DSBs [16]. Throughout the literature, the broad term NHEJ has come to represent the end-joining mechanism associated with the DNA-dependant protein kinase (DNA-PK). DNA-PK consists of a catalytic subunit (DNA-PKcs) and a DNA end-binding component, the heterodimer Ku70/Ku80 [17]. The recruitment of DNA-PKcs to DNA breaks by Ku results in activation of its kinase function that then phosphorylates other proteins as well as itself. DNA-PKcs can then presumably tether the broken ends to facilitate rejoining and also recruit and activate proteins involved in the necessary DNA end-processing and ligation [18]. DNA-end processing enzymes, such as the nuclease Artemis, can modify the end in preparation for ligation [19], and the XRCC4–Ligase IV complex is then recruited to re-ligate the broken DNA ends [20].

In addition, at least one other end-joining mechanism exists in the cell, independent of the DNA-PK complex. For example, a protein complex consisting of MRE11, Rad50, and NBS1 (MRN), which is known to play a role in the processing of DNA ends in order to facilitate HRR, may also facilitate tethering and ligation of DNA ends [21-23]. Since the relative contribution of the various end-joining mechanisms to the repair of DSBs during replication is unclear, NHEJ will be used in this review to describe any mechanism used by the cell to rejoin strand breaks. This broad category is used because many end-joining measurements and end points do not discriminate among the different pathways, except when specifically tested. Importantly, because of the generally error-prone nature of the end-joining mechanisms, any DSBs occurring during replication repaired by these pathways can lead to deletions and rearrangements in chromosomes, such as the rejoining of a one-sided DSB break created during replication of one chromosome with a break from a different chromosome **(Fig. 3)**.

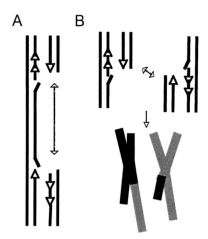

Figure 3. NHEJ can lead to chromosomal rearrangements. (A) End-joining may occur between two one-sided DSBs within the same chromatid, leading to a deletion mutation. Alternately, one-sided DSBs from broken forks on two different chromosomes can be substrates for end-joining mechanisms, causing the recombination of chromatids from different chromosomes (B).

Therefore coordination of these replication-associated events, TLS, HRR, and NHEJ, is undoubtedly important to insure the integrity of the genome, minimizing the impact of suboptimal replication situations, and maintaining the efficiency of damage control when a DNA DSB does occur. A potential candidate group of proteins that may regulate or assist with the cellular decisions needed to cope with replication-associated problems are those that are defective in the disease Fanconi anemia (FA). Studies concerning this chromosomal instability disease are often focused on attempting to connect the activity of these proteins to one of the DSB repair pathways in the cell. However, as discussed below, many apparently conflicting studies are in actuality pointing to a broad role for the FANC proteins in the regulation of TLS, HRR, and NHEJ. Indeed, even early mutagenesis studies on cells from FA patients support this model, which has been revisited by a recent study of mutations in CHO cells.

FA is a rare recessive disorder characterized by developmental abnormalities, progressive bone marrow failure, and cancer susceptibility [24, 25]. FA is genetically heterogeneous, consisting of at least 13 complementation groups for which the genes have been identified: FANCA, B, C, D1 (BRCA2), D2, E, F, G, I, J (BRIP1), L, M, and N (PALB2) [26-28]. Among individuals with FA, complementation groups A, C, and G are the most abundant (representing 65%, 15% and 10% of patients, respectively [29]). Throughout the literature, and this text, the expression "the FA pathway" is used to describe the formation of complexes, post-translational modifications, and the activities of all of the FANC proteins in their entirety. Like any other "pathway" in the cell, the expression is not meant to imply strict linearity to the examined events, but rather to describe an option for the cell that is not available if there is a defect in any of the associated proteins. Therefore, defects in any of the FANC genes result in similar clinical and cellular phenotypes [29].   Though several complexes of the FANC proteins have been identified [30, 31], 10 FANC proteins (including those representing complementation groups A,B,C,E,F,G,L, and M) form what is called the nuclear "core complex" [27, 32]. Two of the proteins (FAAP24 and FAAP100; FANCA-associated proteins) are also integral to the core complex, but defects in the genes encoding these proteins have not been found in FA patients [33, 34]. All of the proteins in the core complex must be present, as the full integrity of the complex is essential for the monoubiquitylation of FANC proteins FANCD2 and FANCI to occur, performed specifically by the core complex protein FANCL, an E3 ubiquitin ligase [35] (Fig. 4). FANCD2 and FANCI are monoubiquitylated in cycling populations of cells as well as following treatment with agents such as mitomycin C (MMC, a DNA interstrand crosslinking agent) or ionizing radiation (IR) [36]. The absence of the ubiquitylated isoform of FANCD2 correlates with the clinical diagnosis of FA, which is based on extreme cellular sensitivity to MMC or diepoxybutane (DEB), another crosslinking agent [37]. Interestingly, the monoubiquitylated FANCD2 and FANCI associate (as a pair) with chromatin during S phase, as does the entire core complex through the DNA binding motifs of FAAP24 and FANCM (Fig. 4). These chromatin associations intensify after DNA damage [38-40], implicating the importance of these proteins in some aspect of replication.

The phenotype of FANC-defective cells provides strong evidence for a role for the FANC proteins in processing DNA damage and maintaining genomic integrity during DNA replication. Traits associated with FANC mutant cells include increased spontaneous chromosomal breakage and exchange, increased apoptosis and high levels of reactive oxygen species (ROS), as well as an increased percentage of cells in the G2 phase [26, 41]. In addition, cells from FA patients consistently show high sensitivity for cell killing in response to DNA crosslinking agents such as MMC and DEB [26, 41].   FA cells also demonstrate a high sensitivity to the induction of chromosomal aberrations with treatment of low doses of agents that induce interstrand crosslinks (ICLs), which serves as a standard diagnostic test for the disease [42].

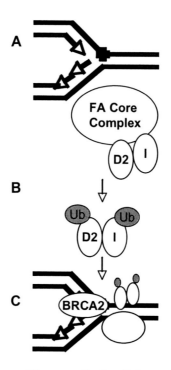

Figure 4. Activation of the FA pathway. When a replication fork arrests and collapses, signaling kinases activate the FANC proteins, several of which form a nuclear core complex (A) necessary for the monoubiquitylation modifications of the FANC proteins FANCD2 and FANCI (B). The core complex, the D2/I heterodimer, BRCA2, and other FA proteins then associate with chromatin at the sites of damage (C).

Though there are disputes over the function and purpose of the activity of the FA pathway, as will be discussed in greater detail below, it is clear that the FANC proteins are functionally linked to DNA damage and replication stress, as they are the targets of post-translation modification by signaling proteins in the cell associated with these responses. The phosphorylation of several FANC proteins plays an important role in activating the FA pathway and associated cell cycle checkpoints [43-45]. The ATM DNA damage-signaling kinase, activated by the presence of DSBs, phosphorylates FANCD2 during S phase, with or without treatment with DNA-damaging agents [45]. After exposure to IR or MMC, ATM phosphorylation of FANCD2 is necessary for activation of the intra-S phase checkpoint but is unnecessary for FANCD2 monoubiquitylation in human cells [43, 45]. In response to crosslink damage, the ATM-related checkpoint kinase, ATR, whose activity is associated with stalled replication forks, phosphorylates FANCD2 either directly or through activation of other kinases, to activate the replication checkpoint [44]. Indeed, the efficient FANCD2 monoubiquitylation after cellular exposure to MMC requires ATR [46]. Other FANC proteins are phosphorylated in an ATR-dependant manner, including FANCG at three different serines [47, 48], as well as FANCA, FANCE, and FANCM, all modifications that are necessary for normal FA pathway function [49, 50].

Links between the FA Pathway and TLS, HRR, and NHEJ

The link between FA and Translesion Synthesis. The role of the FA pathway in assuring the efficient bypass of polymerase-blocking DNA lesions, thereby reducing the occurrences of collapsed and broken replication forks and preventing aberrant chromosomal recombination events, has been minimally explored. Evidence for the involvement of FANC protein FANCC has been convincingly seen in a few studies using chicken DT40 pre-B lymphocytes, in which the high efficiency of gene targeting has allowed for the creation of cells defective in one, or multiples, of many DNA repair, replication, and damage response genes. FANCC was found to promote cell survival and TLS during crosslink repair after exposure to the interstrand crosslinking agent cisplatin [51]. This was determined by epistasis analysis, showing no additional sensitization to exposure to the drug in cells defective in both Rev3 (the catalytic subunit of the TLS polymerase Pol ζ) and FancC, compared to either single mutant, therefore illustrating that the two mutants work in the same pathway. These results were verified by another group in an epistasis study using DT40 mutants in the same genes, but altering the cisplatin exposure conditions [52]. However, an additional study in DT40 cells, measuring epistatic relationships between of FancC and Rad18, suggest a TLS-independent role for FancC in response to alkylation and interstrand crosslink damage [53]. Rad18 is the E3 ubiquitin ligase necessary for monoubiquitylation of the polymerase processivity factor PCNA. This event serves as a signal for recruitment of the error-prone translesion polymerases Pol ζ and Pol η, thus allowing bypass of polymerase-blocking lesions [54, 55]. The double mutant (*fancc/rad18*) showed increased sensitivity to killing after treatment with methyl methanesulfonate or cisplatin compared to either single mutant. In addition, the double mutant showed a clear increase in the formation of spontaneous sister chromatid exchanges, indicative of increased replication fork breaks during DNA synthesis, thus separating the functions of the FA pathway and Rad18-associated TLS in maintaining continuity of replication forks. Though these initial studies provide some indication of the importance of an intact FA pathway in TLS during crosslink repair, which is an essential part of the repair of such DNA lesions, they do not address the role of the FA pathway during S-phase replication and the TLS associated with spontaneously occurring, endogenous DNA damage.

Cellular phenotype similarities between the FA pathway and HRR. The most obvious indication of a role for the FANC proteins in HRR comes from the similarity in phenotypes between FA cells and cells with defects in HRR. Like FA cells, increased chromosomal aberrations are seen consistently in cells with defects in BRCA2 and the Rad51 paralogs [56]. In addition, HRR-defective cells are extremely sensitive to agents that induce ICLs [56], the diagnostic characteristic of cells from FA patients. Therefore it is a common supposition to place the FA pathway as a branch of the HRR pathway responsible for ICL repair, but this is likely an oversimplification. Though there is little compelling evidence for a direct role of the FANC proteins in the enzymatic repair of ICLs or DSBs, conflicting reports, discussed below, show evidence for decreased repair capacity in FANC mutant cells.

Genetic and physical links between the FA pathway and HRR. The sharing of proteins between two pathways is generally strong evidence for at least the coordinated control, if not coordinated activities between them. As mentioned above, FANCD1 was identified as being the essential HR protein BRCA2 [57] and is mutated in patients who have an especially severe clinical phenotype including an earlier onset of solid tumors and leukemias [58-60]. Notably, the activity of BRCA2 is not necessary for the monoubiquitylation of FANCD2 [61]. Thus, BRCA2's activity appears to be "downstream" or independent of the FANCD2 monoubiquitylation pathway. The binding partner of BRCA2, PALB2, is also known as FANCN, another protein shared between HRR and FA [62].

Physical interactions and co-localization of FANC and HRR proteins also help to define a connection between the two pathways. The finding that FANCG interacts directly with BRCA2 and the Rad51 paralog XRCC3, as shown by the yeast two-hybrid system and co-immunoprecipitation, suggests that the FANC core complex may also help facilitate HRR [30, 63]. These results are consistent with the finding that during S-phase GFP-tagged FANCG forms nuclear foci that co-localize with both Rad51 and BRCA2 nuclear foci in response to MMC treatment in HeLa cells [63]. Monoubiquitylated FANCD2 and FANCI localize at sites of putative DSBs with nuclear foci of Rad51 and the BRCA1 breast cancer susceptibility protein [36, 38]. BRCA1 itself appears to play a key role in promoting HRR, as it too is required for Rad51 focus formation after exposure to DNA crosslinking agents [64], and BRCA1 mutant cells show reduced HRR capacity in transfected model plasmid constructs compared to control cells [65, 66]. BRCA1 is also intimately associated with the FA pathway through its binding partner BRIP1, the FANC protein FANCJ [67].

Functional links between the FA pathway and HRR.   There are a number of inconsistencies among studies attempting to assess a functional requirement for FANCD2 monoubiquitylation in HRR.   Reduced homology-directed repair was measured in a *GFP* reporter gene transfected into FA cell lines from three complementation groups (A, D2, and G) relative to complemented controls [68]. Using a modified FANCD2, which mimics the protein in its monoubiquitylated form, they also showed the importance of the modified FANCD2 for its role in promoting HRR activity [68]. Despite these apparently convincing results, the biological relevance of HRR events occurring in artificial direct-repeat recombination substrates such as the one used in this study is not clear, as they do not provide normal chromosomal substrates for HRR. Chicken DT40 cells, though not an ideal model system for HRR studies due to their notoriously high rates of recombination relative to mammalian cells, also show evidence that an intact FA pathway facilitates HRR in reporter gene substrates in response to I-*Sce*I-induced chromosomal DSBs [69, 70]. DT40 knockout mutants disrupted in either *Fancc* or *Xrcc3* show sensitivity to cisplatin, and the double knockout shows the same degree of cisplatin sensitivity as the *fancc* mutant cells, suggesting an overlapping function between Fancc and Xrcc3 proteins [53]. In addition, *fancd2* mutant DT40 cells show reduced gene conversion at the immunoglobulin light chain locus, an event driven by HRR [70].

Conflicting results have also been seen when assessing the necessity of an intact FA pathway for formation of nuclear foci of Rad51 and BRCA2 proteins, an immunochemical indication of the formation of the Rad51 nucleoprotein filament necessary for HRR. Whereas a few studies show the need for an intact FA pathway, including monoubiquitylation of FANCD2 [71, 72], other studies have clearly demonstrated normal Rad51 foci formation in nearly all of the FA complementation groups, except, of course, BRCA2-defective cells [61, 73-75]. Since many of these studies reporting defective focus formation used IR, rather than interstrand crosslinking damage, the relevance of the findings to the FA pathway is unclear as

IR-induced foci likely occur at sites of direct DSBs, rather than at broken replication forks where the FANC proteins act.

An isogenic model system in CHO AA8 cells has been extremely useful in distinguishing the roles of the FA and HRR pathways in genomic stability. Isogenic knockout mutants of *FANCG* [75] and the Rad51 paralog *RAD51D* [8] have allowed for a direct comparison of the contributions of each defect with respect to cell survival and genomic stability after genetic insult. Consistent with cells from FA patients [32], the *fancg* CHO cells are more sensitive (3- to 4-fold) to DNA agents that induce ICLs, such as MMC. However, *fancg* cells are also sensitive (3- to 4-fold) to mono-functional alkylating agents such as ethylnitrosourea and methyl methanesulfonate (MMS), highlighting the important point that these *fancg* cells are not exclusively sensitive to crosslinking agents [75]. Like the *fancg* cells, the HRR-defective *rad51d* cells also showed increased sensitivity to a variety of DNA damaging agents [8], including MMS (3.8-fold) and UV-C (1.5-fold), thus highlighting the roles of both HRR and the FA pathway in coping, although not identically, with a variety of DNA lesions. Intriguingly, there is a profound difference in the sensitivity of the *rad51d* cells to MMC (80-fold) compared with *fancg* cells (3-fold) [8]. This very high level of sensitivity to MMC is consistent with that seen in mutants of other Rad51 paralogs including the *xrcc3* mutant irs1SF in CHO (~90-fold) [76] and the *xrcc2* irs1 mutant of V79 hamster cells (~90-fold) [77], underscoring the well-accepted understanding of the requirement for HRR in resolving cross-linked DNA [78]. Moreover, the lack of extremely high MMC sensitivity of the *fancg* cells argues for an ancillary, rather than intimate, role for the FA pathway in crosslink repair.

The role of FANC proteins in non-homologous end joining. Evidence for a functional link between the FA pathway and NHEJ comes primarily from studies utilizing cells and cell extracts. In general, FA cells do not exhibit the high sensitivity to ionizing radiation (IR) associated with a gross NHEJ deficiency, likely excluding the FA pathway from a direct role in the repair of the frank DSBs that contribute to the lethality of that DNA-damaging agent. However, cells from FA patients representing three different complementation groups (A, C, and F) were reported to have a defect in rejoining of breaks after exposure to IR, as determined by pulse-field gel electrophoresis [79], though none of these cells displayed sensitivity to killing. FANC mutant cells from some complementation groups (A, C, D2, and G) do show decreased survival after electroporation with restriction enzymes, indicative of a defect in the rejoining of frank DSBs induced by enzymatic cleavage [80], but this finding was not reproduced in the isogenic *fancg* CHO cells [81]. Transfection of linearized reporter plasmids into FA cells (of complementation groups A, C, D2, and G) also showed reduced efficiency for end joining [80] relative to retrovirus-complimented controls. In agreement with this, cell extracts from FANC mutant cells also showed reduced end-joining activity of linearized plasmids [82]. Though the aforementioned studies all appear to consistently show a decreased end-joining efficiency in all FA cells examined, other published studies show results to the contrary. Earlier studies found only decreased fidelity (not efficiency) of end joining of blunt ends in cells from FA complementation groups B and C [83, 84], not eliminating a possible role for the FA pathway in end joining, but suggesting its role as that of a qualitative, rather than a quantitative, factor in repairing DSBs. In addition, studies based on chromosomally integrated reporter constructs, containing the I-*Sce*I restriction site for targeted enzymatic cleavage (used in the studies mentioned previously to verify the role of the FA pathway in HRR), found no reduction in NHEJ activity in FANC defective cells from the three complementation groups studied (A, D2, and G) [68]. A few studies have highlighted the possibility that the FA pathway may be functioning in a particular end-joining mechanism. For example, studies in mouse fibroblasts showed an increased sensitivity to killing by IR and transfection of restriction endonucleases in cells defective in both FancD2 and DNA-PKcs ($Fancd2^{-/-}/Prkdc^{sc/sc}$) relative to the already-sensitive DNA-PKcs-defective cells ($Prkdc^{sc/sc}$) [85]. Consistent with previous studies, they see a lack of IR sensitivity in the $Fancd2^{-/-}$ cells. These results suggest that removal of the highly efficient DNA-PKcs-associated NHEJ mechanism, which plays the predominant role in rescuing cells from the lethality of IR exposure, reveals a second end-joining mechanism, independent of DNA-PKcs, that requires the presence of FancD2. This "second" end-joining mechanism may be that which is associated with the MRN complex, as it has been reported that the increase in sensitivity of cells to either DEB or restriction enzyme electroporation, after introduction of anti-Rad50 antibody, was not detected in fibroblasts from FA patients [86]. Thus, the sensitivity to DNA breaks, due to the reduction of the MRN-associated end-joining capacity of the cell, was already present in the FA cells tested. In a related study, this same group has since determined a role for MRN proteins in the fidelity of V(D)J signal joint recombination at immunoglobulin loci, which requires the enzyme-catalyzed formation of DSBs and the activity of DNA-PK-associated NHEJ to rejoin the ends [87]. The study also revealed a tenfold reduction in the fidelity of V(D)J recombination in FA fibroblasts from complementation groups A and C [87], consistent with results seen almost ten years earlier in FA lymphoblasts showing an increase in the aberrant rearrangements of signal joint formation relative to control cell lines [88]. Ultimately, there does appear to be some role for activity of the FA pathway in maintaining the efficiency or fidelity of some aspect of the end-joining capacity in cells, despite the various inconsistencies in the literature. Perhaps creating the greatest limitation on the interpretation of these end-joining results is that the artificial plasmid constructs and/or restriction-enzyme-generated DSBs used in the above studies are likely inappropriate model systems to assess the impact of the FA pathway in regard to the resolution of replication-associated breaks to which the FANC proteins are relevant.

Early mutagenesis studies in human FA cells suggested complexity of FA pathway in maintaining genome stability. The spontaneous and induced mutation rates of a cell population provide a measure of genomic stability based on a surrogate gene and are particularly revealing when combined with an analysis of the nature of the mutations. The X-linked *hprt* gene has been widely used for loss-of-function mutation studies because it detects both point mutations and deletions/insertions, it can be used in many types of cells, and the (functional) hemizygous nature of the locus means disruption of only one locus is necessary for survival of mutant clones in selection [89]. Notably, mutagenesis studies found that cells from FA patients show **hypo**-mutability for induced mutations at the *hprt* locus. Treatment with mutagenic monofunctional or bifunctional psoralens also reduced the recovery of mutants at the *hprt* locus in FA-A fibroblasts [90]. When assessing the spectrum of *hprt* mutations occurring spontaneously, the FA cells showed a much higher percentage of deletions than did the normal fibroblasts [91]. Consistent with these results, a higher proportion of deletions was also found in circulating T-lymphocytes of FA patients relative to lymphocytes of unrelated controls [92]. Upon psoralen treatment, decreased frequencies of mutants were also recovered at the Na^+/K^+ *ATPase* locus in FA-A fibroblasts relative to unrelated non-FA cells [90]. This locus can detect only point mutations, specifically those that prevent binding of the pump-inhibitor drug ouabain, and other types of mutations, including deletions and frame shifts, make the ion pump non-functional and kill the cell.

Though the reduced *hprt* frequencies and deletion-rich mutation spectra seen in human FA cells implies a role for the FANC proteins in some facet of DNA replication integrity, interpretation of the mutagenesis data is complicated by the lack of comparisons made between isogenic cell lines, and the absence of spontaneous mutation rate measurements. In addition, there is an apparent inconsistency with data at other genetic loci in which FA patients had markedly increased mutation frequencies, including the glycophorin A (*GPA*) locus in erythrocytes [93, 94] and the *PIG-A* locus measured in lymphoblasts [95].

Mutagenesis in CHO cells reveals a multifaceted role for the FA pathway when coping with replication-associated problems. Though the specific-locus mutation studies in human FA cells have produced seemingly conflicting results, determination of the rate at which spontaneous mutations occur at the *hprt* locus in the isogenic CHO *fancg* [75] and *rad51d* [8] mutants have helped to clarify and unify all of the previous FA-associated mutagenesis data [81]. Rates of mutagenesis for loss-of-function mutations, by fluctuation analysis [96], which can be measured with high precision in CHO cells, revealed distinct differences between the FA and HRR mutants in mutation rate at the *hprt* locus. Consistent with mutant frequency measurements in human cells, the rate of spontaneous mutagenesis was actually **reduced** in *fancg* cells (>3-fold) compared to control cell lines [75]. Reduced recovery of *hprt* mutants is indicative of a shift in the types of mutagenic events occurring during DNA synthesis, from mutations that allow cell survival during selection of the disrupted locus, such as base substitutions and small deletions, to large deletions and rearrangements that are lethal, compromising essential genes flanking the hemizygous *hprt* locus (**Fig. 5**). The reduced mutation rate at the *hprt* locus in the *fancg* cells, which suggests the inability of the cells to appropriately deal with the types of DNA damage that occur spontaneously (primarily oxidative lesions), was also seen after exposure to exogenous agents producing other types of DNA damage. The *fancg* CHO cells showed a reduced induction of *hprt* mutants at each of two doses of ethyl nitrosourea and UV-C radiation, indicating that the *fancg* cells are somewhat defective in processing a wide variety of lesions (other than DNA crosslinks) [81]. The significance of these findings in the *fancg* cells is highlighted by the comparison to the results of *hprt* mutagenesis in the isogenic *rad51d* CHO cells, which showed a very high rate (~12-fold increase) [8]. The clear divergence between the *fancg* and *rad51d* mutants in spontaneous *hprt* mutagenesis suggests a distinct difference in the roles of the FANC and HRR pathways in preventing loss-of-function mutations. Sequence analysis of the spectrum of *hprt* mutations in these mutants provided additional insight as to the distinct role each of these two pathways play in maintaining the genomic integrity of the cell. The HRR-defective *rad51d* cells had a dramatic shift towards deletion mutations (from 36% to 86%), relative to their gene-complemented control [97] implying that *rad51d* cells have efficient NHEJ activity that can act on DSBs arising from broken replication forks that are not restarted by HRR. With respect to the high rate at which these mutations arise [8], these results clearly show that HRR frequently rescues broken replication forks during DNA synthesis, thereby preventing the low fidelity end-joining mechanisms from resolving the replication-associated DSBs (**Fig. 5**). The reduced mutagenesis in *fancg* cells showed a minor shift in the mutation spectrum towards deletions, but not to the same extent as the *rad51d* mutants. Importantly, the >3-fold decreased rate indicates an extensive shift in the spectrum toward large deletions and rearrangements that cannot be recovered using the *hprt* locus. Thus, these data support the idea that the *fancg* cells have an overall reduced recovery of *hprt* mutations arising through all mechanisms causing mutation, including mutations associated with TLS (resulting in reduced base substitutions) and some component of end-joining (resulting in reduced recoverable deletion mutations).

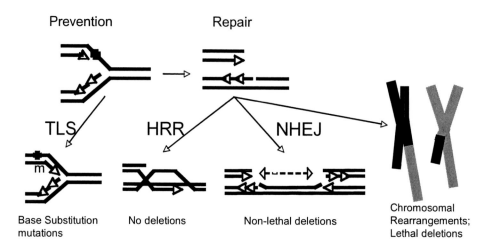

Figure 5. Potential fates of an arrested replication fork in cells with defects in HRR or the FA pathway. In HRR-defective cells, broken replication forks must use NHEJ to resolve the DSBs, but a high proportion are repaired conservatively, leading to non-lethal deletions. In FA cells, reduced TLS leads to reduced base substitution mutations, and reduction in NHEJ capacity leads to fewer deletion mutations that are recovered at the *hprt* locus, shifting the repair events to more chromosomal rearrangements and lethal deletions.

Conclusion

Numerous studies have concluded a role for the FA pathway in maintaining the integrity of the genome through a particular mechanism used by the cell to deal with the various difficulties associated with DNA replication, including the bypass of chemically modified bases via TLS, or the repair of DSBs by either of the major DSB repair pathways, HRR and NHEJ. However, the literature is dominated by biochemical studies and use of artificial plasmid constructs that show a range of somewhat contradictory results regarding the importance of the FANC proteins in any one aspect of cellular damage response during replication.

It is the studies measuring spontaneous and induced mutagenesis in intact cells that have helped reconcile some of the seemingly conflicting observations reported for the role of the FA pathway in promoting DSB repair, as well as helping shed light on the underlying cause of genetic instability and cancer associated with the disease FA. Detailed analyses of the mutagenesis results at model gene loci in FA and HRR-defective CHO cells, and in the human cells derived from FA patients, not only suggest distinct roles for the FA and HRR pathways in preventing mutations, but most importantly, imply a role for an intact FA pathway in coordinating or maintaining the efficiency of all three cellular processes needed to cope with blocked and broken replication forks: TLS, HRR, and DNA end joining.

References

[1]   O'Driscoll M., Jeggo P. A. (2006). The role of double-strand break repair - insights from human genetics. *Nat Rev Genet*, **7**, 45-54.

[2]   Chen X. B., Melchionna R., Denis C. M., Gaillard P. H., Blasina A., Van de Weyer I., et al. (2001). Human Mus81-associated endonuclease cleaves Holliday junctions in vitro. *Mol Cell*, **8**, 1117-27.

[3]   Di Noia J., Neuberger M. S. (2002). Altering the pathway of immunoglobulin hypermutation by inhibiting uracil-DNA glycosylase. *Nature*, **419**, 43-8.

[4]   Lee H., Larner J. M., Hamlin J. L. (1997). A p53-independent damage-sensing mechanism that functions as a checkpoint at the G1/S transition in Chinese hamster ovary cells. *Proc Natl Acad Sci USA*, **94**, 526-31.

[5]   Rothkamm K., Kruger I., Thompson L. H., Löbrich M. (2003). Pathways of DNA double-strand break repair during the mammalian cell cycle. *Mol Cell Biol*, **23**, 5706-15.

[6]   Hinz J. M., Yamada N. A., Salazar E. P., Tebbs R. S., Thompson L. H. (2005). Influence of double-strand-break repair pathways on radiosensitivity throughout the cell cycle in CHO cells. *DNA Repair*, **4**, 782-92.

[7]   Helleday T. (2003). Pathways for mitotic homologous recombination in mammalian cells. *Mutat Res*, **532**, 103-15.

[8]   Hinz J. M., Tebbs R. S., Wilson P. F., Nham P. B., Salazar E. P., Nagasawa H., et al. (2006). Repression of mutagenesis by Rad51D-mediated homologous recombination. *Nucleic Acids Res*, **34**, 1358-68.

[9]   McGlynn P., Lloyd R. G. (2002). Recombinational repair and restart of damaged replication forks. *Nat Rev Mol Cell Biol*, **3**, 859-70.

[10] Thacker J. (2005). The RAD51 gene family, genetic instability and cancer. *Cancer Lett*, **219**, 125-35.

[11] Davies A. A., Masson J. Y., McIlwraith M. J., Stasiak A. Z., Stasiak A., Venkitaraman A. R., et al. (2001). Role of BRCA2 in control of the RAD51 recombination and DNA repair protein. *Mol Cell*, **7**, 273-82.

[12] Kowalczykowski S. C. (2002). Molecular mimicry connects BRCA2 to Rad51 and recombinational DNA repair. *Nat Struct Biol*, **9**, 897-9.

[13] Pellegrini L., Venkitaraman A. (2004). Emerging functions of BRCA2 in DNA recombination. *Trends Biochem Sci*, **29**, 310-6.

[14] Kowalczykowski S. C. (2005). Cancer: catalyst of a catalyst. *Nature*, **433**, 591-2.

[15] Pellegrini L., Yu D. S., Lo T., Anand S., Lee M., Blundell T. L., et al. (2002). Insights into DNA recombination from the structure of a RAD51-BRCA2 complex. *Nature*, **420**, 287-93.

[16] Lieber M. R., Ma Y., Pannicke U., Schwarz K. (2004). The mechanism of vertebrate nonhomologous DNA end joining and its role in V(D)J recombination. *DNA Repair (Amst)*, **3**, 817-26.

[17] Smith G. C., Jackson S. P. (1999). The DNA-dependent protein kinase. *Genes Dev*, **13**, 916-34.

[18] Cary R. B., Peterson S. R., Wang J., Bear D. G., Bradbury E. M., Chen D. J. (1997). DNA looping by Ku and the DNA-dependent protein kinase. *Proc Natl Acad Sci U S A*, **94**, 4267-72.

[19] Ma Y., Pannicke U., Schwarz K., Lieber M. R. (2002). Hairpin opening and overhang processing by an Artemis/DNA-dependent protein kinase complex in nonhomologous end joining and V(D)J recombination. *Cell*, **108**, 781-94.

[20] Martin I. V., MacNeill S. A. (2002). ATP-dependent DNA ligases. *Genome Biol*, **3**, REVIEWS3005.

[21] Moreno-Herrero F., de Jager M., Dekker N. H., Kanaar R., Wyman C., Dekker C. (2005). Mesoscale conformational changes in the DNA-repair complex Rad50/Mre11/Nbs1 upon binding DNA. *Nature*, **437**, 440-3.

[22] Hopfner K. P., Craig L., Moncalian G., Zinkel R. A., Usui T., Owen B. A., et al. (2002). The Rad50 zinc-hook is a structure joining Mre11 complexes in DNA recombination and repair. *Nature*, **418**, 562-6.

[23] de Jager M., van Noort J., van Gent D. C., Dekker C., Kanaar R., Wyman C. (2001). Human Rad50/Mre11 is a flexible complex that can tether DNA ends. *Mol Cell*, **8**, 1129-35.

[24] Tischkowitz M. D., Hodgson S. V. (2003). Fanconi anaemia. *J Med Genet*, **40**, 1-10.

[25] Mathew C. G. (2006). Fanconi anaemia genes and susceptibility to cancer. *Oncogene*, **25**, 5875-84.

[26] D'Andrea A. D., Grompe M. (2003). The Fanconi anaemia/BRCA pathway. *Nat Rev Cancer*, **3**, 23-34.

[27] Thompson L. H. (2005). Unraveling the Fanconi anemia–DNA repair connection. *Nat Genet*, **37**, 921-2.

[28] Reid S., Schindler D., Hanenberg H., Barker K., Hanks S., Kalb R., et al. (2006). Biallelic mutations in PALB2 cause Fanconi anemia subtype FA-N and predispose to childhood cancer. *Nat Genet*, in press.

[29] Kennedy R. D., D'Andrea A. D. (2005). The Fanconi Anemia/BRCA pathway: new faces in the crowd. *Genes Dev*, **19**, 2925-40.

[30] Hussain S., Wilson J. B., Blom E., Thompson L. H., Sung P., Gordon S. M., et al. (2006). Tetratricopeptide-motif-mediated interaction of FANCG with recombination proteins XRCC3 and BRCA2. *DNA Repair*, **5**, 629-40.

[31] Medhurst A. L., Laghmani el H., Steltenpool J., Ferrer M., Fontaine C., de Groot J., et al. (2006). Evidence for subcomplexes in the Fanconi anemia pathway. *Blood*, **108**, 2072-80.

[32] Niedernhofer L. J., Lalai A. S., Hoeijmakers J. H. (2005). Fanconi Anemia (Cross)linked to DNA Repair. *Cell*, **123**, 1191-8.

[33] Ciccia A., Ling C., Coulthard R., Yan Z., Xue Y., Meetei A. R., et al. (2007). Identification of FAAP24, a Fanconi anemia core complex protein that interacts with FANCM. *Mol Cell*, **25**, 331-43.

[34] Ling C., Ishiai M., Ali A. M., Medhurst A. L., Neveling K., Kalb R., et al. (2007). FAAP100 is essential for activation of the Fanconi anemia-associated DNA damage response pathway. *Embo J*.

[35] Meetei A. R., De Winter J. P., Medhurst A. L., Wallisch M., Waisfisz Q., Van De Vrugt H. J., et al. (2003). A novel ubiquitin ligase is deficient in Fanconi anemia. *Nat Genet*, **35**, 165-70.

[36] Garcia-Higuera I., Taniguchi T., Ganesan S., Meyn M. S., Timmers C., Hejna J., et al. (2001). Interaction of the Fanconi anemia proteins and BRCA1 in a common pathway. *Mol Cell*, **7**, 249-62.

[37] Shimamura A., De Oca R. M., Svenson J. L., Haining N., Moreau L. A., Nathan D. G., et al. (2002). A novel diagnostic screen for defects in the Fanconi anemia pathway. *Blood*, **100**, 4649-54.

[38] Taniguchi T., Garcia-Higuera I., Andreassen P. R., Gregory R. C., Grompe M., D'Andrea A. D. (2002). S-phase-specific interaction of the Fanconi anemia protein, FANCD2, with BRCA1 and RAD51. *Blood*, **100**, 2414-20.

[39] Mi J., Kupfer G. M. (2005). The Fanconi anemia core complex associates with chromatin during S phase. *Blood*, **105**, 759-66. Epub 2004 Jul 15.

[40] Smogorzewska A., Matsuoka S., Vinciguerra P., McDonald E. R., 3rd, Hurov K. E., Luo J., et al. (2007). Identification of the FANCI protein, a monoubiquitinated FANCD2 paralog required for DNA repair. *Cell*, **129**, 289-301.

[41] Joenje H., Patel K. J. (2001). The emerging genetic and molecular basis of Fanconi anaemia. *Nat Rev Genet*, **2**, 446-57.

[42] Sasaki M. S., Tonomura A. (1973). A high susceptibility of Fanconi's anemia to chromosome breakage by DNA cross-linking agents. *Cancer Res*, **33**, 1829-36.

[43] Taniguchi T., Garcia-Higuera I., Xu B., Andreassen P. R., Gregory R. C., Kim S. T., et al. (2002). Convergence of the Fanconi anemia and ataxia telangiectasia signaling pathways. *Cell*, **109**, 459-72.

[44] Pichierri P., Rosselli F. (2004). The DNA crosslink-induced S-phase checkpoint depends on ATR-CHK1 and ATR-NBS1-FANCD2 pathways. *EMBO J*, **23**, 1178-87.

[45] Ho G. P., Margossian S., Taniguchi T., D'Andrea A. D. (2006). Phosphorylation of FANCD2 on two novel sites is required for mitomycin C resistance. *Mol Cell Biol*, **26**, 7005-15.

[46] Andreassen P. R., D'Andrea A. D., Taniguchi T. (2004). ATR couples FANCD2 monoubiquitination to the DNA-damage response. *Genes Dev*, **18**, 1958-63.

[47] Mi J., Qiao F., Wilson J. B., High A. A., Schroeder M. J., Stukenberg P. T., et al. (2004). FANCG is phosphorylated at serines 383 and 387 during mitosis. *Mol Cell Biol*, **24**, 8576-85.

[48] Qiao F., Mi J., Wilson J. B., Zhi G., Bucheimer N. R., Jones N. J., et al. (2004). Phosphorylation of Fanconi Anemia (FA) Complementation Group G Protein, FANCG, at Serine 7 Is Important for Function of the FA Pathway. *J Biol Chem*, **279**, 46035-45.

[49] Wang X., Kennedy R. D., Ray K., Stuckert P., Ellenberger T., D'Andrea A. D. (2007). Chk1-mediated phosphorylation of FANCE is required for the Fanconi anemia/BRCA pathway. *Mol Cell Biol*, **27**, 3098-108.

[50] Meetei A. R., Medhurst A. L., Ling C., Xue Y., Singh T. R., Bier P., et al. (2005). A DNA translocase homologous to an ancient DNA repair protein is defective in Fanconi anemia. *Nat Genet*, **37**, 958-63.

[51] Niedzwiedz W., Mosedale G., Johnson M., Ong C. Y., Pace P., Patel K. J. (2004). The Fanconi anaemia gene FANCC promotes homologous recombination and error-prone DNA repair. *Mol Cell*, **15**, 607-20.

[52] Nojima K., Hochegger H., Saberi A., Fukushima T., Kikuchi K., Yoshimura M., et al. (2005). Multiple Repair Pathways Mediate Tolerance to Chemotherapeutic Cross-linking Agents in Vertebrate Cells. *Cancer Res*, **65**, 11704-11.

[53] Hirano S., Yamamoto K., Ishiai M., Yamazoe M., Seki M., Matsushita N., et al. (2005). Functional relationships of FANCC to homologous recombination, translesion synthesis, and BLM. *EMBO J*, **24**, 418-27.

[54] Stelter P., Ulrich H. D. (2003). Control of spontaneous and damage-induced mutagenesis by SUMO and ubiquitin conjugation. *Nature*, **425**, 188-91.

[55] Ulrich H. D., Jentsch S. (2000). Two RING finger proteins mediate cooperation between ubiquitin-conjugating enzymes in DNA repair. *Embo J*, **19**, 3388-97.

[56] Thompson L. H., Schild D. (2001). Homologous recombinational repair of DNA ensures mammalian chromosome stability. *Mutat Res*, **477**, 131-53.

[57] Howlett N. G., Taniguchi T., Olson S., Cox B., Waisfisz Q., De Die-Smulders C., et al. (2002). Biallelic inactivation of BRCA2 in Fanconi anemia. *Science*, **297**, 606-9.

[58] Hirsch B., Shimamura A., Moreau L., Baldinger S., Hag-Alshiekh M., Bostrom B., et al. (2004). Association of biallelic BRCA2/FANCD1 mutations with spontaneous chromosomal instability and solid tumors of childhood. *Blood*, **103**, 2554-9.

[59] Wagner J. E., Tolar J., Levran O., Scholl T., Deffenbaugh A., Satagopan J., et al. (2004). Germline mutations in BRCA2: shared genetic susceptibility to breast cancer, early onset leukemia, and Fanconi anemia. *Blood*, **103**, 3226-9. Epub 2004 Jan 8.

[60] Reid S., Renwick A., Seal S., Baskcomb L., Barfoot R., Jayatilake H., et al. (2005). Biallelic BRCA2 mutations are associated with multiple malignancies in childhood including familial Wilms tumour. *J Med Genet*, **42**, 147-51.

[61] Ohashi A., Zdzienicka M. Z., Chen J., Couch F. J. (2005). FANCD2 functions independently of BRCA2 and RAD51 associated homologous recombination in response to DNA damage. *J Biol Chem*, **280**, 14877-83.

[62] Reid S., Schindler D., Hanenberg H., Barker K., Hanks S., Kalb R., et al. (2007). Biallelic mutations in PALB2 cause Fanconi anemia subtype FA-N and predispose to childhood cancer. *Nat Genet*, **39**, 162-4.

[63] Hussain S., Witt E., Huber P. A., Medhurst A. L., Ashworth A., Mathew C. G. (2003). Direct interaction of the Fanconi anaemia protein FANCG with BRCA2/FANCD1. *Hum Mol Genet*, **12**, 2503-10.

[64] Bhattacharyya A., Ear U. S., Koller B. H., Weichselbaum R. R., Bishop D. K. (2000). The breast cancer susceptibility gene BRCA1 is required for subnuclear assembly of Rad51 and survival following treatment with the DNA cross-linking agent cisplatin. *J Biol Chem*, **275**, 23899-903.

[65] Moynahan M. E., Chiu J. W., Koller B. H., Jasin M. (1999). Brca1 controls homology-directed DNA repair. *Mol Cell*, **4**, 511-8.

[66] Snouwaert J. N., Gowen L. C., Latour A. M., Mohn A. R., Xiao A., DiBiase L., et al. (1999). BRCA1 deficient embryonic stem cells display a decreased homologous recombination frequency and an increased frequency of non-homologous recombination that is corrected by expression of a Brca1 transgene. *Oncogene*, **18**, 7900-7.

[67] Levitus M., Waisfisz Q., Godthelp B. C., de Vries Y., Hussain S., Wiegant W. W., et al. (2005). The DNA helicase BRIP1 is defective in Fanconi anemia complementation group J. *Nat Genet*, **37**, 934-5.

[68] Nakanishi K., Yang Y. G., Pierce A. J., Taniguchi T., Digweed M., D'Andrea A. D., et al. (2005). Human Fanconi anemia monoubiquitination pathway promotes homologous DNA repair. *Proc Natl Acad Sci USA*, **102**, 1110-5.

[69] Yamamoto K., Ishiai M., Matsushita N., Arakawa H., Lamerdin J. E., Buerstedde J. M., et al. (2003). Fanconi anemia FANCG protein in mitigating radiation- and enzyme-induced DNA double-strand breaks by homologous recombination in vertebrate cells. *Mol Cell Biol*, **23**, 5421-30.

[70] Yamamoto K., Hirano S., Ishiai M., Morishima K., Kitao H., Namikoshi K., et al. (2005). Fanconi anemia protein FANCD2 promotes immunoglobulin gene conversion and DNA repair through a mechanism related to homologous recombination. *Mol Cell Biol*, **25**, 34-43.

[71] Digweed M., Rothe S., Demuth I., Scholz R., Schindler D., Stumm M., et al. (2002). Attenuation of the formation of DNA-repair foci containing RAD51 in Fanconi anaemia. *Carcinogenesis*, **23**, 1121-6.

[72] Wang X., Andreassen P. R., D'Andrea A. D. (2004). Functional interaction of monoubiquitinated FANCD2 and BRCA2/FANCD1 in chromatin. *Mol Cell Biol*, **24**, 5850-62.

[73] Godthelp B. C., Wiegant W. W., Van Duijn-Goedhart A., Scharer O. D., Van Buul P. P., Kanaar R., et al. (2002). Mammalian Rad51C contributes to DNA cross-link resistance, sister chromatid cohesion and genomic stability. *Nucleic Acids Res*, **30**, 2172-82.

[74] Godthelp B. C., Wiegant W. W., Waisfisz Q., Medhurst A. L., Arwert F., Joenje H., et al. (2006). Inducibility of nuclear Rad51 foci after DNA damage distinguishes all Fanconi anemia complementation groups from D1/BRCA2. *Mutat Res*, **594**, 39-48.

[75] Tebbs R. S., Hinz J. M., Yamada N. A., Wilson J. B., Salazar E. P., Thomas C. B., et al. (2005). New insights into the Fanconi anemia pathway from an isogenic FancG hamster CHO mutant. *DNA Repair*, **4**, 11-22.

[76] Tebbs R. S., Zhao Y., Tucker J. D., Scheerer J. B., Siciliano M. J., Hwang M., et al. (1995). Correction of chromosomal instability and sensitivity to diverse mutagens by a cloned cDNA of the XRCC3 DNA repair gene. *Proc Natl Acad Sci USA*, **92**, 6354-8.

[77] Liu N., Lamerdin J. E., Tebbs R. S., Schild D., Tucker J. D., Shen M. R., et al. (1998). XRCC2 and XRCC3, new human Rad51-family members, promote chromosome stability and protect against DNA crosslinks and other damages. *Mol Cell*, **1**, 783-93.

[78] Dronkert M. L., Kanaar R. (2001). Repair of DNA interstrand cross-links. *Mutat Res*, **486**, 217-47.

[79] Casado J. A., Nunez M. I., Segovia J. C., Ruiz de Almodovar J. M., Bueren J. A. (2005). Non-homologous end-joining defect in fanconi anemia hematopoietic cells exposed to ionizing radiation. *Radiat Res*, **164**, 635-41.

[80] Donahue S. L., Campbell C. (2002). A DNA double strand break repair defect in Fanconi anemia fibroblasts. *J Biol Chem*, **277**, 46243-7.

[81] Hinz J. M., Nham P. B., Salazar E. P., Thompson L. H. (2006). The Fanconi anemia pathway limits the severity of mutagenesis. *DNA Repair (Amst)*, **5**, 875-84.

[82] Lundberg R., Mavinakere M., Campbell C. (2001). Deficient DNA end joining activity in extracts from fanconi anemia fibroblasts. *J Biol Chem*, **276**, 9543-9.

[83] Escarceller M., Rousset E., Moustacchi E., Papadopoulo D. (1997). The fidelity of double strand breaks processing is impaired in complementation groups B and D of Fanconi anemia. *Somat Cell Mol Genet*, **23**, 401-11.

[84] Escarceller M., Buchwald M., Singleton B. K., Jeggo P. A., Jackson S. P., Moustacchi E., et al. (1998). Fanconi anemia C gene product plays a role in the fidelity of blunt DNA end-joining. *J Mol Biol*, **279**, 375-85.

[85] Houghtaling S., Newell A., Akkari Y., Taniguchi T., Olson S., Grompe M. (2005). Fancd2 functions in a double strand break repair pathway that is distinct from non-homologous end joining. *Hum Mol Genet*, **14**, 3027-33.

[86] Donahue S. L., Campbell C. (2004). A Rad50-dependent pathway of DNA repair is deficient in Fanconi anemia fibroblasts. *Nucleic Acids Res*, **32**, 3248-57.

[87] Donahue S. L., Tabah A. A., Schmitz K., Aaron A., Campbell C. (2007). Defective signal joint recombination in fanconi anemia fibroblasts reveals a role for Rad50 in V(D)J recombination. *J Mol Biol*, **370**, 449-58.

[88] Smith J., Andrau J. C., Kallenbach S., Laquerbe A., Doyen N., Papadopoulo D. (1998). Abnormal rearrangements associated with V(D)J recombination in Fanconi anemia. *J Mol Biol*, **281**, 815-25.

[89] Thompson L. H., Fong S., Brookman K. (1980). Validation of conditions for efficient detection of HPRT and APRT mutations in suspension-cultured Chinese hamster cells. *Mutat Res*, **74**, 21-36.

[90] Papadopoulo D., Porfirio B., Moustacchi E. (1990). Mutagenic response of Fanconi's anemia cells from a defined complementation group after treatment with photoactivated bifunctional psoralens. *Cancer Res*, **50**, 3289-94.

[91] Papadopoulo D., Guillouf C., Mohrenweiser H., Moustacchi E. (1990). Hypomutability in Fanconi anemia cells is associated with increased deletion frequency at the HPRT locus. *Proc Natl Acad Sci USA*, **87**, 8383-7.

[92] Laquerbe A., Sala-Trepat M., Vives C., Escarceller M., Papadopoulo D. (1999). Molecular spectra of HPRT deletion mutations in circulating T-lymphocytes in Fanconi anemia patients. *Mutat Res*, **431**, 341-50.

[93] Sala-Trepat M., Boyse J., Richard P., Papadopoulo D., Moustacchi E. (1993). Frequencies of HPRT- lymphocytes and glycophorin A variants erythrocytes in Fanconi anemia patients, their parents and control donors. *Mutat Res*, **289**, 115-26.

[94] Evdokimova V. N., McLoughlin R. K., Wenger S. L., Grant S. G. (2005). Use of the glycophorin A somatic mutation assay for rapid, unambiguous identification of Fanconi anemia homozygotes regardless of GPA genotype. *Am J Med Genet A*, **135**, 59-65.

[95] Araten D. J., Golde D. W., Zhang R. H., Thaler H. T., Gargiulo L., Notaro R., et al. (2005). A quantitative measurement of the human somatic mutation rate. *Cancer Res*, **65**, 8111-7.

[96] Luria S. E., Delbrück M. (1943). Mutations of bacteria from virus sensitivity to virus resistance. *Genetics*, **28**, 491-511.

[97] Hinz J. M., Nham P. B., Urbin S. S., Jones I. M., Thompson L. H. (2007). Disparate contributions of the Fanconi anemia pathway and homologous recombination in preventing spontaneous mutagenesis. *Nucleic Acids Res*, **35**, 3733-40.

In: Genetic Recombination Research Progress
Editor: Jacob H. Schulz, pp. 265-278

ISBN: 978-1-60456-482-2
© 2008 Nova Science Publishers, Inc.

Chapter 10

Extent and Limits of Genetic Recombination since the Origin of Life

Jean-Pierre Gratia[*]

Pasteur Institute of Brussels, 642, Rue Engeland, B-1180 Brussels

Abstract

Recombination might be an evolutionary development as ancient as the origin of life. In spite of numerous data on the mechanisms of genetic recombination in living organisms since the discovery of Rec mutations and of the existence of genetic exchanges in bacteria, some genetic mechanisms in these organisms remain not fully understood, notably in relation to the genetic diversity of bacteria and to their position with respect to the very first organisms having appeared on earth, ancestors of both eukaryotes and prokaryotes. Spontaneous zygogenesis (or Z-mating), recently discovered in *Escherichia coli*, is intriguing, as it resembles fusion of gametes in eukaryotes in that it involves complete genetic mixing. It is also question on how genetic exchanges can enable bacteria having undergone severe deletions or temporary chromosomal inactivation to survive and overcome what could be a selective disadvantage.

Introduction

Achievements made possible by artificial recombination, notably thanks to the availability of complete genome sequences for organisms such as *Escherichia coli*, yeast, and the human species, may have led some, and particularly research funders, to believe that the story of recombination has reached such a lofty height that there is little point in supporting basic research into the mechanisms of recombination in nature. Yet the relatively recent discovery of the interdependence of DNA replication, homologous recombination, and DNA repair shows that there is still much to be learned, and future data in this area may contribute

---

[*] E-mail address: jpgratia@yahoo.fr. Telephone (private number) 00 32.2.673.88.24, Fax : 00 32 2 373 71 14

importantly to maintaining biodiversity, unravelling complex diseases mechanisms (*e.g.*, carcinogenesis), or even predicting human evolution.

Although bacteria are perhaps the most privileged tool of geneticists today, the mere idea that these organisms might have a genetic apparatus was hardly accepted until the discovery of bacterial transformation, conjugation, and transduction around 1950. Recombination has become the principal tool of genetic analysis or manipulation. Yet molecular biology, despite its achievements, has done little to illuminate what Hayes described in 1968 as one of the few dark corners of cell behaviour [1]. Efforts to lift the veil are leading scientists to challenge accepted views regarding bacteria, their organization, their origin, and their evolution. Evidence also suggests that diploidy is neither a prerequisite for sexuality nor a feature specific to eukaryotes alone.

When Did Recombination First Arise?

In 1964, Muller proposed that sex can occur in asexual populations, enabling them to counter the irreversible accumulation of deleterious mutations [2]. Very recently, Lehman went a step further. On the basis of four lines of evidence, he suggested that recombination is an evolutionary development as ancient as the origin of life, and that sexual reproduction was contemporaneous with asexuality [3]. It is indeed hard to imagine how life could have developed on earth if deleterious mutations accumulating over time were not corrected by dominant positive alleles provided either by complementation or by homologous recombination. Furthermore, the recently established fact that DNA double strand breaks, which occur frequently at replication forks, are corrected by homologous recombination answers the question of how replication can restart after DNA damage. In the absence of such a mechanism, it seems that DNA synthesis would stop and the cell be prevented from dividing and reproducing. It thus seems unimaginable that the first protocells did not possess a minimal set of proteins required for polymerisation, DNA repair, and resolution of replication forks through homologous recombination.

Survival and Evolution through Recombination

At least 25 genes are involved in homologous recombination, which was long erroneously considered not to generate genetic diversity in a population. The first gene identified, *recA*, was isolated by Clark and Margulies in 1971 as a mutation that abolished conjugal recombination in *E. coli* without disrupting the process of conjugation [4]. The RecA protein later turned out to be essential to survival after DNA damage, because it maintains the integrity of the arrested replication fork and signals upregulation of over 40 gene products, most of which are required to restore the genomic template and to facilitate resumption of processive replication [5]. Despite a considerable focus on the interdependence of homologous recombination, replication, and DNA repair since the relatively recent studies on phages and bacteria.[6-11], the roles of many recombination functions are still imperfectly understood [12].

In *Saccharomyces cerevisiae* as well as in bacteria, the mechanisms involved in homologous recombination in double-strand break (DSB) repair require the existence, in the cell, of sequences homologous to those affected by the DSB. Non-homologous end-joining also repairs DSBs in *S. cerevisiae*, but less efficiently than homologous recombination [13-15]. In contrast, non-homologous end joining appears to be of greater importance in mammalian cells [16]. The yeast Rad52 protein, interacting with Rad51 homologous to RecA [15], plays a crucial role in recombination. When it was labelled with the green fluorescent protein, it was found to relocalize from a diffuse nuclear distribution to distinct foci, exclusively during the S phase of mitotic cells. This is consistent with coordinated recombinational repair and DNA replication [17].

E. coli possesses an inducible "SOS response", important in DNA repair and responsible for induced mutagenesis. Induced or activated following damage to DNA, this response requires de novo protein synthesis and several genetic functions, including $recA^+$ and lex^+ [18]. It increases the survival of bacteria exposed to DNA-damaging agents by increasing the capacity of error-free and error-prone DNA repair systems. Inactive during normal aerobic growth, it is induced in a variety of situations likely to occur in nature, and notably in resting bacterial populations [19]. Replication forks are routinely inactivated under such conditions, as shown by Cox et al who note that "If any one of the events (identified through a long series of studies on the interdependence between homologous recombination, replication and DNA repair) fails to take place, the affected cell will either die or undergo an aberrant cell division event. This potential catastrophe may have provided the selective pressure needed for the evolution of homologous recombination systems and other enzymatic components needed for fork reactivation, a critical step paving the way for the evolution of organisms with larger genomes" [20].

Since the SOS system is induced in stressed, non-growing cells, mutations appear which are different from those found during exponential growth. They are called adaptive mutations. They are important, because stationary phase mechanisms may provide models for mutational escape from growth control (*e.g.*, in tumour progression). Their existence implies that evolution-fuelling genetic changes may be accelerated during stress [21, 22].

DNA synthesis associated with recombination may be an important source of spontaneous mutations in non-proliferating cells. The movement of insertion elements, involving site-specific recombination, not only inactivates genes but also provides sequence homology allowing large-scale genomic rearrangements [23]. Moreover, some conjugative plasmids can recombine with the host chromosome and acquire chromosomal genes that can then spread through the population and even to other species (see below).

Insertions and Deletions: Gene Rearrangements Responsible for the Genetic Diversity during Evolution

The linear insertion of prophages into the host chromosome was the first identified example of site-specific recombination. Campbell's model, proposed in 1962 for lambda in lysogens or for the F factor in Hfr strains, received support from P1 transduction analysis of λ lysogens and from more demonstrative deletion mapping of lysogens of Ø80 or Ø80λ hybrids located near the tryptophan operon (see [24, 25, 33] for references). I extended this study to phage Øγ, which shares with Ø80 the same attachment site *att80* on the bacterial chromosome but is differently organised (circular permutation and terminal redundancy of the DNA molecule instead of a *cos*-ended chromosome like that of λ or Ø80) [23]. Surprisingly, Øγ can pick up at a high rate a relatively long segment of the bacterial chromosome, preferentially to the right of *att80*, even upon lytic development in RecA⁻ cells, by a still unknown recombination mechanism. It resembles the *E. coli* phage P1 or Salmonella phage P22 in this respect, except that it is a specialized, not a generalized, transducer.

Phages λ, Ø80, Øγ, and Mu all insert into the host chromosome by means of tyrosine recombinases [27, 28]. Yet, Mu and transposable elements replicate in the transposition process, remaining at their original site while spreading to new sites. This differs from insertion-excision of lambda [29].

Any repeats on a chromosome can induce homologous recombination. The most common are insertion sequences, a major source of mutations (deletions, inversions, duplications) [29-32]. Another deletion-generating process is illegitimate recombination, which does not require pairing of homologous sequences. Illustrative of this mechanism are deletions of the *tonB trp* segment near 27 min on the *E. coli* chromosome [25, 33] .The end points are located anywhere within the tryptophan operon, within and over the *tonB* locus, between this locus and - or over - *att80*. If a prophage is present at this site, the deletion may end within it. As pointed out by Franklin [25], these observations make unlikely the existence of extensive nucleotide sequence complementarity between the ends of deleted segments. The recombination event responsible for such deletions is also involved in the formation of transducing phages, since their genomic composition corresponds with the segment deleted in the chromosome. Deletion does not seem to depend on DNA repair or homologous recombination functions as TonB trp deletions are detectable in RecA⁻ strains.

The acquisition of DNA and the loss of genetic information are two important mechanisms that contribute to strain-specific differences in genome content. A comparative analysis of strains of *Mycobacterium smegmatis,* performed in this Institute by Wang *et al.* (submitted), revealed that the ATCC 607 genome of *M. smegmatis* contains eleven insertion sequences IS*1096* and even more in its derivative mc²155. As mc²155 evolved, there was a considerable expansion in the copy number of IS*1096* (+14) and a large duplication of a 56-kb fragment flanked on both sides by IS*1096*; while concurrently one IS*1096* element and its flank were deleted. This study confirms that IS expansion and IS-induced rearrangements such as duplication and deletion are major forces driving genomic evolution.

Genome Evolution through Multi-step Co-integration Mechanisms in Cell Chromosomes and Plasmids

It is clear that an increasing DNA content in a cell is an important mechanism of evolution. The following are two examples among many that illustrate how the analysis of various forms of genetic recombination contributes to understanding genome enlargement.

Figure 1 illustrates how viruses might participate in the evolution of the organisms they infect. Deletion lysogens of Ø80 have been isolated, where the deletion spans the terminal sequence of the prophage without reaching the phage genome essential to lytic development

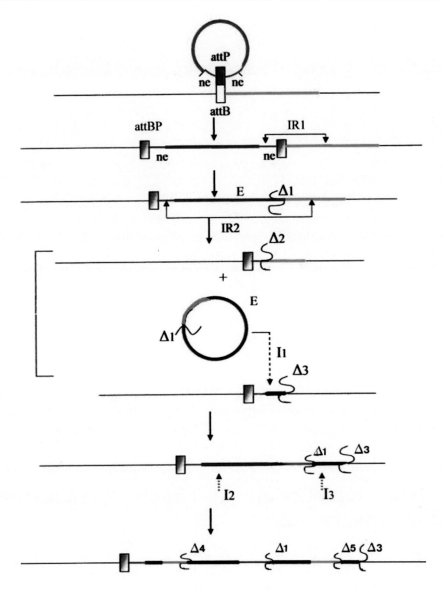

Figure 1. Schematic representation of multistep transduction of *trp* genes in a strain subjected to the complete deletion of the tryptophan operon, using non defective transducing Ø80ptrp phages carrying various segments of that operon (see reference [33] for details and references). IR1 means illegitimate recombination event in a lysogen consisting in a crossover between some point in the non essential region of the Ø80 prophage (ne) and some point within the *trp* operon, generating deletion Δ1. The resulting deletion lysogen becomes unable to produce anymore a high amount of phage particles after induction. An infrequent abnormal excision IR2 generates defective lysogeny by deletion Δ2 and production at a low frequency of plaque forming transducing phage carrying a segment of the *trp* operon (circular form converted to linear form through terminase activity at the *cos* site at the centre of the intact essential phage genome (E)). The genome of the produced phage superinfects a defective lysogen lacking by deletion Δ3 part of the essential phage genome and all the tryptophan genes. This superinfecting phage genome, then re-circularized, is inserted in the residual prophage at a site of genetic homology (I1). I2 and I3: stepwise insertion of other ptrp phages carrying complementary and overlapping segments of the *trp* operon.

(IR1 for illegitimate recombination event 1 in Fig. 1). Such deletion lysogens can produce phage at a frequency of ca.10^{-5} per induced cell through illegitimate recombination (excision by site-specific recombination being prevented by lack of the right side prophage end) involving a crossover at some point on the right side of the distal *attBP* end of the prophage and within the tryptophan operon. Plaque-forming transducing phages Ø80ptrp, generated in different deletion lysogens, can be used to transduce tryptophan-biosynthetic genes into a long deletion lysogen having lost the whole *trp* operon [33]. Provided each transduced segment has a promoter (which is possible, since the *trp* operon possesses at least two promoters [34]) and, if the required tryptophan precursors are supplied, one can introduce the genes in a stepwise fashion, through successive insertions. This generates mosaic structures where chromosomal genes alternate with prophage genes in various orders, according to the sites where a crossover is possible for insertion of the transducing genomes.

There are also some fascinating studies on plasmids that occasionally insert into the bacterial chromosome. The mechanism of linear insertion through a unique crossover between circular genomes was imagined by Fredericq, already one year after Campbell proposed his model, to explain recombinational events in a partial diploid where fertility in conjugation was associated with colicinogeny and chromosomal markers in the tryptophan region [35]. I remember him drawing a series of small circles around a larger one. In crosses involving various R factors, he obtained a strain carrying a long plasmid composed of several Col and R factors associated with a chromosomal segment including the prophage Ø80. He could map the order of the various functions by conjugation and P1 transduction [36], and results obtained by Øγ-mediated transduction confirmed linkage of the various genetic determinants on a single genetic structure [37]. This "instructive example of the recombinational gymnastics possible between heterogenic plasmids" [38] represents an important step in evolutionary genetics.

Do a True Sexuality and Epigenetic Phenomena Exist in Prokaryotes?

Not long after the period just mentioned, Schaeffer achieved something quite revolutionary: induced fusion of bacteria, a kind of "assisted procreation". He converted *Bacillus subtilis* cells to protoplasts, which were able to fuse to form diploid bacteria [39]. Still more remarkable was the finding that these "mating products" were diploids expressing only one chromosome, with possible switching between complete sets of markers [40]. Comparable or related findings were obtained with other bacteria, such as *Bacillus megatherium, Providencia alcalifaciens*, or even *Streptomyces coelicolor* (see [41] for references).

The only known process of spontaneous genetic exchange between two bacterial chromosomes is conjugation. Yet as early as 1949, Lederberg interpreted his discovery of genetic recombination in *E. coli* as resulting from fusion between parental cells. He found in certain crosses an appreciable proportion of prototrophs behaving like stable heterozygotes [42]. These prototrophic strains were confirmed as diploid heterozygotes by pedigree analysis of single cells isolated with a micromanipulator [43]. It is too bad one was unable to repeat these observations and confirm such interpretation.

In my opinion, cell fusion is the most likely explanation of a phenomenon I call spontaneous zygogenesis or "Z-mating" [44-47]. It differs from F-plasmid-promoted conjugation in several ways:

Z-mating occurs between F⁻ *E. coli* K12 strains. The first Szp⁺ clone (thus named because it possesses the "spontaneous zygogenesis property") was isolated from an F⁻ strain incapable of classical genetic transfer. As the history of this strain suggested that a lysogenic clinical isolate used in the initial experiments might contain the genetic determinant of Z-mating (most likely on a phage), it was attempted to create Szp⁺ strains from scratch by exposing various F- strains to a lysate of this isolate. These attempts proved successful.

Z-mating results in complete genetic mixing under conditions that preclude plasmid-mediated conjugation.

In Z-mating, there is no evidence of directional transfer. Products possessing all markers of both parental strains appear almost immediately (parental cells can be co-incubated for as little as 1 minute before vortexing and plating), whatever their position on the bacterial chromosome. Longer incubation fails to increase the frequency of any marker. This contrasts sharply with Hfr-mediated conjugation, as illustrated in Fig. 2.

Immunofluorescence and electron microscopy after single and double labelling of DNA indicate that Szp⁺ bacilli join through their poles [45]. Accordingly, coccal forms of such bacteria, produced by exposure to the drug mecillinam (which specifically blocks penicillin-binding protein 2, involved in lateral cell elongation), attach to each other at any place on their surface. They end up forming giant syncytia-like spheres possessing a single outer envelope (as evidenced notably by the appearance of giant "ghosts" when these structures lyse) [47].

As in the case of the *B. subtilis* exfusants [41, 48-50], diploidy is unstable and most often non-complementing in *E. coli* Z-mating products [44, 46]. This has raised a controversy [51, 52] regarding the diploid status of such cells, as it is not easy to distinguish a non-complementing diploid (which may express the same chromosome for many generations) from a haploid (and indeed, it seems that both *B. subtilis* exfusants and Z-mating products can evolve to true haploidy). Non-complementing diploidy is revealed by phenotype switching, when cells expressing one set of markers yield progeny expressing a whole different set.

Fig. 2 Illustration of events in crosses between an Hfr strain and either an F⁻ or an isogenic Szp⁺ recipient. In A and B, homologous recombination in both crosses gives rise to stable recombinants when streptomycin is used to kill the donor (Hfr P4X transferring the chromosome counterclockwise from *pro*). In B, selection on minimal agar (MA) in the absence of streptomycin gives rise to the formation of recombinants at low frequency, through rare crossover between close markers. In C, are represented the products of the cross between Hfr and Szp⁺ cells, not found in a classical cross between Hfr and F⁻ cells, *i.e.*, completely diploid heterozygotes expressing either parental chromosome

This epigenetic silencing of a whole chromosome in a diploid prokaryotic cell is reminiscent of X chromosome silencing in mammals, discovered in 1962 by Mary Lyon. Although each female mammal inherits an active X chromosome from each parent, by the thousand-cell stage of embryogenesis, transcription of one X chromosome in each cell has become inactivated. Adult female mammals are thus mosaics of cell lineages in which either the maternal or the paternal X chromosome is active. Inactivation of the X chromosome is fully reversible [53]. It is noteworthy that unstable expression of all the dominant markers of both parents is observed also with *B. subtilis* exfusants and *E. coli* Z-mating products. These similarities between mammals and prokaryotes raise questions about the nature of bacteria, their genetics, and their place in evolution.

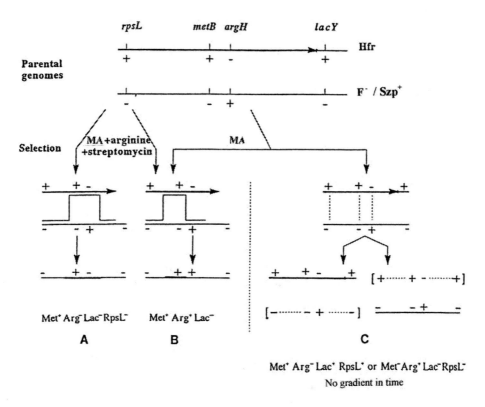

Figure 2. Illustration of events in crosses between an Hfr strain and either an F⁻ or an isogenic Szp⁺ recipient. In A and B, homologous recombination in both crosses gives rise to stable recombinants when streptomycin is used to kill the donor (Hfr P4X transferring the chromosome counterclockwise from *pro*). In B, selection on minimal agar (MA) in the absence of streptomycin gives rise to the formation of recombinants at low frequency, through rare crossover between close markers. In C, are represented the products of the cross between Hfr and Szp⁺ cells, not found in a classical cross between Hfr and F⁻ cells, *i.e.,* completely diploid heterozygotes expressing either parental chromosome

B. subtilis exfusants and *E. coli* Szp⁺ strains differ in the fate of diploid cells in a population. The latter strains can mate with any other strain. Thus, whereas *B. subtilis* exfusants evolve on their own after the initial fusion event, addition of an Szp⁺ strain to a bacterial population should give rise to a succession of matings. It does not seem that these cells accumulate, through successive matings, a large number of parental genomes. Polyploidy is observed only in the large syncytial cells mentioned above, which do not live long. A rare case of triploidy, where only one chromosome was expressed, was identified by chance [46]. Could it be that haploidization is the rule before remating? If so, this would suggest that chromatic reduction, viewed as characteristic of sexual reproduction, might in fact be ancient.

What about recombination and segregation in such haploidization? Unlike the products of meiotic segregation after double cross-overs in eukaryotes, reciprocal recombinants in the descent of each unstable *E. coli* clone selected for complementing diploidy were never found.

In the course of extensive segregation analyses of Z-mating products after many different crosses, I have consistently found either two parental types or one parental type and one recombinant type, never two reciprocal recombinants [45, 46]. This is not the case with *B. subtilis* exfusants [40; L. Hirschbein, personal communication]. Could it be that induced fusion of *B. subtilis* protoplasts and the hypothetical fusion of outer membranes at the poles of Szp$^+$ *E. coli* cells promote different molecular mechanisms at the level of DNA?

In the analysis of *B. subtilis* exfusants, haploid cells with a phenotype corresponding to the inactive chromosome were never observed [41]. In the descent of segregating non complementing diploid *E. coli* Z-mating products, a higher amount of lethal cells was recorded [46]. Therefore, in both cases, irreversible inactivation would prevent the formation of viable haploid cells. The great advantage of the Z-mating system is that it might contribute to help some of the potential haploid cells with a silent chromosome to survive and overcome what could be a selective disadvantage through introduction of a complete active chromosome.

These results contradict the view that sexuality is absent in bacteria and that it did not appear until the emergence of different sexes or mating types in eukaryotes. As true sexuality is linked to diploidy, i.e. coordinate replication of both parental genomes, genetic exchanges in bacteria have always been viewed as parasexual events preventing bacteria from becoming diploid. Yet the above findings clearly show that even if a recessive mutation is expressed as soon as it marks a bacterial chromosome, this does not necessarily mean that the bacterial cell is always haploid. Such a cell may multiply while maintaining two chromosomes, one of which is silenced. If the recessive mutation is on the expressed chromosome, it will be expressed even if a dominant allele is present on the silenced chromosome.

Wide-Ranging Horizontal Gene Transfer: Implications in Evolution and Phylogeny

The discovery of horizontal DNA transfer between unrelated organisms has important implications in evolutionary genetics. Plasmids have been constructed that are capable of conjugative transfer from *E. coli* to Gram-positive *Bacillus, Listeria, Staphylococcus* and *Streptococcus* [54] or from *Enterococcus* to *E. coli* [55], even from *E. coli* to the Actinomycetales, including *Streptomyces* [56].

Conjugative plasmids from *E. coli* are also even transferable to the budding yeast *S. cerevisiae*, via mechanisms similar to those involved in bacterial conjugation and requiring the participation of plasmid genes *tra* and *mob* [57]. *Agrobacterium tumefaciens* causes crown gall in plants by insertion of the Ti plasmid at a semi-random location in the plant genome. In the presence of opines, it produces a diffusible conjugation signal which, by activating transcription factor TraR, upregulates transcription of genes required for conjugation [58, 59].

Because such observations concern only extrachromosomal DNA (phages, plasmids, transposons), the ability to exchange and integrate chromosomal DNA via homologous recombination has been proposed as a basis for species definition in bacteria as well as eukaryotes. DNA sequence analysis indicates that there has been little transfer of chromosomal genes between *E. coli* and *Salmonella enterica* [60].

Recombination between two different species displaying partial genetic homology appears to depend largely on the degree of homology. In crosses between *E. coli* and *Salmonella enterica,* for example, the frequency of crossovers can be several orders of magnitude higher in regions of high homology than in regions of low homology. This effect is largely mismatch repair activity, since recombination between quasi-homologous sequences is increased by *mutL* and *mutS* mutations in the Salmonella recipient [61]. The MutS and MutL proteins appear to block branch migration, presumably in response to mispairing within the newly formed heteroduplex [62].

Enzymatic restriction of foreign DNA at unmethylated sites is another barrier to inter-species recombination. Yet protection against infecting DNA is limited and short-lived on account of the modification system. On the other hand, single-stranded DNA is refractory to most restriction systems, which may explain the success of transformation with chromosomal markers from various bacterial species [63]. Fragmentation of phage or plasmid DNA leads to its complete elimination, whereas DNA fragments generated from a bacterial chromosome subjected to restriction after conjugational transfer may be rescued by homologous recombination with the host genome.

The chromosomal DNA transferred by transduction or conjugation is unable to replicate and has therefore no fate in the descent of the transductant or conjugant in the absence of homologous recombination. Such problem does not concern Z-mating, which can occur between *E. coli* and other bacterial species and where restriction seems to have no effect. After crosses between auxotrophic *E. coli* Szp$^+$ and various F$^-$ Hsd$^+$ or Hsd$^-$ auxotrophic strains of *Salmonella enterica,* no significant difference was observed in the frequency of prototrophic colonies, whether or not the Salmonella partner was restriction-less [64]. This argues in favour of cytoplasmic mixing and cell fusion.

Conservation of gene order in prokaryotes has become important in predicting protein function. Local gene-order conservation reflects functional constraints within the protein. Despite differences in DNA replication mechanisms between Bacteria and Archaea [65], the mode of replication in pyrococci resembles that of eubacteria in that replication begins at a defined single origin and proceeds bidirectionally [66]. This suggests that the proteins involved in replication and recombination might be involved in genome rearrangement in the Archaea.

Genetic transfer in a halophilic archaebacterium occurs through a conjugational mechanism differing from bacterial conjugation in that both parental types can act as donor or recipient. Transfer occurs through cytoplasmic bridges, but the cytoplasmic continuity between mated cells is restricted to certain molecules and there is no general cytoplasmic mixing [67]. The thermophilic bacterium *Aquifex* appears to have acquired hyperthermophilic traits through lateral gene transfer from Archaea, since its reverse gyrase gene is clustered on the chromosome with other archaeal-like genes [68].

The finding of a eukaryotic protein, actin, in both *E. coli* and an archaeon (see [69] for references) suggests that this protein was already present in a hypothetical protoeukaryotic last common ancestor. This and the discovery that Archaea form a third domain [65] render questionable the long-held view that prokaryotes preceded eukaryotes and represent the protocells. They favour, instead, the hypothesis of a regressive evolution from proto-eukaryotes to bacteria [70, 71]. As mentioned above about serial Z-mating in a bacterial population, chromatic reduction may be ancient. Perhaps proto-eukaryotes had their own version of meiosis.

Conclusion

Many questions remain to be answered about recombination and its bearing on evolution. Too many notions, considered established by the scientific community, appear to me as dogmas that should be revisited. This applies notably to the idea that sex in bacteria is limited to plasmid-promoted conjugation. The existence of Z-mating, notably, raises questions regarding the origin of sexuality and challenges the concept of absolute haploidy (with merodiploidy just after mating). It also seems quite reasonable to reject the idea that today's prokaryotes are representatives of protocells. Petes and Hill [30] wrote with good reason that "studies are sufficiently time-consuming to keep geneticists happily employed almost indefinitely".

References

[1]   Hayes W. (1970). *The genetics of bacteria and their viruses* (2nd edition). Oxford & Edinburgh: Blackwell Sci. Publ.

[2]   Muller, H.J. (1964). The relation of recombination to mutational advance. *Mutat. Res.*, **1**, 2-9.

[3]   Lehman, N. (2003). A case for the extreme antiquity of recombination. *J. Mol. Evol.*, **56**, 770-777.

[4]   Clark, A.J. & Margulies, A.D. (1965). Isolation and characterization of recombination-deficient mutants of *Escherichia coli* KProc. *Natl. Acad. Sci. USA*, **53**, 451-459.

[5]   Courcelle, J. & Hanawalt, P.C. (2003). RecA-dependent recovery of arrested DNA replication forks *Annu. Rev. Genet.*, **37**, 611-646.

[6]   Skalka, A.M. (1974). A replicator's view of recombination (and repair). In R.F. Grell (ed.), *Mechanisms in recombination* (pp.421-432). New York: Plenum Publish. Corp.

[7]   Mosig, G. (1987). The essential role of recombination in phage T4 growth. *Annu. Rev. Genet.*, **21**, 347-371.

[8]   Kogoma, T. (1997). Stable DNA replication: interplay between DNA replication, homologous recombination, and transcription. *Mol. Biol. Rev.*, **61**, 212-238.

[9]   Kowalczykowski, S.C. (2000). Initiation of genetic recombination and recombination-dependent replication, *Trends Biochem. Sci.*, **25**, 156-164.

[10] Michel, B., Flores, M-J., Viguera, E., Grompone, G. Seigneur, M. & Bidnenko, V. (2001). Rescue of arrested replication forks by homologous recombination *Proc. Natl. Acad. Sci. USA*, **98**, 8181-8188.

[11] Kuzminov, A. (2001). DNA replication meets genetic exchanges: chromosomal damage and its repair by homologous recombination. *Proc. Natl Acad. Sci. USA*, **98**, 8461-8468

[12] Cox, M. M. (2001). Recombinational DNA repair of damaged replication forks in *Escherichia coli*: questions. *Annu. Rev. Genet.*, **35**, 53-82.

[13] Florers-Rozas, H. & Kolodner, R.D. (2000). Links between replication, recombination, and genome instability in eukaryotes. *Trends Biochem. Sci.*, **25**, 196-2

[14] Borde, V., Golman, A.H. & Lichten, M. (2000). Direct coupling between meiotic DNA replication and recombination initiation. *Science*, **290**, 806-807.

[15] Shinohara, A., Ogawa, H. & Ogawa, T. (1992) Rad51 protein involved in repair and recombination in *S. cerevisiae* is a RecA-like protein. *Cell*, **69**, 457-470.

[16] Critchlow, S. E. & Jackson, S.P. (1998) DNA end-joining: from yeast to man. *Trends Biochem. Sci.*, **23**, 269-275

[17] Lisby, M., Rothstein, R. & Mortensen, W.H. (2001). Rad52 forms DNA repair and recombination centers during S phase. *Proc. Natl. Acad. Sci. USA*, **98**, 8276-8282.

[18] Radman, M. (1975). SOS repair hypothesis: phenomenology of an inducible DNA repair which is accompanied by mutagenesis. In P.H. Hanawalt, & R.C Setlow (Eds), Molecular Mechanisms for Repair of DNA (pp.355-367). New-York: Plenum Press.

[19] Taddei, F. Matic, I.& Radman, M. (1995). cAMP-dependent SOS induction and mutagenesis in resting bacterial populations. *Proc. Natl. Acad. Sci. USA*, **92**, 11736-11740.

[20] Cox, M.M., Goodman, M.F., Kreuzer, K.N., Sherratt, D.J., Sandler, S.J., & Marians, K.K. (2000). The importance of repairing stalled replication forks. *Nature*, **404**, 37-41.

[21] McKenzie, G.J., Harris, R.S., Lee, P.L. & Rosenberg, S.M. (2000). The SOS Response regulates adaptative mutation. *Proc. Natl Acad. Sci. USA*, **97**, 6646-6651.

[22] Bull, H.J., Lombardo, M-J. & Rosenberg, S.M. (2001). Stationary-phase mutation in the bacterial chromosome: recombination protein and DNA polymerase IV-dependence. *Proc. Natl. Acad. Sci. USA*, **98**, 8334-8341.

[23] Foster, P. (1999). Mechanisms of stationary phase mutation: a decade of adaptative mutation. *Annu. Rev. Genet.*, **33**, 57-88.

[24] Campbell, A. (1971) Genetic structure. In A.D. Hershey (Ed.), The Bacteriophage lambda (pp.13-44). New-York: Cold Spring Harbor Lab. Cold Spring Harbor.

[25] Franklin, N.C. (1971). Illegitimate recombination. In A.D. Hershey (Ed.), The Bacteriophage lambda (pp. 175-194). New-York: Cold Spring Harbor Laboratory Press

[26] Gratia, J.P. (1989). Genome organization in hybrids between prophage Ø80 and *Escherichia coli* virus Øγ. *Res. Virol.*, **140**, 373-388.

[27] Esposito, D. & Scocca, J.J. (1997). The integrase family of tyrosine recombinases: evolution of a conserved active site domain. *Nucleic Ac. Res.*, **25**, 3605-3614.

[28] Grindley, N.D.F., Whiteson, K.L., & Rice P.A. (2006). Mechanisms of site-specific recombination. *Annu. Rev. Biochem.*, **75**, 567-605

[29] Shapiro, J.A. (1979). Molecular model for the transposition and replication of bacteriophage Mu and other transposable elements. *Proc. Natl Acad. Sci. USA*, **76**, 1933-1937

[30] Petes, T.D & Hill, C.W. (1988). Recombination between repeated genes in microorganisms. *Annu. Rev. Genet.*, **22**, 147–168.

[31] Shyamala,V., Schneider, E. & Ames, G.F-L. (1990). Tandem chromosomal duplications: role of REP sequences in the recombination event at the joint-point. *EMBO J.*, **9**, 939-946.

[32] Albertini, A.M, Hofer, M., Calos, M.P. & Miller, J.H. (1982). On the formation of spontaneous deletions: the importance of short sequence homologies in the generation of large deletions. *Cell*, **29**, 319-328.

[33] Gratia J.P. (1975). Etude des mécanismes de la transduction par le bactériophage Ø80 d'*Escherichia coli*. *Arch. Biol.*, **86**, 1-44.

[34] Yanofsky, C. & Crawford, I.P. (1988). The tryptophan operon. In F.C. Neidhardt *et al.* (Eds.), *Escherichia coli* and *Salmonella typhimurium* cellular and molecular biology (pp. 1453-1472). Washington, D.C.: American Society for Microbiology.

[35] Fredericq, P. (1963). Linkage of colicinogenic factors with an F agent and with nutritional markers in the chromosome and in an episome of *Escherichia coli*. In Proc. XIth int. Congr. Genetics, The Hague, p. 42.

[36] Fredericq, P. & Delhalle, E. (1974). Recombinaison entre facteurs R et facteurs colicinogènes chez *Escherichia coli* KAnn. *Inst. Pasteur*, **125B**, 3-12

[37] Delhalle, E. & Gratia, J. P. (1976). Transduction de facteurs de résistance aux antibiotiques et de propriétés colicinogènes par le bactériophage Øγ. *Ann. Microbiol* (Inst. Pasteur), **127B**, 447-451.

[38] Falkow, S. (1975) Infectious multiple drug resistance. London: Pion Ltd.

[39] Schaeffer, P., Cami, B. & Hotchkiss, R.D. (1976). Fusion of bacterial protoplasts. *Proc. Natl. Acad. Sci. USA*, **73**, 2151-2155.

[40] Hotchkiss, R.D. & Gabor, M. (1980). Biparental products of bacterial protoplast fusion showing unequal parental chromosome expression. *Proc. Natl; Acad. Sci. USA*, **77**, 3553-3557.

[41] Grandjean, V., Le Hegarat, F. & Hirschbein, L. (1996). Prokaryotic model of epigenetic inactivation: chromosomal silencing in *Bacillus subtilis* fusion products. In V. Russo, R.A. Martiensen, & A.D. Riggs (Eds), Epigenetic Mechanisms of Gene Regulation (pp. 361-376) New-York: Cold Spring Harbor Laboratory Press.

[42] Lederberg, J. (1949). Aberrant heterozygotes in *Escherichia coli*. *Proc. Natl. Acad. Sci. USA*, **35**, 178-184.

[43] Zelle, M.R. & Lederberg, J. (1951). Single-cell isolations of diploid heterozygous *Escherichia coli*. *J. Bacteriol.*, **61**, 351-355.

[44] Gratia, J.P. (1994). Ufr/S variation in *Escherichia coli* K12: a reversible double-mutation or alternate chromosome expression in non-complementing diploids? *Res. Microbiol.*, **145**, 309-325.

[45] Gratia, J.P. & Thiry, M. (2003). Spontaneous zygogenesis in *Escherichia coli*, a form of true sexuality in prokaryotes. *Microbiol.*, **149**, 2571-2584.

[46] Gratia J.P. (2005) Non complementing diploidy resulting from spontaneous zygogenesis in *Escherichia coli*. *Microbiol.*, **151**, 2947-2959

[47] Gratia, J.P. (2007) Spontaneous zygogenesis (Z-mating) in mecillinam-rounded bacteria. *Arch. Microbiol.*, **188** (DOI: 10.1007/s00203-007-0277-y)

[48] Lévi-Meyrueis, C. & Sanchez-Rivas, C. (1984). Complementation and genetic inactivation: two alternative mechanisms leading to prototrophy in bacterial clones. *Mol. Gen. Genet.*, **196**, 488-493.

[49] Guillén, N., Amar, N. & Hirschbein, L. (1985). Stabilized non-complementing diploids (Ncd) from fused protoplast products of *Bacillus subtilis*. *EMBO J.*, **4**, 1333-1338.

[50] Le Dérout, J., Thaler, D.S., Guillén, N. & Hirschbein, L. (1992). The spoOA gene is implicated in the maintenance of non-complementing diploids in *Bacillus subtilis*. *Mol. Microbiol.*, **6** (11), 1495-1505.

[51] Hauser, P. & Karamata, D. (1992). Ploidy of *Bacillus subtilis* exfusants: the haploid nature of cells forming colonies with biparental or prototrophic phenotypes. *J. Gen. Microbiol.*, **138**, 1077-1088.

[52] 52.-Sanchez-Rivas, C. & Levi-Meyrueis, C. (1994). Ploidy of *Bacillus subtilis* exfusants: a controversy. I. The case for diploidy. *Microbiol.*, **140**, 1-2.

[53] Grant, S.G& Chapman, V.M. (1988). Mechanisms of X-chromosome regulation. *Annu. Rev. Genet.*, **22**, 199-233.

[54] Trieu-Cuot, P., Carlier, C., Martin, P. & Courvalin, P. (1987). Plasmid transfer by conjugation from *Escherichia coli* to Gram-positive bacteria. *FEMS Microbiol. Lett.*, **48**, 289-294.

[55] Trieu-Cuot, P., Carlier, C. & Courvalin, P. (1988). Conjugative plasmid transfer from *Enterococcus faecalis* to *Escherichia coli*. *J. Bacteriol.*, **170**, 4388-4391.

[56] Mazodier, P., Petter, R. & Thompson, C. (1989). Intergeneric conjugation between *Escherichia coli* and *Streptomyces* species. *J. Bacteriol.*, **171**, 3583-3585.

[57] Heinemann, J.A. & Sprague, F. Jr (1989). Bacterial conjugative plasmids mobilize DNA transfer between bacteria and yeast. *Nature*, **340**, 205-209.

[58] Schell, J & Van Montagu M. (1977) The Ti-plasmid of *Agrobacterium tumefaciens,* a natural vector for the introduction of *nif* genes in plants? *Basic Life Sci.*, **9**, 159-1

[59] Zupan, J., Muth, T.R., Draper, O. & Zambryski, P. (2000). The transfer of DNA from *Agrobacterium tumefaciens* into plants: a feast of fundamental insights. *Plant J.* **23**, 11-28.

[60] Maynard-Smith, J., Dowson, C.G. & Spratt, B.G. (1991). Localized sex in bacteria. *Nature*, **349**, 29-31.

[61] Rayssiguier, C., Thaler, D.S. & Radman, M. (1989). The barrier to recombination between *Escherichia coli* and *Salmonella typhimurium* is disrupted in mismatch-repair mutants. *Nature*, **342**, 396-401.

[62] Worth, L., Clark, S., Radman, M. & Modrich, P. (1994). Mismatch repair proteins MutS and MutL inhibit RecA-catalyzed strand transfer between diverged DNAs. *Proc. Natl. Acad. Sci. USA*, **91**, 3238-3241.

[63] Lorenz, M.G. & Wackernagel, W. (1994). Bacterial gene transfer by natural genetic transformation in the environment. *Microbiol. Rev.*, **58**, 563-602.

[64] Gratia, J.P. (2007). Spontaneous zygogenesis, a wide-ranging mating process in bacteria. Res. Microbiol. (DOI 10.1016/j.resmic.2007.07.003).

[65] Woese, C.R. & Fox, G.E. (1977). Phylogenetic structure of the prokaryotic domain: the primary kingdoms. *Proc. Natl. Acad. Sci. USA*, **74**, 5088-5090.

[66] Suyama, M. & Bork, P. (2001). Evolution of prokaryotic gene order: genome rearrangements in closely related species. *Trends Genet.*, **17**, 10-13.

[67] Rosenshine, I., Tchelet, R. & Mevarech, M. (1989). The mechanism of DNA transfer in the mating system of an archaeobacterium. *Science*, **245**, 1387-1389.

[68] Forterre, P., Bouthier de la Tour, C., Philippe, H. & Duguet, M. (2000). Reverse gyrase from hyperthermophiles, probable transfer of a thermoadaptation trait from Archaea to Bacteria. *Trends Genet.,* **16**, 152-154.

[69] Xu, Y. & Glansdorff, N. (2002). Was our ancestor a hyperthermophilic prokaryote? (Review) Comp. *Biochem. Physiol.*, **A133**, 677-688.

[70] Forterre, P. (1995). Thermoreduction, a hypothesis for the origin of prokaryotes. *C.R. Acad. Sci. Paris*, **318**, 415-422.

[71] Glansdorff, N. (2000). About the last common ancestor, the universal life-tree and lateral gene transfer: a reappraisal. *Mol. Microbiol.*, **38**, 177-185.

In: Genetic Recombination Research Progress
Editor: Jacob H. Schulz, pp. 279-290

ISBN: 978-1-60456-482-2
© 2008 Nova Science Publishers, Inc.

Chapter 11

In Silico Genetic Recombination an Advanced Biotechnology Tool in Molecular Biology

Viroj Wiwanitkit

Department of Laboratory Medicine, Faculty of Medicine, Chulalongkorn University,
Bangkok Thailand 10330

Abstract

In the past, studying of the gene was very difficult. Genetic laboratory seems to be a complicated and mysterious field. However, the blooming of molecular biology leads to many simplified techniques for genetic studying. Finding the amino acid sequence of a gene can be easily performed by basic sequencing technique. At the end of the 20th century, the completion of the genome project lead to a new era of post genomics. The in silico laboratory there is an advent in the post genomics era. Based on in silico techniques, manipulation on a genetic sequence is easy. This can help us to better understand the genetic recombination phenomenon. To study a genetic recombination, in silico mutating and docking can help create a genetic recombinant. In addition, in silico gene expression analysis such as gene ontology techniques can help identify the changes in phenotypic expression of a designed recombinant. In this article, principles and interesting examples of in silico genetic recombinant studies will be briefly discussed.

Brief History of Genetics

A gene is a union of genomic sequences encoding a coherent set of potentially overlapping functional products [1]. Mendel's original 1865 article is the first publication in the field of genetics. Johann Gregor Mendel (1822-1884) is a very important scientist of the world. In 1865, after 12 years of systematic investigations on peas, he presented his results in the famous paper "Versuche über Pflanzenhybriden." Three years after his return from Vienna he failed to attain his teaching certification a second time [2]. He created his theory of inheritance to explain the results of his experimental hybridizations of peas [3]. Others have proposed that he designed and carried out his experiments to demonstrate the correctness of a theory of inheritance he had already developed [3]. Bailey cited Mendel's 1865 and 1869 papers in the bibliography that accompanied his 1892 paper, Cross-Breeding and Hybridizing [4].

The usual account is that when Mendel gave his paper, no one understood what he said and there were no questions and no discussion [5]. Examination of available evidence indicates that this is not true [5]. Fisher's subsequent research led him to study the work of (Johann) Gregor Mendel, the 19th century monk who first developed the basic principles of heredity with experiments on garden peas [6]. R. A. Fisher (1890-1962) is a professor of genetics, and many of his statistical innovations found expression in the development of methodology in statistical genetics [7]. For instance, his 1930 text on The Genetical Theory of Natural Selection remains a watershed contribution in that area [8]. Although there are many continuous publications against the Mendel theory, the Mendel's pioneering studies of hybridization in the pea continues to influence the way we understand modern genetics.

At the beginning of this century genetics arose out of developmental history as the science of the causal understanding of development [9]. The concept of the difference between the potential for a trait and the trait proper, between the genotype and the phenotype, became clear only during the first decade of the century [10]. After Spemann's epochal discovery (justifiably rewarded with the Nobel Prize in 1935) of the organizer and the beginning of the experimental analysis of developmental fields, little or no progress was made until the last few years when a virtual revolution occurred in developmental biology [9]. The classical view prevailed into the 1930s, and conceived the gene as an indivisible unit of genetic transmission, recombination, mutation, and function [11]. The discovery of intragenic recombination in the early 1940s and the establishment of DNA as the physical basis of inheritance led to the neoclassical concept of the gene, which prevailed until the 1970s [11].

In the past, studying of the gene was difficult. Genetic laboratory seems to be a complicated and mysterious field. The discoveries of DNA technology, beginning in the early 1970s, have led to the second revolution in the concept of the gene in which none of the classical or neoclassical criteria for the definition of the gene hold strictly true [12]. These are the discoveries concerning gene repetition and overlapping, movable genes, complex promoters, multiple polyadenylation sites, polyprotein genes, editing of the primary transcript, pseudogenes and gene nesting [12]. Introduction of Southern, Northern and dot blotting, and DNA sequencing later in the 1970s considerably advanced the diagnostic capabilities [13]. Nevertheless, it was the discovery of the polymerase chain reaction (PCR) in 1985 that led to an exponential growth in molecular biology and the introduction of practicable nucleic acid tests in the routine laboratory [13]. Blooming of molecular biology led to many simplified techniques for genetic studies. Finding the amino acid sequence of a gene can be easily performed by a basic sequencing technique.

The ending of the gene era is the success of genome project. Building on a debate that dates back to 1985, several genome projects are now in full stride around the world, and more are likely to form in the next several years [14]. Italy began its genome program in 1987, and the United Kingdom and U.S.S.R. in 1988 [14]. The European communities mounted several genome projects on yeast, bacteria, Drosophila, and *Arabidospis thaliana* in 1988, and in 1990 commenced a new 2-year program on the human genome [14]. At the end of the 20's century, the completion of the genome project led to a new era of post genomics. The in silico laboratory is an advent in the post genomics era. Recent years have seen a dramatic increase in genomic and proteomic data in public archives [15]. Based on in silico techniques, manipulation on a genetic sequence is easy. Biochemoinformatics can help in the study of pathogenesis, diagnosis and treatment of many disorders in medicine. Now with the complete genome sequences of human and other species in hand, detailed analyses of the genome sequences will undoubtedly improve our understanding of biological systems and at the same time require sophisticated bioinformatic tools [15]. In the post-genomic era, the new discipline of functional genomics is now facing the challenge of associating a function to many thousands of microbial, plant or animal genes of known sequence but unknown function [16]. Bioinformatics now has an essential role both in deciphering genomic, transcriptomic and proteomic data generated by high-throughput experimental technologies, and in organizing information gathered from traditional biology [17]. Sequence-based methods of analyzing individual genes or proteins have been elaborated and expanded, and methods have been developed for analyzing large numbers of genes or proteins simultaneously, such as in the identification of clusters of related genes and networks of interacting proteins [17]. Besides the design of databases, computational methods are increasingly becoming intimately linked with the various experimental approaches. This can help us to better understand the genetic recombination phenomenon.

In Silico Gene Expression Analysis

Gene expression is the showcase of the gene. The function of the gene cannot be known if there is no proven expression. The phenotype is a good explanation for expression of the gene. Basically, genes are located within the nucleus. The human DNA will pass the translation process via RNA from nucleus to protein synthesis in the cytoplasm and pass to the membrane. A main scientific interest in the post-genomic era is gene expression analysis. The exponential growth in the volume of gene information has generated a confusion of voices surrounding the annotation of molecular information about genes and their expressions. To understand a biological process it is clear that a single approach will not be sufficient, just like a single measurement on a protein, such as its expression level, does not describe protein function [18]. Using reference sets of proteins as benchmarks different approaches can be scaled and integrated [18].

Function and other information concerning genes are to be captured, made accessible to biologists or structured in a computable form as a new focus [19]. Gene ontology is the new "logy" for this propose. Gene ontology is a scientific term used to describe the biology of a gene product in any organism. It also means the description of the molecular functions of gene products, the corresponding placement in, and as cellular components, and the participation in biological processes [20]. Since much of biology works by applying prior known knowledge to an unknown entity, the application of a set of axioms that will elicit knowledge and the complex biological data stored in bioinformatics databases are necessary [21]. These often require addition of knowledge to specify and constrain the values held in that database and a way of capturing knowledge within bioinformatics applications and databases is the use of ontologies [21]. In the early part of this century, the Gene Ontology (GO) Consortium was founded. The aim of the GO Consortium is to provide a framework for both the description and the organization of such information [19]. The Gene Ontology (GO) project seeks to provide a set of structured vocabularies for specific biological domains that can be used to describe gene products in any organism [22]. The work includes building three extensive ontologies to describe molecular function, biological process, and cellular component, and providing a community database resource that supports the use of these ontologies [22]. Firstly, the GO Consortium was initiated by scientists associated with three model organism databases: SGD, the Saccharomyces Genome database; FlyBase, the Drosophila genome database; and MGD/GXD, the Mouse Genome Informatics databases then other database were joined [22]. The GO Consortium supports the development of the GO database resource and provides tools enabling curators and researchers to query and manipulate the vocabularies [22]. Two new databases providing new resources for gene annotation: the InterPro database of protein domains and motifs, and the Gene Ontology database for terms that describe the molecular functions and biological roles of gene products, were launched [23]. Presently, gene ontology is applied for advance research in medicine. The gene expressions in many diseases are analyzed based on gene ontology principle. A group of diseasess, which is widely investigated, is malignancy. There are many recent publications in medical research based on the advance in gene ontology.

In addition to gene ontology, transcriptomics is another interesting bioinformatics tool. Basically, gene-specific transcription activators are among the main factors which specifically shape the transcriptome profiles [24]. It is tempting to take advantage of their properties to decipher the genome expression circuitry [24]. Functional genomics refers to the comprehensive analysis, at the protein level (proteome) and at the mRNA level (transcriptome) of all events associated with the expression of whole sets of genes [25]. There are many techniques for transciptome analysis. The advent of microarray technology has offered fantastic opportunities to quickly analyze the expression profiles dictated by specific transcription factors [24]. New technologies designed to facilitate the comprehensive analyses of genomes, transcriptomes and proteomes in health and disease are poised to exert a dramatic change on the pace of research and to impact significantly on the care of patients [26].

The ability to monitor the expression levels of thousands of genes in a single microarray experiment is a huge progression from conventional Northern blot analysis or PCR-based techniques [27]. Microarrays can play a pivotal role in the mass screening of genes in a wide range of fields [27]. It is now possible to examine all the expressed genes present in a cell simultaneously using microarray [28]. For example,

Bahou and Gnatenko said that microarray analysis had demonstrated a clear and reproducible molecular signature unique to platelets [29]. In 2004, Watanabe showed transcriptomes obtained from 10 species of cells infected with human cytomegalovirus, as a model virus, by a synthetic DNA microarray system that he had established previously [30]. Watanabe said that their system provided simultaneous and parallel description on alteration of expression of viral and host genes that were represented within a single area on a slide glass [30]. In addition, Watanabe proposed a project entitled 'comparative virology on cellular responses of infected hosts' that consists of multiple acquisition and integration of transcriptomes from a combination of several cells and viruses as a panel on the identical platform [30].

In addition to microarray, serial Analysis of Gene Expression (SAGE) is another technique based on the massive sequential analysis of short cDNA sequence tags. Each tag is derived from a defined position within a transcript [25]. Its size (14 bp) is sufficient to identify the corresponding gene and the number of times each tag is observed provides an accurate measurement of its expression level [25]. Since tag populations can be widely amplified without altering their relative proportions, SAGE may be performed with minute amounts of biological extract [25]. Combining the analysis of both the transcriptome and proteome in tandem allows changes in RNA transcripts to be followed right through to changes in the level of protein expression [31]. The application of transcriptomics to study host-pathogen interactions has already brought important insights into the mechanisms of pathogenesis, and is expanding further keeping pace with the accumulation of genomic sequences of host organisms and their pathogens [32]. Many new improved SAGE techniques are proposed within a few recent years. Super SAGE is a new potent tool for the transcriptomics of host-pathogen interactions [32]. Super SAGE array combines the advantages of the highly quantitative Super SAGE expression analysis with the high-throughput microarray technology [33]. Notably, the generation of 26 bp tags in the SuperSAGE procedure helps easily show the interaction transcriptome [32]. Deep SAGE, digital transcriptomics with pyrophosphatase based ultra-high throughput DNA sequencing of di-tags, provides high sensitivity and cost-effective gene expression profiling [34]. Sample preparation and handling are greatly simplified compared to simple SAGE [34]. Long Sage is another technique produces tags that are sufficiently long to be reliably mapped to a whole-genome sequence [35]. This can help improve simple SAGE, which has been used widely to study the expression of known transcripts, but limit to annotate new transcribed regions [35].

In Silico Genetic Recombination

To study a genetic recombination, in silico mutating and docking can help create a genetic recombinant. The creation of new genetic recombinant can be performed at either genome or proteome levels. In addition, in silico gene expression analysis such as gene ontology techniques can help identify the changing in phenotypic expression of a designed recombinant.

Molecular Docking

In silico molecular docking is a useful technique for simulation test on the combination of two molecules. Docking involves the development of computer algorithms that evaluate the binding modes of putative ligands in receptor sites [36]. This technique can be used for designing of a combination between molecules. Over the past year there have been some interesting and significant advances in computer-based ligand-protein docking techniques and related rational drug-design tools, including flexible ligand docking and better estimation of binding free energies and solvation energies [37]. There are many techniques for molecular docking. An interesting computational molecular technique is PatchDock [38], which can be used for modeling of the recombination. PatchDock is a computational molecular technique for molecular docking based on shape complementarity principles [38]. The input is two molecules of any type: proteins, DNA, peptides, drugs [38]. The output or result can be further process to be in the format of three - dimension (3D) molecular structure by Swiss-Pdb Viewer (GlaxoSmithKline R&D & the Swiss Institute of Bioinformatics). The property as well as geometry of the derived complex can also studied by the Swiss-Pdb Viewer. Example of the study that used this technique for molecular docking is a structure of human BRCA2-RAD51 by molecular docking study by Wiwanitkit [39]. The derived model from this study provides the basis for further study of local structural changes, which can be introduced by amino acid mutation [39]. In addition, it can be served as the basis for other protein-recombinants production, which can be applied for preventive purpose [39].

In addition to PacthDock, there are also many other tools. SymmDock is another method predicting the structure of a homomultimer with cyclic symmetry given the structure of the monomeric unit [40]. The inputs to the servers are either protein PDB codes or uploaded protein structures. The services are available at http://bioinfo3d.cs.tau.ac.il [40]. The methods behind the servers are very efficient, allowing large-scale docking experiments [40]. FireDock is another efficient method for the refinement and rescoring of rigid-body docking solutions [41]. The refinement process consists of two main steps: (1) rearrangement of the interface side-chains and (2) adjustment of the relative orientation of the molecules [41]. Andrusier et al said that this method accounted for the observation that most interface residues that were important in recognition and binding do not change their conformation significantly upon complexation [41]. Andrusier et al mentioned that FireDock's prediction results were comparable to current state-of-the-art refinement methods while its running time was significantly lower [41]. FireDock is available at http://bioinfo3d.cs.tau.ac.il/FireDock/. 2007 [41]. ZDOCK is another interesting docking tool proposed by Chen et al [42]. It is is freely available to academic users at http://zlab.bu.edu/~ rong/dock [42]. Yi et al used this tool to study the specific binding of maurotoxin (MTX) peptide to Kv1.2 channel [43]. According to this study, the starting conformation of MTX, the side-chain conformation of the most important residue Tyr(32), and proper introduction of flexibility for candidate complexes were demonstrated to be considerably important factors for obtaining the final reasonable complex structure model [43]. RDOCK is a simple and effective algorithm for refining unbound predictions generated by a rigid-body docking algorithm ZDOCK [44]. The main component of RDOCK is a three-stage energy minimization scheme, followed by the evaluation of electrostatic and desolvation energies [44]. Ionic side chains are kept neutral in the first two stages of minimization, and reverted to their full charge states in the last stage of brief minimization [44]. Without side chain conformational search or filtering/clustering of resulting structures, RDOCK represents the simplest approach toward refining unbound docking predictions [44]. These methods can be used for molecular docking to generate a new genetic recombinant.

There are some interesting reports on the in silico genetic recombination. Good examples will be hereby presented. In 2001, Kim et al reported the functional studies of recombinant human Bcl-2 with the deletion of 22 residues at the C-terminal membrane-anchoring region (rhBcl-2Delta22) [45].

In this work, the optimal binding conformation of antimycin A was predicted from molecular docking of antimycin A with the hBcl-2 model created by homology modeling [45]. Kim et al proposed that the finding that antimycin A selectively induces apoptosis in cells overexpressing Bcl-2 suggested that hydrophobic groove-binding compounds might act as selective apoptotic triggers in tumor cells [45]. Deacon et al used computer-aided docking experiments to develop a plausible model for interactions between galactose oxidase and the D-galactose substrate [46]. Deacon et al described an improved expression system for recombinant galactose oxidase in the methylotrophic yeast *Pichia pastoris* [46]. They used this system to express variant proteins mutated at Arg330 and Phe464 to explore the substrate binding model and demonstrated that the Arg330 variants displayed greater fructose oxidase activity than does wild-type galactose oxidase [46].

In Silico Mutating

In silico mutating is another interesting technique for designing of new genetic recombinant. The basic mutating based on the knowledge on the coding for nucleic acids and amino acids. Simulating manipulation on the wild type codes can be easily performed and the mutant can be further used for studies of the gene expression. Gene expression can be further tested by protein structure prediction and functional analysis. This technique seems to be a simple method but requires good basic knowledge on the nomination system. There are some interesting studies on genetic recombinant based on in silico mutating. Recently, Wiwanitkit studied the structure of hemoglobin (Hb) Suan-Dok based on in silico mutating technique. In this study the amino acid sequence of human alpha globin was extracted using ExPASY and compared with that obtained from the Hb Suan-Dok disorder [47]. The derived sequences, alpha globin chains in both the normal and Hb Suan-Dok disorder, were used for further investigation of the tertiary structures [47]. Modeling these proteins for the tertiary structure was performed using the CPHmodels 2.0 Server [47]. For comparison the tertiary structure of human alpha globin chains in normal and hemoglobin Suan-Dok are calculated and presented [47]. Based on this information, there was no significant difference between the predicted alpha globin tertiary structures of normal hemoglobin and Hb Suan-Dok [47]. Therefore, from this study we can state that the tertiary structure of alpha globin is not significantly affected by the mutation in the Hb Suan-Dok disorder and that the effect of this hemoglobin abnormality may be silent [47]. In 2005, Wiwanitkit studied the elongation part of Hb Geneva. In this work, a bioinformatic analysis was performed to study the secondary and tertiary structures of the abnormal amino acid sequences in elongated part of Hb Geneva [48]. A computer-based study for protein structure modeling was performed. According to this study, the secondary structure analysis of the Hb Geneva showed many defects in helix and strand of the Hb Geneva compared with normal beta-globin chains [48]. On the basis of this information, the main alteration in the Hb Geneva might be due to these aberrations [48]. With regard to the tertiary structure, the deterioration of folds, accompanied by the aberration in secondary structure of globin in Hb Geneva can be identified [48]. Wiwanitkit also performed a similar study on Hb India [49]. In this study, the secondary and tertiary structures of human alpha globin chains of normal and hemoglobin Q-India disorder were calculated and presented [49]. Based on this information, the main difference between the predicted alpha globin secondary structures of normal and Hb Q-India is an extra helix in the Hb Q-India [49]. The predicted tertiary structure also supports this finding. The results from this study can be good data for further study on Hb Q-India disorder, which can bring to the further understanding on this hemoglobinopathy [49]. According to another study by Wiwanitkit [50], a functional analysis was performed on 4 important beta hemoglobinopathies (hemoglobin C, D, E, and S) using PolyPhen, a novel bioinformatic tool. The mutations Hb C (beta 6, Glu --> Lys), Hb D (beta 121, Glu --> Gln), Hb E (beta 26, Glu --> Lys), and Hb S (beta 6, Glu --> Val) were selected for further study. According to the in silico mutation study, the functional change in the studied hemoglobinopathies was variable

[50]. The position-specific independent counts (PSIC) difference score ranged from 1.362 (Hb D) to 2.986 (Hb S). Regarding the degree of damage, all had probable damage [50]. This analysis demonstrated that the functional aberration in the hemoglobinopathy was based on complex pathogenesis [50]. Identifying only the structural aberration in a hemoglobinopathy is not sufficient; additional functional analysis is recommended [50].

Example of in Silico Genetic Recombination Study

A. Introduction

Here, the author will present an example of in silico genetic recombination study on Hb Abruzzo. Hb Abruzzo is an example of hemoglobinopathy firstly described in Italy. It is an unstable hemoglobin variant associated with beta-thalassemia. It is first described in Abruzzo, Italy by Tentori *et al* in 1972 [1]. The main abnormality is the the histidine residue 143 of the beta-globin chain was replaced by arginine [51]. It usually presents as an erythrocytosis in the heterozygote [51 - 53]. Sometimes it can also be found in in combination with beta-thalasssemia [51 – 53]. This hemoglobin it moves between Hb F and Hb S at alkaline pH [51 – 53]. By chromatography, it can be isolated on can be separated by cation and anion exchange chromatography [51 - 53]. Pathophysiologically, peptide maps of tryptic digests of the abnormal beta chain show a single amino acid mutation. Although the primary structure of Hb Abruzzo disorder is well-known the secondary structure of Hb Abruzzo is not well documented. The study on the secondary of the haemoglobin Abruzzo, which can help explain more in the pathogenesis of the Hb Abruzzo disorder, is needed. Here, the author performs a bioinformatic analysis to study the secondary and tertiary structures of those elongated amino acid sequence. Answering this question, a computer-based study for protein structure modeling is performed.

B. Materials and Methods

The database ExPASY [54] was used for data mining of the amino acid sequence for human beta globin chain. Then the mutation beta 143, His-Arg was experimentally performed. Concerning secondary structure modeling, yhe author performs protein secondary structure predictions of beta globins in both normal and hemoglobin Abruzzo disorder from its primary sequence using NNPREDICT server [55]. The calculated secondary structures were presented and compared.

C. Results

Using NNPREDICT server, the calculation for secondary structure of the elongated part of hemoglobin Abruzzo disorder performed (Figure 1). Comparing Hb Abruzzo to normal beta globin, a more helix can be detectd.

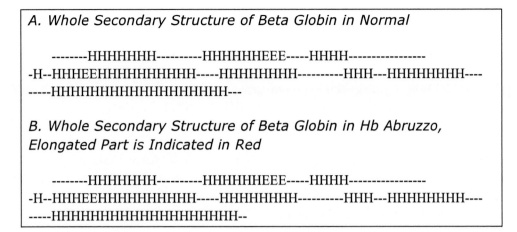

Figure 1. Calculated secondary structures of the elongated part of hemoglobin Abruzzo (Secondary structure prediction: H = helix, E = strand, - = no prediction).

D. Discussion

Here, the author performed a structural analysis for the Hb Abruzzo. According to this study, the secondary structure analysis of the Hb Abruzzo showed additional helices to the normal beta globin chains. Based on this information, the main alteration in the Hb Abruzzo might be due to the additional helices in the elongated part. Indeed, the structural aberration relating to the helix part of the globin chain seems to show possible correlation to erythrocytosis such as in Hb Great Lakes [56]. In addition, the residue 153 in the beta-chain is mentioned as a 2,3- -diphosphoglycerate (DPG) binding site [57 - 58]. Therefore, increased oxygen affinity can be expected. Based on this information, a significant difference in structure between the predicted alpha globin secondary structures of normal and Hb Abruzzo is clarified. In addition, results from this study can be good data for further study on the phylogenetic and tertiary structure of globin chain in Hb Abruzzo disorder, which can lead to the further understanding of this hemoglobinopathy.

References

[1]   Gerstein MB, Bruce C, Rozowsky JS, Zheng D, Du J, Korbel JO, Emanuelsson O, Zhang ZD, Weissman S, Snyder M. What is a gene, post-ENCODE? History and updated definition. *Genome Res.* 2007 Jun;17(6):669-81.

[2]   Weiling F. Historical study: Johann Gregor Mendel 1822-1884. *Am J Med Genet.* 1991 Jul 1;40(1):1-25

[3]   Monaghan FV, Corcos AF. Mendel, the empiricist. *J Hered.* 1985 Jan-Feb;76(1):49-54.

[4]   MacRoberts MH. L. H. Bailey's citations to Gregor Mendel. *J Hered.* 1984 Nov-Dec;75(6):500-1.

[5]   Monaghan FV, Corcos AF. Reexamination of the fate of Mendel's paper. *J Hered.* 1987 Mar-Apr;78(2):116-8.

[6]   Piegorsch WW. Fisher's contributions to genetics and heredity, with special emphasis on the Gregor Mendel controversy. *Biometrics.* 1990 Dec;46(4):915-24.

[7]   Thompson EA. R.A. Fisher's contributions to genetical statistics. *Biometrics.* 1990 Dec;46(4):905-14.

[8]   Piegorsch WW. Fisher's contributions to genetics and heredity, with special emphasis on the Gregor Mendel controversy. *Biometrics.* 1990 Dec;46(4):915-24.

[9]   Opitz JM. Developmental disorders of man. Part 2. *Monatsschr Kinderheilkd.* 1992 May;140(5):264-72.

[10] Falk R. The gene in search of an identity. Hum Genet. 1984;68(3):195-204.

[11] Portin P. The concept of the gene: short history and present status. *Q Rev Biol.* 1993 Jun;68(2):173-223.

[12] Portin P. Historical development of the concept of the gene. J Med Philos. 2002 Jun;27(3):257-86.

[13] Csako G. Present and future of rapid and/or high-throughput methods for nucleic acid testing. *Clin Chim Acta.* 2006 Jan;363(1-2):6-31.

[14] Watson JD, Cook-Deegan RM. Origins of the Human Genome Project. *FASEB J.* 1991 Jan;5(1):8-11.

[15] Yu U, Lee SH, Kim YJ, Kim S. Bioinformatics in the post-genome era. *J Biochem Mol Biol.* 2004 Jan 31;37(1):75-82.

[16] Claverie JM, Abergel C, Audic S, Ogata H. Recent advances in computational genomics. *Pharmacogenomics.* 2001 Nov;2(4):361-72

[17] Kanehisa M, Bork P. Bioinformatics in the post-sequence era. *Nat Genet.* 2003 Mar;33 Suppl:305-10.

[18] Jensen LJ, Steinmetz LM. Re-analysis of data and its integration. *FEBS Lett.* 2005 Mar 21;579(8):1802-7.

[19] Ashburner M, Lewis S. On ontologies for biologists: the Gene Ontology--untangling the web. *Novartis Found Symp.* 2002; 247:66-80

[20] Takai T, Takagi T. Introduction to gene ontology. Tanpakushitsu Kakusan Koso. 2003; 48(1):79-85.

[21] Stevens R, Goble CA, Bechhofer S. Ontology-based knowledge representation for bioinformatics. *Brief. Bioinform.* 2000; 1(4):398-414.

[22] Gene Ontology Consortium. Creating the gene ontology resource: design and implementation. *Genome Res.* 2001; 11(8):1425-33.

[23] Lewis S, Ashburner M, Reese MG. Annotating eukaryote genomes. *Curr Opin Struct Biol.* 2000; 10(3):349-54.

[24] Devaux F, Marc P, Jacq C. Transcriptomes, transcription activators and microarrays. *FEBS Lett.* 2001 Jun 8;498(2-3):140-4.

[25] Marti J, Piquemal D, Manchon L, Commes T. Transcriptomes for serial analysis of gene expression. *J Soc Biol.* 2002;196(4):303-7.

[26] Martin DB, Nelson PS. From genomics to proteomics: techniques and applications in cancer research. *Trends Cell Biol.* 2001 Nov;11(11):S60-5.

[27] Gobert GN, Moertel LP, McManus DP. Microarrays: new tools to unravel parasite transcriptomes. *Parasitology.* 2005 Oct;131(Pt 4):439-48.

[28] Saito H. Microarray as a standard laboratory technique and as an unprecedented tool for understanding systems biology. *Rinsho Byori.* 2006 Jul;54(7):732-7.

[29] Bahou WF, Gnatenko DV. Platelet transcriptome: the application of microarray analysis to platelets. *Semin Thromb Hemost.* 2004 Aug;30(4):473-84.

[30] Watanabe S. Transcriptome analysis of virus-infected cells. *Uirusu.* 2004 Jun;54(1): 23-31.

[31] Gare DC. Analysis of differentially expressed parasite genes and proteins using transcriptomics and proteomics. *Methods Mol Biol.* 2004;270:203-18.

[32] Matsumura H, Ito A, Saitoh H, Winter P, Kahl G, Reuter M, Kruger DH, Terauchi R. SuperSAGE. *Cell Microbiol.* 2005 Jan;7(1):11-8.

[33] Matsumura H, Bin Nasir KH, Yoshida K, Ito A, Kahl G, Krüger DH, Terauchi R. SuperSAGE array: the direct use of 26-base-pair transcript tags in oligonucleotide arrays. *Nat Methods.* 2006 Jun;3(6):469-74.

[34] Nielsen KL, Høgh AL, Emmersen J. DeepSAGE--digital transcriptomics with high sensitivity, simple experimental protocol and multiplexing of samples. *Nucleic Acids Res.* 2006;34(19):e133.

[35] Keime C, Semon M, Mouchiroud D, Duret L, Gandrillon O. Unexpected observations after mapping LongSAGE tags to the human genome. *BMC Bioinformatics.* 2007 May 15;8:154.

[36] Jones G, Willett P. Docking small-molecule ligands into active sites. *Curr Opin Biotechnol.* 1995 Dec;6(6):652-6.

[37] Schneidman-Duhovny D, Inbar Y, Polak V, Shatsky M, Halperin I, Benyamini H, Barzilai A, Dror O, Haspel N, Nussinov R, Wolfson HJ. Taking geometry to its edge: fast unbound rigid (and hinge-bent) docking. *Proteins* 2003; 52: 107-12.

[38] Lybrand TP. Ligand-protein docking and rational drug design. *Curr Opin Struct Biol.* 1995 Apr;5(2):224-8.

[39] Wiwanitkit V. Structure of human BRCA2-RAD51 by molecular docking study. *Arch Gynecol Obstet.* 2007 Jun 5;

[40] Schneidman-Duhovny D, Inbar Y, Nussinov R, Wolfson HJ. PatchDock and SymmDock: servers for rigid and symmetric docking. *Nucleic Acids Res.* 2005 Jul 1;33(Web Server issue):W363-7.

[41] Andrusier N, Nussinov R, Wolfson HJ. FireDock: fast interaction refinement in molecular docking. *Proteins.* 2007 Oct 1;69(1):139-59.

[42] Chen R, Li L, Weng Z. ZDOCK: an initial-stage protein-docking algorithm. *Proteins.* 2003 Jul 1;52(1):80-7.

[43] Yi H, Qiu S, Cao Z, Wu Y, Li W. Molecular basis of inhibitory peptide maurotoxin recognizing Kv1.2 channel explored by ZDOCK and molecular dynamic simulations. *Proteins.* 2007 Aug 29

[44] Li L, Chen R, Weng Z. RDOCK: refinement of rigid-body protein docking predictions. Proteins. 2003 Nov 15;53(3):693-707.

[45] Kim KM, Giedt CD, Basanez G, O'Neill JW, Hill JJ, Han YH, Tzung SP, Zimmerberg J, Hockenbery DM, Zhang KY. Biophysical characterization of recombinant human Bcl-2 and its interactions with an inhibitory ligand, antimycin *A. Biochemistry.* 2001 Apr 24;40(16):4911-22.

[46] Deacon SE, Mahmoud K, Spooner RK, Firbank SJ, Knowles PF, Phillips SE, McPherson MJ. Enhanced fructose oxidase activity in a galactose oxidase variant. *Chembiochem.* 2004 Jul 5;5(7):972-9.

[47] Wiwanitkit V. Modeling for tertiary structure of globin chain in Hemoglobin Suan-Dok disorder. *Hematology*. 2005 Apr;10(2):163-5.

[48] Wiwanitkit V. Structural analysis on the abnormal elongated hemoglobin "hemoglobin Geneva". *Nanomedicine*. 2005 Sep;1(3):216-8.

[49] Wiwanitkit V. Secondary and tertiary structure aberration of alpha globin chain in haemoglobin Q-India disorder. *Indian J Pathol Microbiol*. 2006 Oct;49(4):491-4.

[50] Wiwanitkit V. Analysis of functional aberration of some important beta hemoglobinopathies (hemoglobin C, D, E, and S) from nanostructures. *Nanomedicine*. 2005 Sep;1(3):213-5.

[51] Tentori L, Sorcini MC, Buccella C. Hemoglobin Abruzzo: beta 143 (H 21) his to arg. *Clin Chim Acta* 1972; 38: 258-262.

[52] Chiancone E, Norne JE, Bonaventura J, Bonaventura C, Forsen S. Nuclear magnetic resonance quadrupole relaxation study of chloride binding to hemoglobin Abruzzo (beta 143 his-to-arg). *Biochim Biophys Acta* 1974; 336: 403-406.

[53] Zhao W, Wilson JB, Webber BB, Huisman TH, Sciarratta GV, Ivaldi G, Ripamonti M. A second observation of Hb Abruzzo [beta143(H21)his-to-arg] in an Italian family. *Hemoglobin* 1990; 14: 463-466.

[54] Gasteiger E, Gattiker A, Hoogland C, Ivanyi I, Appel RD, Bairoch A. ExPASy: The proteomics server for in-depth protein knowledge and analysis. *Nucleic Acids Res* 2003;31:3784-8.

[55] Kneller DG, Cohen FE, Langridge R. Improvements in Protein Secondary Structure Prediction by an Enhanced Neural Network. *J Mol Biol* 1990; 214: 171-182.

[56] Rahbar S, Winkler K, Louis J, Rea C, Blume K, Beutler E. Hemoglobin Great Lakes (beta 68 [E12] leucine replaced by histidine): a new high-affinity hemoglobin. *Blood* 1981;58:813-7.

[57] Cheng Y, Lin H, Xue D, Li R, Wang K. Lanthanide ions induce hydrolysis of hemoglobin-bound 2,3-diphosphoglycerate (2,3-DPG), conformational changes of globin and bidirectional changes of 2,3-DPG-hemoglobin's oxygen affinity. *Biochim Biophys Acta* 2001;1535:200-16.

[58] Cheng Y, Shen TJ, Simplaceanu V, Ho C. Ligand binding properties and structural studies of recombinant and chemically modified hemoglobins altered at beta 93 cysteine. *Biochemistry* 2002;41:11901-13.

In: Genetic Recombination Research Progress
Editor: Jacob H. Schulz, pp. 291-305

ISBN: 978-1-60456-482-2
© 2008 Nova Science Publishers, Inc.

Chapter 12

Gene Transfer in Human Skin with Different Pseudotyped HIV-Based Vectors

A. Hachiya,[1, 2,] P. Sriwiriyanont,[3] A. Patel,[4, 5] N. Saito,[1] A. Ohuchi,[1] T. Kitahara,[1] Y. Takema,[1] R. Tsuboi,[2] E. Boissy,[3, 6] M.O. Visscher,[3] J.M. Wilson[7] and G.P. Kobinger[4, 5]*

[1]Kao Biological Science Laboratories, Haga, Tochigi, Japan.
[2]Department of Dermatology, Tokyo Medical University, 6-7-1 Nishishinjuku
Shinjuku-ku, Tokyo, Japan.
[3]The Skin Sciences Institute, Division of Neonatology, Cincinnati Children's Hospital
Medical Center, Cincinnati, OH, USA.
[4]Special Pathogens Program, National Microbiology Laboratory, Public Health Agency
of Canada, Department of Medical Microbiology, University of Manitoba,
Winnipeg, Canada.
[5]Department of Medical Microbiology, University of Manitoba, Winnipeg, Canada.
[6]Department of Dermatology, University of Cincinnati College of Medicine
Cincinnati, OH, USA.
[7]Gene Therapy Program, Division of Medical Genetics, University of Pennsylvania
Health System, Philadelphia, PA, USA.

Abstract

Pseudotyping lentiviral vector with other viral surface proteins could be applied for treating genetic anomalies in human skin. In this study, the modification of HIV vector tropism by pseudotyping with the envelope glycoprotein from vesicular stomatitis virus (VSV), the Zaire Ebola (EboZ) virus, murine leukemia virus (MuLV), lymphocytic choriomeningitis virus (LCMV), Rabies, or the rabies-related Mokola virus encoding *LacZ* as a reporter gene was

* E-mail address: hachiya.akira@kao.co.jp, Phone: (513) 475-6637. Fax: (513) 263-7500

evaluated qualitatively and quantitatively in human skin xenografts. High transgene expression was detected in dermal fibroblasts transduced with VSV-G-, EboZ-, or MuLV-pseudotyped HIV vector with tissue irregularities in the dermal compartments following repeated injections of EboZ- or LCMV-pseudotyped vectors. Four weeks after transduction, double-labeling immunofluorescence of β-galactosidase and involucrin or integrin β1 demonstrated that VSV-G-, EboZ-, or MuLV-pseudotyped HIV vector effectively targeted quiescent epidermal stem cells and their progenies, which were expressed dorsally and underwent terminal differentiation. Among the six different pseudotyped HIV-based vectors evaluated, VSV-G-pseudotyped vector was found to be the most efficient viral glycoprotein for cutaneous transduction as demonstrated by the highest level attained in β-galactosidase activity and genome copy number evaluated by *Taq*Man PCR.

Key Words: lentivirus, gene therapy, pseudotype, vesicular stomatitis virus glycoprotein, stem cells

Introduction

The skin, the largest and most accessible organ in the body, is an attractive target for somatic gene therapy because it is readily available for topical application and intradermal injection of genetic materials is a simple procedure.[1-5] In general, genetic modification of the skin can be performed *ex vivo* or *in vivo*. For *ex vivo* gene transfer, cells are grown from skin biopsies and genetically engineered *in vitro* prior to grafting, while for *in vivo* transduction, the genetic material is administered directly into intact skin. Gene transfer technology for intact skin or cells is an important approach that can be applied to several fields ranging from basic research to innovation in the development of novel clinical modalities.

The use of viral vectors in meditating gene transfer have been examined intensively for therapeutical usage of inherited skin diseases such as Dystrophic Epidermolysis Bullosa,[6-9] Junctional Epidermolysis Bullosa (JED),[10-12] Lamellar ichthyosis,[13] and X-linked ichthyosis.[14,15] Several types of viral vectors such as retrovirus, adenovirus, adeno-associated virus (AAD), and lentivirus have been evaluated for epidermal gene transfer *in vivo*.[16-21] Adenovirus vector can efficiently transduce keratinocytes, however, transgene expression is transient due to the loss of unintegrated vector during epithelium differentiation as well as immunodestruction of target cells.[22] Retroviral vectors based on murine leukemia virus (MuLV) can lead to long term expression of the transgene in keratinocytes both in *in vitro*[23] and *in vivo*[24] probably due to integration of the genetic material into the host genome. However, transduction is limited to proliferating cells which required intense tissue manipulation such as dermabrasion to create an adequate population of dividing cells.[18,25]

Lentivirus vectors such as HIV-based vector can integrate into the genome of quiescent as well as proliferating cells resulting in stable and long-term expression of the transgene.[26] In order to optimize transduction of target tissue, the incorporation of the various viral envelope proteins onto the viral membrane of lentiviral vector (i.e. pseudotyping) has been developed. Early pseudotyping mainly involved the use of the vesicular stomatitis virus envelope glycoprotein (VSV-G) which conferred both broadened tropism and increased stability facilitating concentration and storage.[26-28] It has been reported that a single intradermal injection of VSV-G-pseudotyped HIV-1-based lentivirus into human skin grafted on immunodeficient mice could efficiently transduce most major skin cell types including keratinocytes, fibroblasts, endothelial cells and macrophages within one month and persisted for at least 6 months.[20] Pseudotyping lentiviral vector with different envelope proteins has been shown a valuable strategy for optimizing vector tropism to a specific tissue of interest. Up to date, several pseudotyped lentiviral vectors using surface glycoproteins from various viruses were re-targeted or targeted with improved efficiency to different organs or tissues such as liver, lung, muscle and islets.[29-38]

In the present study, HIV-based vector was pseudotyped with the envelope glycoprotein of VSV, Zaire Ebola virus (EboZ), MuLV, lymphocytic choriomeningitis virus (LCMV), Rabies virus or the rabies-related Mokola virus and evaluated for efficacy and cell type specificity following transduction of human skin grafted onto severe combined immunodeficient (SCID) mice. Transduction efficiency was evaluated qualitatively and quantitatively within the time frame of epidermal renewal (4 weeks), where the replication of stem cells and transit amplifying cells undergo differentiation and desquamation. EboZ-, MuLV- and VSV-G-pseudotyped lentiviral vectors effectively targeted quiescent epidermal stem cells as well as the proliferating daughter cells. The quantitative analyses of β-galactosidase expression and numbers of integrated transgene in genomic DNA and the level of transgene mRNA expression indicate that VSV-G-pseudotyped lentiviral vector is more suitable for cutaneous transduction of human skin in the xenograft model.

Results

Production of Adenovirus and pseudotyped HIV-based vectors. The following viral envelope glycoproteins were used for pseudotyping HIV-1 based vector: envelope glycoprotein of VSV-G, EboZ, MuLV, LCMV, Rabies, or Mokola virus. All pseudotyped viruses were produced in parallel under the same conditions for every experiment. The average titers of VSV-G, EboZ, MuLV, LCMV, Rabies, or Mokola-pseudotyped HIV vector preparations established on 293T cells by limiting dilutions were respectively 3.5×10^9, 2.6×10^8, 4.7×10^8, 6.0×10^8, 8.0×10^7, and 4.2×10^8 transducing units (TU)/ml with a variability not exceeding 40%. Each transduction of human skin xenograft was performed under the same conditions with the same amount of virus particles per injection (based on p24 Gag; 10 μg of p24 equivalent to 5×10^7 TU of VSV-G pseudotyped vector). The preparation of adenovirus vector encoding the *LacZ* gene (Ad-LacZ) used in this study was titrated at 3.2×10^8 plaque forming units (pfu)/ml by immunostaining of the hexon protein (see Materiels and Methods for more details).

Localization of transgene expression during epidermal turnover. The cutaneous transduction patterns of lentiviral vector pseudotyped with VSV-G, EboZ, MuLV, LCMV, Rabies, or Mokola glycoproteins were qualitatively evaluated by immunocytochemistry with specific antibodies against β-gal, integrin β1 (marker for stem cell and basal cell), involucrin (differentiation marker of keratinocyte) or fibroblast. Human skin xenograft transplants were injected twice with 25 μl (10 μg of p24) at 1 week of interval and excised tissues were analyzed one week after the second injection. Results show that β-galactosidase expression was detected at variable levels in the stratum corneum, epidermal keratinocytes and dermal fibroblasts in the skin treated with each pseudotyped HIV vector evaluated (Fig. 1). Notably,

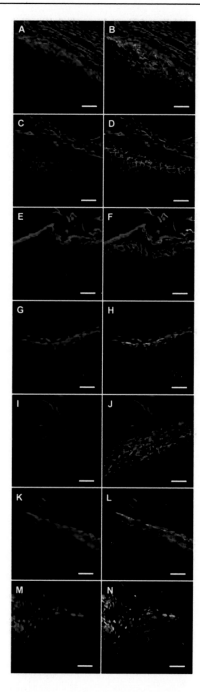

Figure 1. Immunocytochemistry on human epidermis 7 days after transduction with pseudotyped HIV-based vectors expressing β-galactosidase. Each vector was administered twice and xenografts were analyzed by immunocytochemistry 7 days after the last injection. (A) and (B): VSV-G, (C) and (D): EboZ, (E) and (F): MuLV, (G) and (H): LCMV, (I) and (J): Rabies, (K) and (L): Mokola, and (M) and (N) Ad-LacZ. immunocytochemistry staining of the grafted human skin was performed with a rabbit polyclonal antibody recognizing the *E. coli* β-galactosidase (red) (A, C, E, G, I, K and M) and merged with a monoclonal antibody specific for human involucrin (green) (B, D, F, H, J, L and N) as detailed in "MATERIALS AND METHODS". Bar = 50 μm.

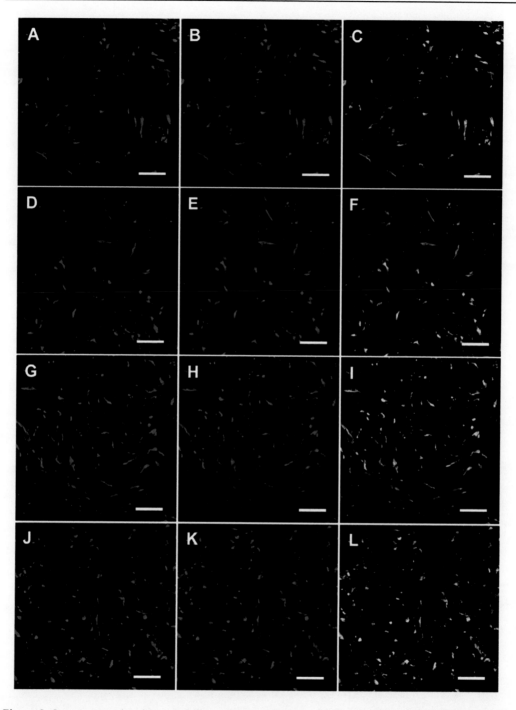

Figure 2. Immunocytochemistry on human dermis 7 days post-transduction with pseudotyped HIV-based vectors expressing β-gal. Each vector was administered twice and xenograft was analyzed by immunocytochemistry 7 days following the last injection. (A-C): VSV-G, (D-F): EboZ, (G-I): MuLV, and (J-L) Adenovirus. Immunocytochemistry staining of grafted human skin was performed with a rabbit polyclonal antibody recognizing the *E. coli* β-galactosidase (red) and a monoclonal antibody specific for human fibroblast (green) as detailed in "MATERIALS AND METHODS". Merging images were presented in (C), (F), (I) and (L). Bar = 50 μm.

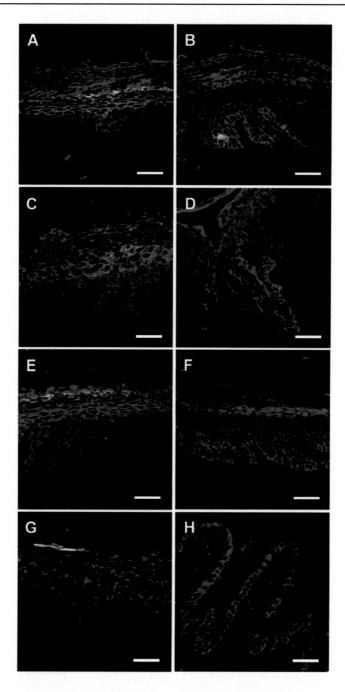

Figure 3. β-galactosidase expression is observed in column in epidermis treated with VSV-G-, EboZ-, or MuLV-pseudotyped lentiviral vectors 4 week after the 2nd injection. (A) and (B): VSV-G, (C) and (D): EboZ, (E) and (F): MuLV, and (G) and (H): Adenovirus. Immunofluorescence staining of grafted human skin intradermally injected with pseudotyped HIV vectors encoding *E. coli* β-galactosidase 4 week after the 2nd injection was performed with a rabbit polyclonal antibody that recognizes *E. coli* β-galactosidase (red) and a monoclonal antibody specific for human involucrin (green) (A, C, E and G) or a monoclonal antibody specific for human integrin β1 (green) (B, D, F and H) as detailed in "MATERIALS AND METHODS". Bar = 50 μm.

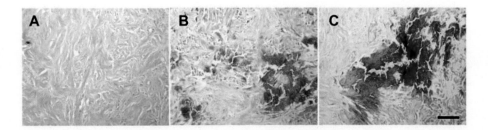

Figure 4. Hematoxylin nuclei staining of EboZ- or LCMV-pseudotyped vector treated human skin. Grafted human skin intradermally injected with HIV vector pseudotyped with VSV-G (A), EboZ (B) or LCMV (C) expressing β-gal were embedded in paraffin and analyzed following hematoxylin and eosin staining 28 days after the last injection. Tissue degenerations in the architecture of the dermis are visible as dark red areas. Bar = 100 μm.

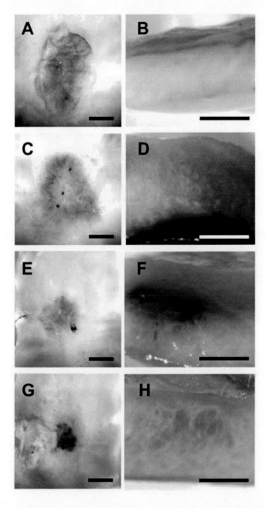

Figure 5. β-galactosidase staining of human skin xenografts transduced with HIV-based vector pseudotyped with different envelope glycoproteins. (A) and (B): PBS, (C) and (D): VSV-G, (E) and (F): EboZ, and (G) and (H): MuLV. β-gal staining of grafted human skin was performed 24 h post-intradermal injection. Bars = 3 mm in A, C, E, and G (top views) and bars = 1 mm in B, D, F, and H (side views).

VSV-G, EboZ or MuLV-pseudotyped vector consistently showed β-gal expression comparable to that of adenoviral vector in human skin epidermis and dermis (Fig. 1 and 2). HIV-based vector-transduced skin showed normal keratinocyte differentiation patterns similar to that of PBS injected control (data not shown) whereas adenoviral vector induced evident epidermal hyperplasia (Fig. 1 M and N). Four weeks post-transduction, β-gal expression was observed in a columnar pattern with proper epidermal differentiation in the skin treated with VSV-G, EboZ, and MuLV-pseudotyped vectors (Fig. 3 A to F). β-gal expression was not detected in epidermis treated with adenoviral vector 28 days post-treatment (Fig. 3 G and H), consistent with the previous report indicating that adenoviral vector does not efficiently transduce skin progenitor cells.[16] Expression of β-gal persisted for 28 days in VSV-G, EboZ, or MuLV-pseudotyped vector transduced skin fibroblasts, the longest time point analyzed. However, tissue degeneration was observed in the dermis of skin xenografts treated with EboZ- or LCMV-pseudotyped vector (Fig. 4).

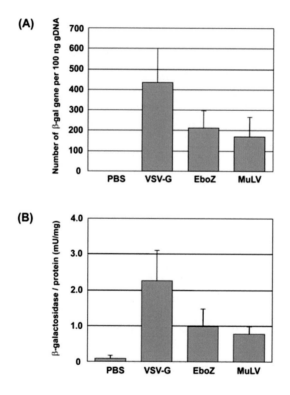

Figure 6. The integrated *E. coli* β-galactosidase copy number and β-galactosidase activity is the highest in the skin treated with VSV-G-pseudotyped lentiviral vector compared to the rest of viral vectors. (A) The integrated *E. coli* β-galactosidase copy number was analyzed after genomic DNA was extracted from each tissue using a DNeasy kit as detailed in "MATERIALS AND METHODS". The values represent mean ± SD of integrated copy number per 100 ng of total DNA from three to four individual grafted skins. (B) β-galactosidase activity was evaluated after pulverized biopsies were lysed in buffer for β-galactosidase assay as detailed in "MATERIALS AND METHODS". The values represent the mean ± SD from three to four individual grafted skins. The activity in the skin treated with VSV-G-pseudotyped lentiviral vector was significantly ($p<0.05$) higher than those of the rest of viral vectors and the activities in the skin treated with EboZ- and MuLV-pseudotyped lentiviral vector were also significantly ($p<0.05$ and $p<0.01$, respectively) higher compared to the PBS control.

Transduction of human skin after a single injection. *En face* visualization of the human skin xenograft transplant after a single injection of VSV-G-, EboZ-, or MuLV-pseudotyped vector demonstrated comparable levels of positive staining for β-gal (Fig. 5 C, E and G). Interestingly, intense staining was observed in the underlying dermal region when VSV-G-pseudotyped vector was administered (Fig 5 D). To rule out pseudotransduction of β-gal protein generated during vector production, VSV-G- or MuLV-pseudotyped vector was injected in the presence of 100 μM of AZT (zidovudine) which inhibits reverse transcription and thus *de novo* synthesis of the transgene product. Transduction of human skin with either VSV-G- or MuLV-pseudotyped vector in the presence of AZT resulted in β-gal staining comparable to the untreated negative control xenograft indicating that pseudotransduction was not responsible for β-gal activity in transduced human skin (data not shown).

Quantitative evaluation of transduction efficiencies by pseudotyped HIV-based vectors in human skin xenografts. Gene transfer efficiency of VSV-G-, EboZ- or MuLV-pseudotyped vector was further analyzed by *Taq*Man PCR to identify the most efficient vector in human skin xenografts. To determine the copy numbers of reverse transcribed vector transgene, *TaqMan* PCR was performed with primers specific to two regions inside the *LacZ* gene. The *LacZ* gene was detected in skin xenografts after administration of VSV-G-, EboZ-, or MuLV-vector, but not in PBS injected control xenografts. VSV-G-pseudotyped vector demonstrated the highest copy number of *LacZ* DNA followed by EboZ and MuLV pseudotyped vectors (Fig 6 A). Comparable results were observed in the level of β-galactosidase activity, which was measured enzymatically by spectrophotometry after incubation with the chemiluminescent substrate CPRG. (see Materials and Methods for details). The overall level of β-galactosidase activity was higher in human skin transduced with VSV-G than with EboZ or MuLV-pseudotyped vector (Fig. 6 B). Overall, these results indicate that VSV-G-pseudotyped HIV vector is the most efficient and the less toxic vector analyzed for transduction of human skin.

Discussion

Retargeting or expanding tropisms by changing the coat protein responsible for attachment and entry of viral vectors into target cells has been successfully applied to several gene transfer vector systems such as lentiviral and AAV vectors.[29-38] The present study evaluated tropism of HIV-based vector pseudotyped with envelope glycoproteins from VSV, EboZ, MuLV, LCMV, Rabies or Mokola virus for different cell types present in differentiated human skin. Two to four square centimeter sections of human skin were transplanted onto the back of SCID mice and injected intradermally with each pseudotyped vector preparation. This study shows that among all the pseudotyped HIV vectors evaluated, VSV-G, EboZ-, and MuLV vectors consistently transduced fibroblasts in human skin xenografts, at levels comparables to that obtained with others and adenovirus. In addition, VSV-G, EboZ-, and MuLV-pseudotyped HIV vectors efficiently transduced epidermal progenitor cells with subsequent expression in the epidermis following normal terminal differentiation. This finding is consistent with a recent report indicating that transgene expression from a VSV-G-pseudotyped HIV vector can be detected along the basal layer and in columns of differentiated epidermal cells in human foreskin xenografts.[39] Long-term expression of a transgene of interest in several cell types of the skin, especially in epidermal stem cells, would be of therapeutic value for inherited skin diseases such as JED which manifest by a separation of the epidermis from the basal lamina forming a blister cavity in the plane of the lamina lucida.[10-12]

In this study, adenoviral vector induced epidermal hyperplasia possibly by inducing secretion of cytokines that regulate epidermal differentiation and proliferation such as IL-1 as previously described.[40] Expression of β-gal persisted for 28 days in keratinocytes and skin fibroblasts following transduction with VSV-G, EboZ-, or MuLV pseudotyped HIV vector. However, anomalies reminiscent of tissue degeneration were observed in the dermis treated with EboZ- pseudotyped HIV vector. Interestingly, skin transduced with LCMV-pseudotyped HIV vector also showed similar tissue degeneration despite low levels of β-gal expression. Toxicity associated with intracellular expression of the Ebola glycoprotein has been reported for different cell type *in vitro*.[41,42] However, our observation is the first report indicating that incoming envelope glycoproteins in the context of a HIV vector preparation can be toxic to a human tissue. The mechanism has yet to be elucidated and experiments are currently evaluating the development of human skin xenograft after treatment with a HIV vector pseudotyped with the envelope NTD4L, a deletion mutant of the Ebola glycoprotein which has demonstrated undetectable toxicity *in vitro*.[42]

In the study outlined here, we demonstrated how pseudotyping lentiviral vector could influence the transduction efficacies of epidermal and dermal skin cells as well as the ability to target the quiescent stem cells. Among six types of envelopes we investigated, VSV-G-pseudotyped vector is the most efficient vehicle for cutaneous gene transfer into keratinocyte stem cells and fibroblasts. In addition, no signs of toxicity were detected following transduction with VSV-G-pseudotyped-HIV vector. These findings provide new insights for the development of an efficient gene transfer strategy for cutaneous gene therapy.

Materials and Methods

Grafting Human Skin onto SCID Mice and Injection of Pseudotyped-Lentiviral Vectors

Animals were handled according to the guidelines of the Institutional Animal Care and Use Committee and Cincinnati Children's Hospital Research Foundation-approved protocol (Cincinnati, OH). ICR-SCID mice 4-6 weeks old (Taconic, NY) were kept under pathogen-free condition (Children's Hospital Research Foundation, Cincinnati, OH) throughout the experiments. After one week of acclimation, they were shaved with an electric clipper to remove the dorsal hair. Prior to the surgery, mice were anesthetized with isofluorane/oxygen (3%/0.8 liter) and were maintained using isofluorane/oxygen (2%/0.7 liter) throughout the surgery. The shaved dorsal skin was treated with a non-staining surgical scrub upon confirmation of anesthesia. The dorsal site was cut to produce a wound bed of approximately 2.0-3.0 cm in diameter. Cadaveric skin (U.S. Tissue and cells, Cincinnati, OH), which were obtained one day before the surgery and maintained in DMEM with L-glutamine and antibiotic/antimycotic (Invitrogen, Carsbad, CA) at 2-4°C, was sutured in place with a reverse cutting precision monofilament PS-3, 6-0 (Ethicon Endosurgery, Cincinnati, OH). The edges of the graft bed were treated with sensorcaine to provide analgesia. Following the surgery, the mice were kept in an incubator at 37°C either for an hour or until recovered from anesthesia.

The viral stocks of pseudotyped-lentiviral vectors with different type of glycoprotein envelopes were RCL-free certified by a vector core at Indiana University. After complete healing, a 25 μl (10 μg of p24 equivalent to 5×10^7 TU of VSV-G pseudotyped vector) of each viral stock encoding *LacZ* gene was intradermally injected onto grafted human skin on ICR-SCID mice once a weeks for 2 weeks. Two groups of control were conducted using either adenoviral vector encoding the same reporter gene or phosphate buffer saline (PBS).

DNA Construction and Vector Production

The helper packaging construct pCMVΔR8.2 encoding the HIV helper function, the transfer vector pHx*LacZ*WP encoding *E. coli* β-galactosidase gene, and plasmids encoding for envelope proteins were used for triple transfection as described elsewhere.[43] Titers of pseudotyped lentiviruses were examined by serial dilution on 293T cells seeded as described elsewhere.[44]

Adenovirus vector was prepared using Adeno-X™ Adenoviral Expression Systems (BD Clontech, Mountain View, CA), according to the manufacture's instructions. The titer of adenoviral stock was determined using Adeno-X™ Rapid Titer Kit (BD Clontech). In brief, HEK 293 cells were transfected with several dilutions of adenoviral stocks. After 48 hrs in culture, cells were fixed in methanol and infected cells were detected using an antibody against adenoviral hexon protein.

All experiments involving the production and functional analysis of replication-incompetent HIV-based pseudotyped vectors were performed under biosafety level 2+ containment as approved by the Institutional Biosafety Commitees of both the Wistar Institute and the Cincinnati Children's Hospital.

X-gal Staining, Immunohistochemistry and Confocal Microscopy

One and four weeks after the 2nd injection, harvested biopsy specimen was immediately fixed in 0.2% glutaraldehyde for 10 minutes, followed by the incubation in 1mg/ml X-gal diluted in 0.1M sodium phosphate buffer (pH 7.4) with 5 mM $K_3Fe(CN)_6$, 5 mM $K_4Fe(CN)_6$, 1 mM $MgCl_2$, 0.02% NP-40 and 0.01% DOC at 37°C overnight. Tissue was trimmed around the blue green staining of X-Gal hydrolyses, post fixed in 2% *p*-formaldehyde and embedded in paraffin and either counterstained with hematoxylin and eosin or proceeded with immunofluorescence. In another set of experiment, the expression level of reporter gene post-transduction was visualized by X-gal staining following 24 hour after a single injection.

To examine the transfection efficacy of each vectors in targeting skin cells, the immunoreactivities of *E. coli* β-galactosidase and several skin markers were investigated using rabbit polyclonal antibody against *E. coli* β-galactosidase (Abcam; Ab4761, Cambridge, MA), mouse monoclonal antibodies against human CD29 (integrin β1; clone MAR4, BD Pharmingen, San Diego, CA), involucrin (Clone SY5, Labvision, Fremont, CA), and CD90/Thy1 (BD Pharmingen, San Jose, CA). To avoid skin autofluorescence, paraffin sections were deparaffinized and free aldehyde groups from glutaraldehyde/formaldehyde fixation were saturated with 0.25% ammonia in 70% ethanol for 30 minutes, and detected with Quantum Dot™ nanocrystal (QDot 605 goat anti mouse, biotinylated rabbit antibody and QDot 655 streptavidin conjugated, Hayward, CA), which emits fluorescence in the near infared region. Normal mouse IgG, and rabbit IgG (Vector Laboratories, Burlingame, CA) were used as negative controls. Double-labeling were analyzed using Zeiss LSM confocal laser scanning microscope (Zeiss, Germany) with an excitation wavelength of 488 nm, and emission filter BP 585-615 nm and LP 650 nm for QDot 605 and QDot 655, respectively.

Genomic DNA PCR

Half of a 6 mm punch biopsy collected at the injection site 4 weeks after the 2nd injection was minced and used for genomic DNA extraction using a DNeasy tissue kit (Qiagen) according to the manufacture's instructions.

Viral cDNA integrated host genome was quantified using ABI PRISM 7300 sequence detection system (Applied Biosystems) as described elsewhere.[44] Briefly, each 50-µl reaction mixture contained 22.5 µl of gDNA, 0.9 µM each primer, 0.2 µM labeled probe, and 25 µl of *Taq*Man Universal PCR Master Mix (Applied Biosystems). The *Taq*Man probe for detection of the integrated pHx*LacZ*WP was labeled with the fluorescent reporter dye 6FAM at the 5' end and the quencher dye TAMRA at the 3' end (6FAM-AGCTCTCTCGACGCAGGACTCGGC- TAMRA). The following *Taq*Man primer sets for amplification of the region encompassing the packaging signal of pHx*LacZ*WP were used in this study: sense, 5'-TGA-AAG -CGA-AAG-GGA-ACC-A-3', and antisense, 5'-CCG-TGC-GCG-CTT-CAG-3'. All reactions were set up in MicroAmp Optical 96-well reaction plates (Applied Biosystems). Amplification conditions in the ABI PRISM 7300 sequence detection system were 55°C for 2 min and 95°C for 10 min, followed by 95°C for 15 sec and 60°C for 1 min and repeated for 50 cycles. Data was collected from the ABI PRISM 7300 sequence detection system and transferred directly to a computer and analyzed using Sequence Detection Software (Applied Biosystems). Unknown samples, standards, and template negative controls were run in duplicate. The frequencies of unknown samples were interpolated from a standard curve derived from the simultaneous amplification of linearized pHx*LacZ*WP plasmid DNA standards. The quantity of vector transgene copies was calculated per 100 ng of genomic DNA.

β-Galactosidase Assay

Another half of a 6 mm punch biopsy specimen that was collected at the injection site of each animal 4 weeks after the 2nd injection was pulverized in liquid nitrogen, and transfer to a glass vial. Lysis buffer was added in each vial after liquid nitrogen was completely evaporated. A quick freeze and thaw was done to break up cells. The supernatant was collected after centrifugation at 4°C for 5 minutes. Relative β-galactosidase expression in the supernatant was determined using enhanced β-galactosidase assay kit with chlorophenol red-β-D-galactopyranoside; CPRG (Gene Therapy Inc., San Diego, CA) and BCA protein assay (Pierce, Rockford, IL) according to the manufacture's instructions.

Statistics

The level of significance of the difference was calculated by the Student t test. The differences in the mean or raw values among treatment groups were considered significant when $p < 0.05$.

References

[1] Cao T, Wang XJ, Roop DR. Regulated cutaneous gene delivery: the skin as a bioreactor. *Hum Gene Ther* 2000; 11: 2297-2300.

[2] Ghazizadeh S, Taichman LB. Virus-mediated gene transfer for cutaneous gene therapy. *Hum Gene Ther* 2000; 11: 2247-2251.

[3] Khavari PA. Genetic correction of inherited epidermal disorders. *Hum Gene Ther* 2000; 11: 2277-2282.

[4] Uitto J, Pulkkinen L. The genodermatoses: candidate diseases for gene therapy. *Hum Gene Ther* 2000; 11: 2267-2275.

[5] Vogel JC. Nonviral skin gene therapy. *Hum Gene Ther* 2000; 11: 2253-2259.

[6] Chen M, O'Toole EA, Mullenhoff M, Medina E, Kasahara N, Woodley DT. Development and characterization of a recombinant truncated type VII collagen "minigene". Implication for gene therapy of dystrophic epidermolysis bullosa. *J Biol Chem* 2000; 275: 24429-24435.

[7] Chen M, Kasahara N, Keene DR, Chan L, Hoeffler WK, Finlay D *et al*. Restoration of type VII collagen expression and function in dystrophic epidermolysis bullosa. *Nat Genet* 2002; 32: 670-675.

[8] Mecklenbeck S, Compton SH, Mejia JE, Cervini R, Hovnanian A, Bruckner-Tuderman L *et al*. Microinjected COL7A1-PAC vector restores synthesis of intact procollagen VII in a dystrophic epidermolysis bullosa keratinocyte cell line. *Hum Gene Ther* 2002; 13: 1655-1662.

[9] Ortiz-Urda S, Thuagarajan B, Keene DR, Lin Q, Fang M, Calos MP *et al*. Stable nonviral genetic correction of inherited human skin disease. *Nat Med* 2002; 8: 1166-1170.

[10] Dellambra E, Vailly J, Pellegrini G, Bondanza S, Golisano O, Macchia C *et al*. Corrective transduction of human epidermal stem cells in laminin-5-dependent junctional epidermolysis bullosa. *Hum Gene Ther* 1998; 9: 1359-1370.

[11] Vailly J, Gagnoux-Palacios L, Dell'Ambra E, Romero C, Pinola M, Zambruno G *et al*. Corrective gene transfer of keratinocytes from patients with junctional epidermolysis bullosa restores assembly of hemidesmosomes in reconstructed epithelia. *Gene Ther* 1998; 5: 1322-1332.

[12] Seitz CS, Giudice GJ, Balding SD, Marinkovich MP, Khavari PA. BP180 gene delivery in junctional epidermolysis bullosa. *Gene Ther* 1999; 6: 42-47.

[13] Choate KA, Medalie DA, Morgan JR Khavari PA. Corrective gene transfer in the human skin disorder lamellar ichthyosis. *Nat Med* 1996; 2: 1263-1267.

[14] Boyce ST, Ham RG. Calcium-regulated differentiation of normal human epidermal keratinocytes in chemically defined clonal culture and serum-free serial culture. *J Invest Dermatol* 1983; 81: 33s-40s.

[15] Freiberg RA, Choate KA, Deng H, Alperin ES, Shapiro LJ, Khavari PA. A model of corrective gene transfer in X-linked ichthyosis. *Hum Mol Genet* 1997; 6: 927-933.

[16] Setoguchi Y, Jaffe HA, Danel C, Crystal RG. Ex vivo and in vivo gene transfer to the skin using replication-deficient recombinant adenovirus vectors. *J Invest Dermatol* 1994; 102: 415-421.

[17] Tang DC, Shi Z, Curiel DT. Vaccination onto bare skin. *Nature* 1997; 388: 729-730.

[18] Ghazizadeh S, Harrington R, Taichman L. In vivo transduction of mouse epidermis with recombinant retroviral vectors: implications for cutaneous gene therapy. *Gene Ther* 1999; 6: 1267-1275.

[19] Hengge UR, Mirmohammadsadegh A. Adeno-associated virus expresses transgenes in hair follicles and epidermis. *Mol Ther* 2000; 2: 188-194.

[20] Baek SC, Lin Q, Robbins PB, Fan H, Khavari PA. Sustainable systemic delivery via a single injection of lentivirus into human skin tissue. *Hum Gene Ther* 2001; 12: 1551-1558.

[21] Kuhn U, Terunuma A, Pfutzner W, Foster RA, Vogel JC. In vivo assessment of gene delivery to keratinocytes by lentiviral vectors. *J Virol* 2002; 76: 1496-1504.

[22] Niwa H, Yamamura K, Miyazaki J. Efficient selection for high-expression transfectants with a novel eukaryotic vector. *Gene* 1991; 108: 193-199.

[23] Mathor MB, Ferrari G, Dellambra E, Cilli M, Mavilio F, Cancedda R *et al*. Clonal analysis of stably transduced human epidermal stem cells in culture. *Proc Natl Acad Sci USA* 1996; 93: 10371-10376.

[24] Kolodka TM, Garlick JA, Taichman LB. Evidence for keratinocyte stem cells *in vitro*: Long term engraftment and persistence of transgene expression from retrovirus-transduced keratinocytes. *Proc Natl Acad Sci USA* 1998; 95: 4356-4361.

[25] Anderson WF. Human gene therapy. Nature1998; 392: 25-30.

[26] Naldini L, Blomer U, Gallay P, Ory D, Mulligan R, Gage FH *et al*. In vivo gene delivery and stable transduction of nondividing cells by a lentiviral vector. *Science* 1996; 272: 263-267.

[27] Burns JC, Friedmann T, Driever W, Burranscano M, Yee JK. Vesicular stomatitis virus G glycoprotein pseudotyped retroviral vectors: concentration to very high titer and efficient gene transfer into mammalian and non-mammalian cells. *Proc Natl Acad Sci USA* 1993; 90: 8033-8037.

[28] Akkina RK, Walton RM, Chen ML, Li QX, Planelles V, Chen IS. High-efficiency gene transfer into CD34[+] cells with a human immunodeficiency virus type 1-based retroviral vector pseudotyped with vesicular stomatitis virus envelope glycoprotein G. *J Virol* 1996; 70: 2581-2585.

[29] Reiser J, Harmison G, Kluepfel-Stahl S, Brady RO, Karlsson S, Scubert M. Transduction of nondividing cells using pseudotyped defective high-titer HIV type 1 particles. *Proc Natl Acad Sci U S A* 1996; 93: 15266-15271.

[30] Mochizuki H, Schwartz JP, Tanaka K, Brady RO, Reiser J. High-titer human immunodeficiency virus type 1-based vector systems for gene delivery into nondividing cells. *J Virol* 1998; 72: 8873-8883.

[31] Wool-Lewis RJ, Bates P. Characterization of Ebola virus entry by using pseudotyped viruses: identification of receptor-deficient cell lines. *J Virol* 1998; 72: 3155-3160.

[32] Mitrophanous K, Yoon S, Rohll J, Patil D, Wilkes F, Kim V *et al*. Stable gene transfer to the nervous system using a non-primate lentiviral vector. *Gene Ther* 1999; 6: 1808-1818.

[33] Chan SY, Speck RF, Ma MC, Goldsmith MA. Distinct mechanisms of entry by envelope glycoproteins of Marburg and Ebola (Zaire) viruses. *J Virol* 2000; 74: 4933-4937.

[34] Stitz J, Buchholz CJ, Engelstadter M, Uckert W, Bloemer U, Schmitt I *et al*. Lentiviral vectors pseudotyped with envelope glycoproteins derived from gibbon ape leukemia virus and murine leukemia virus 10A1. *Virology* 2000; 273: 16-20.

[35] Desmaris N, Bosch A, Salaun C, Petit C, Prevost MC, Tordo N *et al*. Production and neurotropism of lentivirus vectors pseudotyped with lyssavirus envelope glycoproteins. *Mol Ther* 2001; 4: 149-156.

[36] Kobinger GP, Weiner DJ, Yu QC, Wilson JM. Filovirus-pseudotyped lentiviral vector can efficiently and stably transduce airway epithelia in vivo. *Nat Biotechnol* 2001; 19: 225-230.

[37] Lewis BC, Chinnasamy N, Morgan RA, Varmus HE. Development of an avian leukosis-sarcoma virus subgroup A pseudotyped lentiviral vector. *J Virol* 2001; 75: 9339-9344.

[38] Beyer WR, Westphal M, Ostertag W, von Laer D. Oncoretrovirus and lentivirus vectors pseudotyped with lymphocytic choriomeningitis virus glycoprotein: generation, concentration, and broad host range. *J Virol* 2002; 76: 1488-1495.

[39] Ghazizadeh S, Taichman LB. Organization of stem cells and their progeny in human epidermis. *J Invest Dermatol* 2005; 124: 367-372.

[40] Shayakhmetov DM, Li ZY, Ni S, Lieber A. Interference with the IL-1-signaling pathway improves the toxicity profile of systemically applied adenovirus vectors. *J Immunol* 2005; 174: 7310-7319.

[41] Yang ZY, Duckers HJ, Sullivan NJ, Sanchez A, Nabel EG, Nabel GJ. Identification of the Ebola virus glycoprotein as the main viral determinant of vascular cell cytotoxicity and injury. *Nat Med* 2000; 6: 886-889.

[42] Medina MF, Kobinger GP, Rux J, Gasmi M, Looney DJ, Bates P *et al*. Lentiviral vectors pseudotyped with minimal filovirus envelopes increased gene transfer in murine lung. *Mol Ther* 2003; 8: 777-789.

[43] Watson DJ, Kobinger GP, Passini MA, Wilson JM, Wolfe JH. Targeted transduction patterns in the mouse brain by lentivirus vectors pseudotyped with VSV, Ebola, Mokola, LCMV, or MuLV envelope proteins. *Mol Ther* 2002; 5: 528-537.

[44] Croyle MA, Callahan SM, Auricchio A, Schumer G, Linse KD, Wilson JM *et al*. PEGylation of a vesicular stomatitis virus G pseudotyped lentivirus vector prevents inactivation in serum. *J Virol* 2004; 78: 912-921.

In: Genetic Recombination Research Progress
Editor: Jacob H. Schulz, pp. 307-322

ISBN: 978-1-60456-482-2
© 2008 Nova Science Publishers, Inc.

Chapter 13

Genomic DNA Rearrangement in Brain

Toyoki Maeda[*]

Department of Molecular and Cellular Biology, Kyushu University, 4546, Tsurumihara,
Beppu, Oita, 874-0838, Japan

Abstract

The central nervous system comprises neuron networks, in which enormously diversified neurons connect and interact with each other. For decades, the diversity of such neurons has been hypothetically ascribed to a gene diversification mechanism similar to that of the antigen receptor genes in the immune system. Synaptogenesis, memory, and odoreceptor diversification have been raised as candidate representatives of the neuronal functions involved in DNA rearrangements mediated by the hypothetical somatic DNA recombination mechanism in the brain. Some reports have described the somatic DNA recombination activity and the possible rearranging gene loci in association with these neuronal functions in the brain. In spite of every effort to search for the associated gene loci for traces of gene rearrangement, no physiologically functional gene rearrangement has yet been identified. Two rearranged genomic regions were, however, identified in the brain by a rearranged genomic region-oriented approach. A repetitive genomic region (LINE) and a non-repetitive genomic region (BC-1) are the only known examples to undergo genomic rearrangement in the brain, thus far. Both of the regions have been observed to undergo DNA rearrangement not only in the brain and but also in the lens, thus implying that these DNA rearrangements are associated with ectodermal development. The relationship between the neighboring gene function and the rearrangement of the non-repetitive genomic region is herein discussed.

Introduction

The somatic recombination mechanism is considered to play a role in controlling gene functions in higher vertebrates, mainly in the regulation of gene expression and somatic cell diversification. The unique mechanisms of somatic DNA recombination in vertebrates have been exclusively identified in the immune system, and they are known as antigen receptor

---

[*] E-mail address: maedat@beppu.kyushu-u.ac.jp, Telephone: 81-977-27-1681, Fax: 81-977-27-1682

gene rearrangements. From a simple suspicion that the somatic DNA recombination mechanism functions not only for the immune system, the exploration for the somatic DNA recombination system in the extra-immune system was initiated about 20 years ago. As one of the possible organs to bear the somatic DNA recombination, the brain has been an attractive candidate for a long time. The brain has been regarded as possibly bearing somatic DNA recombination activity to develop and maintain its function. The brain consists of 10^{11}~10^{12} cells derived from only one source of neural stem cells, which are characterized by immensely diversified protein molecules yielded in each brain cell to process various kinds of information and memorize them [1, 2]. It seems very natural that a somatic DNA recombination mechanism functions also in the central nervous system similar to that in the immune system which yields 10^7~10^9 diversified antigen receptor molecules (immunoglobulin and T-cell receptors). Although the immune system has been believed to be the only system bearing physiological somatic DNA recombination systems (antigen receptor gene rearrangement and immunoglobulin class-switch rearrangement), many reports have proposed the existence of some hypothetical functions in the brain possibly established by somatic DNA recombination. The possible somatic DNA recombination mechanisms have been proposed to mainly establish memory function, synaptogenesis, or odorant receptor diversity. The following section introduces some representative examples for these brain functions in terms of the putative somatic DNA recombination mechanism.

Brain Functions Expected to be Associated with Genomic Rearrangement

-Memory Consolidation-

The brain senses and recognizes various kinds of environmental changes occurring outside of the brain, and stores them as memories. The environmental information is processed electro-physiologically in the neuron-networks in a very short time (mostly within a second or shorter), but memory storage for a long period may require maintenance of diversified protein molecules for a longer time (from several hours to a lifetime) in the brain. For long-time memory, the hippocampal formation has been regarded as a potential candidate region undergoing genomic recombination to yield the vast diversity of proteins. It was recently shown that non-homologous endjoining (NHEJ), which is one of the steps of the DNA recombination process, in the hippocampus was activated in the process of learning and memory consolidation [3]. The intraperitoneal injection of cytosine arabinoside (Ara-C), which is an inhibitor of NHEJ, blocked memory consolidation without neurotoxicity in mice [4]. These observations imply that somatic DNA recombination containing NHEJ plays a role in memory consolidation in the hippocampus.

-Synaptogenesis-

The central nervous system consists of neuronal synapse networks of 10^{11} neurons. Cell-to-cell interactions are crucial steps for the development of the highly complex nervous system. The synapse is a joint structure between neurons where neuronal stimuli are transmitted from one neuron to another. Synaptogenesis among the 10^{11} neurons can yield greatly diversified neuro-transmitting networks. A variety of cell surface molecules have been found to be expressed at synapses in the developing nervous system. Some of them contribute to the cell-to-cell interaction and connection to form the synapse network, and somatic DNA recombination in the neurons may play a role to diversify combinations of neurons. Among the cell adhesion molecules on the surface of neurons, one of the cadherin family members, protocadherin, can be a candidate gene undergoing DNA rearrangements, because its genomic organization is similar to the immunoglobulin gene structure and the protocadherin gene (*Pcdh*) is expressed in the central nervous system at a neuronal developmental stage [5, 6]. The genomic structure of protocadherin consists of cluster regions of variable exons and constant exons in mice and humans. The cluster of variable region exons consists of ~50 exons and it is divided into three clusters, *Pcdhα*, *Pcdhβ* and *Pcdhγ*, and a constant region exon at the 3' site of *Pcdhα* and *Pcdhγ* This *Pcdh* gene structure presents the possibility of a DNA rearrangement in the exons, although somatic DNA recombination has not been identified on the protocadherin locus, to date.

-Odorant Receptor Diversity-

There are ~1500 odorant receptor genes in the mouse olfactory system. One olfactory neuron expresses one odorant receptor molecule. A somatic DNA recombination system in the odorant receptor gene loci has been hypothesized to produce vastly diversified odorant receptor molecules to correspond to immense kinds of odor discretely, in a somewhat analogous function to that of the immune system. The association between odorant molecules and odorant receptors reminds one of the associations between antigens and antigen receptor molecules. The odorant receptor locus comprises multiple segments of odorant receptor gene tandem arrayed on a chromosome, which resembles the structure of the antigen receptor gene loci. In addition, the allelic exclusive expression of the antigen receptor gene is achieved by a successful antigen receptor gene rearrangement formed on one chromosome, suppressing the following DNA rearrangement on the other allele during lymphocyte differentiation [7]. The same kind of allelic exclusion mechanism with DNA rearrangement may function for a one neuron-one receptor rule in the olfactory system.

The Somatic DNA Recombination Activity in the Brain

There have been some evidence-based reports for somatic DNA recombination activity in brain cells. The enzymatic components for DNA recombination, including immune system-specific and -non-specific components, are expressed in the embryonic brain, and mice lacking the genes coding for these components reveal lethally impaired neuronal development.

There are two kinds of components functioning in the antigen receptor gene rearrangement. One consists of the components functioning almost specifically for antigen receptor gene rearrangement, which are recombination-activating gene products (RAG-1 and RAG-2). The others are Ku70, Ku80, DNA-PK, XRCC4, DNA ligase IV, and Kin, which are not specific for antigen receptor gene rearrangements. *RAG-1* and *RAG-2* play key roles cooperatively in the site-specific antigen receptor gene recombination. The expression of

RAG-1 is widely observed in the neurons of the pre- and postnatal rat brain [8]. On the other hand, *RAG-2* expression is not detectable in the rat brain. *RAG-1* may play a role for somatic DNA recombination in brain in association with an unknown factor expressed specifically for the brain other than *RAG-2*, or may serve to maintain the genomic DNA sequence in long-lived mature neurons lacking mitotic activity in the central nervous system. On the other hand, the expression of both *RAG-1* and *RAG-2* is observed in the olfactory neurons of zebrafish [9]. This observation supports the idea of somatic DNA recombination of odorant receptor genes as described above. Nevertheless, no evidence of a DNA rearrangement of the odorant receptor genes has yet been identified.

The latter components, non-specific for antigen receptor gene rearrangement, are involved in non-homologous end-joining (NHEJ). Mice lacking the gene encoding Ku70, XRCC4, or DNA ligase IV reveal neurodegenerative change in the brain during late embryogenesis [10-12]. These observations suggest that some neurons at the late embryonic stage are highly active in DNA double strand break repair in the brain. It implies that the genomic DNA in the neurons becomes unstable before the neuron-network construction is completed. The apparent instability of genomic DNA of the neurons may be associated with the somatic DNA recombination in the brain. The embryonic neurodegeneration in the brain caused by a lack of the NHEJ component(s) involved in the antigen receptor gene rearrangement can be rescued by a p53- or ATM-deficient condition [13]. On the other hand, the immunodeficiency caused by the similar NHEJ impairment cannot be rescued by the same condition. The impairment of NHEJ results in damaged genomic DNA with accumulated un-repaired double strand breaks (DSBs). Immature neurons bearing the damaged DNA are eliminated only through a p53- and ATM-involving DNA damage monitoring system. The immune system bears another damaged-cell-eliminating system, which can function even under p53- or ATM-deficient conditions. Moreover, an impairment of p53 and ATM cause other types of neuronal abnormality including neurobehavioral disorders or ataxia. Neurons, which successfully undergo somatic DNA recombination, if it exists, may participate in the normal development of the neuronal networks in the brain, and the neurons bearing an incomplete process of the DNA recombination should be removed during the brain development. The hypothetical DNA recombination machinery in brain may be similar to that in the immune system, but not the same.

Previous reports have shown that DNA-repairing enzymes, such as XRCC, DNA polymerase β, Kin, and the genes for DNA excision repair responsible for Xeroderma pigmentosum and Cockayne syndrome, are also involved in the neuronal development of the embryonic brain [14-19].

The disruption of these genes leads to impaired neuronal development. Although these neuronal impairments may possibly be derived from impairment of the brain-specific DNA rearrangement, it is still unclear whether all of these DNA-repairing enzymes are involved in the somatic DNA recombination in brain

Other than DNA repair enzymes, terminal deoxynucleotidyl transferase (TdT) also functions in the antigen receptor gene rearrangement. TdT contributes to diversify antigen receptor genes by adding a very short (1~20 base pairs) random nucleotide sequence at the joining points between the selected V, (D) and J segment. Since TdT is not expressed early in ontogeny, TdT seems to contribute to V-(D)-J diversification mainly in the adult stage. TdT was found to be expressed not only in the thymus but also in the brain of adult mice. In the mouse brain, TdT expression was detected in neurons of the hippocampus, neocortex, cerebellum, amygdala, and olfactory cortex [20]. These observations may support the concept of the hypothetical somatic recombination system functioning in the brain.

Detection of Somatic DNA Recombination Mechanism in the Brain Parallel to That in the Immune System

The detection of somatic DNA recombinational events by using the DNA substrate sequence can be one of the approaches used to capture DNA rearrangements in the brain. Site specificity for the somatic recombination in the immune system is ascribed to the heptamer-nonamer recognition signal sequence in the V-(D)-J joining of the variable region of the antigen receptor genes, and to the switch regions in the class switch rearrangement of the immunoglobulin constant region. The recognition signal sequence (RSS) for the V-(D)-J joining comprises a CACAGTG heptamer and ACAAAAACC nonamer separated by a 12 base pair- or a 23 base pair-spacer, which is located adjacent to every gene segment of the V, (D), and J region. A transgenic experiment using the signal sequence for the V-(D)-J joining was successfully performed to detect the activity for DNA rearrangements in the brain. A transgene construct containing two sets of the RSS DNA substrate was engineered to undergo DNA rearrangement at the RSSs [21]. Indeed, apparent recombinations in the transgene were detected near RSS in the transgenic substrate sequence, which strongly suggested that a V-(D)-J recombination mechanism functioned in the brain. However, another possibility of somatic DNA recombination activity such as DNA recombination activity in the immune cells in the brain, or the random integration of the transgene into genomic DNA resulting in an apparent expression of a reporter gene without DNA recombination can not yet be ruled out [22].

RNA-Mediated Genomic Recombination

DNA rearrangement involving RNA molecules can be one of the possible mechanisms for brain genomic rearrangement. L1 DNA is known as a member of LINE (Long Interspersed Nucleotide Element) and its open reading frames encode an RNA chaperon protein, endonuclease, and reverse transcriptase. These components allow L1 DNA components to be integrated into other loci mediated by self reverse-transcription. Such L1-retroposition was reported in several carcinomas' cells in association with their carcinogenesis [23-25]. Moreover, a recent report showed that L1-retrotransposition affected neural progenitor cell differentiation diversely [26]. These observations suggest that L1-retrotransposition in the human gives rise to a positive aspect as achieving cell variation, and a negative aspect as an impairment of a normally functioning gene, thus resulting in pathological cell death or carcinogenesis. The vertebrate must have gained evolutionally genetic mechanisms selecting the locus and the period of L1-retrotransposition in favor of normal cell growth and differentiation, although the mechanisms have not yet been clearly elucidated.

DNA-Rearranged Locus-Directed Approaches for Genomic DNA Recombinations in the Brain

In an early stage of the research for identification of somatic DNA recombination in brain, a cytogenetical analysis implied the possibility of it. The somatic pairing of chromosome 1 centromeres in the interphase nuclei in the human cerebellum has been reported [27]. This observation implies that certain genomic events, possibly including DNA recombinations, at satellite DNA sequence locations around the centromere occurs in the cerebellum. In addition, the hippocampal neurons exhibiting long-term potentiation were reported later to reveal the rearrangement of centromeric satellite DNA [28]. The latter report seems to confirm that somatic DNA recombination in a centromeric satellite DNA region occurs in the central nervous system. It is, however, still unclear as to which gene is involved in the centromeric satellite DNA rearrangement in association with the brain function. A more specific locus-oriented approach seems preferable to identify brain-specific gene rearrangements.

Two methods have been reported to direct the loci undergoing somatic DNA recombination in brain. One is to clone a rearranged locus and the other is to clone side products derived from somatic DNA recombination with DNA deletion. Regarding the former method, a restriction-digested genomic DNA subtraction method is a very reasonable way to clone locus- and tissue-specific rearranged genomic regions. A genomic fragment was cloned by the subtraction of an excess amount of genomic liver DNA from genomic brain DNA on gel electrophoresis after restriction digestion. A LINE (Long Interspersed Nucleotide Element) repetitive DNA fragment was cloned by this method [29]. The rearranged LINE DNA is also identified in the lens. These observations suggest that this LINE DNA rearrangement is thus related to ectodermal development, thus possibly affecting the functions of unknown neighboring gene(s).

For the other method which focuses on the side products of somatic DNA recombination, it is hypothesized that genomic structural changes occur in the brain in a similar way to that observed in the antigen receptor gene rearrangements. In the immune system, the antigen receptor gene segments arrayed on a chromosome combine at RSSs to produce a functional gene for an antigen receptor, and this rearrangement accompanies a deletion of the intervening genomic DNA lying between the gene segments. The intervening DNA sequence is deleted as a circular DNA molecule (Fig. 1). If the somatic DNA recombination in the brain occurs in a similar way, a circular DNA molecule would be formed. Therefore, circular DNA molecules can be used as a tool to detect site-specific DNA recombinations.

Analysis of Circular DNA Population in the Mouse Brain

Circular DNA molecules in the nucleus should be analyzed, because the cytoplasm contains many mitochondria, the DNA of which is in circular form and would disturb the analysis of the circular DNA derived from nuclear DNA. We recently reported the analyses of nuclear circular DNA from a 16-day-old embryonic brain. The brain tissue at this embryonic age is expected to bear genomic recombinational events as it is reported that the enzymatic components function during the late embryonic stage on brain development [10-12]. Nuclear circular DNA was extracted from an embryonic (16-day-old) Balb/c mouse brain [30]. From the analysis of one hundred and fifty clones from the 16-day-old embryonic brain circular

DNA library, the clones were all different from each other. About 30% of the clones contained L1 DNA sequence. The non-repetitive DNA clones did not contain any recombinational joining point, implying that many of the circular DNA bear joining points at a repetitive DNA sequence, probably mainly at the L1 DNA sequence. The L1-containing circular DNA molecules can be candidates for intermediate molecules for L1 retrotransposition. A quantitative comparison of the L1 DNA-containing circular DNA molecules between embryonic and postnatal brain and liver is shown in Table 1.

Table 1. Quantitative estimation of the circular DNA clones in pre- and postnatal brain tissue and liver tissue. (per 10^8 cells)

Tissue	Brain		Liver	
Age	16E	P21	16E	P21
Total recombinants	1.3×10^4	7.2×10^2	3.8×10^4	2.5×10^5
L1-bearing plaques (%)	27.5	0.5	44.0	80.7

E; embryonic day-old. P; postnatal day-old.

The results show that all the tissues analyzed bear nuclear circular DNA and the content of L1 DNA varies among the tissues. The postnatal brain bore much less of the circular DNA and much less of the L1 DNA than liver and embryonic brain. These observations suggest that the postnatal brain has very low activity for circular DNA formation, and the circular DNA formation involves less L1 DNA in the postnatal brain than in the others. The L1 sequence may be preferentially involved in circular DNA formation in tissues except in postnatal brain tissue, and the L1 circular DNA formation may be a major mechanism for circular DNA formation in the tissues. This mechanism may be suppressed or be almost lacking in the postnatal brain, and the circular DNA not involving L1 DNA may be formed by a mechanism other than the L1 circular DNA formation, for example, illegitimate somatic DNA recombination. The circular DNA population in brain was also analyzed quantitatively in the pre- and postnatal periods (Table 2).

Table 2. Quantitative estimation of the circular DNA clones in pre- and postnatal brain tissue (per 10^8 cells)

Age	14E	16E	P0	P8	P21
Total recombinants	1.4×10^4	1.3×10^4	2.7×10^4	6.4×10^3	7.2×10^2
L1-bearing clones (%)	5.9	27.5	14.4	17.2	0.5

E; embryonic day-old. P; postnatal day-old.

The brain tissue of the embryonic 14-day-old through postnatal 8-day-old bears many more circular DNA clones than the postnatal 21-day-old brain. This observation suggests that postnatal brain tissue loses its DNA recombination activity accompanying circular DNA formation. As observed above, the L1 DNA recombinational activity seems to be different among the organs and to change with aging. The L1 DNA sequence locates dispersedly all over genome DNA, and so the quantitative changes of the L1-circular DNA population of organs may reflect the ubiquitous and non-functional DNA rearrangement in somatic DNA.

From this view, the postnatal brain has less L1-rearranging activity, and a mechanism suppressing the activity may contribute to protecting brain-specific gene functions against detrimental DNA rearrangements in non-dividing cells, such as neurons, during the normal lifetime; only at the developmental stage the L1-retrotransposition may contribute to neuron diversification as described in a previous section.

Somatic DNA Recombination Activity in the Aging Brain

Our observation of the postnatal loss of DNA recombinational activity in brain is supported by previous reports, which showed that aged brain bears less genomic DNA rearrangement than aged liver by analyzing a transgene locus comprised of tandem arrayed β-Lac Z-containing plasmid DNA of aged and young transgenic mice [31, 32]. The observed transgene rearrangements are mediated by deletion, inversion, or transposition, and involve various loci ranging over all genomic regions, but no site-specific DNA recombination was detected. The authors speculated that the accumulation of the randomly occurring genomic DNA rearrangements possibly resulted in an inappropriate expression of genes located near the rearranged regions as detrimental genetic changes with aging; however, no evidence was shown to support that the observed DNA recombinational activity contributed to a site-specific DNA rearrangement in the brain or in other organs. Either the analysis of the transgene rearrangement or our L1-circular DNA analysis does not seem to lead to loci undergoing functional DNA rearrangements. The next question is whether there are L1- or non-L1-circular DNA clones produced by a site-specific recombination mechanism.

Identification of Site-Specific Genomic Recombination in the Brain

In order to search circular DNA molecules in the brain for one derived from site-specific genomic DNA recombination, we developed the external-region directed polymerase chain reaction method (ED-PCR) (Figure 1) [30].

External-region directed primers complementary to the DNA sequence at both terminals for 100 non-repetitive clones randomly isolated from the brain circular DNA library were synthesized. Only one of the primer pairs yielded an amplified product. All others seemed to be derived from illegitimate somatic DNA recombination, because these products could not be amplified from embryonic brain DNA reproducibly. The only reproducibly amplified ED-PCR product contained a recombinational joining point. The original non-repetitive circular DNA fragment yielding the ED-PCR product reproducibly was designated as BC-1 DNA fragment (Brain Circular-1 DNA). Circular DNA molecules were produced from around the BC-1 fragment region at the late embryonic stage. A longer than 100kb genomic region containing the BC-1 fragment turned out to be segmentally orthologous to the human genomic DNA sequence (Figure 2) [33]. The recombinational joining points of BC-1 circular DNA locate in non-human-orthologous sequence and this reminds us of the antigen receptor gene rearrangement editing the gene segments inside to produce functional genes. However, there was no reasonable open reading frame identified in the BC-1 region. The BC-1 deletion occurs at recombinational joining points quite different from those for BC-1 circular DNA formation.

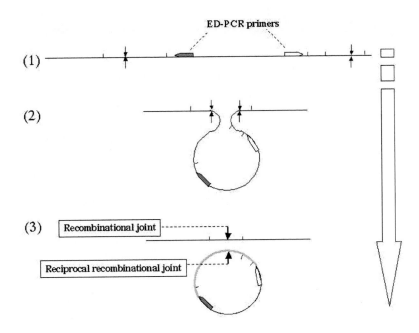

Figure 1. Schematic drawing demonstrating the principle of ED-PCR applied for DNA deletion with circular DNA formation in DNA rearrangement. A white and a gray arrow toward opposite directions are PCR primers for ED-PCR. Vertical arrows are recombinational joining points. Note that the ED-PCR primers are facing to each other to amplify the PCR product (thick pale gray line) after the circularization of the sequence containing the primer regions. The ED-PCR product contains a reciprocal recombinational joint. Short vertical bars represent restriction sites. In the original report, Eco RI was used to digest the circular DNA molecules [30].

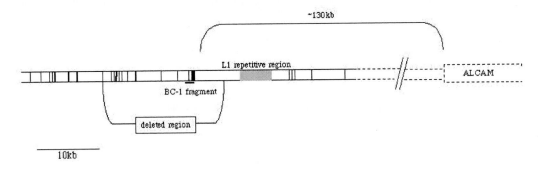

Figure 2. Genomic region surrounding the BC-1 fragment sequence on mouse chromosome 16.Black vertical bars represent human-chromosome 3-orthologous regions. The gray zone represents an L1 repetitive region.

From these observations, BC-1 circular DNA molecules (with a size of 0.5-2kb) are probably intermediate side products which follow large (about 20kb or longer) genomic deletions. The detected deletion does not seem to contribute to the assembly of neighboring gene segments, because no conservative DNA sequence, which can be a marker for a functional gene segment, was identified in the combined region formed after the rearrangement. The BC-1 deletion may have an effect for gene expression rather on a long distant gene than on an adjacent gene. BC-1-20kb-deletion can be detected in the newborn brain within a few days and in the lens until at least 3 weeks after birth (Table 3).

Table 3. Circular DNA formation, DNA rearrangement, and RNA expression of the BC-1 region in the brain, lens, and spleen with aging

	Age	16E	18E	P0	P8	P21	P100
Brain	Circular DNA	(+)	(-)	(-)	(-)	(-)	(-)
	DNA rearrangement	(-)	(+)	(+)	(-)	(-)	(-)
	RNA expression	(-)	(-)	(-)	ND	ND	ND
Lens	Circular DNA	ND	ND	(+)	ND	(-)	(-)
	DNA rearrangement	ND	ND	(+)	ND	(+)	(-)
	RNA expression	(-)	(+)	(+)	(-)	ND	ND
Spleen	Circular DNA	ND	ND	ND	ND	(-)	(-)
	DNA rearrangement	ND	ND	ND	ND	(+)	(-)
	RNA expression	(-)	ND	ND	ND	(-)	ND

RNA expression was analyzed by *in situ* hybridization. 16E and 18E denote embryonic 16- and 18-day-old. P0, P8, P21, and P100 denote postnatal 0-, 8-, 21-, 100-day-old. Brain does not contain olfactory bulb in this analysis. ND is 'not done'.

However, a genomic Southern blot probed with the BC-1 fragment showed no difference of density among the liver DNA, brain DNA, and lens DNA of the postnatal mouse. These indicate that only a few brain cells undergo the BC-1 deletion and the rearrangement disappears within a short period after birth. *In situ* hybridization experiments showed that the BC-1 region was expressed transiently in the newborn lens but not expressed in the brain (Figure 3). The possibility of BC-1 RNA expression in brain, however, could not be excluded, because the *in situ* hybridization experiment did not detect signals in brain tissue between slices. It seemed that many of the lens cells expressed BC-1 RNA and a few lens cells underwent the BC-1 deletion, and they maintained it for a longer period than in brain. The BC-1-associated genomic DNA events and RNA expression with aging is summarized in Table 3. The cells bearing BC-1 rearrangement seemed to appear transiently and disappear later. There are several possibilities of the cells' fate applicable to these observations. Firstly, larger rearrangements occurred in the cells and the earlier rearranged regions deleted. In this case, the detected rearranged regions were regarded as intermediate products for further rearrangements. Secondly, the cells bearing the rearranged BC-1 might migrate to other organs. The other central nervous tissues, including the olfactory bulb and spinal cord, may therefore be candidates for the destination of the migration. Finally, the cells bearing the rearranged BC-1 lead to cell death. The BC-1 rearrangement might cause harmful effects for cell survival or lead to programmed cell death. One of these events, or a combination of these events, might cause the transient detection of the BC-1 rearrangement. It is unclear whether BC-1 RNA expression is associated with BC-1 deletion in the BC-1 RNA-expressing cells.

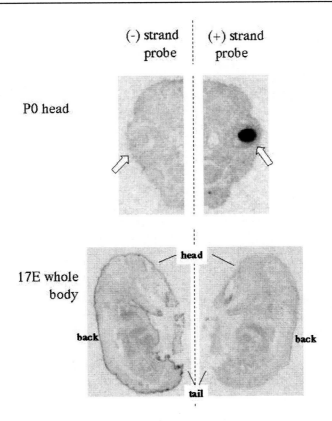

Figure 3. Detection of BC-1 RNA in the late embryonic and newborn stage by *in situ* hybridization probed with a human-orthologous region in the BC-1 fragment. The strand of the probes was assigned to (+) or (-) depending on whether it yielded a positive or negative signal at the lens, respectively [31]. The upper panels depict coronal sections of the 0-day-old newborn head portion. White arrows depict the lens. The lower panels are saggital sections of the central part of the whole body of a 17-day-old embryo. Note that the 17-day-old embryo revealed no significant signal in the section and the 0-day-old postnate revealed a positive signal only at the lens. No significant signals were detected in the brain at both stages.

The BC-1 region does not seem to contain a standard gene structure, such as an exon-intron structure rimmed by a donor and acceptor site of an RNA splicing signal, and a definite length of BC-1 transcript was not detected, although BC-1 RNA expression in the mouse newborn lens was reconfirmed by observing that the *in situ* hybridizing BC-1 signal was erased by RNase-pretreatment. It is also possible that BC-1 RNA expression and BC-1 deletion are associated with controlling the expression of other genes or of BC-1 itself. One of the gene candidates affected by the BC-1 expression or deletion is the *ALCAM* (activated leukocyte cell adhesion molecule) gene which is the nearest gene locus to the BC-1 region on mouse chromosome 16 and on human chromosome 3 (Figure 2) [33, 34]. The *ALCAM* gene is well conserved among vertebrates (Table 4). Orthologues of ALCAM has been identified in the mammalian, avian, reptilian, and amphibian species. *ALCAM* gene function has been found to be associated with neuronal development but *ALCAM* expression was not identified in the lens, so far [35]. No BC-1 RNA expression was detected in brain (Figure 3), but BC-1 expression may occur in a very limited small cell population in brain, if it occurs. The BC-1 deletion or expression is not likely to be linked to the *ALCAM* gene expression control in association with neuronal development.

Contrary to the *ALCAM* gene conserved widely in the vertebrates, the BC-1 region reveals a limited evolutional conservation. The BC-1-homologous region is well conserved only among the placental mammals but not in the monotreme, marsupial, or bird (Table 4).

Table 4. Homology of the orthogues of BC-1 and ALCAM between the mouse and other animal species

Animal species	Homology to mouse BC-1(%)	Homology to mouse ALCAM(%)
Homo sapiens (human)	87	86
Pan troglodytes (chimpanzee)	87	87
Macaca mulatta (rhesus monkey)	86	86
Equus caballus (horse)	84	85
Bos taurus (bovine)	84	86
Canis lupus familiaris (dog)	86	88
Myotis lucifugus (bat)	86	88
Spermophilus tridecemlineatus (squirrel)	87	87
Mus musculus (mouse)	100	100
Rattus norvegicus (rat)	92	91
Cavia porcellus (guinea pig)	86	88
Dasypus novemcinctus (armadillo)	80	84
Tupaia belangeri (tree shrew)	87	86
Sorex araneus (shrew)	80	91
Echinops telfairi (tenrec)	79	88
Monodelphis domestica (opposum)	(-)	86
Ornithorhynchus anatinus (duck-billed platypus)	(-)	84
Gallus gallus (chicken)	(-)	86

Genomic sequence of underlined species is completed. (-); No significant orthologous sequence was found in an NCBI BLAST search.

From the evolutional aspect of the BC-1 appearance in the placental mammals, the BC-1 RNA or the BC-1 deletion might contribute to establish a new biological function in placental mammals. And the genes functioning both in the lens and in the brain could achieve this. From this view, the pineal gland, which evolves from the parietal eye of submammalian species, can be a candidate organ bearing functioning genes in the brain commonly expressed in the lens. In addition, the functions of the pineal gland are different between mammal and non-mammals. For example, the pacemaker of the biological clock locates in the pineal gland in birds but it locates in the suprachiasmatic nucleus of the hypothalamus in mammals [36]. The evolutional altered function of the pineal gland may be attributed to mammals gaining the BC-1 genomic region. This is one of the attractive hypotheses about the physiological function of BC-1, but it is unclear as to which gene function is associated with the BC-1 deletion in the pineal gland. Moreover, as the spleen also reveals the BC-1 deletion, biological function commonly observed in the lens, spleen, and brain (possibly in pineal gland) should involve the BC-1 deletion. Further study is necessary to confirm this hypothesis and elucidate the physiological significance of the BC-1 deletion. For example, it will be interesting to study the phenotypical alterations with the pineal gland, lens, and spleen in mouse artificially lacking the BC-1 genomic region systemically.

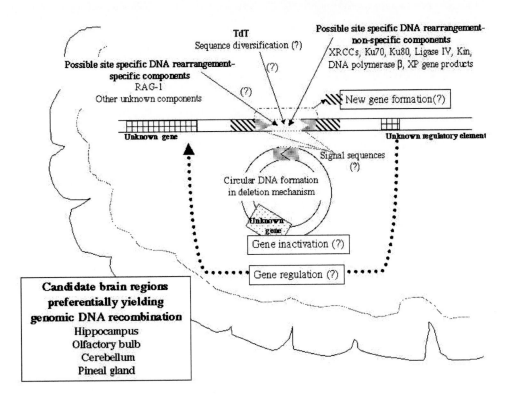

Figure 4. Schematic drawing demonstrating the hypothetical mechanisms for somatic DNA recombination and hypothetical functions derived from it in the brain.

There are only sporadic reports indicating the existence of functional somatic DNA recombination in brain. The possible mechanism of somatic DNA recombination in brain with the associated components and the hypothetical genetic functions of the DNA recombination are schematically summarized in Figure 4. The relationship among the reported hypothesized somatic DNA recombinations is not known. The proposed mechanisms appear to differ with respect to and independent of each other. If the different mechanisms function in the brain, how the mechanisms differ and whether the mechanisms are associated with each other, or not, are also interesting problems to be clarified in further investigations.

References

[1] O'Kusky J, Colonnier M. A laminar analysis of the number of neurons, glia, and synapses in the adult cortex (area 17) of adult macaque monkeys. *J. Comp. Neurol.* **210**: 278–290, 1982.

[2] Dietrich A and Been W. Memory and DNA. *J. theor. Biol.* **208**: 145-149, 2001

[3]   Colón-Cesario M, Wang J, Ramos X, García HG, Dávila JJ, Laguna J, Rosado C, Peña de Ortiz S. An inhibitor of DNA recombination blocks memory consolidation, but not reconsolidation, in context fear conditioning. *J Neurosci* **26**: 5524 –5533, 2006

[4]   Zou C, Huang W, Ying G, Wu Q. Sequence analysis and expression mapping of the rat clustered protocadherin gene repertoires. *Neuroscience* **144**: 579–603, 2007

[5]   Noonan JP, Grimwood J, Schmutz J, Dickson M, Myers RM. Diversity Gene Conversion and the Evolution of Protocadherin Gene Cluster. *Genome Res.* **14**: 354-366, 2004

[6]   Yagi T. Diversity of the cadherin-related neuronal receptor/protocadherin family and possible DNA rearrangement in the brain. *Genes to Cells.* **8**:1-8, 2003.

[7]   Serizawa S, Miyamichi K, Sakano H. One neuron-one receptor rule in the mouse olfactory system. *Trends Genet.* **20**:648-653, 2004

[8]   Chun JJ, Schatz DG, Oettinger MA, Jaenisch R, Baltimore D. The recombination activating gene-1 (RAG-1) transcript is present in the murine central nervous system. *Cell.* **64**: 189-200, 1991.

[9]   Jessen JR, Jessen TN, Vogel SS, Lin S. Concurrent expression of recombination activating genes 1 and 2 in zebrafish olfactory sensory neurons. *Genesis* **29**: 156-162, 2001.

[10] Barnes DE, Stamp G, Rosewell I, Denzel A, Lindahl T. Targeted disruption of the gene encoding DNA ligase IV leads to lethality in embryonic mice. *Curr Biol.* **8**:1395-1398, 1998.

[11] Gao Y, Sun Y, Frank KM, Dikkes P, Fujiwara Y, Seidl KJ, Sekiguchi JM, Rathbun GA, Swat W, Wang J, Bronson RT, Malynn BA, Bryans M, Zhu C, Chaudhuri J, Davidson L, Ferrini R, Stamato T, Orkin SH, Greenberg ME, Alt FW. A critical role for DNA end-joining proteins in both lymphogenesis and neurogenesis. *Cell.* **95**: 891-902, 1998.

[12] Gu Y, Sekiguch J, Gao Y, Dikkes P, Frank K, Ferguson D, Hasty P, Chun JJ, and Alt FW. Defective embryonic neurogenesis in Ku-deficient but not DNA-dependent protein kinase catalytic subunit-deficient mice. *Proc. Natl. Acad. Sci. USA.* **97**: 2668-2673, 2000.

[13] Orii KE, Lee Y, Kondo N, McKinnon PJ. Selective utilization of nonhomologous end-joining and homologous recombination DNA repair pathways during nervous system development. *Proc Natl Acad Sci USA.* **103**:10017-10022, 2006

[14] Zhu C, Mills KD, Ferguson DO, Lee C, Manis J, Fleming J, Gao Y, Morton CC, Alt FW. Unrepaired DNA breaks in p53-deficient cells lead to oncogenic gene amplification subsequent to translocations. *Cell* **109**: 811–821, 2002

[15] Deans B, Griffin CS, Maconochie M. Thacker J. Xrcc2 is required for genetic stability, embryonic neurogenesis and viability in mice. *EMBO J.* **19**: 6675–6685, 2000

[16] Gao Y, Ferguson DO, Xie W, Manis JP, Sekiguchi J, Frank KM, Chaudhuri J, Horner J, DePinho RA, Alt FW. Interplay of p53 and DNA-repair protein XRCC4 in tumorigenesis, genomic stability and development, *Nature* **404**: 897–900, 2000

[17] Murai M, Enokido Y, Inamura N, Yoshino M, Nakatsu Y, van der Horst GT, Hoeijmakers JH, Tanaka K, Hatanaka H. Early postnatal ataxia and abnormal cerebellar development in mice lacking Xeroderma pigmentosum Group A and Cockayne Syndrome Group B DNA repair genes. *Proc Natl Acad Sci USA.* **98**: 13379–13384, 2001

[18] Sugo N, Aratani Y, Nagashima Y, Kubota Y, Koyama H. Neonatal lethality with abnormal neurogenesis in mice deficient in DNA polymerase beta. *EMBO J.* **191**:397-404, 2000

[19] Araneda S, Angulo J, Touret M, Sallanon-Moulin M, Souchier C, Jouvet M. Preferential expression of kin, a nuclear protein binding to curved DNA, in the neurons of the adult rat. *Brain Res.* **762**:103-113, 1997

[20] Peña De Ortiz S, Colón M, Carrasquillo Y, Padilla B, Arshavsky YI. Experience-dependent expression of terminal deoxynucleotidyl transferase in mouse brain. *Neuroreport.* **14**:1141-1144, 2003.

[21] Matsuoka M, Nagawa F, Okazaki K, Kingsbury L, Yoshida K, Müller U, Larue DT, Winer JA, Sakano H. Detection of somatic DNA recombination in the transgenic mouse brain. *Science.* **254**: 81-86, 1991.

[22] Schatz DG, Chun JJ. V(D)J recombination and the transgenic brain blues. *New Biol.* **4**:188-196, 1992

[23] Morse B, Rotherg PG, South VJ, Spandorfer JM, Astrin SM. Insertional mutagenesis of the myc locus by a LINE-1 sequence in a human breast carcinoma. *Nature (London)* **333**:87–90, 1988

[24] Miki Y, Nishisho I, Horii A, et al. Disruption of the APC gene by a retrotransposal insertion of L1 sequence in a colon cancer. *Cancer Research.* **52**:643–645, 1992

[25] Liu J, Nau MM, Zucman-Rossi J, Powell JI, Allegra CJ, Wright JJ. LINE-I element insertion at the t(11;22) translocation breakpoint of a desmoplastic small round cell tumor. Genes, Chromosomes and *Cancer.* **18**:232–239, 1997

[26] Alysson R. Muotri, Vi T. Chu, Maria C. N. Marchetto, Wei Deng, John V. Moran and Fred H. Gage. Somatic mosaicism in neuronal precursor cells mediated by L1 retrotransposition. *Nature* **435**: 903-910, 2005.

[27] Arnoldus EP, Peters AC, Bots GT, Raap AK, van der Ploeg M. Somatic pairing of chromosome 1 centromeres in interphase nuclei of human cerebellum. *Hum. Genet.* **83**: 231-234, 1989.

[28] Billia F, Baskys A, Carlen PL, De Boni U. Rearrangement of centromeric satellite DNA in hippocampal neurons exhibiting long-term potentiation. *Brain Res Mol Brain Res.* **14**:101-108, 1992.

[29] Yokota H, Iwasakii T, Masayoshi Takahashi, And Michio Oishi. A tissue-specific change in repetitive DNA in rats *Proc. Nati. Acad. Sci. USA.* **86**: 9233-9237, 1989.

[30] Maeda T, Chijiiwa Y, Tsuji H, Sakoda S, Tani K, Suzuki T. Somatic DNA recombination yielding circular DNA and deletion of a genomic region in embryonic brain. *Biochem. Biophys. Res. Commun..* **319**:1117-1123, 2004.

[31] Dollé ME, Giese H, Hopkins CL, Martus H-J, Hausdorff JM, Vijg J. Rapid accumulation of genome rearrangements in liver but not in brain old mice. *Nat. Genet.* **17**: 431-41997

[32] Dollé ME, Vijg J. Genome Dynamics in Aging Mice. *Genome Res.* **12**: 1732-1738, 2002.

[33] Maeda T, Mizuno R, Sugano M, Satoh S, Oyama J, Sakoda S, Suzuki T, Makino N. Somatic DNA recombination in a mouse genomic region, BC-1, in brain and non-brain tissue. *Can J Physiol Pharmacol.* **84**: 443-449, 2006

[34] Maeda T, Sakoda S, Suzuki T, Makino N. Somatic DNA recombination in the brain. *Can J Physiol Pharmacol.* **84**:319-324, 2006

[35] Weiner JA, Koo SJ, Nicolas S, Fraboulet S, Pfaff SL, Pourquié O, Sanes JR. Axon fasciculation defects and retinal dysplasias in mice lacking the immunoglobulin superfamily adhesion molecule BEN/ALCAM/SCMol. *Cell. Neurosci.* **27**: 59–69, 2004

[36] Csernus V, Mess B. Biorhythms and pineal gland. *Neuroendocrinology Letters.* **24**: 404-411, 2003.

In: Genetic Recombination Research Progress
Editor: Jacob H. Schulz, pp. 323-333

Chapter 14

Homologous Recombination and Innocuous Intron Elimination[1]

Kejin Hu[*]

The Wisconsin National Primate Research Center, University of Wisconsin, Madison, Wisconsin, 53715

Abstract

Innocuous loss of intron or *in situ* loss of intron is common during evolution. Together with intron gain, loss of intron has re-shaped many genomes dramatically, and has changed many gene structures together with intron gain. Five modes of intron loss can be defined in a multiple-intron gene: complete loss of all introns, 3′-biased loss, concerted loss of several internal adjacent introns, intron exclusion and multiple intron exclusion. The cDMHR/DSBR model, which we recently established, can accommodate all the five patterns of intron losses. In this model, cDNA undergoes homologous recombination (HR) with its parent intron-containing genomic copy. This cDNA recombination is facilitated by the HR repair machinery of DSB in the cell. This process can be triggered by a DSB in a specific intron, which results in the loss of the very intron suffering the DSB. The reverse transcriptase activity could be from retrotransposon such as the yeast Ty1 element and possibly the mammalian LINE. DSB in intron, retrotransposon-encoded reverse transcriptase and homologous repair machinery might be the long searched driving forces for *in situ* intron elimination. This model is strongly supported by the independent experimental data from yeast in which cDNA recombination with the corresponding genomic copy is directly demonstrated, and this process is dependent on the HR protein RAD52. The widely used gap-repair technique also supports this model.

Introduction

Thirty years ago, a breakthrough finding was presented at a conference at Cold Spring Harbor, that a gene is interrupted by intervening sequences, which is now well known as

---

[1] This article is written to celebrate the 30th anniversary of the finding of intron.
[*] E-mail address: hukejin@gmail.com or khu3@wisc.edu

intron [Sambrook, 1977]. The spliceosomal intron is now demonstrated to be a common feature of eukaryotic genomes. Spliceosomal intron is the sequence in a gene in which the corresponding RNA portion is removed from the primary RNA during the splicing process carried out by the spliceosome. Unlike an exon, an intron generally lacks homology in sequence and length. A most useful feature of an intron is the conservation of the location of a specific intron in a given gene even among distant lines. This conservation provides a convenient avenue for uncovering the evolutionary history of a gene. With the accumulation of sequence data and the completion of many genome projects, it is now apparent that genes/genomes undergo extensive loss and gain of introns during their evolution [Roy and Gilbert, 2006; Rogozin et al., 2003]. These two opposite processes greatly shaped different genomes. For example, yeast is believed to experience extensive loss of introns, and this intron elimination process in yeast has resulted in a paucity of introns in the entire genome, and resulted in the overwhelming biased location of the rare introns in the 5′ region of the genes that do retain some of the introns [Fink, 1987]. The intronless prokaryote is also proposed to be the result of such an intron elimination process [Hu, 2006]. A new concept about bacteria is that bacteria are not primitive to eukaryotes, and that they are most likely the derivative of a common ancestor, the Last Universal Cellular Ancestor (LUCA), which is thought to be an intron-rich organism, through a complete elimination of introns during a fitness process [Poole et al., 1999; Forterre and Philippe, 1999; Hu, 2006]. Eukaryotic genomes differ significantly in intron density [Roy and Gilbert, 2006], and this difference is believed to be a result of the different extents of intron gain and loss in different genomes [Jeffares et al., 2006]. The mechanisms for both processes are not well elucidated. The molecular mechanism for intron gain is more   elusive although more hypotheses have been formulated [Roy and Gilbert, 2006]. However, for intron loss, we have a greater degree of consensus in that homologous recombination (HR) between a cDNA and its corresponding genomic DNA is thought to be responsible for the removal of the introns [Hu, 2006; Hu and Leung, 2006; Lewin, 1983; Fink, 1987; Roy and Gilbert, 2006]. Yet, the mechanism of intron loss remains poorly understood because cDNA recombination is more an idea than an established model. Recently, we established a model for cDNA recombination theory of intron elimination [Hu, 2003, Hu and Leung, 2006; Hu, 2006]. In this chapter, I will summarize the model of cDNA homologous recombination in the intron removal process, and the most recent insights into this model, including the newly recognized pattern of intron loss which strongly points to our model, classification of intron losses, driving forces for and trigger to this process. It must be kept in mind that this chapter deals with the innocuous intron loss in active genes only. For the intron loss in retrotransposed gene, readers are referred to other elegant reviews [Mighell et al., 2000; Vanin, 1985; Weiner at el., 1986].

A Refined Model for Innocuous Intron Elimination in the Currently Active Genes

Several years after the finding of introns, it was noticed that an intron can be eliminated without negative effect on the host gene [Lewin, 1983]. cDNA recombination has been a dominant theory for such an intron elimination since then although genomic deletion is also proposed by some authors [Lewin, 1983; Fink, 1987; Robertson, 2000]. However, cDNA recombination is more an idea than a model/theory for almost 30 years. More insights into this theory have been gained based on the patterns of intron loss in a gene, and a model called cDMHR/DSBR has recently been established, which stands for cDNA-mediated homologous recombination that is facilitated by DSB (double-strand break of DNA) repair machinery [Hu,

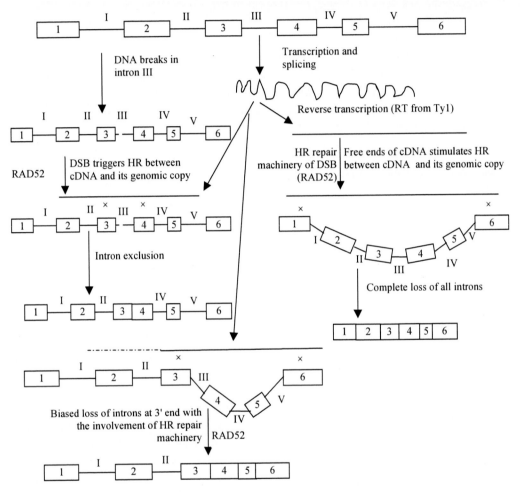

Figure 1. A Model for intron elimination. A 5-intron gene is used as an example for the illustration of intron elimination. It is transcribed, and then spliced. The resulted mature mRNA is subsequently reverse transcribed into cDNA. Reverse transcriptase activity is provided by the Ty1 transposon in the case of budding yeast (Derr. *et al*, 1991). The cDNA can undergo homologous recombination with its genomic locus in different ways depending on the status of cDNA and its genomic copy. Full length cDNA undergoes homologous recombination by double crossover at both terminal regions through stimulation by both free ends which mimic DSB. A DSB within an intron triggers homologous recombination between flanking coding sequences and the corresponding regions in cDNA, and this gives rise to specific loss of the broken intron. Truncated cDNA undergoes homologous recombination with genomic copy stimulated by free ends as does full length cDNA. RAD52 is involved in this process (Derr and Strathern, 1993; Derr, 1998). Numbered box signifies exon; linking thin line is intron. Free thin line stands for cDNA. Broken line indicates truncation of cDNA. The gap in intron 3 denotes a double-strand-break of DNA. Scribbled line designates mRNA. Roman numerals by intron are the intron number. Combined exons mean precise excision of intron from genomic copy. DNA, mRNA and cDNA are all oriented in a 5'-to-3' direction from left to right.

2003; Hu and Leung, 2006; Hu, 2006]. This model is depicted in Figure 1, and is composed of the following components. 1) Pre-mRNA is spliced out of introns, and is subsequently reverse transcribed into cDNA. 2) A cDNA undergoes homologous recombination with its parent intron-containing genomic copy. 3) HR repair machinery for DSB facilitates such a cDNA recombination. 4) A DSB in a specific intron triggers such a cDNA recombination. 5) Free ends of cDNA mimic DSB and promote homologous recombination with the parent intron-containing genomic copy. 6) The reverse transcriptase activity for this cDNA recombination process is retrotransposon-encoded, Ty1 in the case of yeast. 7) This cDMHR/DSBR model deduced from the natural data (patterns of intron loss) is supported by the laboratory demonstration with yeast as a model genome. A brief description of the cDMHR/DSBR model is given in the legend to Figure 1. The following sections will highlight some aspects of this model.

Patterns of Intron Loss in an Active Gene

There are two types of intron losses mediated by an RNA intermediate: inactivation loss of introns in retropseudogene and the innocuous intron loss that does not result in the inactivation of the processed gene. Inactivation loss of intron is so called because it usually coincides with the inactivation of the transposed gene. Inactivation loss is a result of retrotransposition with the involvement of retroelements such as integrase activity in addition to the reverse transcriptase activity, while innocuous loss of intron occurs at the original locus mediated by other cellular processes other than retrotransposition (see below) [Hu, 2006; Hu and Leung, 2006]. In this communication, discussed is only the innocuous loss of introns in a currently active gene. It occurs at the original locus. There are many reports of individual cloning of intronless genes while the intron-containing orthologues exist, implying a loss of introns [Hu and Leung, 2004; Bolland and Hewitt, 2001; Kato et al., 1999]. Comparative studies display differential loss of some introns in a gene [Hu and Leung, 2006; Frugoli et al., 1998]. In the post-genome era, there are many genome-wide analyses that revealed extensive loss of introns [Roy et al., 2003; Nielsen et al., 2004; Robertson, 1998 and 2000; Bryson-Richardson et al., 2004]. Nowadays, innocuous loss of introns is recognized as common. Generally, there are five types of innocuous losses of introns in a multiple-intron gene as illustrated in Figure 2, which fall into two categories: simultaneous and individual losses of introns. Simultaneous loss can be divided into three types: complete loss of all introns, biased loss of several introns in the 3' region and simultaneous loss of several adjacent internal introns. Individual loss includes loss of a single intron anywhere, and double or/and multiple events of such an individual loss of a single intron. Simultaneous losses of introns are well recognized, especially for complete loss and 3' biased loss. Complete loss of introns in a gene possibly occurs extensively in currently intron-poor genomes such as yeast [Fink, 1987]. Complete loss of intron is also reported by cloning or analysis of a specific gene such as cathepsin L [Hu, 2003, Hu and Leung, 2004 and 2006], plant *adh* genes [Hu, 2006], *Tetraodon SART1* [Bolland and Hewitt, 2001], and human *HMGN4* [Birger et al., 2001]. Biased loss of introns in the 3'-end region is believed to happen in many unicellular organisms and the nucleomorph *Guillardia theta* [Fink, 1987; Jeffares et al., 2006]. Biased intron loss in the 3'-end region is best known as biased location of intron in the 5' region of a gene in many genes and genomes. Individual loss of introns or intron exclusion is a recently defined mode of intron loss to describe the precise elimination of a single intron anywhere with the surrounding introns retained [Hu, 2006]. It can occur multiple times in a gene. Intron exclusion is a random process in terms of the location of introns in a gene. Plant alcohol dehydrogenase provides an example of intron exclusion [Hu, 2006]. Large-scale comparison of intron positions uncovered six individual losses of introns in rodent genomes [Roy et al., 2003]. Individual elimination of intron (intron exclusion) is common in the three large families of chemoreceptor genes str, stl and srh in C. elegans [Robertson, 1998 and 2000].

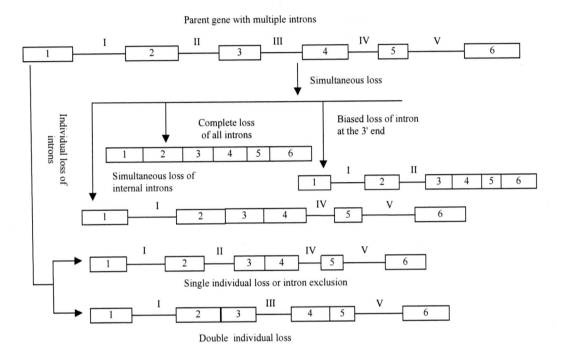

Figure 2. Patterns of intron loss. Numbered box is exon while the thin line linking adjacent exons represents intron. Combined exons indicate precise loss of intron between them. The Roman numerals above intron are the intron numbers. Sequences are placed in a 5'-to-3' orientation from left to right.

The patterns of intron loss within a gene provide useful clues to how introns are lost. First of all, the precise loss of introns in all five modes implies an RNA-mediated process. Pre-mRNA is spliced out of introns by spliceosomal machinery, and then the mature RNA is reverse transcribed. The cDNA is involved in the subsequent recombination process. Simultaneous complete loss of all introns in a currently active gene is an apparent and strong evidence for cDNA-mediated homologous recombination with its genomic locus. This is more obvious when the intronless active gene is single-copied in the genome [Hu and Leung, 2006; Bolland and Hewitt, 2001]. Simultaneous loss of introns in the 3' region is a conventional molecular fossil in support of cDNA homologous recombination. Apart from the features provided by the complete loss of introns, 3' loss directly shows that a cDNA with its 5' portion truncated recombines with the 5' intron-containing portion of its genomic counterpart. Therefore, loss of intron occurs in situ, not through retrotransposition which is a random integration into the genome. Simultaneous loss of all internal introns clearly supports the assumption of in situ recombination at the parent intron-containing locus. For example, maize catalase 3 lost all its 6 internal introns leaving only the two terminal introns intact on each end [Frugoli et al., 1998].

This reasoning applies to the intron exclusion as well (see below). Taken together, precise elimination of some of the intron complement in a gene indicates that the elimination process occurs at the original locus because there are two smoking guns at the same venue in these cases: intronless cDNA relics and intron-containing fragments of the same gene. Individual loss of intron, especially such a loss in the middle region of a gene is thought to be inconsistent with cDNA recombination mechanism. Several authors suggest that genomic deletion is a possible mechanism of individual loss of intron. However, new interpretation finds that intron exclusion (individual loss) is an added evidence because a DSB in a specific intron can give rise to an individual loss other than simultaneous loss of introns. The DSB theory [Hu, 2006] is in nice agreement with gap repair phenomenon in yeast and bacteria. A gap or a DSB within homologous sequence can stimulate homologous recombination, and limited nonhomologous sequence around DSB is permitted (see more detailed discussion below). Therefore, all modes of intron losses are pointing to the above mentioned cDNA recombination model.

Experimental Evidence for cDMHR/DSBR Model of Intron Loss

The above model is experimentally demonstrated in yeast. To distinguish the fates of the cDNA sequence through RNA-mediated recombination, Derr and Strathern [Derr and Strathern, 1993; Derr, 1998] designed an elegant experiment as depicted in Figure 3. Although their focus was not to address the intron elimination, they experimentally proved a mechanism of intron loss. Therefore, their experiment is worth detailed discussion here. In their experiments, they clearly showed that a free cDNA can undergo homologous recombination with its chromosomal edition in addition to conversion with plasmid copy, and retrotransposition. They made two mutant *his3* copies as a reporter gene in a yeast strain with *his3* deletion (*his3-Δ200*), one of which is on a plasmid and the other of which is chromosomal. The plasmid provides the *his3* cDNA donor for homologous recombination with the chromosomal copy. The plasmid *his3* (*pGAL1-his3-ΔATG*) lacks upstream promoter and has a deletion of the first 27 bp coding for the first 9 amino acids of HIS3. To generate *his3* cDNA, a *GAL1* promoter is placed downstream of the *HIS3* sequence, so that an antisense *his3* mRNA is produced. An artificial intron is also engineered into *his3* sequence on the *MscI* site in a spliceable orientation in respect to the *GAL1* promoter, and unspliceable in the native sense mRNA of *HIS3* gene. The chromosomal *his3* contains a native *HIS3* promoter, and is rendered null by a deletion of 34 bp spanning the *MscI* site (*his3-ΔMscI*). This design ensures that the His+ prototroph is only from the homologous recombination between the *HIS3* cDNA and its mutant chromosomal version. Direct homologous recombination between the plasmid *his3* and the chromosomal *his3* cannot generate *His+* prototroph because the presence of the unspliceable artificial intron. And, homologous recombination between the *his3* cDNA and the plasmid *his3* allele cannot generate His+ prototroph due to the lack of a promoter in the sense direction (A in Figure 3) although such recombination removes the unspliceable artificial intron. However, homologous recombination between the *his3* cDNA resulted from the antisense *his3* mRNA driven by the *GAL1* promoter, and the null chromosomal *his3* is expected to reconstitute the fully functional *HIS3*, and therefore render yeast His[+] prototroph. Acutally, this is the only recombination in their design that is expected to render yeast His[+] prototroph. Derr and Strathern did observe His[+]

prototroph by transformation of the *his3-ΔMscI* yeast with the *pGAL1-his3-ΔATG* plasmid. From this experiment, one can make the following conclusions: 1) cellular mRNA can be reverse transcribed by cellular reverse transcriptase (see below also); 2) A free cDNA can undergo homologous recombination with its genomic copy; 3) Intron can be eliminated precisely by such a recombination process. This is in great agreement with our model based on patterns of intron losses in a gene.

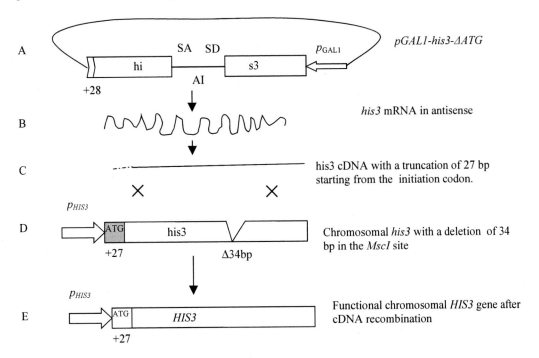

Figure 3. Experimental demonstration of intron elimination by cDNA homologous recombination with genomic locus (Derr, 1998). A. the plasmid construct for the *HIS3* cDNA donor. The *his3* coding sequence with a truncation of the first 27 bp from the initiation code is driven by *GAL1* promoter, but in an antisense direction relative to *HIS3* gene. Therefore, the transcript is an antisense mRNA without *HIS3* coding function. Apart from a deletion of 27 bp from initiation codon, the *his3* sequence lacks upstream promoter element. In addition, the coding sequence is interrupted by an artificial intron (AI) that is in a spliceable orientation in respect to the *GAL1* promoter, but unspliceable if a *HIS3* sense mRNA be made. B. Antisense *his3* mRNA with a truncation of the first 27 bp starting from the initiation codon corresponding to the sense strand. C. The *HIS3* cDNA with a 27 bp deletion from the initiation codon. D. Chromosomal copy of *HIS3* gene with the *HIS3* native promoter, but deleted of 34 bp which renders *HIS3* null. E. Fully functional chromosomal *HIS3* gene after gene conversion. Boxes are *HIS3* coding sequence. Arrows are the *GAL1* or *HIS3* promoters as indicated. The notch in D indicates a 34-bp deletion of *HIS3*. The chevron in A shows a 27-bp deletion of *HIS3* from its initiation codon which corresponds to the shaded box in D. × marks region where crossovers occur. SA, splice acceptor; SD, splice donor; *pGAL1*, *GAL1* promoter; *pHIS3*, *HIS3* promoter.

Source of Reverse Transcriptase

The current model for the common innocuous loss of introns requires the existence of the cellular activity of a reverse transcriptase. The strong evidence, albeit indirect, for the existence of cellular reverse transcriptase is the ubiquitous occurrence of retropseudogene [Mighell *et al.*, 200]. The three major hallmarks of retropseudogene is the lack of introns, the absence of a promoter (upstream sequence) and the presence of a poly(A) tract at the 3' end, indicative of an RNA intermediate. The spliced RNA is transcribed, and the cDNA is hitch-hiked and integrated back randomly into genome by retro-element. Using *HIS3* as reporter gene with an artificial intron in an unspliceable direction, it is directly demonstrated that this RT activity is from the Ty1 element in yeast because expression of Ty1 can facilitate the generation of *HIS3* "retropseudogene" [Derr *et al.*, 1991]. Retrotranscribed cDNA has two fates, retrotransposition randomly into the genome and homologous recombination precisely back onto the parent locus [Hu and Leung, 2006, Derr and Strathern, 1993]. Therefore, the RT source could be the same for these two processes. Using the same assay system, it was directly demonstrated that cDNA generated by Ty1 RT can undergo homologous recombination with the chromosomal homologous counterpart [Derr and strathern, 1993; Derr, 1998]. Retrotransposon is common in cell, for example, in human, LINE retrotransposons are responsible for the generation of the retropseudogene [Esnault *et al.*, 2000]. However, it remains to clarify if mammalian LINE RT activity is involved in cDNA recombination with the parent genomic locus. It is also unclear if other cellular RT such as reverse transcriptase encoded by group II intron, can act *in trans* and contributes to cDNA homologous recombination.

The Driving Forces for Intron Loss

A key component of the cDMHR/DSBR model of intron elimination in a currently active gene is that the homologous repair machinery for DSB is involved in this process [Hu, 2003, Hu and Leung, 2006; Hu, 2006]. First of all, the intron elimination is a faithful recombination. There is not any indel at the junction between the intron-containing portion and the intronless portion of a gene in the biased loss of introns. In addition, no indel is found on both ends at the site of intron exclusion. This faithful recombination is reminiscent of homologous recombination in DSB repair which represents the most faithful mechanism for DSB repair. A strong support is that RAD52 is proved to be involved in the cDNA-mediated removal of intron at the homologous locus [Derr and Strathern, 1993; Derr, 1998]. RAD52 is a key and conserved enzyme involved in HR repair of DSB [Dudáš and Chovanec, 2004].

Another key point in the current model of intron elimination is that a DSB in a specific intron triggers the cDNA-mediated homologous recombination. The clue is from the phenomenon of intron exclusion (individual loss of a single intron). Intron exclusion is best explained by cDNA recombination carried out by the DSB repair machinery that is triggered by a DSB in the very intron (see Figure 1). This model is in agreement with the commonly used gap repair technique in yeast [Ma, *et al.*, 1987; Kostrub *et al.*, 1998] and in bacteria [Lee *et al.*, 2001; Zhang *et al.*, 200]. In the gap repair process, a DSB can stimulate homologous recombination using the homologous sequence flanking the gap. However, adjacent intron, given it is small, can also be eliminated by the homologous recombination triggered by a DSB in a specific intron. Ma *et al.*, [1987] showed that both proximal and distal homologous region to the DSB can undergo homologous recombination. Therefore, DSB activation of HR repair machinery not only explains intron exclusion, but also concerted loss of adjacent internal introns. Finally, free ends of cDNA mimic DSB and can stimulate cDNA homologous recombination (Figure 1). This is analogous to the gene knock-out reaction in yeast in which the homologous recombination with target locus can be facilitated by the homologous free ends that encompass a selection marker [Rothstein, 1983; Baudin et al., 1993]. In summary, cellular mRNA, reverse transcriptase, DSB in introns, and HR repair machinery form joint forces to drive the loss of introns in genomes.

cDNA recombination triggered by a DSB, and facilitated by the HR repair machinery of DSB causes loss of introns. With the current model, we can better understand the many phenomena we observed. This process has reshaped dramatically the unicellular genomes, and makes them intron-poor due to the selection pressure for an economic and compact genome [Hu, 2006]. Secondly, recombination with cDNA truncated from the 5' end might result in the 5' biased location of introns. Thirdly, cDNA recombination might have totally eliminated all introns in prokaryotic genomes. Lastly, cDNA recombination has changed the gene structures by elimination of one or several internal introns. The gene structure is further complicated by an opposite process, intron gain.

cDMHR/DSBR model also implies that loss and gain of introns might not be balanced in some genomes. First of all, intron loss through cDNA homologous recombination is relatively innocuous, but intron insertion into coding sequence is highly detrimental. A genome that favors cDMHR/DSBR process finally ends up with an intron-poor genome, or an intronless genome. Bacteria and yeast genomes use homologous recombination for DSB repair more frequently than the human genome which uses non-homologous end joining as a default process for DSB repair [Dudáš and Chovanec, 2004]. Therefore, bacteria and yeast are more prone to loss of introns due to the powerful HR repair machinery. Another factor that might affect loss of intron through cDNA homologous recombination might be the intron density. Human genome has more introns, and these introns are generally longer. This decreases the homology between a cDNA and its genomic counterpart, and therefore prevents cDNA homologous recombination. Therefore, the current model can also explain why different genomes have different intron density and dynamics.

Conclusion

Genomes undergo loss and gain of introns during the evolution. Loss of intron in a currently active gene is found to have five patterns: complete loss of all introns in a multiple-intron gene, biased loss of introns in the 3' portion of a gene, simultaneous loss of several internal introns, intron exclusion, and multiple intron exclusion. Intron exclusion is the random loss of a single intron in a multiple-intron gene with its surrounding introns unchanged. A cDNA recombines with the parental intron-containing genomic counterpart, and therefore eliminates the introns. The reverse transcriptase activity comes from the wide spread retrotransposons, Ty1 in the case of yeast, and likely LINE in mammalian genomes. This cDNA-mediated loss of intron can be triggered by a DSB in a specific intron. Homologous repair machinery of DSB is involved in this process which is supported by the fact that RAD52 is involved in the RNA-mediated HR process. This can also be inferred by the fact that the recombination process in the intron removal is a faithful process. The reverse transcriptase activity from retrotransposon, a DSB in an intron and the DSB repair machinery combine their forces to drive the losses of introns. This cDNA-mediated process of intron elimination has greatly changed the genome profile and gene structures. My model has the power to accommodate all modes of intron losses, and the different intron profiles among genomes.

References

[1] Baudin, A., Ozier-Kalogeropoulos, O., Denouel, A., Lacroute, F. and Cullin, C. (1993) A simple and efficient method for direct gene deletion in *Saccharomyces cerevisiae*. *Nucleic Acids Research*, **21**, 3329-3330.

[2] Bolland, D. J., and Hewitt, J. E. (2001) Intron loss in the *SART1* genes of *Fugu rubripes* and *Tetraodon nigroviridis*. *Gene*, **271**, 43-49.

[3] Bryson-Richardson, R. J., Logan, D. W., Currie, P. D., and Jackson, I. J. (2004) Large-scale analysis of gene structure in rhodopsin-like GPCRs: evidence for widespread loss of an ancient intron. *Gene*, **338**, 15-23.

[4] Derr, L. K., Strathern, J. N., and Garfinkel, D. J. (1991) RNA-mediated recombination in *S. cerevisiae*. *Cell*, **67**, 355-364.

[5] Derr, L. K. and Strathern, J. N. (1993) A role for reverse transcripts in gene conversion. *Nature*, **361**, 170-173.

[6] Derr, L. K. (1998) The involvement of cellular recombination and repair genes in RNA-mediated recombination in *Saccharomyces cerevisiae*. *Genetics*, **148**, 937-945.

[7] Dudáš, A., and Chovanec, M. (2004) DNA double-strand break repair by homologous recombination. *Mutation Research*, **566**, 131-167.

[8] Esnault, C., Maestre, J., and Heidmann, T. (2000) Human LINE retrotransposons generate processed pseudogenes. *Nature Genetics*, **24**, 363-367.

[9] Fink, G. R. (1987) Pseudogenes in yeast? *Cell*, **49**, 5-6.

[10] Forterre, P., and Philippe, H. (1999) Where is the root of the universal tree of life? *BioEssays*, **21**, 871-879.

[11] Frugoli, J. A., McPeek, M. A., Thomas, T. L., and McClung, C. R. (1998) Intron loss and gain during evolution of the catalase gene family in angiosperms. *Genetics*, **149**, 355-365.

[12] Birger, Y., Ito, Y., West, K. L., Landsman, D., and Bustin, M. (2001) HMGN4, a newly discovered nucleosome-binding protein encoded by an intronless gene. *DNA and Cell Biology,* **5**, 257-264.

[13] Hu, K. Molecular cloning and characterization of the cathepsin L gene from the marine shrimp *Metapenaeus ensis*. Ph.D. thesis. University of Hong Kong, Hong Kong, China. 2003.

[14] Hu, K. (2006) Intron exclusion and the mystery of intron loss. *FEFS Letters*, **580**, 6361-6465.

[15] Hu, K., and Leung, P. (2004) Shrimp cathepsin L encoded by and intronless gene has predominant expression in hepatopancreas, and occurs in the nucleus of oocyte. *Comparative Biochemistry and Physiology, B*, **137**, 21-33.

[16] Hu, K. and Leung, P. (2006) Complete, precise, and innocuous loss of multiple introns in the currently intronless, active cathepsin L-like genes, and inference from this event. *Mol. Phylogenet. Evol.*, **38**, 685-696.

[17] Jeffares, D. C., Mourier, T. and Penny, D. (2006) The Biology of intron gain and loss. *Trends in Genetics*, **22**, 16-22.

[18] Kato, H., Shintani, A., and Minamikawa, T. (1999) The structure and organization of two cysteine enopeptidase genes from rice. *Plant Cell Physiol.*, **40**, 462-467.

[19] Kostrub, C. F., Lei, E. P., and Enoch, T. (1998) Use of gap repair in fission yeast to obtain novel alleles of specific genes. *Nucleic Acids Research*, **26**, 4783-4784.

[20] Lee, E., Yu, D., De Velasco, J., Tessarollo, L., Swing, D. A., Court, D. L., Jenkins, N. A., and Copeland, N. G. (2001) A highly efficient *Escherichia coli*-based chromosome engineering system adapted for recombinogenic targeting and subcloning of BAC DNA. *Genomics*, **73**, 56-65.

[21] Ma, H., Kunes, S., Schatz, P. J., and Botstein, D. (1987) Plasmid construction by homologous recombination in yeast. *Gene*, **58**, 201-216.

[22] Mighell, A. J., Smith, N. R., Robinson, P. A., and Markham, A. F. (2000) Vertebrate Pseudogenes. *FEBS Letters*, **468**, 109-114.

[23] Mourier, T., and Jeffares, D. C. (2003) Eukaryotic intron loss. *Science*, **300**, 1393.

[24] Poole, A., Jeffares, D., and Penny, D. (1999) Early evolution: prokaryotes, the new kids on the block. *BioEssays*, **21**, 880-889.

[25] Lewin, R. (1983) How mammalian RNA returns to its genome. *Science*, **219**, 1052-1054.

[26] Nielsen C. B., Friedman, B., Birren, B., Burge, C. B., and Galagan. (2004) Patterns of intron gain and loss in fungi. *PLos Biology*, **2**, 1-9.

[27] Robertson, H. M. (1998) Two large families of chemoreceptor genes in the nematodes *Caenorhabditis elegans* and *Caenorhabditis briggsae* reveal extensive gene duplication, diversification, movement and intron loss. *Genome Research*, **8**, 449-463.

[28] Robertson, H. M. (2000) The large *srh* family of chemoreceptor genes in *Caenorhabditis* nematodes reveals processes of genome evolution involving large duplications and deletions and intron gains and losses. *Genome Research*, **10**, 192-203.

[29] Rogozin, I. B., Wolf, Y. I., Sorokin, A. V., MIrkin, B. G., and Koonin, E. (2003) Remarkable interkingdom conservation of intron positions and massive, lineage-specific intron loss and gain in eukaryotic evolution. *Current Biology*, **13**, 1512-1517.

[30] Rothstein, R. J. (1983) One-step gene disruption in yeast. *Methods in Enzymology*, **101**, 202-211.

[31] Roy, S.W., Fedorov, A., and Gilbert, W. (2003) Large-scale comparison of intron positions in mammalian genes shows intron loss but no gain. *PNAS*, **100**, 7158-7162.

[32] Roy, S. W., and Gilbert, W. (2006) The evolution of spliceosomal introns: patterns, puzzles and progress. *Nature Reviews, Genetics.*, **7**, 211-221.

[33] Sambrook, J. (1977) Adenovirus amazes at Cold Spring Harbor. *Nature*, **268**, 101-104.

[34] Vanin, E. F. (1985) Processed pseudogenes: characteristics and evolution. *Ann. Rev. Genet.*, **19**, 253-272.

[35] Weiner, A. M., Deininger, P. L., and Efstratiadis, A. (1986) Nonviral retroposons: genes, pseudogenes, and transposable elements generated by the reverse flow of genetic information. *Ann. Rev. Biochem.*, **55**, 631-661.

[36] Zhang, Y., Muyrers, J. P. P., Testa, G., and Stewart, A. F. (2000) DNA cloning by homologous recombination in *Escherichia coli. Nature Biotechnology*, **18**, 1314-1317.

In: Genetic Recombination Research Progress
Editor: Jacob H. Schulz, pp. 335-343

ISBN: 978-1-60456-482-2
© 2008 Nova Science Publishers, Inc.

Chapter 15

Natural Genetic Recombination of Pathogens

Viroj Wiwanitkit

Department of Laboratory Medicine, Faculty of Medicine, Chulalongkorn University,
Bangkok Thailand 10330

Abstract

Genetic recombination is an important phenomenon in gene medicine. This phenomenon can lead to a new genotype and phenotype. Genetic recombination can be either natural or artificial. Natural genetic recombination is an important contributing factor to genetic shift and drift. In medicine, the genetic recombination in pathogenesis of an existing disease is important. Considering three epidemiological determinants, genetic recombination of agent is easier than that of host and environment. For pathogens, natural genetic recombination can result in a change in virulence and susceptibility. Natural genetic recombination in the virus is well characterized in clinical microbiology. In this article, the author performed a literature review on the natural genetic recombination of pathogens. Examples of important genetic recombination of pathogens and their clinical correlations are presented.

Introduction to Genetic Recombination

Genetic recombination is an important phenomenon in gene medicine. This phenomenon can lead to new genotype and phenotype. Basically, different types of nucleotide sequences (repetitions, palindromes, homopolymers) represented in the regions of genome transcription, as well as in mRNA's of the cytoplasm and nuclear messenger-like RNA's are mentioned [1]. It is noted that this population of molecules is heterogeneous not only by the molecular weight, but also by some other parameters [1 - 2]. A number of genes in higher organisms and in lower particles as viruses appear to be split [3]. That is, they have "nonsense" stretches of DNA interspersed within the sense DNA [3]. The cell produces a full RNA transcript of this DNA, nonsense and all, and then appears to splice out the nonsense sequences before sending the RNA to the cytoplasm [3 - 4]. It is suggested that development is essentially a simple process, the cells having a limited repertoire of overt activities and interacting with each other by means of simple signals, and that general principles may be discerned [5]. The complexity lies in the specification of the internal state which may be described in terms of a gene-switching network. Pattern formation is a central feature in development; it is the process whereby states are assigned to the cells according to their position such that the appropriate type of cytodifferentiation is selected from the repertoire [5]. It is accepted that there are many thousands of genetic translation from nucleic codes of genes to proteins for expressions [6]. A defect can be expected in any step of the process and this might lead to a new expression. Some new expressions can bring advantages while others may not. With many alterations, genetic recombination between two sets of codons can usually bring a new expression. Genetic recombination is the transmission-genetic process by which the combinations of alleles observed at different loci in two parental individuals become shuffled in a new resulting expression.

Genetic recombination can be either natural or artificial. Natural genetic recombination is an important contributing factor to genetic shift and drift. In a living organism level, many researches show that homologous recombination plays an important role in the reparation of DNA in various groups of organisms, irrespective of the way they reproduce [7]. Involvement of recombination in meiosis, however, is impossible to explain only by DNA repair functions [7]. The hypothesis, that a recombination in the course of sexual process is a source of variability, and is not capable of explaining the existence of this process well [7]. There is also evidence that recombination at meiosis is largely confined to structural genes or adjacent DNA. It is proposed that the absence of a functionally important methyl group in a promotor or operater region produces a recombinator or signal for the initiation of recombination [8]. The formation of hybrid DNA in this region then allows the lost methyl groups to be replaced by maintenance methylase activity [8]. The removal of epigenetic defects by recombination during meiosis therefore becomes an essential part of a reprogramming and rejuvenation process [8]. Homologous recombination is an important process that occurs between homologous chromosomes during meiotic prophase I [9]. Formation of chiasmata, which hold homologous chromosomes together until the metaphase I to anaphase I transition, is critical for proper chromosome segregation [9]. Indeed, mutation plays the primary role in evolution [10]. Homologous recombination, as in sex, is important for population genetics shuffling of minor variants, but relatively insignificant for large-scale evolution [10]. Major evolutionary innovations depend much more on illegitimate recombination, which makes novel genes by gene duplication and by gene chimaerisation [10].

Theories on the evolution of recombination in regard to its ability to increase mean fitness require a consistent source of negative linkage disequilibrium among loci affecting fitness to show an advantage to recombination [11]. Gessler and Xu presented that, at least theoretically, genetic variation for recombination can spread in very large populations under a strictly multiplicative-fitness, deleterious-allele model [11]. It is stated that double-strand breaks of chromosomal DNA, which can be successfully repaired during meiosis, are also the cause of meiotic conjugation of homologous chromosomes and of their genetic recombination (crossing-over) [12]. Formation of chiasmata, which hold homologous chromosomes together until the metaphase I to anaphase I transition, is critical for proper chromosome segregation. Recent studies have suggested that the SPO11 proteins have conserved functions in a number of organisms in generating sites of double-stranded DNA breaks (DSBs) that are thought to be the starting points of homologous recombination [13]. Processing of these sites of DSBs requires the function of RecA homologs, such as RAD51, DMC1, and others, as suggested by mutant studies; thus the failure to repair these meiotic DSBs results in abnormal chromosomal alternations, leading to disrupted meiosis [13]. Lichten said that recent studies point to a universal mechanism for initiating meiotic recombination: the formation of double-strand DNA breaks by Spo11p [14]. Blueyard et al recently reported that the absence of AtXrcc3, an Arabidopsis Rad51 paralogue, leads to extensive chromosome fragmentation during meiosis, first visible in diplotene of meiotic prophase I [14]. This study clearly shows that this fragmentation results from un- or mis-repaired AtSpo11-1 induced double-strand breaks and is thus due to a specific defect in the meiotic recombination process [15]. Indeed, for every eukaryote that has been tested, spo11 mutants are deficient for meiotic recombination and are partially or completely sterile [16]. Depending on the species, this reduced fertility reflects either a defect in chromosome segregation, or an arrest response in germ cell differentiation [16]. Similarities and differences from species to species uncover a complex set of regulations that coordinate recombination with other events of meiotic prophase, such as chromosome pairing and meiotic cell cycle [16]. These described processes are the basis of the organism or cellular level natural recombination. In sexually reproducing organisms, homologous recombination increases genetic diversity in gametes and ensures proper chromosome segregation [17]. This is a classical example of an expression of natural genetic recombination.

In the molecular level, the recombination is also concerned. Genetic change often occurs via natural genetic engineering systems including cellular biochemical functions, such as recombination complexes, topoisomerases, and mobile elements, capable of altering the DNA sequence information and joining together different genomic components [18]. Levels of diversity vary across the human genome [19]. This variation is caused by two forces: differences in mutation rates and the differential impact of natural selection [19]. Pertinent to the question of the relative importance of these two forces is the observation that both diversity within species and interspecies divergence increase with recombination rates [19]. Hellmann et al noted that current recombination rates are a better predictor of diversity than of divergence [19]. Qualitative disturbances with synthesis of abnormal proteins, or protein diseases of protein structure join together with their consequences [20]. Quantitative disorders, with modified synthesis of normal proteins, result very often from abnormalities of structural genes, but also from abnormalities of transcription or translation [20]. In medicine, natural molecular genetic recombination in pathogenesis of existing disease is important. Shapiro noted that prokaryotic genetic determinants are organized as modular composites of coding sequences and protein-factor binding sites joined together during evolution [21]. Studies of genetic change have revealed the existence of biochemical functions capable of restructuring the bacterial genome at various levels and joining together different sequence elements [21]. Defenses against most specialized pathogens are often initiated by a disease resistance gene [22]. Genomes encode several classes of genes that can function as resistance genes [22]. Many of the mechanisms that drive the molecular evolution of these genes are now becoming clear [22]. The processes that contribute to the diversity of resistance genes include tandem and segmental gene duplications, recombination, unequal crossing-over, point mutations, and diversifying selection [22].

In addition to natural genetic recombination, with advent of biotechnology, artificial genetic recombination can be developed. Genetic engineering is the baseline for artificial recombinant generation. Gene engineering is a new method of operating directly with genes.

It permits constructing in vitro any hybrid genomes desirable [23]. There is no limitation of combining ability for gene engineering [23]. Three main stages of constructing hybrid genomes should be taken into account for the proper determination of gene engineering as a method of genome constructing: 1) the gene isolation; 2) their cross-linking in vitro; 3) the transfer of hybrid DNA into recipient cell or its genome [23 - 24]. In genetic engineering, a wide range of techniques is now available for the construction of hybrid DNA molecules comprising components from disparate species [25]. Transfer of segments of DNA from other organisms, and especially eukaryotes permits their preparation in quantities sufficient for detailed analysis of their structure and mechanism of expression [25]. At first, the study of *Escherichia coli* and its plasmids and bacteriophages had provided a vast body of genetical information, much of it relevant to the whole of biology [26]. This was true even before the development of the new techniques, for cloning and analysing DNA, that have revolutionized biological research [26 - 27]. Presently, more advent techniques are in used. Gene transfer is an example. Gene transfer is a useful technique to analyse the gene-controlled mechanism of cellular expression [28]. Using various oncogenes (DNA tumor virus oncogenes: adenovirus 12 E1 and its transcriptional subunits, myc and SV40 T antigens; RNA tumor retrovirus oncogens: fos, src and H-ras) for the transfer, it was demonstrated that these genes invariably caused remarkable and characteristic changes of the cellular expression [28]. For gene transfer, electroporation of cells in the presence of DNA is widely used for the introduction of transgenes either stably or transiently into bacterial, fungal, animal, and plant cells [29]. Recently, in vivo electroporation has emerged as a leading technology for developing nonviral gene therapies and nucleic acid vaccines [30]. Electroporation (EP) involves the application of pulsed electric fields to cells to enhance cell permeability, resulting in exogenous polynucleotide transit across the cytoplasmic membrane [30]. Similar pulsed electrical field treatments are employed in a wide range of biotechnological processes including in vitro EP, hybridoma production, development of transgenic animals, and clinical electrochemotherapy [30]. Electroporative gene delivery studies benefit from well-developed literature that may be used to guide experimental design and interpretation [30]. There are also array of other methods available to move DNA into the nucleus provides the flexibility necessary to transfer genes into cells as physically diverse as sperm and eggs [31 - 32]. Some of the more promising alternative strategies such as sperm-mediated gene transfer, restriction enzyme-mediated integration, metaphase II transgenesis, and a new technique on retrovirus-mediated gene transfer [31]. A method for the production of transgenic animals has been available, namely sperm-mediated gene transfer, based on the intrinsic ability of sperm cells to bind and internalise exogenous DNA molecules and to transfer them into the oocyte at fertilization since 1989 [33 - 34].The major benefits of the sperm-mediated gene transfer technique were found to be its high efficiency, low cost and ease of use compared with other methods [33]. Furthermore, sperm-mediated gene transfer does not require embryo handling or expensive equipment [33]. Given the potential impact of this method for the generation of transgenic animals, for both mammalian and non-mammalian species this technique is widely used [35]. Considering restriction enzyme-mediated integration, it is a powerful tool to investigate the molecular basis of most genetically determined processes. An improved version of this method is the insertional mutagenesis via restriction enzyme mediated integration (REMI) [36]. Transformation efficiency and mode of vector integration are species dependent and further influenced by vector conformation, restriction enzyme activity, and transformation protocol [36]. However, the REMI is limited used in general gene transfer.

Concerning metaphase II transgenesis, it is a method describing by microinjection of membrane-depleted sperm heads and tg DNA into metaphase II oocytes [37]. It is noted that the efficiency of metaphase transgenesis is at least as high as that of the well-established and prevailing alternative, pronuclear microinjection [37]. While retrovirus-mediated gene transfer is the most advanced technology. Several viruses are now used for vector for gene transfer. Examples of those viruses are Sindbis virus and adenovirus.

Natural Recombinations of Pathogens

Considering three epidemiological determinants, genetic recombination of agent is easier than that of host and environment. Concerning the basic principles, human diseases do not occur by chance and factors that cause or contribute to diseases and injuries can be identified by means of systematic investigation. Epidemiologic Triad [38], a traditional model of disease causation is the necessary basis in epidemiology. The epidemiologic triangle recognized three factors in the pathogenesis of disease: host, agent and environment. Considering the molecular epidemiology, host molecular biology (resistance, susceptibility) and pathogen molecular biology (resistance, virulence) should be focused. The main important approaches in molecular biology include a) incidence and prevalence, b) pathogen population and clustering and c) host genetic background [39 - 41]. For pathogens, natural genetic recombination can result in change in virulence and susceptibility. Natural genetic recombination in the virus is well characterized in clinical microbiology. In this article, the author performed a literature review on natural genetic recombination of pathogens. Example of important genetic recombination of pathogens and their clinical correlations are hereby presented.

Important Natural Genetic Recombination of Bacteria

Natural genetic transformation can facilitate gene transfer in many genera of bacteria and requires the presence of extracellular DNA [41]. Although cell lysis can contribute to this extracellular DNA pool, several studies have suggested that the secretion of DNA from living bacteria may also provide genetic material for transformation [41]. On the basis of established knowledge of microbial genetics one can distinguish three major natural strategies in the spontaneous generation of genetic variations in bacteria [43]. Arber said that these strategies were: (1) small local changes in the nucleotide sequence of the genome, (2) intragenomic reshuffling of segments of genomic sequences and (3) the acquisition of DNA sequences from another organism [43]. Arber noted that the three general strategies differ in the quality of their contribution to microbial evolution [43]. Arber also mentioned that genetic determinants of variation generators as well as of modulators of the frequency of genetic variation were defined as evolutionary genes [44]. Arber noted that this postulate was consistent with the notion that spontaneous mutagenesis is in general not adaptive and that the direction of evolution depends on natural selection exerted on populations of genetic variants [44].

There are many important natural genetic recombinations in bacteria. For example, gonococci undergo frequent and efficient natural transformation [45]. Hamilton and Dillard said that gonococcal transformation occured so often that the population structure is panmictic, with only one long-lived clone having been identified [45]. They noted that this high degree of genetic exchange was likely necessary to generate antigenic diversity and allow the persistence of gonococcal infection within the human population [46]. Hamilton et al said that *Neisseria gonorrhoeae* secreted DNA via a specific process and donated DNA might be used in natural transformation, contributing to antigenic variation and the spread of antibiotic resistance, and it might modulate the host immune response [46].

Fussenegger et al reported that the processes of type-4 pilus biogenesis and DNA transformation were functionally linked and play a pivotal role in the life style of *Neisseria gonorrhoeae* [47]. They concluded that PilE was essential for the first step of transformation, such as DNA uptake, and was itself also subject to transformation-mediated phase and antigenic variation [47].

In addition to *Neisseria gonorrhoeae*, *Pseudomonas aeruginosa* is another bacteria that is widely mentioned for its natural recombination. Basically, the large *Pseudomonas aeruginosa* pathogenicity island PAPI-1 of strain PA14 is a cluster of 108 genes that encode a number of virulence features [47]. Qui et al found that demonstrate that, in a subpopulation of cells, PAPI-1 could exist in an extrachromosomal circular form after precise excision from its integration site within the 3' terminus of the tRNA(Lys) gene [48]. Qui et al noted that PAPI-1 played an important role in the evolution of *Pseudomonas aeruginosa*, by expanding its natural habitat from soil and water to animal and human infections [48]. Kurasekara et al noted that comparison of genomes of various *Pseudomonas aeruginosa* strains showed that that the ExoU determinant was found in the same polymorphic region of the chromosome near a tRNA(Lys) gene, suggesting that exoU was a horizontally acquired virulence determinant. [49].

Important Natural Genetic Recombination of Virus

Natural genetic recombination of virus can be expected. Due to the fact that many millions of virus exist, a number of natural recombination can be expected per day. For example, classical swine fever (CSF) virus, one member of the family Flaviviridae is the pathogen of CSF, an economically important and highly contagious disease of pigs is well described for natural genetic recombination [50]. To detect possible recombination events, He et al performed a phylogenetic analysis of 25 full-length CSFV strains isolated all over the world [50]. According to this work, recombination events were confirmed by bootscaning. A mosaic virus, CSFV 39 (AF407339) isolated in China was found [50]. Infectious bursal disease virus (IBDV), a double-stranded RNA virus and member of the Birnaviridae family, is another virus that is documented for the natural genetic recombination [51]. Wei et al suggested that VP2 was not the sole determinant of IBDV virulence, and that the RNA-dependent RNA polymerase protein, VP1, might play an important role in IBDV virulence [51]. Wei et al noted that the discovery of reassortant viruses in nature suggested an additional risk of using live IBDV vaccines, which could act as genetic donors for genome reassortment [51]. Natural recombinant between equine herpesviruses 1 and 4 in the ICP4 gene is also reported [52]. According to a recent report by Pagamjav et al [52], The 3 ' -end and downstream of ICP4 gene of EHV-1 B were found to be replaced by the corresponding region of EHV-4, indicating that EHV-1 B is a naturally occurring recombinant virus between progenitors of EHV-1 P and EHV-4. Recently, homologous recombination happens between strains of chicken anemia virus (CAV), the only member of the genus Gyrovirus in the family Circoviridae and the pathogen of chicken infectious anemia, was also reported [53]. An important point of natural recombination of viral pathogen is the expectation of the recombination to form a new high virulence strain that can cause pandemic around the world such as in the present situation of surveillance of bird flu (H5N1) virus.

References

[1] Gazarian KG, Tarantul VZ. Molecular organization and genome expression in eukaryotes. *Ontogenez.* 1978;9(1):20-38.

[2] Boczkowski K. Modern concept of the gene. *Ginekol* Pol. 1972 May;43(5):625-7

[3] Crick F. Split genes and RNA splicing. *Science.* 1979 Apr 20;204(4390):264-71.

[4] Sharp PA. On the origin of RNA splicing and introns. *Cell.* 1985 Sep;42(2):397-400.

[5] Wolpert L, Lewis JH. Towards a theory of development. *Fed Proc.* 1975 Jan;34(1): 14-20.

[6] Winter WP, Hanash SM, Rucknagel DL. Genetic mechanisms contributing to the expression of the human hemoglogin loci. *Adv Hum Genet.* 1979;9:229-91, 361-7.

[7] Babynin EV. Molecular mechanism of homologous recombination in meiosis: origin and biological significance. *Tsitologiia.* 2007;49(3):182-93.

[8] Holliday R. The biological significance of meiosis. *Symp Soc Exp Biol.* 1984;38:381-94.

[9] Li W, Ma H. Double-stranded DNA breaks and gene functions in recombination and meiosis. *Cell Res.* 2006 May;16(5):402-12.

[10] Cavalier-Smith T. Origins of the machinery of recombination and sex. *Heredity.* 2002 Feb;88(2):125-41.

[11] Gessler DD, Xu S. On the evolution of recombination and meiosis. *Genet Res.* 1999 Apr;73(2):119-31.

[12] Gershenzon SM. The role of double-stranded DNA breaks in the mechanism of meiosis. *Tsitol Genet.* 1994 Jan-Feb;28(1):83-9.

[13] Li W, Ma H. Double-stranded DNA breaks and gene functions in recombination and meiosis. *Cell Res.* 2006 May;16(5):402-12.

[14] Lichten M. Meiotic recombination: breaking the genome to save it. *Curr Biol.* 2001 Apr 3;11(7):R253-6.

[15] Bleuyard JY, Gallego ME, White CI. The atspo11-1 mutation rescues atxrcc3 meiotic chromosome fragmentation. *Plant Mol Biol.* 2004 Sep;56(2):217-24.

[16] Baudat F, de Massy B. SPO11: an activity that promotes DNA breaks required for meiosis. *Med Sci (Paris).* 2004 Feb;20(2):213-8.

[17] Smith KN, Nicolas A. Recombination at work for meiosis. *Curr Opin Genet Dev.* 1998 Apr;8(2):200-11.

[18] Shapiro JA. Genome system architecture and natural genetic engineering in evolution. *Ann N Y Acad Sci.* 1999 May 18;870:23-35.

[19] Hellmann I, Prufer K, Ji H, Zody MC, Paabo S, Ptak SE. Why do human diversity levels vary at a megabase scale? *Genome Res.* 2005 Sep;15(9):1222-31.

[20] Biserte G. Natural history of molecular pathology. *Ann Biol Clin (Paris).* 1975;33(4): 261-80

[21] Shapiro JA. Genome organization, natural genetic engineering and adaptive mutation. *Trends Genet.* 1997 Mar;13(3):98-104.

[22] Meyers BC, Kaushik S, Nandety RS. Evolving disease resistance genes. *Curr Opin Plant Biol.* 2005 Apr;8(2):129-34.

[23] Alikhanian SI. Successes and prospects for genetic engineering. *Genetika.* 1976; 12(7): 150-73.

[24] Tikhonenko NI. Genetic engineering. *Patol Fiziol Eksp Ter.* 1977 Jan-Feb;(1):5-13.

[25] Murray K. Genetic engineering: possibilities and prospects for its application in industrial microbiology. *Philos Trans R Soc Lond B Biol Sci.* 1980 Aug 11;290(1040): 369-86.

[26] Datta N. Escherichia coli as a genetic tool. *J Hyg (Lond).* 1985 Dec;95(3):611-8.

[27] Imanaka T. Application of recombinant DNA technology to the production of useful biomaterials. *Adv Biochem Eng Biotechnol.* 1986;33:1-27.

[28] Nagai Y, Nakaishi H, Sanai Y. Gene transfer as a novel approach to the gene-controlled mechanism of the cellular expression of glycosphingolipids. *Chem Phys Lipids.* 1986 Dec 15;42(1-3):91-103

[29] Lurquin PF. Gene transfer by electroporation. *Mol Biotechnol.* 1997 Feb;7(1):5-35.

[30] Somiari S, Glasspool-Malone J, Drabick JJ, Gilbert RA, Heller R, Jaroszeski MJ, Malone RW. Theory and in vivo application of electroporative gene delivery. *Mol Ther.* 2000 Sep;2(3):178-87.

[31] Wall RJ. New gene transfer methods. *Theriogenology.* 2002 Jan 1;57(1):189-201.

[32] Miyazaki M, Obata Y, Abe K, Furusu A, Koji T, Tabata Y, Kohno S. Gene transfer using nonviral delivery systems. *Perit Dial Int.* 2006 November-December;26(6):633-640.

[33] Lavitrano M, Busnelli M, Cerrito MG, Giovannoni R, Manzini S, Vargiolu A. Sperm-mediated gene transfer. *Reprod Fertil Dev.* 2006;18(1-2):19-23.

[34] Chan AW, Luetjens CM, Schatten GP. Sperm-mediated gene transfer. *Curr Top Dev Biol.* 2000;50:89-102.

[35] Gandolfi F. Sperm-mediated transgenesis. *Theriogenology.* 2000 Jan 1;53(1):127-37.

[36] Maier FJ, Schafer W. Mutagenesis via insertional- or restriction enzyme-mediated-integration (REMI) as a tool to tag pathogenicity related genes in plant pathogenic fungi. *Biol Chem.* 1999 Jul-Aug;380(7-8):855-64

[37] Perry AC. Metaphase II transgenesis. *Reprod Biomed Online.* 2002 May-Jun;4(3): 279-84.

[38] Thristed RD. Are there social determinants of health and disease? *Perspect Biol Med.* 2003 Summer;46(3 Suppl):S65-73.

[39] Foxman B, Riley L. Molecular epidemiology: focus on infection. *Am J Epidemiol.* 2001 Jun 15;153(12):1135-41.

[40] Lipuma JJ. Molecular tools for epidemiologic study of infectious diseases. *Pediatr Infect Dis J.* 1998 Aug;17(8):667-75.

[41] Barata Rde C. The challenge of emergent diseases and the return to descriptive epidemiology. *Rev Saude Publica.* 1997 Oct;31(5):531-7.

[42] Draghi JA, Turner PE. DNA secretion and gene-level selection in bacteria. *Microbiology*. 2006 Sep;152(Pt 9):2683-8.

[43] Arber W. Genetic variation: molecular mechanisms and impact on microbial evolution. *FEMS Microbiol Rev*. 2000 Jan;24(1):1-7.

[44] Arber W. Involvement of gene products in bacterial evolution. *Ann N Y Acad Sci*. 1999 May 18;870:36-44.

[45] Hamilton HL, Dillard JP. Natural transformation of Neisseria gonorrhoeae: from DNA donation to homologous recombination. *Mol Microbiol*. 2006 Jan;59(2):376-85.

[46] Hamilton HL, Dominguez NM, Schwartz KJ, Hackett KT, Dillard JP. Neisseria gonorrhoeae secretes chromosomal DNA via a novel type IV secretion system. *Mol Microbiol*. 2005 Mar;55(6):1704-21.

[47] Fussenegger M, Rudel T, Barten R, Ryll R, Meyer TF. Transformation competence and type-4 pilus biogenesis in Neisseria gonorrhoeae--a review. *Gene*. 1997 Jun 11;192(1):125-34.

[48] Qiu X, Gurkar AU, Lory S. Interstrain transfer of the large pathogenicity island (PAPI-1) of Pseudomonas aeruginosa. *Proc Natl Acad Sci U S A*. 2006 Dec 26;103(52): 19830-5.

[49] Kulasekara BR, Kulasekara HD, Wolfgang MC, Stevens L, Frank DW, Lory S. Acquisition and evolution of the exoU locus in Pseudomonas aeruginosa. *J Bacteriol*. 2006 Jun;188(11):4037-50.

[50] He CQ, Ding NZ, Chen JG, Li YL. Evidence of natural recombination in classical swine fever virus. *Virus Res*. 2007 Jun;126(1-2):179-85.

[51] Wei Y, Li J, Zheng J, Xu H, Li L, Yu L. Genetic reassortment of infectious bursal disease virus in nature. *Biochem Biophys Res Commun*. 2006 Nov 17;350(2):277-87.

[52] Pagamjav O, Sakata T, Matsumura T, Yamaguchi T, Fukushi H. Natural recombinant between equine herpesviruses 1 and 4 in the ICP4 gene. *Microbiol Immunol*. 2005;49(2): 167-79.

[53] He CQ, Ding NZ, Fan W, Wu YH, Li JP, Li YL. Identification of chicken anemia virus putative intergenotype recombinants. *Virology*. 2007 Jul 4;

In: Genetic Recombination Research Progress
Editor: Jacob H. Schulz, pp. 345–357
ISBN 978-1-60456-482-2
© 2008 Nova Science Publishers, Inc.

Chapter 16

Existence and Uniqueness of Positive Solutions of a Randomized Spruce Budworm Model[*]

Yanqiu Li[1][†]*and Hailong Gao*[2]
[1]Department of fundamental sciences,
JiLin Teacher's Institute of Engineering And Technology,
Changchun 130052, Jilin, P. R. China
[2] Department of Basic Sciences,
JiLin Architectural and Civil Engineering Institute,
Changchun 130021, Jilin, P. R. China

Abstract

This paper discusses a randomized Spruce Budworm model with Holling III Functional Response . We show that the positive solution of the associated stochastic differential equation does not explode to infinity in a finite time. We proof the existence and uniqueness of the positive solutions.In addition, Uniformly Continuous of solution is studied.

Keywords: Itô's formula; Existence; Uniqueness; Uniformly Continuous.

2000 MR Subject Classification. 34F05, 34E10.

---

[*]The work was supported by NNSF of China(10571021)
[†]E-mail address: liyanqiusd@yahoo.com.cn

1. Introduction

A practical model which exhibits two positive linearly stable steady state populations is that for the sparuce budworm which can, with ferocious efficiency, defoliate the balsam hr: it is a major problem in Canada. Ludwig et.al.(1978)considered the budworm population dynamics to be governed by the equation:

$$\frac{dN(t)}{dt} = N(t)r_B(1 - \frac{N}{K_B}) - p(N).$$

Spruce Budworm model with Holling III Functional Response

Here r_B is the linear birth rate of the budworm and K_B is the carrying capacity which is related to the density of foliage available on the trees. The p(N)-teerm reoresents predation, generally by birds; the qualitative form of it id important . Predation usually saturates fot latge enough N. There is an approximate threshold balue N_c, below which the predation is small. while above it the predation is close to its saturation value: such a funjctional form densities N, the birda tend to seck food elsewhere and so the predation term p(N) drops more rapidly, as $N \rightarrow 0$, than a linear rate proportional to N.To be spedifid we take the form for p(N) suggested by Ludwig et.al.(1978)namely $\frac{BN^2(t)}{1+N^2(t)}$,wher A and B are positive constants, and the dynamics of N(t) is then governed by :

$$\frac{dN(t)}{dt} = N(t)r_B(1 - \frac{N}{K_B}) - \frac{BN^2(t)}{1+N^2(t)}. \tag{1.1}$$

Now, we consider a special Spruce Budworm model with Holling III Functional Response

$$\frac{dx(t)}{dt} = x(t)(a - bx(t)) - \frac{x^2(t)}{1+x^2(t)}. \tag{1.2}$$

It is importanct,in practice,that Spruce Budworm model is often subject to environmental noise. Moreover, Mao et al. [11] considered the following stochastically perturbed system

$$dx(t) = diag(x_1(t),\cdots,x_n(t))[(b + Ax(t))dt + \sigma x(t)dW(t)], \tag{1.3}$$

Bahar and Mao [13] considered the stochastically perturbed delay system

$$dx(t) = diag(x_1(t),\cdots,x_n(t))[(b + Ax(t-\tau))dt + \sigma x(t)dW(t)], \tag{1.4}$$

Jiang and Shi [19] considered the stochastically perturbed Competition System

$$dx(t) = diag(x_1(t),\cdots,x_n(t))[(b - Ax(t))dt + \sigma dW(t)], \tag{1.5}$$

where

$$x = (x_1,\cdots,x_n)^T, \quad b = (b_1,\cdots,b_n)^T, \quad A = (a_{ij})_{n\times n},$$

and Jiang and Gao [20] considered the stochastically perturbed delay Competition System

$$dx_i(t) = x_i(t)[(b_i - a_{ii}x_i(t) - \sum_{j=1,j\neq i}^{n} a_{ij}x_j(t-\tau))dt + \sigma_i dW_i(t)], \tag{1.6}$$

and

$$\sigma_{ii} > 0 \quad if \ 1 \leq i \leq n \quad whilst \quad \sigma_{ij} \geq 0.$$

They all have revealed an important fact that environmental noise can suppress a potential population explosion. Moreover, Bahar and Mao [13] have shown that the above simple hypothesis on the noise is enough to guarantee the stochastically ultimate boundedness of the solutions of the stochastic Lotka-Volterra delay model (1.4). We therefore wonder whether different form of environmental noise perturbations will lead to different results and the presence of a such noise will affect the results which have been known about Eq. (1.2). One of the aim of this paper is to show that the presence of such a noise can not affect the main qualitative properties of Eq. (1.2) provided the intensity of the noise is small enough.

Suppose that a and f in Eq. (1.2) is stochastically perturbed, with

$$a \rightarrow a + \sigma \dot{W}(t) .$$

where $\dot{W}(t)$ are independent white noises with $W(0) = 0$, $t \geq 0$, and σ^2 represents the intensity of the noise. Then this environmentally perturbed system may be described by the Itô equation

$$dx(t) = [x(t)(a - bx(t)) - \frac{x^2(t)}{1 + x^2(t)}]dt + \sigma_1 x(t)dW_1(t). \tag{1.7}$$

with initial data $x(0) \in R_+$, where $W(t)$ are independent standard Brownian motions, $a > 0$ and $b > 0$.

In order that a stochastic delay differential equation has a unique global (i.e. no explosion in a finite time) solution for any initial data $x(0) \in R_+$, the coefficients of the equation are usually required to satisfy the linear growth condition and local Lipschitz condition (cf. Arnold [1] and Friedman [2]). However, the coefficients of Eq. (1.7) do not satisfy the linear growth condition, though they are local Lipschitz continuous, so the solution of Eq. (1.7) may explode at a finite time. We shall show that the solution will not explode in a finite time which is uniqueness and global positive. In addition, we obtain the upper bound of p-th moment . The significant contributions of this paper are therefore clear. For example, consider the corresponding version of Eq. (1.7), with the initial data $x(0)$ in the positive cone R_+, where $W(t)$ are dependent on standard Brownian motions. Suppose that

(A) $b > 0$;

(B)   There exists $p \geq 1$ such that

$$x(0) < \frac{a + \frac{1}{2}(p-1)\sigma_1^2}{b}. \tag{1.8}$$

CONCLUSION. Our theory shows that under simple assumption A the solution of Eq. (1.7) is not only positive but will also not explode to infinity at any finite time, see Sections 2 . In addition, we obtain Stochastically ultimate boundedness and moment average in time under assumptions A and B, see Section 3. the upper bound of p-th moment by comparison principle ,see Section 4 . We also obtain the solutiom is uniformly continuous see Section 5.

2. Positive and Global Solutions

Throughout this paper, unless otherwise specified, we let $(\Omega, \mathcal{F}, \{\mathcal{F}_t\}_{t \geq 0}, P)$ be a complete probability space with a filtration $\{\mathcal{F}_t\}_{t \geq 0}$ satisfying the usual conditions (i.e. it is right continuous and $\mathcal{F}_0$ contains all P-null sets). Let $W_i(t)$ denote the independent standard Brownian motions defined on this probability space, $i = 1, \cdots, n$. We introduce the notation $R_+ = \{x \in R : x > 0\}$.

Assumption A. $b > 0$.

As the state $x(t)$ of Eq (1.7) is the size of x species in the system, it should be nonnegative. In this section we shall show under simple Assumption A the solution of Eq (1.7) is not only positive but will also not explode to infinity at any finite time.

Theorem 2.1. *Let Assumption A hold, and the initial data $x(0) \in R_+$. Then there exists a unique solution $x(t)$ to Eq. (1.7) on $t \geq 0$ and the solution will remain in R_+ with probability 1, namely $x(t) \in R_+$ for all $t \geq 0$ almost surely.*

Proof. Since the coefficient of the equation are locally Lipschitz continuous, for any given the initial data $x(0) \in R_+$, where τ_e is the explosion time (cf. Arnold [1] and Friedman [2]). To show this solution is global, we need to show that $\tau_e = \infty$ a.s. Let $k_0 > 0$ be sufficiently large for every component of $x(0) \in R_+$ lying within the interval $[\frac{1}{k_0}, k_0]$. For each integer $k \geq k_0$, define the stopping time

$$\tau_k = \inf\{t \in [0, \tau_e) : x(t) \notin (\frac{1}{k}, k) \text{ or } y(t) \notin (\frac{1}{k}, k)\},$$

where throughout this paper we set $\inf \emptyset = \infty$ (as usual $\emptyset$ denotes the empty set). Clearly, τ_k is increasing as $k \to \infty$. Set $\tau_\infty = \lim\limits_{k \to \infty} \tau_k$, whence $\tau_\infty \leq \tau_e$ a.s. If we can show that $\tau_\infty = \infty$ a.s., then $\tau_e = \infty$ a.s. and $(x(t), y(t)) \in R_+^2$ a.s. for all $t \geq 0$. In other words, to complete the proof all we need to show is that $\tau_\infty = \infty$ a.s. For if this statement is false, then there is a pair of constants $T > 0$ and $\varepsilon \in (0, 1)$ such that

$$P\{\tau_\infty \leq T\} > \varepsilon.$$

Hence there is an integer $k_1 \geq k_0$ such that

$$P\{\tau_k \leq T\} \geq \varepsilon \qquad for \ all \quad k \geq k_1. \tag{2.1}$$

Define a C^2-function $V : R_+ \to R_+$ by

$$V(x) = x - 1 - \log(x).$$

The nonnegativity of this function can be seen from

$$u - 1 - \log(u) \geq 0 \qquad \text{on } u > 0.$$

If $x(t) \in R_+ (t > 0)$, the Itô's formula shows that

$$dV(x(t))$$

$$= (1 - x(t)^{-1})[[x(t)(a - bx(t)) - \frac{x^2(t)}{1 + x^2(t)}]dt + \sigma x(t)dW(t)] + 0.5(x^{-2}(t))x^2(t)\sigma^2 dt$$

$$= \{-bx^2(t) + ax(t) + bx(t) - \frac{x^2(t)}{1 + x^2(t)} - a + 0.5\sigma^2 + \frac{x(t)}{1 + x^2(t)}\}dt + (\sigma x(t) - \sigma)dW(t)$$

$$= F(x(t))dt + (\sigma x(t) - \sigma)dW(t),$$

where

$$F(x(t)) : = -bx^2(t) + ax(t) + bx(t) - \frac{x^2(t)}{1 + x^2(t)} - a + 0.5\sigma^2 + \frac{x(t)}{1 + x^2(t)}$$

And we write $x(t) = x$. Compute

$$F(x) \leq -bx^2 + ax + bx - \frac{x^2}{1 + x^2} - a + 0.5\sigma^2 + \frac{x}{1 + x^2}$$

which is upper bounded, say by K, in R. We therefore obtain

$$dV(x) \leq Kdt + (\sigma x - \sigma)dW(t)$$

Intergrating bothsides from 0 to $\tau_k \wedge T$, yields

$$\int_0^{\tau_k \wedge T} d(V(x(t))) \leq \int_0^{\tau_k \wedge T} Kdt + \int_0^{\tau_k \wedge T} (\sigma x - \sigma)dW(t),$$

since $x(T \wedge \tau_k) \in R_+$. Whence taking expectations, yields

$$E[V(x(\tau_k \wedge T))] \leq V(x(0), y(0)) + KE(\tau_k \wedge T). \tag{2.2}$$

Here, and in the sequel, "$E(f)$" shall mean the mathematical expectation of f. Consequently,

$$EV(x(\tau_k \wedge T)) \leq V(x(0)) + KT.$$

Set $\Omega_k = \{\tau_k \leq T\}$ for $k \geq k_1$ and, by (2.1), $P(\Omega_k) \geq \varepsilon$. Note that for every $\omega \in \Omega_k$, there is $x(\tau_k, \omega)$ equals either k or $\frac{1}{k}$, and hence $V(x(\tau_k, \omega))$ is no less than either

$$k - 1 - \log(k)$$

or

$$\frac{1}{k} - 1 - \log(\frac{1}{k}) = \frac{1}{k} - 1 + \log(k).$$

Consequently,

$$V(x(\tau_k, \omega)) \geq [k - 1 - \log(k)] \wedge [\log(k) - 1 + \frac{1}{k}].$$

It then follows from (2.2) that

$$\begin{aligned} V(x(0)) + KT &\geq E[1_{\Omega_k}(\omega)V(x(\tau_k, \omega))] \\ &\geq \varepsilon([k - 1 - \log(k)] \wedge [\log(k) - 1 + \frac{1}{k}]), \end{aligned}$$

where 1_{Ω_k} is the indicator function of Ω_k. Letting $k \to \infty$ leads to the contradiction

$$\infty > V(x(0)) + KT = \infty.$$

So we must therefore have $\tau_\infty = \infty$ a.s. This completes the proof of Theorem 2.1.

3. Stochastically Ultimate Boundedness and Moment Average in Time

Definition 3.1. Equation (2.2) is said to be Stochastically ultimate bounded if for any $\varepsilon \in (0,1)$,there exists a positive constant $H = H(\varepsilon)$ such that for any initial date $x(0)$ in R_+,the solution $x(t)$ of Eq. (1.7) has the property that

$$\lim_{t \to +\infty} \sup P\{x(t) > H\} < \varepsilon. \tag{3.1}$$

Lemma 3.1. *Let hypothesis (A)hold and $\theta > 0$.Then there is a positive constant $K = K(\theta)$,which is independent of the initial date $x(0)$ in R_+,such that the solution $x(t)$ of Eq. (1.7) has the property that*

$$\lim_{t \to +\infty} \sup E(x(t)^\theta) \leq K. \tag{3.2}$$

Proof. Define

$$V(x) = x^\theta \quad for \ x \in R_+.$$

By the Itô's formula ,we have

$$dV(x(t)) = LV(x(t))dt + \theta\sigma x(t)^\theta dw(t), \tag{3.3}$$

where $LV : R_+ \to R$ is defined by

$$LV(x(t)) = -\theta \, bx(t)^{\theta+1} + \theta \, ax(t)^\theta - \theta \frac{x(t)^{\theta+1}}{1 + x(t)^2} + \frac{\theta(\theta-1)}{2}\sigma^2 x(t)^\theta,$$

Compute

$$\begin{aligned} LV(x(t)) &\leq -\theta \, bx(t)^{\theta+1} + \theta \, ax(t)^\theta + \frac{\theta(\theta-1)}{2}\sigma^2 x(t)^\theta \\ &= F(x) - V(x), \end{aligned}$$

where

$$F(x) = x(t)^\theta - \theta \, bx(t)^{\theta+1} + \theta \, ax(t)^\theta + \frac{\theta(\theta-1)}{2}\sigma^2 x(t)^\theta,$$

Note that F(x)is bounded in R_+,namely

$$K := \sup_{x \in R_+} F(x) < \infty.$$

We therefore have

$$LV(x(t)) \leq K - V(x(t)).$$

Substituting this into (3.3)gives

$$dV(x(t)) \leq [K - V(x(t))]dt + \theta\sigma x(t)^\theta dw(t).$$

Once again by the Itô's formula we have

$$
\begin{aligned}
d[e^t V(x(t))] &= e^t[V(x(t))dt + dV(x(t))] \\
&\le e^t[V(x(t)) + K - V(x(t))] \\
&= Ke^t.
\end{aligned}
$$

We hence derive that

$$
e^t E(V(x(t))) \le V(x(0)) + Ke^t - K
$$

This implies immediately that

$$
\lim_{t \to +\infty} \sup E(V(x(t))) \le K.
$$

We therefore finally have

$$
\lim_{t \to +\infty} \sup E(x(t)^\theta) \le K,
$$

Remark 3.1. *It is easily known that when $\theta > 0$, then $E(x(t)^\theta) \le K', K' = V(x(0)) + K_1$*

Theorem 3.1. Under hypothesis (A), Eq. (1.7) is stochastically ultimately bounded.
Proof. By lemma 3.1, there is $K > 0$ such that

$$
\lim_{t \to +\infty} \sup E(x(t)^{\frac{1}{2}}) \le K
$$

Now, for any $\varepsilon > 0$, let $H = \dfrac{K^2}{\varepsilon^2}$. Then by Chebyshev's inequality

$$
P\{x(t) > H\} \le \frac{Ex(t)^{\frac{1}{2}}}{H^{\frac{1}{2}}}.
$$

Hence
$$
\lim_{t \to +\infty} \sup P\{x(t) > H\} \le \frac{K}{H^{\frac{1}{2}}} = \varepsilon.
$$

The result in the previous section shows that the solutions of Eq. (1.7) will be stochastically ultimately bounded. That is, the solutions will be ultimately bounded with large probability. the following result shows that the average in time of the θ-th moment of the solutions will be bounded.

Theorem 3.2. Under hypothesis (A), there is a positive constant K, which is independent of the initial date $x(0) \in R_+$, such that the solution $x(t)$ of Eq. (1.7) has the property that

$$
\lim_{t \to +\infty} \sup \frac{1}{T} \int_0^T E(x(t)^\theta) dt \le K, \quad \theta \in (0,2). \tag{3.4}
$$

Proof. We use the same notations as in the proof of the Theorem 2.1

$$
F(x) = F_1(x) - x(t)^\theta, \quad \theta \in (0,2)
$$

with

$$F(x(t)): \quad = -bx^2(t) + ax(t) + bx(t) - \frac{x^2(t)}{1+x^2(t)} - a + 0.5\sigma^2 + \frac{x(t)}{1+x^2(t)} + x(t)^{\theta}.$$

$$F_1(x) \quad \leq \quad -bx^2(t) + ax(t) + bx(t) + 0.5\sigma^2 + \frac{x(t)}{1+x^2(t)} + x(t)^{\theta}.$$

Clearly, $F_1(x)$ is bounded in R_+, namely

$$K = \max_{x \in R_+} F_1(x) < \infty.$$

So

$$F(x) \leq K - x(t)^{\theta}.$$

Using this estimation, integrating both sides of (2.2) from 0 to $\tau_k \wedge T$, and then taking expectations, we have

$$0 \leq V(x(0)) + KE(\tau_k \wedge T) - E \int_0^{\tau_k \wedge T} x(t)^{\theta} dt.$$

Letting $k \to \infty$ yields

$$E \int_0^T x(t)^{\theta} dt \leq V(x(0)) + KT.$$

Dividing both sides by T and then letting $T \to \infty$ we get

$$\lim_{t \to +\infty} \sup \frac{1}{T} \int_0^T Ex(t)^{\theta} dt \leq K, \quad \theta \in (0,2)$$

4. Upper Bound of P-th Moment

Assumption B. For any initial value $x(0) \in R_+$, there exists $p \geq 1$ such that

$$\left\{ x(0) < \frac{a + \frac{1}{2}(p-1)\sigma_1^2}{b}. \right. \tag{4.1}$$

Remark 4.1 *It is easily known that when p is large enough, then (4.1) holds .*

The following result shows that the p-th moment of positive solution to Eq. (1.7) will be upper bounded.

Theorem 4.1 *Assume that Assumptions A and B hold. Let $x(t)$ be a solution of Eq. (1.7) with initial value $x(0) in R_+$, then*

$$\left\{ E(x^p(t)) \leq k(p). \right. \tag{4.2}$$

Where

$$\left\{ k(p) := \left(\frac{a + \frac{1}{2}(p-1)\sigma_1^2}{b} \right)^p. \right.$$

Proof. we have

$$
\begin{aligned}
dx^p(t) &= px^{p-1}(t)dx(t) + \tfrac{1}{2}p(p-1)x^{p-2}(t)(dx(t))^2 \\
&= px^{p-1}(t)[(x(t)(a-bx(t)) - \frac{x^2(t)}{1+x^2(t)})dt + \sigma x(t)dW(t)] \\
&\quad + \tfrac{1}{2}p(p-1)\sigma^2 x^p(t)dt \\
&= (px^p(t)(a-bx(t)) - px^p(t)\frac{x(t)}{1+x^2(t)} + \tfrac{1}{2}p(p-1)\sigma^2 x^p(t))dt \\
&\quad + p\sigma x^p(t)dW(t) \\
&= px^p(t)(a-bx(t) - \frac{x(t)}{1+x^2(t)} + \tfrac{1}{2}(p-1)\sigma_1^2)dt \\
&\quad + p\sigma_1 x^p(t)dW_1(t).
\end{aligned}
$$

Integrating from 0 to t and taking expectations, yields

$$
E(x^p(t)) - E(x^p(0)) = \int_0^t pE(x^p(s)(a-bx(s) - \frac{x(s)}{1+x^2(s)} + \tfrac{1}{2}(p-1)\sigma^2))ds.
$$

So,

$$
\begin{aligned}
\frac{dE(x^p(t))}{dt} &= pE(x^p(t)(a-bx(t) - \frac{x(t)}{1+x^2(t)} + \tfrac{1}{2}(p-1)\sigma^2)) \\
&\leq paE(x^p(t)) - pbE(x^{p+1}(t)) + \tfrac{1}{2}p(p-1)\sigma^2 E(x^p(t)) \\
&\leq paE(x^p(t)) - pb[E(x^p(t))]^{\frac{p+1}{p}} + \tfrac{1}{2}p(p-1)\sigma^2 E(x^p(t)) \\
&= pE(x^p(t))\{[a + \tfrac{1}{2}(p-1)\sigma^2] - b[E(x^p(t))]^{\frac{1}{p}}\}.
\end{aligned}
$$

Let $X(t) = E(x^p(t))$, then we have

$$
\frac{dX(t)}{dt} \leq pX(t)[a + \tfrac{1}{2}(p-1)\sigma^2 - bX^{\frac{1}{p}}(t)].
$$

From (4.1), we know

$$
0 < bX^{\frac{1}{p}}(0) = bx(0) < a + \tfrac{1}{2}(p-1)\sigma_1^2,
$$

then by the standard comparison argument shows that

$$
[E(x^p(t))]^{\frac{1}{p}} = X^{\frac{1}{p}}(t) \leq \frac{a + \tfrac{1}{2}(p-1)\sigma^2}{b},
$$

i.e.,

$$
E(x^p(t)) \leq (\frac{a + \tfrac{1}{2}(p-1)\sigma_1^2}{b})^p.
$$

Remark 4.2 *If there exists $p \geq 1$ such that*

$$
x(0) < \frac{a + \tfrac{1}{2}(p-1)\sigma_1^2}{b}.
$$

then

$$x(0) < \frac{a + \frac{1}{2}(m-1)\sigma_1^2}{b}.$$

for all $m > p$. By Theorem 3.1,

$$E(x^m(t)) \leq \left(\frac{a + \frac{1}{2}(m-1)\sigma_1^2}{b}\right)^m.$$

5. Uniformly Continuous

Lemma 5.1[14,16]. *Suppose that an n-dimensional stochastic process X(t) on $t \geq 0$ satisfies the condition*

$$E|X(t) - X(s)|^\alpha \leq c|t-s|^{1+\beta}, \qquad 0 \leq s,t < \infty,$$

for some positive constants α, β and c. Then there exists a continuous modification $\tilde{X}(t)$ of X(t), which has the property that for every $\gamma \in (0, \beta/\alpha)$, there is a positive random variable $h(w)$ such that

$$p\left\{\omega: \sup_{0<|t-s|<h(\omega),\ 0\leq s,\ t<\infty} \frac{|\tilde{X}(t,\omega) - \tilde{X}(s,\omega)|}{|t-s|^\gamma} \leq \frac{2}{1-2^{-\gamma}}\right\} = 1.$$

In other word, almost every sample path of $\tilde{X}(t)$ is locally but uniformly Hölder-continuous with exponent γ.

Theorem 5.1 *Assume that Assumptions A and B hold. Let x(t) be a solution of Eq. (1.7) on $t \geq 0$ with initial data $x(0) \in R_+$, then almost every sample path of x(t) is uniformly continuous on $t \geq 0$.*

   Proof. We shall consider the following stochastic integral equation instead of Eq. (1.7)

$$x(t) = x(0) + \int_0^t f_1(s, x(s))ds + \int_0^t g_1(s, x(s))dW(s), \tag{5.1}$$

where

$$\begin{aligned} f(s, x(s)) &= x(s)(a - bx(s)) - \frac{x^2(s)}{1 + x^2(s)}; \\ g(s, x(s), y(s)) &= \sigma x(s). \end{aligned}$$

Then,

$$
\begin{aligned}
E(|f(s,x(s))|^p) &= E(x^p(s)|a - bx(s) - \frac{1}{1+x^2(s)}x(s)|^p) \\
&\leq \frac{1}{2}E(x^{2p}(s)) + \frac{1}{2}E((a - bx(s) - \frac{1}{1+x^2(s)}x(s))^{2p}) \\
&\leq \frac{1}{2}E(x^{2p}(s)) + \frac{1}{2}(n+1)^{2p-1}(E(a - bx(s))^{2p} + E(x^{2p}(s))) \\
&\leq \frac{1}{2}E(x^{2p}(s)) + \frac{1}{2}(n+1)^{4p-2}(a^{2p} + b^{2p}E(x^{2p})) \\
&\quad + \frac{1}{2}(n+1)^{2p-1}c^{2p}E(x^{2p}(s)) \\
&\leq \frac{1}{2}K(2p) + \frac{1}{2}(n+1)^{4p-2}(a^{2p} + b^{2p}K(2p)) \\
&\quad + \frac{1}{2}(n+1)^{2p-1}c^{2p}K(2p) \\
&=: F(p),
\end{aligned}
\tag{5.2}
$$

and

$$
E(|g(s,x(s))|^p) = E(\sigma^p x^p(t)) = \sigma^p E(x^p(t)) \leq \sigma^p K(p) =: G(p). \tag{5.3}
$$

Here we use Remark 4.2, $E(x^m(t)) \leq K(m)$ for all $m > p, t > 0$. Without loss of generality, we assume that $p > 2$. By the moment inequality (cf. Friedman[2] or Gao[20]) for stochastic integrals(5.1), we have that for $0 \leq t_1 < t_2 < \infty$ and $p > 2$,

$$
\begin{aligned}
&E|\int_{t_1}^{t_2} g_1(s,x(s))dW(s)|^p \\
&\leq [\frac{p(p-1)}{2}]^{p/2}(t_2 - t_1)^{(p-2)/2}\int_{t_1}^{t_2} E|g(s,x(s))|^p ds.
\end{aligned}
\tag{5.4}
$$

Let $0 < t_1 < t_2 < \infty, t_2 - t_1 \leq 1, 1/p + 1/q = 1$, then from (5.2), (5.3) and (5.4), we yield:

$$
\begin{aligned}
E|x_1(t_2) - x_1(t_1)|^p &\leq 2^{p-1}E(\int_{t_1}^{t_2} |f(s,x(s))|ds)^p + 2^{p-1}E|\int_{t_1}^{t_2} g(s,x(s))d\omega(s))|^p \\
&\leq 2^{p-1}(\int_{t_1}^{t_2} 1^q ds)^{\frac{p}{q}}E(\int_{t_1}^{t_2} |f(s,x(s))|^p ds) \\
&\quad + 2^{p-1}[\frac{p(p-1)}{2}]^{\frac{p}{2}}(t_2 - t_1)^{\frac{p-2}{2}}\int_{t_1}^{t_2} E|g(s,x(s))|^p ds \\
&\leq 2^{p-1}(\int_{t_1}^{t_2} 1^q ds)^{\frac{p}{q}}E(\int_{t_1}^{t_2} F(p)ds) \\
&\quad + 2^{p-1}[\frac{p(p-1)}{2}]^{\frac{p}{2}}(t_2 - t_1)^{\frac{p-2}{2}}\int_{t_1}^{t_2} G(p)ds \\
&= 2^{p-1}(t_2 - t_1)^{(p-1)+1}F(p) + 2^{p-1}[\frac{p(p-1)}{2}]^{\frac{p}{2}}(t_2 - t_1)^{\frac{p-2}{2}+1}G(p) \\
&\leq 2^{p-1}(t_2 - t_1)^{\frac{p}{2}}\{(t_2 - t_1)^{\frac{p}{2}} + [\frac{p(p-1)}{2}]^{\frac{p}{2}}\}M(p) \\
&\leq 2^{p-1}(t_2 - t_1)^{\frac{p}{2}}\{1 + [\frac{p(p-1)}{2}]^{\frac{p}{2}}\}M(p),
\end{aligned}
$$

where $M(p) := F(p) + G(p)$.

We see from Lemma 5.1 that almost every sample path of $x(t)$ is locally but uniformly Hölder-continuous with exponent γ for every $\gamma \in (0, p - 2/2p)$ and therefore almost every sample path of $x(t)$ is uniformly continuous on $t \geq 0$.

References

[1] L. Arnold, *Stochastic Differential Equations: Theory and Applications*, New York,Wiley, 1972.

[2] A. Friedman, *Stochastic Differential Equations and their Applications*, New York,Academic Press, 1976.

[3] Maynard Smith J. , *Models in ecology,* Cambridge ,Cambridge Univ Press ,1974

[4] He X Z ,(1996). Stability and delays in a predator-prey system. *J. Math. Anal. Appl,* **198** , 355-370.

[5] Li Y K, (1999).Periodic solutions of a periodic delay predator-prey system, *Proc of Amer Math Soc,* **127**(5), 1331-1335.

[6] M. Fan and K. Wang,(2001). Periodicity in a delayed ratio-dependent predator-prey system. J. *Math. Anal. Appl,***262**(1), 179-190.

[7] B. S. Goh,(1997). Global stability in many species systems. *Amer. Nat* **111** , 135-143.

[8] K. Golpalsamy,(1982). Globally asymptotic stability in a periodic Lotka-Volterra system. *J. Austral. Math. Soc. Ser B*, **24** , 160-170.

[9] R. M. L ,(1969). Why the prey curve has a hump .*Amer. Nat,***103**, 81-87.

[10] R. M. L and M. R. H ,(1963). Graphical representation and stability conditions for predator-prey interactions. *Amer, Nat,* **97**, 209-223.

[11] X. R. Mao, G. Marion and E. Renshaw,(2002). Environmental Brownian noise suppresses explosions in Population dynamics.*Stochastic Processes and their Applications,***97**,95-110.

[12] J. A. D. Appleby and X. R. Mao,(2005). Stochastic stabilisation of functional differential equations . *Systems and Control Letters,* **54**(11), 1069-1081.

[13] A. Bahar and X. R. Mao,(2004). Stochastic delay Lotka-Volterra model. *J. Math. Anal. Appl.* **292**,364-380 .

[14] X. Mao,(1999). Stochastic versions of the Lassalle theorem. *J. Differential Equations,* **153** 175-195, .

[15] X. Mao, *Stochastic Differential Equations and Applications*, Chichester,Horwood, 1997.

[16] I. Karatzas and S. E. Shreve, *Brownian Motion and Stochastic Calculus*, Berlin/New York,SpringerVerlag, 1991.

[17] I. Barbalat, (1959).Systems dequations differentielles d'osci d'oscillations nonlineaires. *Rev. Roumaine Math Pures Appl.* **4**,267-270.

[18] D. Q. Jiang and N. Z. Shi, A note on nonautonomous Logistic equation with random perturbation, *J. Math. Anal. Appl.*, (in Press).

[19] D. Q. Jiang and N. Z. Shi, *Existence, Uniqueness and Global Asymptotic Stability of Positive Solutions and MLE of the Parameters of Two Species Lotka-Volterra Competition System with Random Perturbation*, (in Press).

[20] D. Q. Jiang and H. L. Gao, *Existence, uniqueness and global asymptotic stability of positive solutions and MLE of the parameters of n-species delay Lotka-Volterra competition system with random perturbation, (in Press).*

Index

B

D

E

F

G

H

I

M

O

P

S

T

U

V

W

X

X chromosome, 271
xenograft, 293, 295, 299, 300
xenografts, xii, 292, 294, 297, 298, 299
X-linked, 62, 176, 186, 292, 304
x-ray -ray, 3, 23, 25, 40, 54, 55
x-rays, 25, 40

Y

yeast, vii, xiii, 13, 39, 42, 46, 169, 181, 200, 246,
254, 265, 267, 273, 276, 278, 280, 284, 323, 324,
325, 326, 328, 329, 330, 331, 332, 333
yield, 39, 70, 81, 122, 230, 271, 308, 309, 355

Z

zebrafish, 310, 320
zinc, 13, 17, 25, 26, 55, 72, 165, 177, 178, 179, 187,
260